OXFORD MATHEMATICS

Intermediate GCSE for Edexcel

Course Editors

Peter McGuire   Ken Smith

The huge blades on a wind turbine trace out a circle when they are turning.

You can find out more about circles in section 25.

OXFORD
UNIVERSITY PRESS

OXFORD
UNIVERSITY PRESS
Great Clarendon Street, Oxford OX2 6DP

Oxford University Press is a department of the University of Oxford.
It furthers the University's objective of excellence in research, scholarship,
and education by publishing worldwide in

Oxford   New York
Auckland   Bangkok   Buenos Aires   Cape Town   Chennai
Dar es Salaam   Delhi   Hong Kong   Istanbul   Karachi   Kolkata
Kuala Lumpur   Madrid   Melbourne   Mexico City   Mumbai   Nairobi
São Paulo   Shanghai   Taipei   Tokyo   Toronto

Oxford is a registered trade mark of Oxford University Press
in the UK and in certain other countries

Many thanks to the original authors of this series for the use of their material:
Sue Briggs   Peter McGuire   Derek Philpott   Susan Shilton   Ken Smith

The authors would like to thank:
Paul Metcalf for his authoritative coursework guidance; and
Keith Gordon and Trevor Senior for their invaluable contribution to the revised content of this book.

British Cataloguing in Publication Data

Data available

ISBN 0 19 914810 4

The publishers would like to thank Edexcel for their kind permission to reproduce past paper questions. Edexcel
accept no responsibility for the answers to the past paper questions which are the sole responsibility of the publishers.

The publishers and authors are grateful to the following:

Illustrators	Gecko, Oxford Illustrators
Photographers	Mike Dudley, Martin Sookias, Andrew Ward
Photographic Libraries	Ancient Art and Architecture Collection Ltd, Architectural Association, Mary Evans Picture Library
Suppliers	Cornish Seal Sanctuary, Eurostar
Cover artwork	Pictor Photo Library

Typeset in Great Britain by Mathematical Composition Setters Ltd, Hilltop Business Park, Devizes Road, Salisbury, Wiltshire.
Printed in Spain by Edelvives, Zaragoza.

About this book

This book is designed to help you achieve your best possible grade in the **Edexcel Specification A** Intermediate GCSE examination. It can also be used for the **Modular specification B** examination. The book is divided into two parts: a full-colour part containing the GCSE content that you need to learn, and a black-and-white part containing plenty of exam practice and answers.

This is more than just a book of questions: it is a learning package that will help you to make the most of your mathematical talents and expertise. You can be confident that you will be well prepared for your Edexcel GCSE examination.

Wordfinder As well as a detailed **Contents** list and an **index**, the book contains a **Wordfinder** on pages 10 and 11. This provides an alphabetical list of the mathematical terms used in the book, and tells you where to find them.

Starting points The content of the book is divided into sections, each covering a particular topic. Each section begins with **Starting points**, which lists the facts and techniques that you should already know. There are some questions for you to try out before starting the section.

Sections Colour is used in the sections in the following ways:

- Yellow panels contain explanations and worked examples.
- The more difficult questions in an exercise are numbered in blue - for example, on page 84:
  7 Show that $2 (n + 2) + 2n$ and $4n + 4$ are equivalent expressions.
- Blue text is used at times to stress important points – for example, on page 98, the digits 4 and 0 are emphasized in the number 0.00403.
- Words in the margin that link with the main text are coloured red. For example, on page 69: quadrilateral is defined in the margin and referred to in the main text. There are also plenty of exercises to consolidate your learning. A **non-calculator icon** is used to identify exercises where you should not use a calculator.

End points Each section finishes with **Endpoints**, where the main work of the section is summarized. These are excellent for revision purposes.

Skills break At certain points throughout the book there are **Skills breaks**. Each one provides a variety of questions all linked to the same data, and allows you to draw on the skills and techniques that you have learned.

In focus Towards the end of the book are **In focus** pages which relate to a particular section, and you can use these for in-depth revision.

Coursework There is a separate **Coursework Guidance** section, which details what you need to do for your investigative task and your statistical task. It contains sample tasks, with **Moderator comments** to help you get better marks.

Exam questions The **Exam questions** section contains hundreds of past paper questions from Edexcel to help you prepare for your exam. There are also two **practice exam papers**, one calculator and one non-calculator, so that you can gain familiarity and confidence with the new Edexcel examination style.

Answers Numerical **answers** are given at the end of the book.

Good luck in your exams!

Note for students taking the modular GCSE course:

If you are following **Edexcel Specification B (1388) modular**, the symbol will tell you which module the topic is assessed in.

■ Stage 1 (Comprising module 1)
■ ■ Stage 2 (Comprising module 2)
■ ■ ■ Stage 3 (Comprising the terminal examination)

The coursework component (contained within the terminal assessment) is covered in the Coursework Guidance section.

A note on accuracy

Make sure your answer is given to any degree of accuracy stated in the question, for example 2 dp or 1 sf. Where it is not stated, choose a sensible degree of accuracy for your answer, and make sure you work to a greater degree of accuracy through the problem. For example, if you choose to give an answer to 3 sf, work to at least 4 sf through the problem, then round your final answer to 3 sf.

Examination groups differ in their approach to accuracy. Some say that you should not give your final answer to a greater degree of accuracy than that used for the data in the question, but others state answers should be given to 3 sf.

If you are in any doubt, check with your examination group.

Metric and imperial units

	Metric	Imperial	Some approximate conversions
Length	millimetres (mm) centimetres (cm) metres (m) kilometres (km) 1 cm = 10 mm 1 m = 100 cm 1 km = 1000 m	inches (in) feet (ft) yards (yd) miles 1 ft = 12 in 1 yd = 3 ft 1 mile = 1760 yd	1 inch = 2.54 cm 1 foot ≈ 30.5cm 1 metre ≈ 39.4 in 1 mile ≈ 1.61 km
Mass	grams (g) kilograms (kg) tonnes 1 kg = 1000 g 1 tonne = 1000 kg	ounces (oz) pounds (lb) stones 1 lb = 16 oz 1 stone = 14 lb	1 pound ≈ 454 g 1 kilogram ≈ 2.2 lb
Capacity	millilitres (ml) centilitres (cl) litres 1 cl = 10 ml 1 litre = 100 cl = 1000 ml	pints (pt) gallons 1 gallon = 8 pints	1 gallon ≈ 4.55 litres 1 litre ≈ 1.76 pints ≈ 0.22 gallons

Starting points
You need to know about ...

... so try these questions.

A Negative numbers

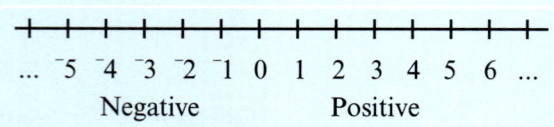

... ⁻5 ⁻4 ⁻3 ⁻2 ⁻1 0 1 2 3 4 5 6 ...

Negative   Positive

- To **add** a negative number:
  - subtract the corresponding positive number.

$7 + {}^-4$	$2 + {}^-5$	${}^-4 + {}^-3$
$= 7 - 4$	$= 2 - 5$	$= {}^-4 - 3$
$= 3$	$= {}^-3$	$= {}^-7$

- To **subtract** a negative number:
  - add the corresponding positive number.

$7 - {}^-4$	$2 - {}^-5$	${}^-4 - {}^-3$
$= 7 + 4$	$= 2 + 5$	$= {}^-4 + 3$
$= 11$	$= 7$	$= {}^-1$

A1 Calculate these missing numbers.
- **a** $1 + {}^-3 = \square$
- **b** $^-5 + {}^-2 = \square$
- **c** $9 + {}^-5 = \square$
- **d** $4 - {}^-6 = \square$
- **e** $^-7 - {}^-3 = \square$
- **f** $^-2 - {}^-7 = \square$

A2 Calculate these missing numbers.
- **a** $3 + {}^-5 = \square$
- **b** $^-5 + 3 = \square$
- **c** $7 - {}^-2 = \square$
- **d** $^-2 - 7 = \square$
- **e** $6 + \square = 3$
- **f** $7 - \square = 5$
- **g** $^-3 - \square = 1$
- **h** $^-4 + \square = {}^-9$
- **i** $^-5 + \square = {}^-1$
- **j** $^-6 - \square = 0$

B Multiples, factors and primes

- The **multiples** of a number can be divided by the number without leaving a remainder.
- The **common multiples** of two numbers are numbers that are multiples of both the numbers.

Multiples of 4:
  4, 8, 12, 16, 20, 24, 28, 32, ...
Multiples of 6:
  6, 12, 18, 24, 30, 36, 42, ...
Common multiples of 4 and 6:
  12, 24, 36, ...

- The **factors** of a number are whole numbers that divide into it without leaving a remainder.
- The **common factors** of two numbers are numbers that are factors of both the numbers.

Factors of 18:
  1, 2, 3, 6, 9, 18
Factors of 42:
  1, 2, 3, 6, 7, 14, 21, 42
Common factors of 18 and 42:
  1, 2, 3, 6

- A number is **prime** if it has only two different factors.

5 is a prime number:
it has factors
1 and 5.

9 is not prime:
it has factors
1, 3 and 9.

B1 Give seven multiples of:
- **a** 5
- **b** 8
- **c** 12

B2 Give three common multiples of:
- **a** 5 and 8
- **b** 8 and 12

B3 Give all the factors of:
- **a** 24
- **b** 30
- **c** 25
- **d** 17

B4 Give the common factors of:
- **a** 24 and 30
- **b** 25 and 30

B5 Explain why 1 is not prime.

B6 Give all the prime numbers less than 20.

C Writing powers using index notation

$3 \times 3 \times 3 \times 3 = 3^4$
$1.6 \times 1.6 = 1.6^2$

4 is the **index**.
4 and 2 are **indices**.

We say 3^4 is:
3 **to the power** 4, or
3 **raised to the power** 4.

C1 Write in index notation:
- **a** 5×5
- **b** $2 \times 2 \times 2 \times 2 \times 2$

C2 Evaluate:
- **a** 4^3
- **b** 3^4
- **c** 1.5^2
- **d** 1.1^5

D Writing a number as a product of primes

- The factors of a number which are prime are called **prime factors**.
- A multiplication of prime factors is called a **product of primes**.
- To write a number as a product of primes:
  - break the number down into pairs of factors until all the factors are prime
  - use index notation to write the powers of each prime.

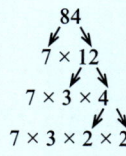

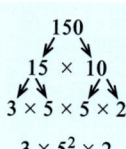

D1 Give the prime factors of:
- **a** 84
- **b** 154

D2 Write these as a product of primes.
- **a** 120
- **b** 350

E Types of fraction

- There are several ways to think of a fraction such as $\frac{3}{5}$.

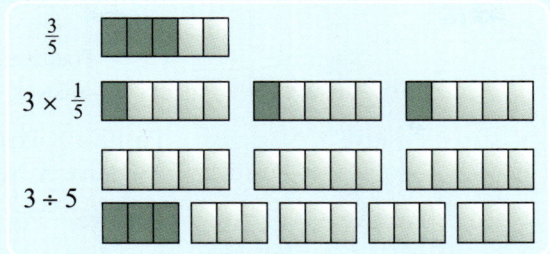

$\frac{3}{5}$

This shows that **multiplying by $\frac{1}{5}$** has the same effect as **dividing by 5**.

$3 \times \frac{1}{5}$

$3 \div 5$

- An **improper fraction** is one where the numerator is larger than the denominator.

$$\frac{9}{7}$$

- An improper fraction is greater than 1, so it can be written as a **mixed number**.

$$\frac{9}{7} = \frac{7}{7} + \frac{2}{7} = 1\frac{2}{7}$$

- A **whole number** can be written as an improper fraction.

$$3 = \frac{3}{1}$$

F Equivalent fractions

- To write an equivalent fraction:
  - ❖ multiply or divide the numerator or denominator by the same number.

$$\overset{\times 5}{\frac{3}{4}} = \frac{15}{20} \qquad \overset{\div 3}{\frac{27}{42}} = \frac{9}{14}$$

G Writing fractions as decimals

- To write a fraction as a decimal:
  - ❖ divide the numerator by the denominator.

$$\frac{3}{8} = 3 \div 8 = \mathbf{0.375}$$

Some fractions give **recurring decimals**.

$$\frac{2}{3} = 0.\dot{6} \qquad \frac{5}{6} = 0.8\dot{3} \qquad \frac{59}{270} = 0.2\dot{1}8\dot{5}$$

H Calculating with fractions

- Multiplying by a fraction less than 1 gives a smaller amount.

- To calculate a fraction of an amount: either
  - ❖ divide the amount by the denominator and multiply by the numerator

or

  - ❖ write the fraction as a decimal and multiply by the amount.

$$\frac{3}{8} \text{ of } \mathbf{£112}$$

$$112 \div 8 \times 3 = \mathbf{£42}$$

$$0.375 \times 112 = \mathbf{£42}$$

- Dividing by a fraction less than 1 gives a larger amount.

$$6 \div \frac{2}{3} = 9$$

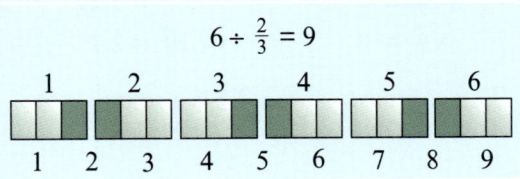

E1 Draw a diagram to show that $\frac{2}{3}$ is the same as $2 \times \frac{1}{3}$ and $2 \div 3$.

E2 Write as a mixed number:

    a $\frac{4}{3}$ **b** $\frac{13}{6}$ **c** $\frac{23}{4}$

E3 Write as an improper fraction:

    a $1\frac{2}{5}$ **b** $3\frac{1}{3}$ **c** 5

    d $2\frac{11}{13}$ **e** 11

F1 Write three equivalent fractions for:

    a $\frac{2}{5}$ **b** $\frac{4}{12}$ **c** $\frac{20}{30}$

G1 Write these fractions as decimals.

    a $\frac{7}{16}$ **b** $\frac{5}{9}$

    c $\frac{7}{11}$ **d** $\frac{23}{54}$

H1 Calculate:

    a $\frac{5}{8}$ of £144

    b $\frac{2}{7}$ of 322 litres

    c $\frac{9}{14}$ of 154 m

H2 Draw a diagram to show that $6 \div \frac{3}{4} = 8$

Multiplying and dividing negative numbers

♦ Multiplying or dividing numbers give either a negative or a positive answer.

Negative answer	Positive answer
Positive × Negative	Positive × Positive
Negative × Positive	Negative × Negative
Positive ÷ Negative	Positive ÷ Positive
Negative ÷ Positive	Negative ÷ Negative

Exercise 1.1
Multiplying and dividing negative numbers

Do not use a calculator for Exercise 1.1.

1 Which of these calculations give a negative answer?

A 4 × ⁻5   B ⁻2.5 × 6   C ⁻3 ÷ ⁻2.8   D ⁻7.1 ÷ 4.6   E ⁻6.9 × ⁻5

F ⁻3.7 × 4.2   G ⁻3 ÷ ⁻4   H ⁻1.8 × 6   I 6.2 ÷ ⁻3.9   J 12.5 ÷ 3

2 Copy and complete these calculations.

a 3 × ⁻5 = ☐   b ⁻2 × 6 = ☐   c 5 × ⁻4 = ☐   d ⁻3 × ⁻6 = ☐
e ⁻12 ÷ ⁻3 = ☐   f ⁻8 ÷ 4 = ☐   g 24 ÷ ⁻6 = ☐   h ⁻40 ÷ 5 = ☐

3 Copy and complete these calculations.

a 4 × ☐ = ⁻28   b ⁻15 ÷ ☐ = 3   c ⁻4 × ☐ = ⁻32   d ☐ × ⁻6 = 18
e ☐ ÷ ⁻7 = 5   f ☐ ÷ ⁻6 = ⁻6   g ⁻3 × ☐ = 36   h ☐ ÷ 7 = 2

4 ⁻2.3 × 4.1 = ⁻9.43   ⁻12.6 ÷ ⁻1.75 = 7.2

Use these two calculations to answer:

a 2.3 × 4.1   b ⁻2.3 × ⁻4.1   c ⁻4.1 × 2.3
d ⁻12.6 ÷ 1.75   e 12.6 ÷ 1.75   f 12.6 ÷ ⁻1.75

5 ⁻1.9 × 7.4 × ⁻5.8   ⁻3.4 × 5.7 × 1.2 × ⁻5.3 × ⁻2.6

Can you predict if these calculations have a negative or a positive answer? If so, explain how.

Squares and square roots

♦ To **square** a number is to multiply the number by itself.

3 × 3 = 9
3 rows of 3 dots gives 9 dots.

9 is the **square** of 3
3 is a **square root** of 9

$3^2 = 9$
$\sqrt{9} = 3$

♦ You can also square negative numbers and decimal numbers.

⁻3 × ⁻3 = 9
$(⁻3)^2 = 9$
$\sqrt{9} = ⁻3$

⁻3 is a square root of 9

2.7 × 2.7 = 7.29
$2.7^2 = 7.29$
$\sqrt{7.29} = 2.7$

2.7 is a square root of 7.29

⁻2.7 × ⁻2.7 = 7.29
$(⁻2.7)^2 = 7.29$
$\sqrt{7.29} = ⁻2.7$

⁻2.7 is a square root of 7.29

This type of curve is called a **parabola**.

- The square roots of 9 are **3** and **¯3**.

- The square roots of 7.29 are **2.7** and **¯2.7**.

- Each positive number has two square roots:
  - a **positive** one
  - a **negative** one.

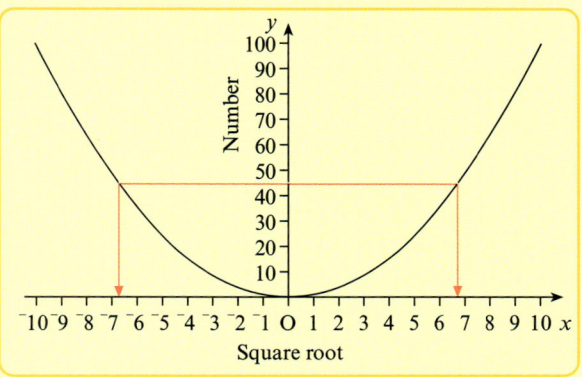

Square roots graph: numbers 0 to 100

The square roots of 45 are about **¯6.7** and **6.7**.

**Exercise 1.2
Square roots**

1   Draw the square roots graph above.

2   Use your graph to estimate the values of each of these square roots.

   a $\sqrt{30}$ **b** $\sqrt{74}$ **c** $\sqrt{12}$ **d** $\sqrt{45}$ **e** $\sqrt{60}$ **f** $\sqrt{95}$

3   Use your calculator to find the values to 2 dp of each of these square roots.

Using the square root key on a calculator only gives the positive square root.

   a $\sqrt{48}$ **b** $\sqrt{72}$ **c** $\sqrt{39.1}$ **d** $\sqrt{26.82}$ **e** $\sqrt{900}$

   f $\sqrt{0.01}$

4   Without a calculator find each of these square roots.

   a $\sqrt{169}$ **b** $\sqrt{196}$ **c** $\sqrt{100}$ **d** $\sqrt{81}$ **e** $\sqrt{225}$

Cubes and cube roots

- To **cube** a number is to multiply the number by itself twice.

The symbol for a cube root is $\sqrt[3]{\ }$.

$$4 \times 4 \times 4 = 64$$
4 layers of 4 rows of 4 cubes gives 64 cubes.

64 is the cube of 4
4 is the cube root of 64

$$4^3 = 64$$
$$\sqrt[3]{64} = 4$$

- You can also cube negative numbers and decimal numbers.

$$^-4 \times ^-4 \times ^-4 = 64$$
$$(^-4)^3 = ^-64$$
$$\sqrt[3]{^-64} = ^-4$$

$$2.7 \times 2.7 \times 2.7 = 19.683$$
$$2.7^3 = 19.683$$
$$\sqrt[3]{19.683} = 2.7$$

- All numbers, positive and negative, have only one cube root.

Exercise 1.3
Cubes and cube roots

1 List the key presses on your calculator to find 6^3.

2 Find:

  a 9^3 **b** 30^3 **c** 3.5^3 **d** $(^-9)^3$ **e** $(^-1.1)^3$ **f** 0.4^3

3 List the key presses on your calculator to find $\sqrt[3]{68.921}$.

4 Find:

  a $\sqrt[3]{8}$ **b** $\sqrt[3]{64}$ **c** $\sqrt[3]{1000}$ **d** $\sqrt[3]{27}$ **e** $\sqrt[3]{729}$

  f $\sqrt[3]{343}$ **g** $\sqrt[3]{15.625}$ **h** $\sqrt[3]{8000}$ **i** $\sqrt[3]{125}$ **j** $\sqrt[3]{0.343}$

> An integer is a positive or negative whole number. 0 is also an integer.

5 **a** Find the cube of each integer from $^-5$ to 5.
  b Draw a graph to show these integers and their cubes.

Reciprocals

> Zero does not have a reciprocal as $\div 0$ is not defined

♦ Two numbers which multiply together to equal 1 are **reciprocals** of each other.

  $$0.8 \times 1.25 = 1$$ 0.8 is the reciprocal of 1.25.
  1.25 is the reciprocal of 0.8.

♦ The reciprocal of any number **n** is: $\dfrac{1}{n}$ or $1 \div n$.

  $$\frac{1}{0.8} = 1.25 \qquad \frac{1}{1.25} = 0.8$$

♦ The reciprocal of any fraction $\dfrac{p}{q}$ is: $\dfrac{q}{p}$.

  Writing 0.8 and 1.25 as fractions gives: $\dfrac{4}{5} \times \dfrac{5}{4} = 1$ $\dfrac{4}{5}$ is the reciprocal of $\dfrac{5}{4}$.
  $\dfrac{5}{4}$ is the reciprocal of $\dfrac{4}{5}$.

Exercise 1.4
Reciprocals

1 Find the reciprocal of:

  a 4 **b** 0.1 **c** 2.5 **d** 3 **e** $0.\dot{6}$ **f** $0.8\dot{3}$ **g** 0

2 Write as a fraction the reciprocal of:

  a $\dfrac{3}{5}$ **b** $\dfrac{7}{4}$ **c** $1\dfrac{1}{2}$ **d** $3\dfrac{1}{3}$ **e** 1.2

3 **a** Copy these axes.
  b Complete the graph by plotting the whole numbers 1 to 10 against their reciprocal.

4 Investigate the relationship between negative numbers and their reciprocals.

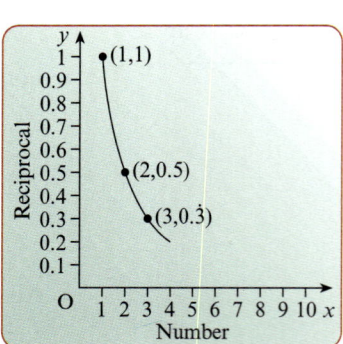

Standard form

♦ A number can be written in several ways using powers of 10. For example:

10^{-3}	10^{-2}	10^{-1}	10^0	10^1	10^2	10^3	10^4
0.001	0.01	0.1	1	10	100	1000	10 000

3600

$0.36 \times 10\,000$
3.6×1000
36×100
360×10
3600×1
$36\,000 \times 0.1$
$360\,000 \times 0.01$
$3\,600\,000 \times 0.001$

0.729

$0.000\,729 \times 1000$
$0.007\,29 \times 100$
0.0729×10
0.729×1
7.29×0.1
72.9×0.01
729×0.001
7290×0.0001

♦ One way of writing a number is called standard form. 3600 and 0.729 in standard form are:

Standard form is also called standard index form.

❖ the first part is a number greater than or equal to 1 and less than 10

3.6×10^3
7.29×10^{-1}

❖ the second part is a power of 10 written in index notation.

♦ To write a number in standard form:

Include any zeros between non-zero digits.
For example:
$40\,900\,000 = 4.09 \times 10^7$

❖ use all the non-zero digits for the first part

43 900 000
4.39

0.000 58
5.8

❖ decide what power of 10 the first part must be multiplied by to give the number

43 900 000
$= 4.39 \times 10\,000\,000$

0.000 58
$= 5.8 \times 0.0001$

❖ write the power of 10 in index notation.

43 900 000
$= 4.39 \times 10^7$

0.000 58
$= 5.8 \times 10^{-4}$

♦ To rewrite a number in standard form as an ordinary number:

❖ write the power of 10 in full and multiply by the first part.

2.4×10^6
$= 2.4 \times 1\,000\,000$
$= \qquad 2\,400\,000$

3.152×10^{-3}
$= 3.152 \times 0.001$
$= \qquad 0.003\,152$

Exercise 1.5
Standard form

1   Write these numbers in standard form.
   a   380 000   **b**   45 100 000   **c**   0.000 92   **d**   0.000 0262

2   Write these as ordinary numbers.
   a 9.42×10^4 **b** 2.5×10^{-3} **c** 7.414×10^8 **d** 6.27×10^{-7}

3
   | **P** 13×10^4 | | **Q** 5.7×1000 | | **R** $6.42 \div 10^3$ | | **S** 0.58×10^{-4} |

   a   Explain why each of these numbers is not written in standard form.
   b   Write each number in standard form.

4   List the key presses on your calculator to find $60 \times (6.19 \times 10^{-11})$.

> Give your answers
> in standard form.

5 Use your calculator to find:

a $365 \times (2.4 \times 10^4)$ **b** $(8.84 \times 10^6) \div 52$

c $(5.5 \times 10^{-4}) \times (3.6 \times 10^7)$ **d** $(8.2 \times 10^8) \div (1.04 \times 10^{-2})$

e $(6.8 \times 10^{-16}) + (7 \times 10^{-15})$ **f** $(4 \times 10^4) - (2.8 \times 10^2)$

Rules of indices

- ◆ To multiply powers of the same number:
  - ❖ add the indices.

  $3^2 \times 3^3$
  $= 3^{2+3}$
  $= 3^5$

  $(3 \times 3) \times (3 \times 3 \times 3)$
  $= 3 \times 3 \times 3 \times 3 \times 3$
  $= 243$

- ◆ To divide powers of the same number:
  - ❖ subtract the indices.

  $3^5 \div 3^2$
  $= 3^{5-2}$
  $= 3^3$

  $(3 \times 3 \times 3 \times 3 \times 3) \div (3 \times 3)$
  $= 243 \div 9$
  $= 27$

- ◆ To raise a power of a number to another power:
  - ❖ multiply the indices.

  $(3^4)^2$
  $= 3^{4 \times 2}$
  $= 3^8$

  $(3 \times 3 \times 3 \times 3) \times (3 \times 3 \times 3 \times 3)$
  $= 81 \times 81$
  $= 6561$

Exercise 1.6
Rules of indices

1 Give the answer to these using index notation.

a $3^2 \times 3^5$ **b** $4^{-3} \times 4^5$ **c** $2^7 \div 2^4$ **d** $7^{-4} \times 7^{-2}$ **e** $6^3 \div 6^5$

f $7^{-4} \div 7^2$ **g** $5^{-1} \times 5^1$ **h** $2^4 \div 2^{-1}$ **i** $6^{-3} \div 6^{-2}$ **j** $7^{-3} \times 7^0$

2 Copy and complete these calculations.

a $2^3 \times 2^\square = 2^7$ **b** $4^5 \times 4^\square = 4^3$ **c** $3^5 \div 3^\square = 3^2$ **d** $7^\square \times 7^{-3} = 7^4$

e $8^4 \div 8^\square = 8^{-2}$ **f** $5^\square \div 5^2 = 5^2$ **g** $2^\square \times 2^5 = 2^{-1}$ **h** $3^4 \div 3^\square = 3^3$

3 Give the answer to these using index notation.

a $(3^2)^4$ **b** $(5^3)^3$ **c** $(4^5)^{-2}$ **d** $(4^{-2})^5$ **e** $(2^4)^{-4}$ **f** $(7^3)^0$

4 Copy and complete these calculations.

a $2^\square \times 2^2 = 32$ **b** $3^6 \times 3^{-3} = \square$ **c** $2^8 \div 2^\square = 64$ **d** $4^2 \div 4^{-1} = \square$

e $(3^2)^\square = 81$ **f** $(5^3)^\square = 125$ **g** $(2^{-3})^\square = 0.125$ **h** $(5^\square)^2 = 0.04$

Standard form without a calculator

- ◆ To multiply: $(6 \times 10^7) \times (3 \times 10^{-2})$
  - ❖ multiply the first parts
  - ❖ multiply the second parts
  - ❖ combine the two answers
  - ❖ write in standard form.

  $6 \times 3 = 18$
  $10^7 \times 10^{-2} = 10^5$
  18×10^5
  1.8×10^6

- ◆ To divide: $(4 \times 10^{-3}) \div (8 \times 10^5)$
  - ❖ divide the first parts
  - ❖ divide the second parts
  - ❖ combine the two answers
  - ❖ write in standard form.

  $4 \div 8 = 0.5$
  $10^{-3} \div 10^5 = 10^{-8}$
  0.5×10^{-8}
  5×10^{-9}

Exercise 1.7
Standard form
without a calculator

1 Give the answer to these in standard form.

a $(6 \times 10^{-4}) \times (4 \times 10^7)$ **b** $(6 \times 10^{-2}) \div (3 \times 10^4)$

c $(5 \times 10^{-9}) \div (2.5 \times 10^{-12})$ **d** $(4 \times 10^{-3}) \times (1.5 \times 10^{-8})$

Least common multiples

♦ To find the least common multiple (LCM) of 24 and 45:

   ❖ write each number as a product of primes

$$24 = 2^3 \times 3$$
$$45 = 3^2 \times 5$$

   ❖ take the highest power of each prime factor to give a new product of primes

$$2^3 \times 3^2 \times 5$$

   ❖ evaluate the new product of primes.

$$8 \times 9 \times 5$$
$$= \mathbf{360}$$

Exercise 1.8
Least common multiples

1   Find the LCM of:

   a 12 and 18   **b** 20 and 42   **c** 28 and 30   **d** 150 and 315
   e 10, 18 and 35   **f** 15, 27 and 70   **g** 66, 135 and 275

2   The LCM of two numbers is 630. One number is 42.
     Give the numbers the other one could be.

Adding and subtracting fractions

♦ You can only add or subtract fractions when the denominators are the same.

♦ To add or subtract fractions:

$$\frac{3}{4} + \frac{1}{6} \qquad \frac{7}{8} - \frac{1}{3}$$

   ❖ find the LCM of the denominators

$$12 \qquad 24$$

   ❖ find equivalent fractions with the LCM as the new denominator

$$\frac{9}{12} + \frac{2}{12} \qquad \frac{21}{24} - \frac{8}{24}$$

   ❖ add or subtract the numerators.

$$\frac{11}{12} \qquad \frac{13}{24}$$

Any common multiple can be used as the new denominator.

Using the LCM will usually give an answer in its lowest terms.

Exercise 1.9
Adding and subtracting fractions

1   Evaluate:

   a $\frac{1}{2} + \frac{1}{3}$ **b** $\frac{2}{5} + \frac{1}{4}$ **c** $\frac{1}{2} - \frac{1}{3}$ **d** $\frac{3}{4} - \frac{1}{5}$

   e $\frac{2}{5} + \frac{3}{8}$ **f** $\frac{1}{3} + \frac{1}{4} + \frac{1}{6}$ **g** $\frac{2}{3} - \frac{1}{6}$ **h** $\frac{3}{5} + \frac{1}{2} - \frac{1}{4}$

2   Write the answer to each of these as a mixed number.

   a $\frac{2}{3} + \frac{1}{2}$ **b** $\frac{1}{4} + \frac{4}{5}$ **c** $\frac{1}{2} + \frac{1}{3} + \frac{1}{4}$ **d** $\frac{3}{8} + \frac{3}{4} + \frac{1}{2}$

3   Evaluate:

   a $1\frac{1}{2} + \frac{4}{5}$ **b** $1\frac{4}{5} - \frac{1}{2}$ **c** $1\frac{1}{3} + 2\frac{1}{2}$

   d $2\frac{2}{5} - 1\frac{1}{2}$ **e** $2\frac{5}{8} + 1\frac{1}{3}$ **f** $2\frac{1}{4} - 1\frac{3}{5}$

Write any mixed numbers as improper fractions before adding or subtracting.

4   Fractions with a numerator of 1 are called **unit fractions**.

   Investigate fractions like this which can be written as the sum of two unit fractions.

$$\frac{2}{5} = \frac{1}{3} + \frac{1}{15}$$

Thinking ahead to ...
multiplying and
dividing fractions

A   Calculate:

a $10 \div 1.25$ **b** 10×0.8 $\boxed{1.25 = \dfrac{5}{4} \quad 0.8 = \dfrac{4}{5}}$

c $16 \div 1.25$ **d** 16×0.8

B   What do your answers to Question **A** tell you about dividing by a fraction?

Multiplying and dividing fractions

◆ To multiply fractions:
  ❖ multiply the numerators
  ❖ multiply the denominators
  ❖ write the answer in its lowest terms.

$$\frac{4}{5} \times \frac{3}{8} = \frac{12}{40} = \frac{3}{10}$$

> The answer may already be in its lowest terms.

◆ Dividing by a fraction has the same effect as multiplying by its reciprocal.

◆ To divide fractions:
  ❖ write the division as a multiplication
  ❖ multiply the fractions.

$$\frac{1}{6} \div \frac{2}{3}$$
$$= \frac{1}{6} \times \frac{3}{2} = \frac{3}{12} = \frac{1}{4}$$

Exercise 1.10
Multiplying and
dividing fractions

1   Evaluate:

a $\dfrac{2}{3} \times \dfrac{5}{8}$ **b** $\dfrac{3}{5} \times \dfrac{5}{6}$ **c** $\dfrac{4}{7} \times \dfrac{3}{5}$ **d** $\dfrac{9}{14} \times \dfrac{2}{3}$

e $\dfrac{1}{4} \div \dfrac{5}{8}$ **f** $\dfrac{2}{3} \div \dfrac{6}{7}$ **g** $\dfrac{2}{5} \div \dfrac{3}{4}$ **h** $\dfrac{1}{4} \div \dfrac{3}{5}$

2   Write the answer to each of these as a mixed number or a whole number.

a $6 \times \dfrac{2}{5}$ **b** $8 \times \dfrac{3}{4}$ **c** $10 \div \dfrac{5}{6}$ **d** $7 \div \dfrac{2}{3}$

e $12 \times 1\dfrac{3}{4}$ **f** $9 \times 3\dfrac{1}{2}$ **g** $5 \div 1\dfrac{1}{6}$ **h** $11 \div 2\dfrac{3}{5}$

> Write any mixed numbers or whole numbers as improper fractions before multiplying or dividing.

3   Use the fraction key on your calculator to evaluate:

a $2\dfrac{1}{2} \times 1\dfrac{4}{5}$ **b** $1\dfrac{1}{6} \div \dfrac{2}{5}$ **c** $4\dfrac{1}{2} \div 1\dfrac{3}{4}$ **d** $2\dfrac{3}{4} \times 2\dfrac{2}{3}$

e $\dfrac{4}{5} \times 1\dfrac{1}{4}$ **f** $2\dfrac{1}{2} \div 2\dfrac{1}{4}$ **g** $1\dfrac{3}{5} \times 3\dfrac{3}{4}$ **h** $\dfrac{5}{6} \div 1\dfrac{3}{8}$

Highest common factors

◆ To find the highest common factor (HCF) of 120 and 252:

  ❖ write each number as a product of primes

$$120 = 2^3 \times 3 \times 5$$

$$252 = 2^2 \times 3^2 \times 7$$

  ❖ take the lowest power of each common prime factor to give a new product of primes

$$2^2 \times 3$$

  ❖ evaluate the new product of primes.

$$4 \times 3$$
$$= 12$$

Exercise 1.11
Highest common factors

1   Use products of primes to find the HCF of:

a   252 and 360   **b**   120 and 315   **c**   693 and 1078

d   48, 66 and 225

End points

You should be able to ...   ... so try these questions.

A   Multiply and divide negative numbers

A1   Copy and complete these calculations.
  a $7 \times {}^-2 = \square$ **b** ${}^-3 \times {}^-5 = \square$ **c** ${}^-12 \div 4 = \square$
  d $20 \div {}^-5 = \square$ **e** $\square \times {}^-3 = {}^-18$ **f** $\square \div {}^-2 = 8$
  g ${}^-4 \times \square = 36$ **h** $\square \div 7 = {}^-3$

B   Find roots, cubes and reciprocals

B1   Use your calculator to find:
  a $\sqrt{1.44}$ **b** 3.3^3 **c** $\sqrt[3]{216}$ **d** $\sqrt[3]{3.375}$

B2   Find the reciprocal of:
  a 8 **b** 0.2 **c** $0.\dot{5}$ **d** $\frac{2}{5}$ **e** $1\frac{1}{4}$

C   Understand and use the rules of indices

C1   Give the answer to these using index notation:
  a $3^3 \times 3^4$ **b** $4^5 \div 4^2$ **c** $(7^2)^3$

C2   Copy and complete these calculations:
  a $2^3 \times 2^\square = 128$ **b** $3^5 \times 3^\square = 27$ **c** $(4^2)^\square = 0.0625$

C3   Give the answer to these using index notation.
  a $2^5 \times 2^{-3}$ **b** $3^{-2} \div 3^7$ **c** $(5^2)^5$ **d** $(4^{-2})^3$

C4   Copy and complete these calculations.
  a $3^4 \times 3^\square = 3^{10}$ **b** $4^3 \div 4^\square = 4^7$
  c $7^\square \times 7^7 = 7^5$ **d** $(6^{-2})^0 = 6^\square$

D   Understand and use standard form

D1   Write these numbers in standard form.
  a $70\,620\,000$ **b** $0.000\,003\,75$

D2   Write these as ordinary numbers.
  a 6.9×10^{-8} **b** 1.03×10^{10}

D3   Use your calculator to find:
  a $52 \times (1.92 \times 10^{13})$ **b** $(6.58 \times 10^{-7}) \div 7$ **c** $(4.1 \times 10^{27}) + (7 \times 10^{26})$

D4   Without using a calculator, give the answer to these in standard form.
  a $(7 \times 10^5) \times (6 \times 10^{-9})$ **b** $(8 \times 10^{-2}) \div (4 \times 10^{-6})$

E   Calculate with fractions

E1   Evaluate:
  a $\frac{2}{3} + \frac{1}{5}$ **b** $\frac{5}{6} - \frac{2}{5}$ **c** $2\frac{3}{8} - 1\frac{3}{4}$ **d** $1\frac{5}{7} + 2\frac{1}{6}$

E2   Evaluate:
  a $\frac{3}{4} \times \frac{2}{5}$ **b** $\frac{1}{6} \div \frac{2}{3}$ **c** $3\frac{1}{3} \times 1\frac{4}{5}$ **d** $2\frac{1}{2} \div 1\frac{3}{4}$

F   Find least common multiples and highest common factors

F1   Find the LCM of:
  a 12 and 21 **b** 24 and 90

F2   Find the HCF of:
  a 14 and 35 **b** 105 and 350

Some points to remember

- Each positive number has two square roots: a positive one and a negative one.
- All numbers, positive and negative, have only one cube root.
- A number to a negative power is a fraction.
- Fractions can only be added or subtracted when the denominators are the same.

Starting points

You need to know about ...

... so try these questions.

A Finding the value of an expression

- Examples of expressions are:

$3n + 1 = (3 \times n) + 1$ $9 - 4n = 9 - (4 \times n)$ $2a - b = (2 \times a) - b$

$k^2 + 1 = (k \times k) + 1$ $2y^2 = 2 \times y \times y$ $g(g + 1) = g \times (g + 1)$

- The value of an expression depends on the value of the letters.

  For example: ❖ when $n = 5$, $3n + 1 = (3 \times n) + 1$
  $$= (3 \times 5) + 1$$
  $$= 16$$

  ❖ when $y = 3$, $2y^2 = 2 \times y \times y$
  $$= 2 \times 3 \times 3$$
  $$= 18$$

B Some types of number sequences

- Even numbers: 2, 4, 6, 8, 10, ...

- Odd numbers: 1, 3, 5, 7, 9, ...

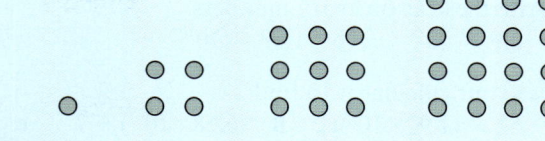

- Square numbers: 1,   4,   9,   16, ...

- Triangle numbers: 1,   3,   6,   10, ...

- Powers of 2:   2, 4, 8, 16, 32, ...

- Powers of 3:   3, 9, 27, 81, 243, ...

- Fibonacci numbers: 1, 1, 2, 3, 5, 8, 13, ...

  The sequence of Fibonacci numbers begins 1, 1, ...
  Further numbers in the sequence are found by adding the previous two numbers together. For example, the next Fibonacci number is $8 + 13 = 21$.

  Start with two different numbers to find other sequences like this, for example: 3, 6, 9, 15, 24, ...

A1 Find the value of these expressions when $n = 4$.
  a $5n + 1$ **b** $7 - n$
  c $2n - 8$ **d** $3(2 + n)$

A2 Find the value of these expressions when $a = 3$.
  a $a^2 + 10$ **b** $4a^2$
  c $a^2 - 5$ **d** $a(a + 1)$

B1 Draw a pattern of dots to show that 25 is a square number.

B2 Find four odd square numbers.

B3 Write down the first six triangle numbers.

B4 List the first four powers of 5.

B5 Which of these are Fibonacci numbers:
  30   34   62   90?

B6 Each number in this sequence is found by adding the previous two numbers together.

  1, 3, 4, 7, 11, 18, ...

Find the next four numbers in this sequence.

C Continuing a sequence

♦ A **sequence** of numbers usually follows a pattern or rule.

♦ Each number in a sequence is called a **term**.

For example,
in the sequence: 3, 5, 7, 9, 11, ...   the 1st term is 3
the 2nd term is 5
the 3rd term is 7.

♦ A sequence can often be continued by finding a pattern in the **differences**.

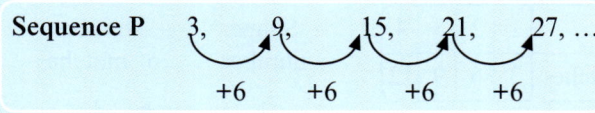

Sequence P   3,   9,   15,   21,   27, ...

+6   +6   +6   +6

❖ In Sequence P, the **first difference** is 6 each time.
So continue the sequence by adding 6.

D A rule for a sequence

These are the first three matchstick patterns in a sequence.

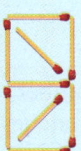 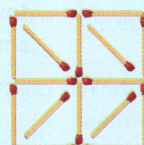 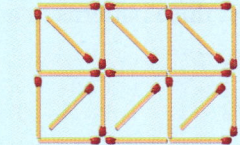

Pattern 1   Pattern 2   Pattern 3

♦ The number of matches in each pattern can be shown in a table.

Pattern number (n)	Number of matches (m)
1	9
2	16
3	23
4	30

+1 (between pattern numbers, +7 between matches

❖ The pattern number goes up by 1 each time.

❖ The number of matches goes up by 7 each time.

❖ So a rule that links the number of matches (m) with the pattern number (n) begins $m = 7n$ ...

❖ A rule that fits all the results in the table is $m = 7n + 2$

♦ This rule can be used to calculate the number of matches in any pattern,
for example: in Pattern 10 there are $(7 \times 10) + 2 = 72$ matches.

C1 Find the 6th and 7th term in sequence P.

C2 What is the 10th term in this sequence?

4, 7, 10, 13, 16, ...

C3 For each sequence, find the next three terms.
 a 6, 7, 10, 15, 22, ...
 b 2, 8, 14, 20, 26, ...
 c 2, 5, 11, 20, 32, ...
 d 2, 3, 9, 20, 36, ...

D1 These are the first three matchstick patterns in a sequence.

Pattern 1

Pattern 2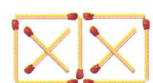

Pattern 3

 a Draw pattern 4 and pattern 5 in this sequence.

 b How many matches are in:
 i pattern 3   **ii** pattern 5?

 c Make a table for the first five patterns in this sequence.

 d Which of these rules fits the results in your table?

$m = 4n + 2$ $m = 5n + 1$

$m = 6n - 1$

Sequences and mappings

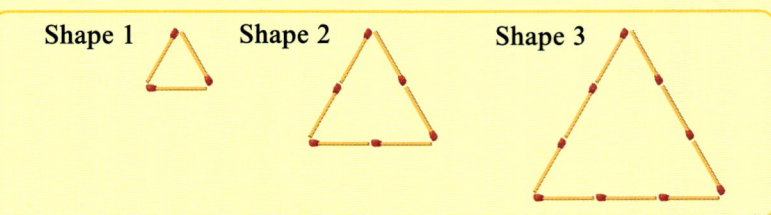

These equilateral triangles are the first three in a sequence of shapes.

Shape 1   Shape 2   Shape 3

◆ Data for this sequence can be shown in a table.

Shape number	1	2	3	4
Number of matches	3	6	9	12

◆ The data can also be shown in a **mapping diagram** like this.

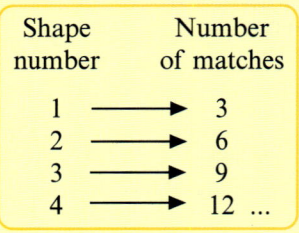

Shape number	Number of matches
1	3
2	6
3	9
4	12 ...

You can choose any letter to stand for the shape number.

For example:
the rule
$n \longrightarrow 3n$
can be written as
$s \longrightarrow 3s$
or
$p \longrightarrow 3p$
or …

◆ The total number of matches is 3 times the shape number. For example, the 50th shape in the sequence uses 150 matches.

Using n to stand for the shape number, the rule for the mapping diagram can be written: $n \longrightarrow 3n$

Exercise 2.1
Sequences and mappings

1 These are the first three patterns in a sequence.

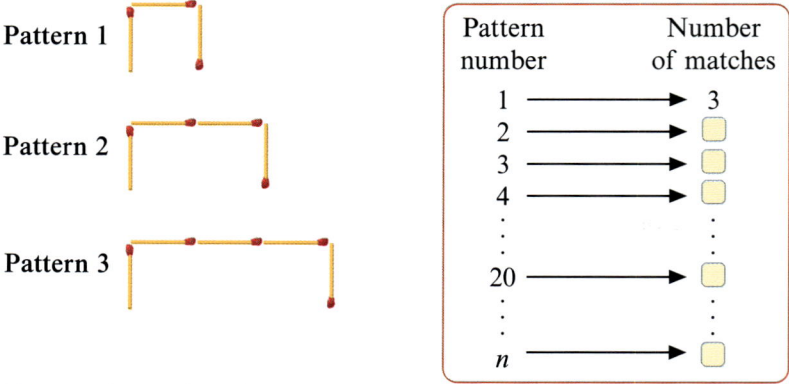

Pattern 1

Pattern 2

Pattern 3

Pattern number	Number of matches
1	3
2	☐
3	☐
4	☐
⋮	⋮
20	☐
⋮	⋮
n	☐

Copy and complete the mapping diagram for the sequence.

2 These patterns of touching squares are the first four in a sequence.

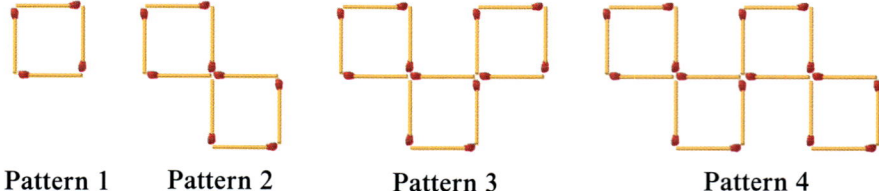

Pattern 1   Pattern 2   Pattern 3   Pattern 4

a Draw a mapping diagram for the first six patterns of touching squares.
b Find a rule for the sequence in the form $n \longrightarrow$ …. , where n is the pattern number.
c Use your rule to calculate the number of matches in the 100th pattern.

Thinking ahead to ...
finding rules

A These are the first three patterns in sequence A.

Sequence A

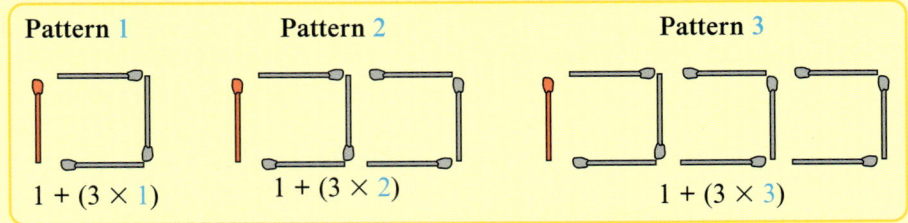

Pattern 1   Pattern 2   Pattern 3

How many matchsticks are in the 100th pattern?
Explain how you worked it out.

Finding rules

To find a rule for the number of matches (*m*) in the *n*th pattern in sequence A above

◆ **Method 1** Look at how the patterns are made.

| Pattern 1 | Pattern 2 | Pattern 3 |

1 + (3 × 1)   1 + (3 × 2)   1 + (3 × 3)

❖ So a rule for the number of matches (*m*) in the *n*th pattern is **$m = 1 + 3n$**

◆ **Method 2** Look at differences.

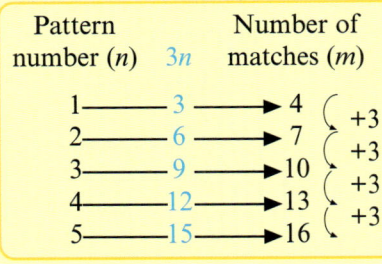

Pattern
number (*n*)   3*n*   matches (*m*)

| 1 — 3 → 4 +3 |
| 2 — 6 → 7 +3 |
| 3 — 9 → 10 +3 |
| 4 — 12 → 13 +3 |
| 5 — 15 → 16 +3 |

❖ The pattern number goes up by 1 each time.

❖ The number of matches goes up by 3 each time so there is a linear rule that begins $m = 3n$ ... .

Examples of linear rules are:

$m = 4n + 3$
$y = 2 - 5x$
$a = 3b - 1$

❖ Compare 3*n* with the number of matches.

❖ The number of matches is 1 more than 3*n* each time.

❖ So a rule for the number of matches (*m*) in the *n*th pattern is **$m = 3n + 1$**.

Exercise 2.2
Finding rules

1 These triangle patterns are the first three in sequence B.

Sequence B

 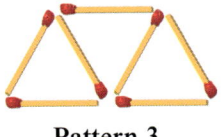

Pattern 1   Pattern 2   Pattern 3

a Find a rule for the number of matches (*m*) in the *n*th triangle pattern. Explain your method.
b Use your rule to find the number of matches in the 8th pattern.
c Check your answer by drawing the 8th pattern and counting the matches.

2   Sequence C

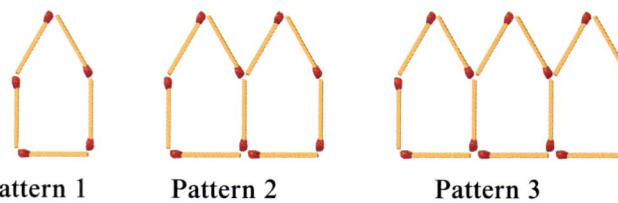

Pattern 1   Pattern 2   Pattern 3

   a   For sequence C, find a rule for the number of matches (m) in the nth pattern.
         Explain your method.
   b   Calculate the number of matches in the 40th pattern.
   c   Which pattern uses exactly 129 matches?

3   Sequence D

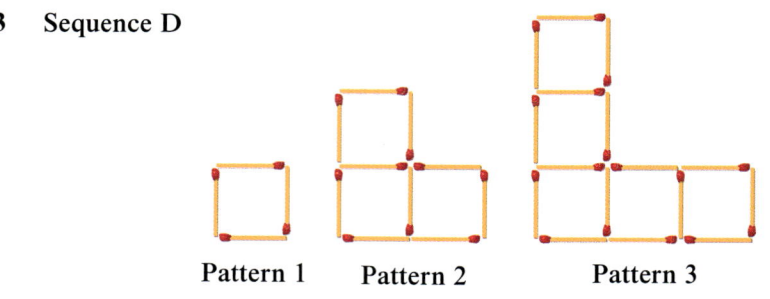

Pattern 1   Pattern 2   Pattern 3

   a   For sequence D, find a rule for the number of matches (m) in the nth pattern.
         Explain your method.
   b   How many of matches are in the 100th pattern?

4   This mapping diagram fits a sequence of matchstick patterns.

Pattern number (n)		Number of matches (m)
1	⟶	6
2	⟶	10
3	⟶	14
4	⟶	18
5	⟶	22

   a   Draw a sequence of matchstick patterns that fits this mapping diagram.
   b   Find a rule for the number of matches (m) in the nth pattern.

5   Copy and complete each mapping diagram.

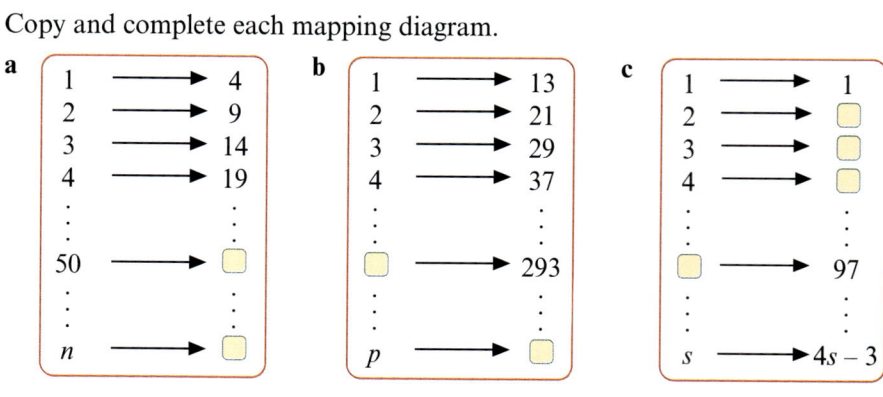

Thinking ahead to ...
finding the *n*th term

A

> 7, 9, 11, 13, 15, 17, ...

a Write the next two terms in this sequence.
b Find the 12th term.
c What is the 50th term?

The *n*th term of a sequence

To find an expression for the *n*th term in the sequence 7, 9, 11, 13, 15, ...

> The 1st term is 7
> The 2nd term is 9
> The 3rd term is 11
> The 4th term is 13
> The 5th term is 15
> . . .

◆ These results can
be shown in
a mapping diagram.

Term number (n)	$2n$	Term
1	2	7
2	4	9
3	6	11
4	8	13
5	10	15

❖ The first difference for the terms is 2 each time.
So there is a linear expression for the *n*th term that begins $2n$ ... .

❖ Each term is 5 more than $2n$.
So an expression for the *n*th term is $2n + 5$.

Exercise 2.3
Finding the *n*th term

1 A 6, 9, 12, 15, 18, ...   B 1, 6, 11, 16, 21, ...
   C 13, 23, 33, 43, 53, ...   D 2, 10, 18, 26, 34, ...

For each of the sequences A to D:
a find an expression for the *n*th term
b use your expression to calculate the 50th term.

2 A student has tried to find the *n*th term of this sequence.

> 5, 8, 11, 14, 17, ...
>
> *n*th term is *n* + 3   ✗

a Explain the mistake you think he has made.
b Find a correct expression for the *n*th term of this sequence.

3 The 2nd term of a sequence is 7.
Which of these could not be an expression for the *n*th term?

> $3n + 1$ $11 - 2n$ $n + 5$ $n + 7$ $5n - 3$

4 Find an expression for the *n*th term of the sequence: 20, 18, 16, 14, 12, ...

Extending number patterns

Exercise 2.4
Extending patterns

1 Morag finds an expression for the *n*th number in this sequence.

4, 10, 18, 28, 40, ...

This is her working:

1st number	4	$= 1 \times 4$
2nd number	10	$= 2 \times 5$
3rd number	18	$= 3 \times 6$
4th number	28	$= 4 \times 7 ...$

So *n*th number $= n \times (n + 3)$

a Show Morag's line of working for the 5th number.
b Find the 10th number in this sequence.
c Explain how Morag's working helps to find an expression for the *n*th number in this sequence.

2 These are the first five triangle numbers:

1, 3, 6, 10, 15, ...

The triangle numbers follow this pattern:

1st triangle number	$1 = \dfrac{1 \times 2}{2}$
2nd triangle number	$3 = \dfrac{2 \times 3}{2}$
3rd triangle number	$6 = \dfrac{3 \times 4}{2}$
4th triangle number	$10 = \dfrac{4 \times 5}{2}$

a What is the next line in this pattern?
b Use the pattern to find the 12th triangle number.
c Find an expression for the *n*th triangle number.

3 These are the first five powers of 2:

2, 4, 8, 16, 32, ...

Powers of 2 follow this pattern:

1st power	2		$= 2^1$
2nd power	4	$= 2 \times 2$	$= 2^2$
3rd power	8	$= 2 \times 2 \times 2$	$= 2^3$
4th power	16	$= 2 \times 2 \times 2 \times 2$	$= 2^4$

a What is the next line in this pattern?
b Use the pattern to find the 8th power of 2.
c Write an expression for the *n*th power of 2.

Writing rules in different ways

Exercise 2.5
Different ways
to write rules

1 These hollow square tile designs are the first three in a sequence.

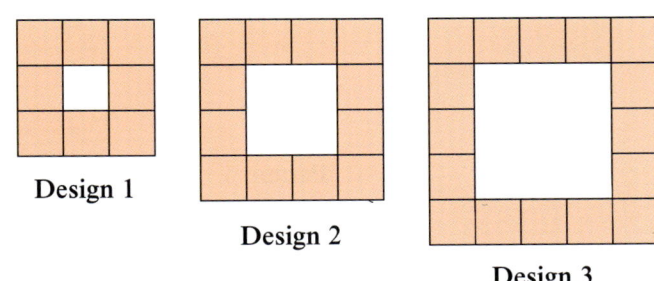

Design 1

Design 2

Design 3

Three students find a rule for the number of tiles in the nth design.
All three students are correct.

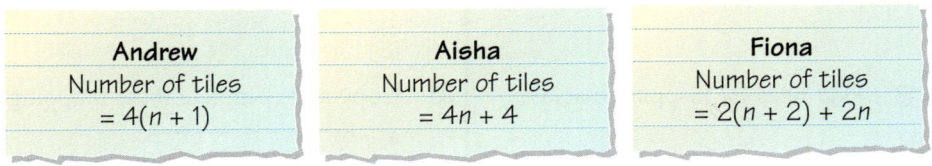

Andrew	Aisha	Fiona
Number of tiles	Number of tiles	Number of tiles
$= 4(n + 1)$	$= 4n + 4$	$= 2(n + 2) + 2n$

The way each student wrote their rule shows how they found it.

Andrew drew this diagram to
show how he found his rule.

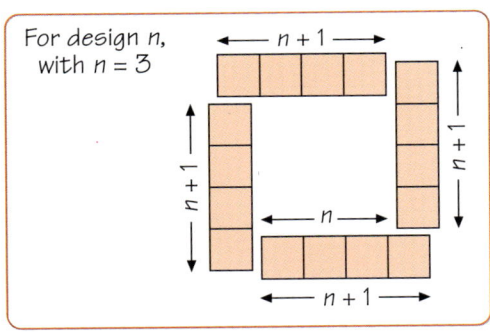

For design n,
with $n = 3$

Fiona's rule is not as easy to
use as the other two rules.

Using Fiona's rule to find
the number of tiles in the
10th design gives:

Number of tiles

$= 2 \times (10 + 2) + (2 \times 10)$
$= 24 + 20$
$= 44$

a Draw a diagram for Aisha's rule.
b Draw a diagram for Fiona's rule.
c Show how you can calculate the number of tiles in the 50th design using:
  i Andrew's rule   **ii** Aisha's rule   **iii** Fiona's rule
d Which design has 324 tiles?
  Explain how you found your result.

2 These designs are the first three in a sequence.

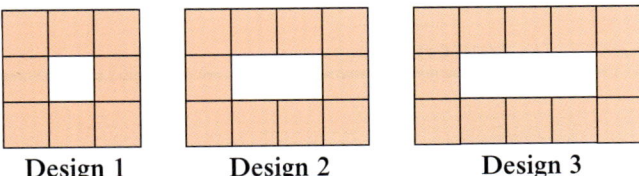

Design 1   Design 2   Design 3

a Find two different ways of writing a rule for the number of tiles in
  the nth design.
b Explain how you found your results.

End points

You should be able to ...   ... so try these questions

A Find a rule that fits a sequence of patterns

A1 These are the first three matchstick patterns in a sequence.

Pattern 1   Pattern 2   Pattern 3

a Find a rule for the number of matches (*m*) in the *n*th pattern. Explain your method.

b Use your rule to find the number of matches in the 100th pattern.

B Find rules for mapping diagrams

B1 Copy and complete these mapping diagrams.

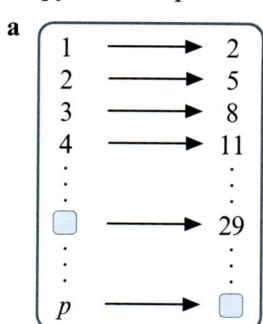

a

1	→	2
2	→	5
3	→	8
4	→	11
⬚	→	29
p	→	⬚

b

1	→	7
2	→	12
3	→	17
4	→	22
20	→	⬚
n	→	⬚

C Find an expression for the *n*th term in a sequence

C1
A 7, 10, 13, 16, 19, ...   B 2, 7, 12, 17, 22, ...
C 5, 9, 13, 17, ...   D 6, 7, 8, 9, 10, ...
E 9, 12, 15, 18, 21, ...   F 11, 18, 25, 32, ...

For each of the sequences A to F:

a find an expression for the *n*th term

b use your expression to calculate the 20th term.

Some points to remember

♦ It is often possible to find a rule for the *n*th pattern in a sequence of patterns by looking at how each pattern can be made.

♦ In a sequence: ❖ If the first differences are *k* each time, there is a simple linear expression for the *n*th term that begins *kn* ... .

Starting points

You need to know about ...   ... so try these questions

A Naming angles and triangles

- Any angle less than 90° is an **acute angle**.
- Any angle equal to 90° is a **right angle**.
- Any angle between 90° and 180° is an **obtuse angle**.
- Any angle between 180° and 360° is a **reflex angle**.

- Any triangle which has:
  - three sides of equal length
  - three equal angles (60°)
  is an **equilateral triangle**.

- Any triangle which has:
  - two sides of equal length
  - two equal angles
  is an **isosceles triangle**.

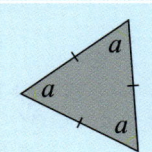
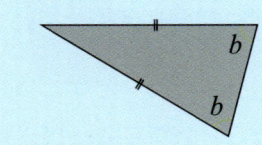

- Any triangle which has no sides of equal length and no equal angles is a **scalene triangle**.
- Any triangle which has one right angle is a **right-angled triangle**.

B Angle sums

- Angles at a point on a straight line add up to 180°.

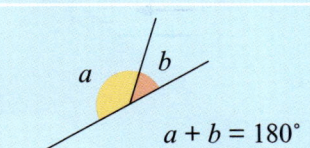

$$a + b = 180°$$

- Angles round a point add up to 360°.

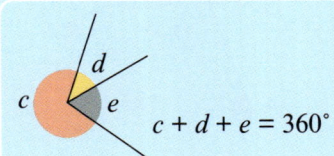

$$c + d + e = 360°$$

- Vertically opposite angles are equal.

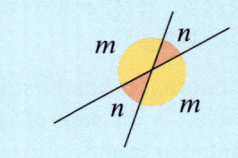

- Angles in a triangle add up to 180°.

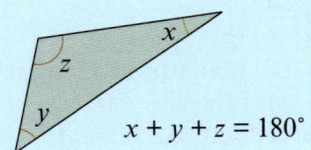

$$x + y + z = 180°$$

C Parallel lines

At each point where a straight line crosses a set of parallel lines there are two pairs of vertically opposite angles.

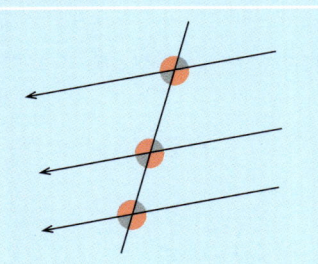

Parallel lines are marked with arrows.

Here equal angles are marked with the same colour.

A1

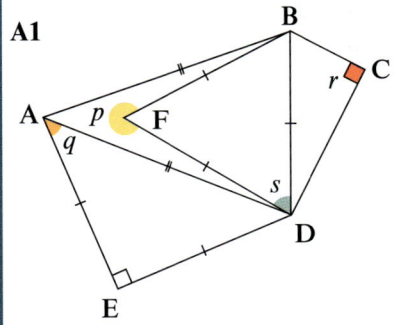

a What type of angle is:
  i p ii q iii r?
b What type of triangle is:
  i BFD   ii BCD?
c Which triangles are isosceles?

B1 In the diagram above calculate:
a the size of angle s
b angle p
c angle $A\hat{D}E$.

B2 On this diagram, angles marked with the same letter are equal in size.

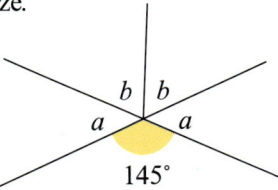

a Work out angles a and b.
b Explain why a triangle can only have one obtuse angle.

C1

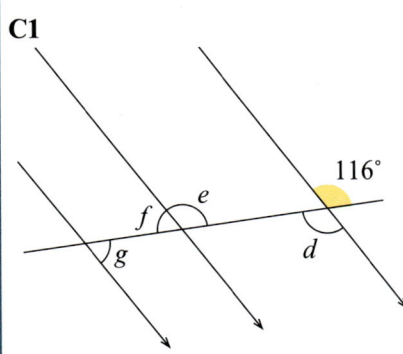

Work out the angles d to g in this diagram.

D Quadrilaterals

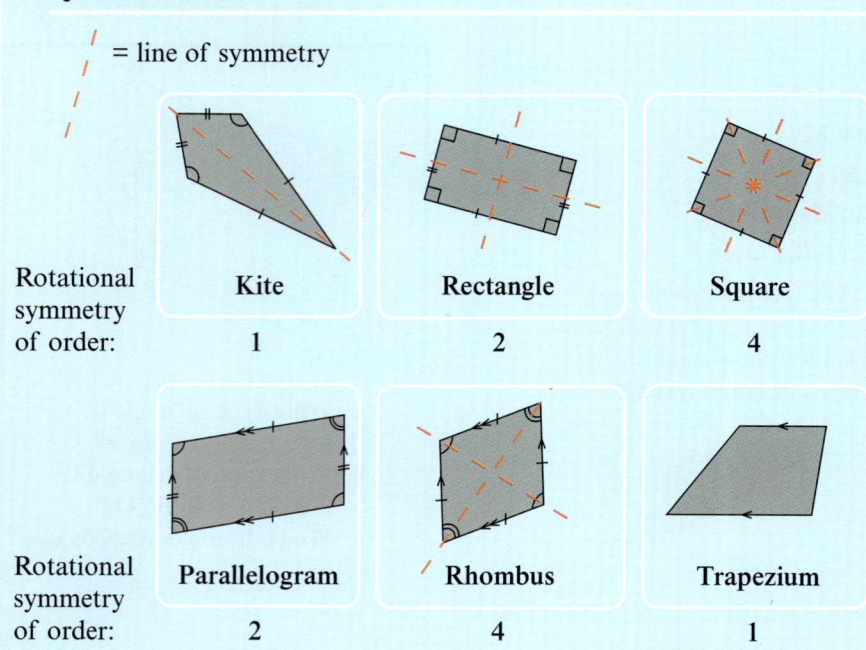

/ = line of symmetry

Kite	**Rectangle**	**Square**
Rotational symmetry of order: 1	2	4
Parallelogram	**Rhombus**	**Trapezium**
Rotational symmetry of order: 2	4	1

E Polygons

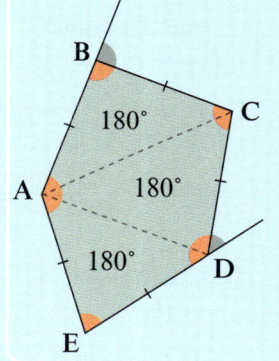

- In ABCDE:
  - the **interior angles** are marked in red
  - the angles marked in blue are not interior angles.

- The **sum of the interior angles** of a polygon with *n* sides is: $(n - 2) \times 180°$
  So for ABCDE the sum of interior angles is:
  $(5 - 2) \times 180°$
  $= 3 \times 180°$
  $= 540°$

- In a **regular polygon** all the sides are equal and all the interior angles are equal.

- ABCDE is an **irregular polygon** The sides are all equal but the interior angles are not.

Name of polygon	Number of sides	Sum of interior angles	Interior angle of a regular polygon
Triangle	3	180° ——— ÷3 ➤	60°
Quadrilateral	4	360° ——— ÷4 ➤	90°
Pentagon	5	540° ——— ÷5 ➤	108°
Hexagon	6	720° ——— ÷6 ➤	120°
Heptagon	7		
Octagon	8		
Nonagon	9		
Decagon	10		

Another expression for the **sum of the interior angles** of a polygon with *n* sides is: $(180° \times n) - 360°$

D1 Name all the quadrilaterals that fit each of these labels.

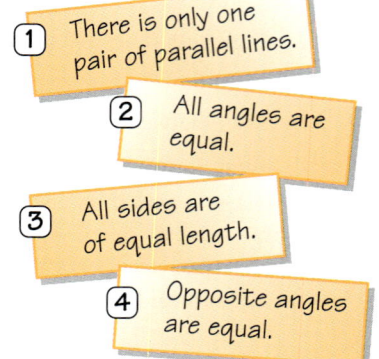

① There is only one pair of parallel lines.

② All angles are equal.

③ All sides are of equal length.

④ Opposite angles are equal.

D2 Draw a trapezium with one line of symmetry.

E1 What is the sum of the interior angles of an octagon?

E2 Calculate the angle *a* in this pentagon.

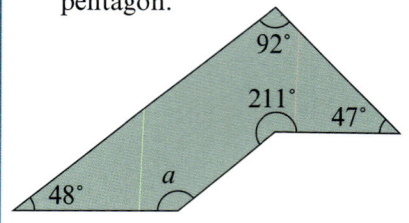

E3 Calculate the interior angle of a regular heptagon to the nearest degree.

E4 A dodecagon has 12 sides.
  a What is the sum of the interior angles of a dodecagon?
  b Calculate the interior angle of a regular dodecagon.

Angles in triangles

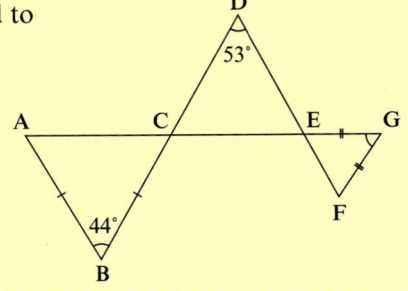

To calculate an angle you may need to work out some other angles first.

Example

Calculate the angle EĜF.

◆ You should sketch a diagram and label each angle that you calculate.

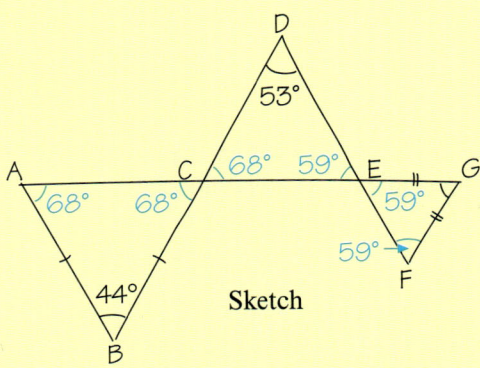

Sketch

To calculate EĜF

Calculation …	**… Reason**
AĈB = CÂB…	…ABC is an isosceles △
= (180° − 44°) ÷ 2	
= 68°	
DĈE = 68°…	…Vertically opposite ACB
DÊC = 180° − (68° + 53°)…	…Angle sum of △
= 59°	
FÊG = 59°…	…Vertically opposite DEC
GÊF = EF̂G…	…EFG is an isosceles △
EĜF = 180° − (59° × 2)…	…Angle sum of △

So the angle EĜF = 62°

△ stands for triangle

An angle can be written in different ways.

For example:
DĈE is the same angle as DĈG and EĈD

DÊC is the same angle as DÊA and CÊD.

You may not need to calculate all the intermediate angles.

Exercise 3.1
Angles in triangles

1 a Which is the easiest angle to calculate in this diagram?
   b Calculate the angles *a* to *f* in this diagram.
   c In what order did you calculate the angles? Explain why.

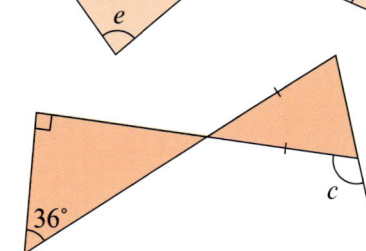

2

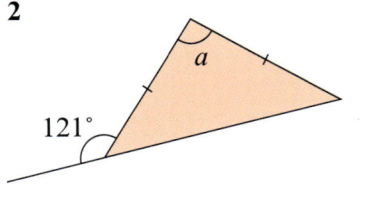

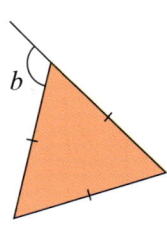

Calculate the angles *a*, *b* and *c* in these diagrams.
Give a reason for each calculation that you do.

You will need to work out some other angles first.

Parallel lines

In each of these diagrams a straight line crosses two parallel lines.

♦ In each diagram a pair of **corresponding angles** is labelled. Corresponding angles are equal.

♦ In each diagram a pair of **alternate angles** is labelled. Alternate angles are equal.

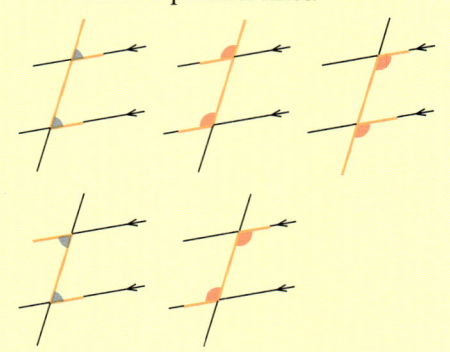

Exercise 3.2
Angles in parallel lines

1 a List five pairs of corresponding angles in this diagram.
b List three pairs of alternate angles.

To find corresponding angles in a diagram you could look for an F shape which may be upside down and/or back to front.

To find alternate angles in a diagram you could look for a Z shape which may be back to front.

2 Sketch these diagrams. Work out the angles *a*, *b* and *c*. You may need to calculate some other angles first.

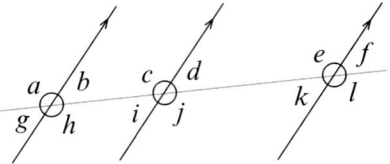

3 In this diagram AS, BR and NQ intersect to make the triangle DPO. CE, FI, JM and NQ are parallel.

a List three pairs of corresponding angles along the line:
  i AS   **ii** BR
b Explain why DĜH and DĤI are not corresponding angles.
c List three pairs of alternate angles in this diagram.
d Calculate each of these angles.
  i HD̂G   **ii** AD̂C
  iii GK̂L   **iv** HL̂K
  v QP̂L   **vi** NÔR

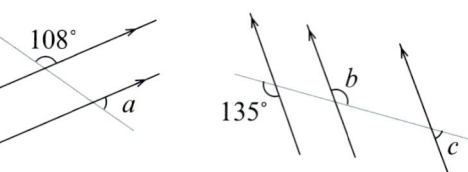

Give a reason for each calculation that you do.

4 In each of these diagrams there is one pair of parallel lines.

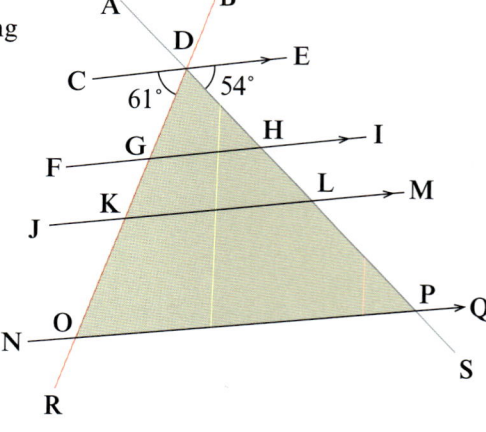

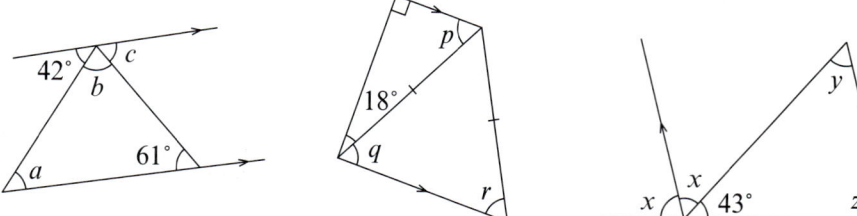

Sketch these diagrams and calculate each of the angles marked with a letter.

Angles in polygons

♦ At each vertex of a polygon the angle between an extended side and the adjacent side is called **an exterior angle**.

In ABCDE:

❖ the exterior angles are marked in orange
❖ the interior angles are marked in blue.

♦ The sum of the exterior angles of any polygon is 360°.

In ABCDE:
$a + b + c + d + e = 360°$

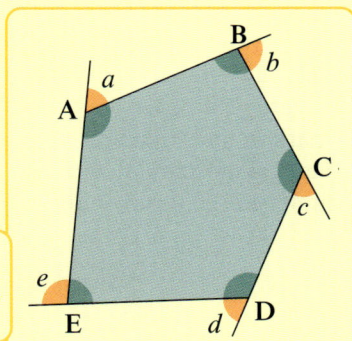

You can show this by tracing the angles and fitting them together round a point.

♦ At each vertex the sum of the interior angle and exterior angle is 180°.

Exercise 3.3
Exterior angles
of polygons

1   For this polygon:

   a   calculate each exterior angle
   b   check that the total of the exterior angles is 360°.

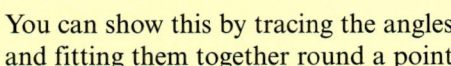

2   A dodecagon has 12 sides.
In this regular dodecagon one side is extended to form the angle p.

   a   Explain why the exterior angles of a regular dodecagon are all equal to 360° ÷ 12.
   b   Calculate the angle p.
   c   Calculate the interior angle of a regular dodecagon.

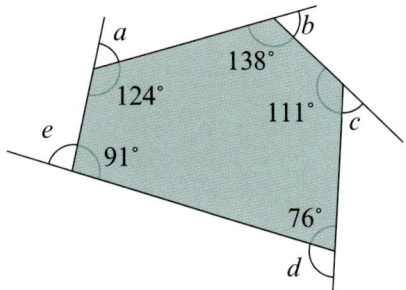

3   This is part of regular nonagon drawn inside a circle with centre C.
One side of the nonagon is extended to form the angle e.
Calculate the angles e to h.

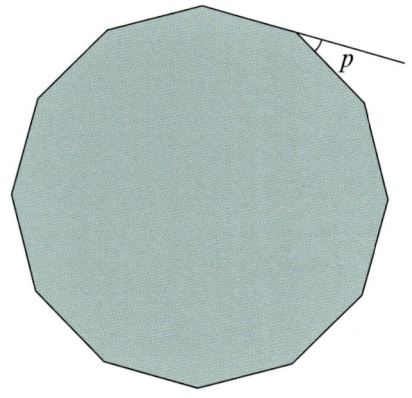

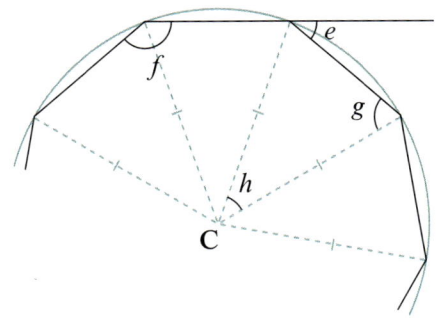

Exercise 3.4
Triangles investigation

You can mark eight points that are equally spaced on the circumference of a circle if you:
- draw a circle on square grid paper
- mark in lines that are vertical, horizontal and at 45° to the horizontal.

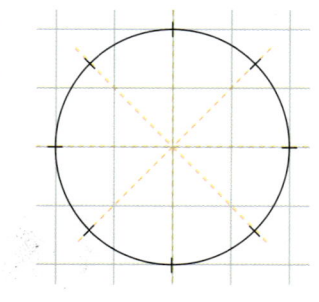

1 This eight-point circle has the points A to H equally spaced on the circumference.

Δ ABD is drawn by joining three of the points.

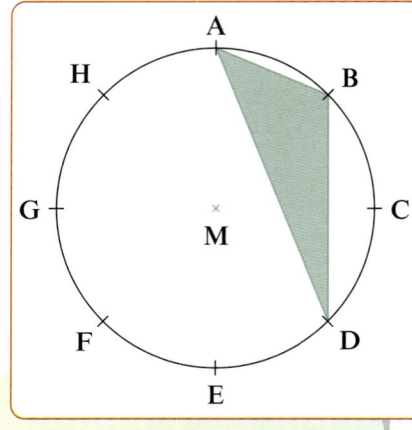

This is how a student calculated the exterior angle at A for Δ ABD.

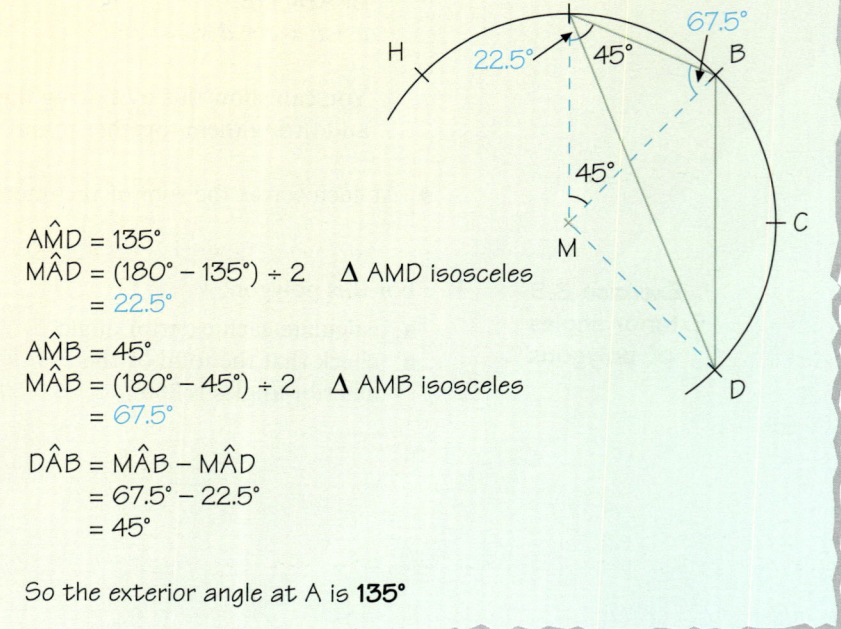

To find the exterior angle at A

$A\hat{M}D = 135°$
$M\hat{A}D = (180° - 135°) \div 2$ Δ AMD isosceles
$\qquad = 22.5°$

$A\hat{M}B = 45°$
$M\hat{A}B = (180° - 45°) \div 2$ Δ AMB isosceles
$\qquad = 67.5°$

$D\hat{A}B = M\hat{A}B - M\hat{A}D$
$\qquad = 67.5° - 22.5°$
$\qquad = 45°$

So the exterior angle at A is **135°**

a Explain why A$\hat{M}$D is 135°.
b For triangle ABD:
  i calculate the exterior angles at B and D
  ii check that the total of the exterior angles is 360°.

2 Triangle ACF is also drawn on an eight-point circle.

a For triangle ACF:
  i calculate each interior angle
  ii calculate each exterior angle.
b How many different triangles is it possible to draw in an eight-point circle?
c What different exterior angles are possible for triangles drawn on an eight-point circle?

Do not count any that are reflections or rotations of another polygon.

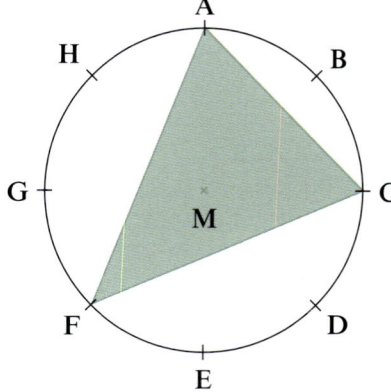

End points

You should be able to ...   ... so try these questions

A Calculate angles in parallel lines	**A1** Calculate the angles *a*, *b* and *c* in this diagram.

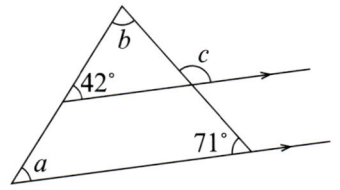

B Use the properties of polygons

B1 Polygons A to E are drawn on an equilateral grid.

Which of these polygons:
a is a regular polygon
b has only one obtuse angle?

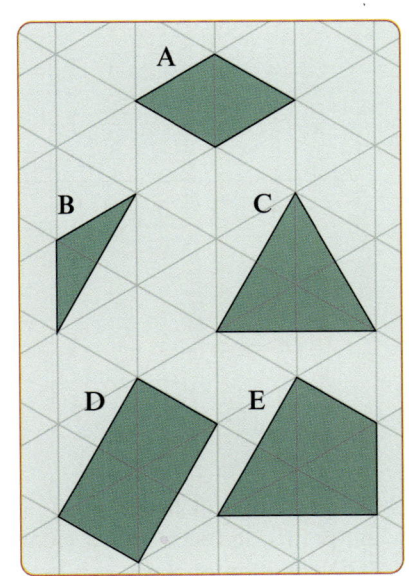

C Calculate angles in polygons

C1 For each of the polygons B and E:

a calculate the interior angles
b calculate the exterior angles.

C2 What is the exterior angle of a regular octagon?

Some points to remember

- For a polygon with *n* sides:

  - the sum of the interior angles is $(n - 2) \times 180°$

  - the sum of the exterior angles is 360°

  - each exterior angle of a regular polygon is $360° \div n$.

- Examples of quadrilaterals

	Square	Rectangle	Kite	Rhombus	Parallelogram
The diagonals:					
◆ bisect the interior angles	✓	✗	✗	✓	✗
◆ bisect each other	✓	✓	✗	✓	✓
◆ intersect at 90°.	✓	✗	✓	✓	✗
Number of lines of symmetry	4	2	1	2	0
Order of rotational symmetry	4	2	1	2	2

Starting points
You need to know about ...

... so try these questions

A Graphs of vertical and horizontal lines

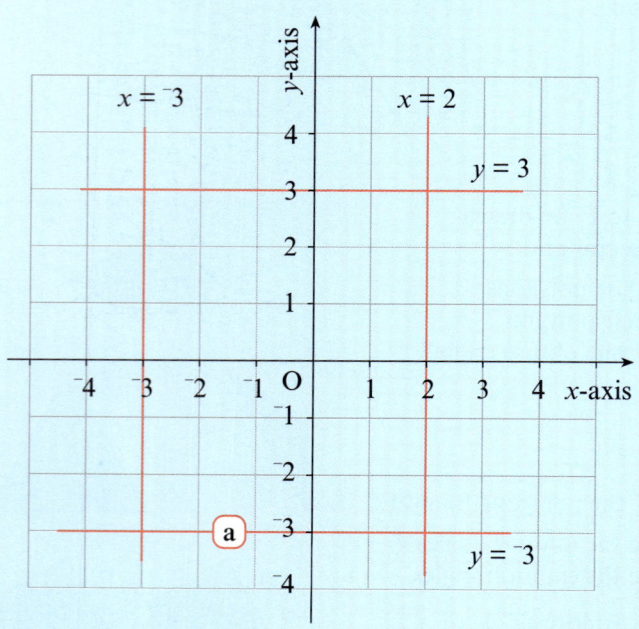

The equation of **line a** is $y = ^-3$ as:
all points on **line a** have a y-coordinate of $^-3$.
The equation of the x-axis is $y = 0$ and
the equation of the y-axis is $x = 0$.

A1 a Draw a pair of axes.
 b Label the x-axis.
 c Label the y-axis.
 d Draw and label the line
   $y = 2$ and the line $x = ^-1$
 e Give the coordinates of
   where lines $y = 2$ and
   $x = ^-1$ cross.

A2 a Give the coordinates of
   where the lines $x = 0$ and
   $y = 2$ cross.
 b Will the line $y = 2$ cross the
   line $y = ^-2$?
   Explain your answer.
 c What can you say about all
   the points on the line $x = 0$?

A3 Will each of these lines be
  vertical or horizontal?

  a $y = 4$ **b** $x = ^-2$
  c $x = 3$ **d** $y = ^-8$
  e $y = 0$ **f** $x = 0$

B A graph from a table of values

This table of values shows how values of
x and y are linked by the equation $y = x + 1$

x	$^-1$	0	1	1.5	2
y	0	1	2	2.5	3

From the table of values:
the points

($^-1$, 0)
(0, 1)
(1, 2)
(1.5, 2.5)
(2, 3)

can be plotted and
joined for the graph.

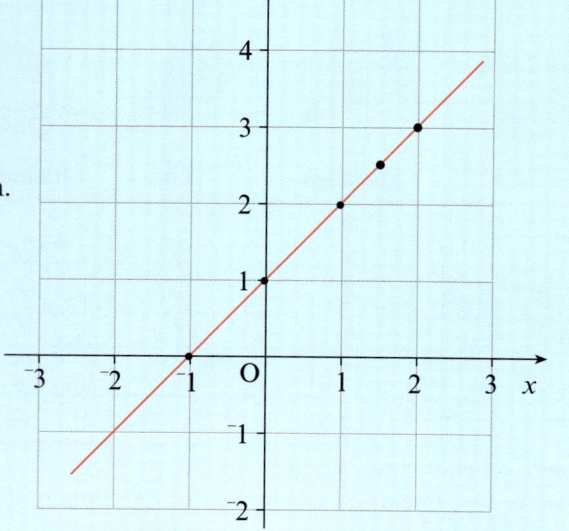

B1

x	$^-2$	$^-1$	0	1	2	3
y						

Copy and complete this table
of values for each equation.

 a $y = 3x - 1$
 b $y = x + 2$
 c $y = x$
 d $y + 3 = x$
 e $y = 1.5x + 1$

B2 For each equation in **B1** draw
  and label a graph.

B3 Do not draw a graph.
  Which of these points lie on
  the line $y = 2x - 3$?

  a ($^-2$, $^-7$)   **b** (0, 3)
  c (2, 1)   **d** (3, 3)
  e (0, 1.5)   **f** ($^-15$, $^-27$)
  g ($^-0.25$, $^-3.5$) **h** (4, 4)

Linear graphs

Exercise 4.1
Interpreting linear graphs

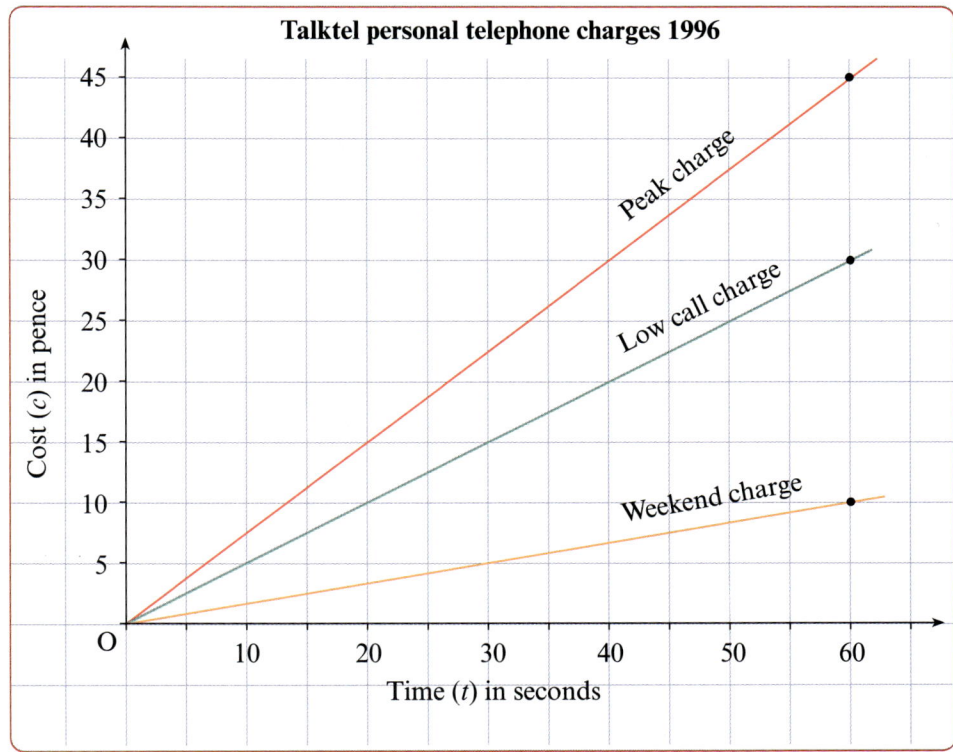

Talktel personal telephone charges 1996

Peak charge

Low call charge

Weekend charge

Cost (*c*) in pence

Time (*t*) in seconds

Linear is another word for straight.

A linear graph is a straight line graph.

Each linear graph shows a different charge made by Talktel.
Each call is timed (in seconds), and the cost is calculated for the customer bill.

1   Talktel's peak charge is 45 pence per minute.
   a   From the graph, what is the low call charge per minute?
   b   What is the weekend charge per minute?

2   **a**   Does a 35 second call at peak charge cost more or less than 25 pence?
   b   Estimate the cost of this call.

3   Estimate the cost of a 45 second call at peak charge.

4   Which is cheaper: 15 seconds at peak or 20 seconds at low call charge?
   Explain your answer.

5   Jess paid 7.5 pence for a 15 second call. Which charge was used for this?

6   At the weekend charge, estimate the cost of a 2 minute 35 second call.

7   At peak charge, how long a call can you make for 75 pence?

8   **a**   Copy the graph for the Talktel charges in 1996.
   b   Add a line to show the Infotel charge of 32 pence per minute.
   c   The line graph for which charge is the steepest?
   d   List the charges in order of the steepness of their line graphs.
   e   Describe any link you can spot between charge rates and steepness.

In the formula $c = t \div 6$:

c is the cost in pence

t is the time in seconds.

---

9   Talktel use the formula $c = t \div 6$ for their weekend charge.
   Use the formula to find the cost of a 96 second call at weekend charge.

10   **a**   Write a formula for the low call charge.
   b   Use your formula to find the cost of $1\frac{1}{2}$ minutes at low call charge.

The gradient of a linear graph

♦ The gradient of a linear graph is a measure of how steep the line is.

♦ The gradient of a linear graph is the same for any part of the line.

♦ The gradient of a linear graph is given by:

Change along the y-axis
Change along the x-axis

The change along an axis can be an **increase** or a **decrease**.

For a gradient we need to look at what happens for an **increase** along the x-axis.

When:
an **increase** along the x-axis gives
an **increase** along the y-axis

We say the gradient of the line is **positive**.

For example:
To find the gradient of this linear graph:

♦ choose two points on the line, e.g. A and B

♦ along the y-axis the change is 6 units (from 8 to 14)

♦ along the x-axis the change is 2 units (from 2 to 4).

The gradient is given by:

$$\frac{6}{2} = 3$$

The gradient of the line is 3.

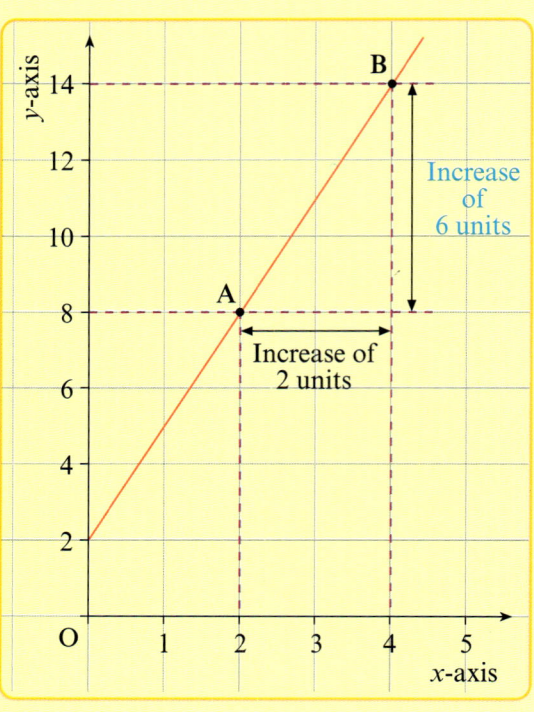

Exercise 4.2
Gradients of linear graphs

1   Calculate the gradients of lines **a** to **h**.

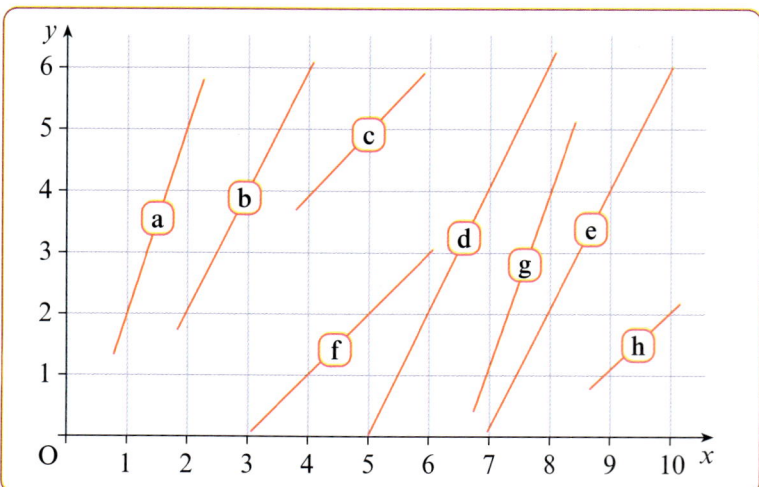

2   **a** Draw a pair of axes: x-axis from 0 to 10, y-axis from 0 to 10.
    b On your axes draw and label lines with each of these gradients:
       5   6   4   1   2   0.5

3   How are the gradient of a linear graph and its steepness linked?

The gradient of a linear graph has a value that can be given:

◆ as a whole number

or ◆ as a fraction (in its lowest terms)

or ◆ as a decimal.

The gradient of a linear graph is given by:

$$\frac{\text{Change in } y\text{-coordinates}}{\text{Change in } x\text{-coordinates}}$$

For example:

The linear graph **a** has a gradient given by:

$\frac{3}{1}$ (increase of 3) (increase of 1)

$= 3$ (whole number)

The linear graph **b** has a gradient given by:

$\frac{5}{2}$

$\frac{5}{2}$ is a fraction in its lowest terms

but $\frac{5}{2} = 2\frac{1}{2}$

The linear graph **c** has a gradient given by:

$\frac{2}{4}$ or 0.5 (decimal)

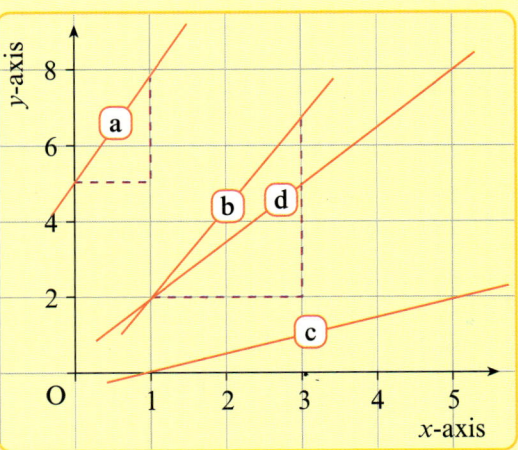

The linear graph **d** has a gradient of:

$\frac{6}{4} = \frac{3}{2}$ (in its lowest terms)

Exercise 4.3
Calculating gradients

1

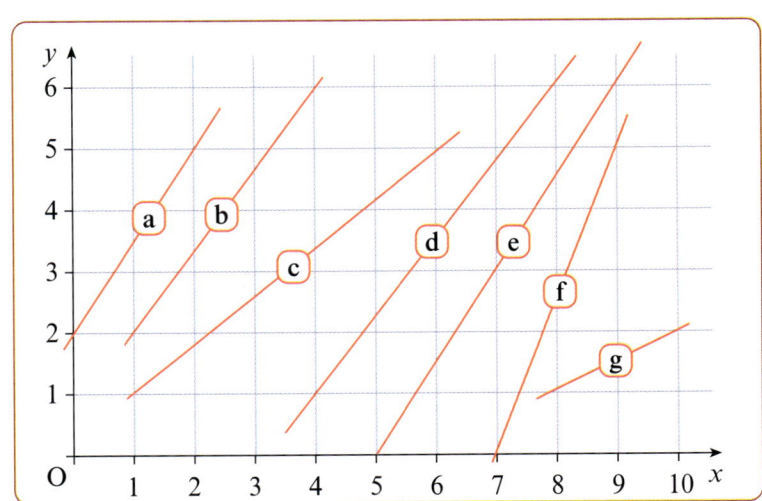

a Give the gradient of each linear graph **a** to **g**, as:
   i a fraction in its lowest terms
   ii a decimal value.

b List the lines with their gradients in order of steepness.
   Start with the steepest.

2 What can you say about the gradients of lines which are parallel?
   Explain your answer with an example and a diagram.

The line CD slopes:
downwards from left to right.

To find the gradient of line CD, use the points C and D:

◆ the change along the *y*-axis is
   5 units   (decrease)

◆ the change along the *x*-axis is
   2 units   (increase)

The gradient of CD is $\frac{^-5}{2}$ or $^-2.5$

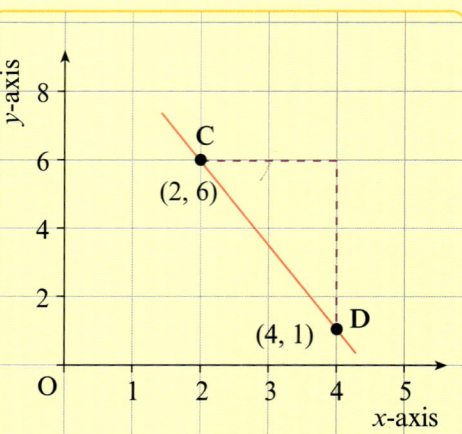

> When:
> an **increase** along the *x*-axis gives
> a **decrease** along the *y*-axis
>
> we say the gradient of the line is **negative**.
>
> A negative sign is used to show a negative gradient.
> For example a gradient of
> $\frac{^-3}{4}$

Gradients of straight lines can be described in this way:

A line which slopes   **upwards from left to right** (╱)
            has   **a positive gradient**.

A line which slopes   **downwards from left to right** (╲)
            has   **a negative gradient**.

Exercise 4.4
Gradients:
positive and negative

1

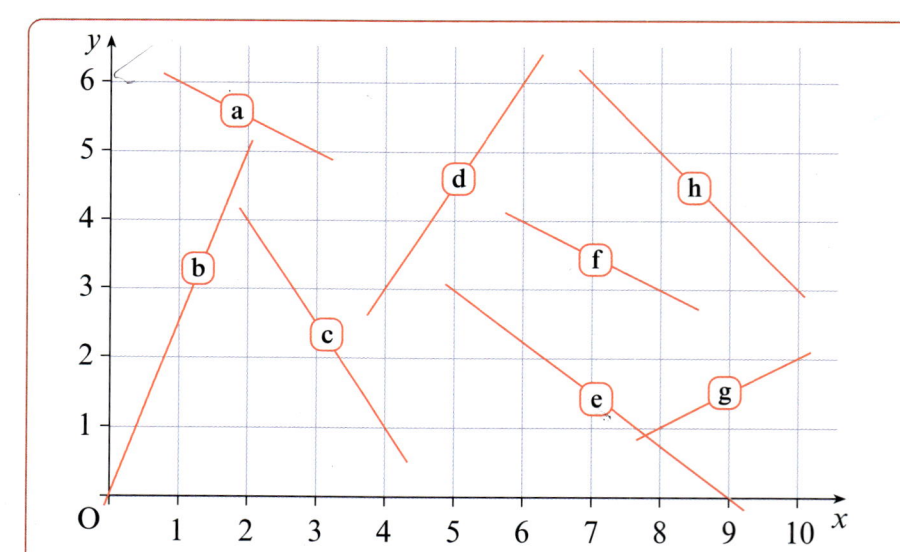

a   Which of the lines, **a** to **h**, has a negative gradient?
b   For each of the lines **a** to **h**, give its gradient as:
   i a fraction in its lowest terms
   ii a decimal value.

2   a   Draw a pair of axes: the *x*-axis from 0 to 10 and the *y*-axis from 0 to 6.
   b   For each of these gradients, draw and label a line:
      gradient of 3, gradient of $^-2$, gradient of $\frac{2}{3}$, gradient of $\frac{^-3}{4}$.

3   A line slopes downwards from right to left. Describe its gradient.

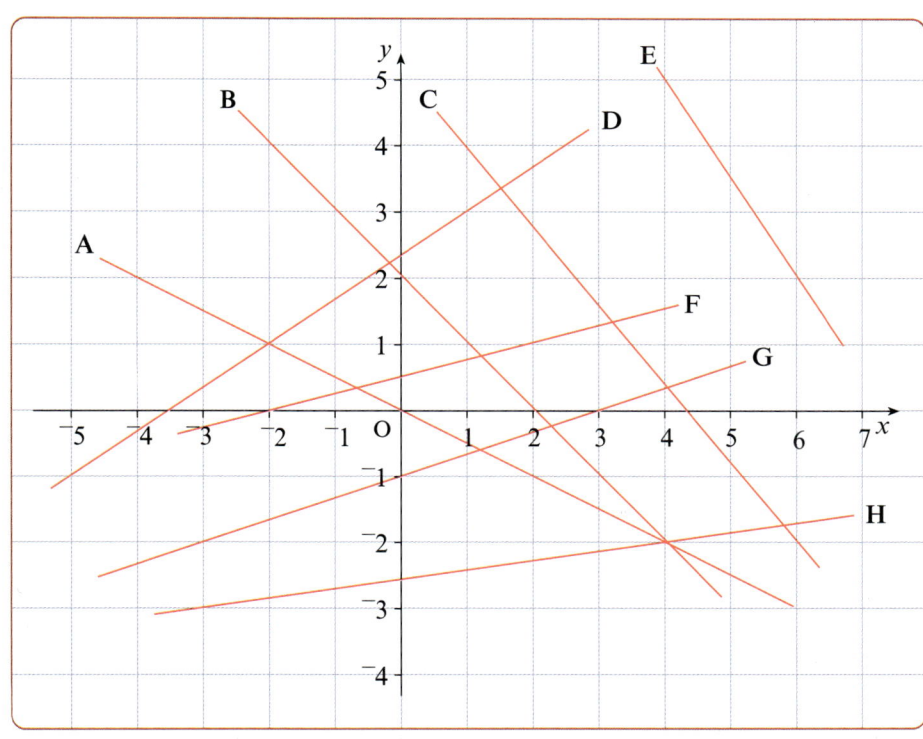

4 Which of these lines has a gradient of $^-\frac{1}{2}$?

5 Which line has a gradient of $\frac{1}{3}$?

6 Which is steeper: line F or line H? Explain your answer using gradients.

7 Give the gradient of line B.

8 Which line has a gradient of $^-1.2$?
   Explain your answer.

9 Asa gave the gradient of line E as $\frac{2}{3}$.
   Do you agree? Explain your answer.

10 Mina gave the gradient of line D as $\frac{3}{2}$.
   Do you agree? Explain your answer.

11 a Draw a pair of axes with:
      values of x from $^-5$ to $^+5$, and values of y from $^-5$ to $^+5$.

    b Plot the point ($^-3$, $^-5$) and label it P, and label ($^-5$, 5) as R.

    c Line T has a gradient of $\frac{4}{3}$, and it passes throught point P.
      On your axes draw and label line T.

    d Line V has a gradient of $^-\frac{1}{4}$, and it passes through point R.
      On your axes draw and label line V.

    e Give the coordinates of where line T and line V cross.

    f Line W joins points R and P.
      What is the gradient of line W?

    g Line Z has a gradient of $\frac{9}{2}$, and passes through point P.
      Give the coordinates of where line Z and line V cross.

The gradient of lines that are parallel or at right angles to each other

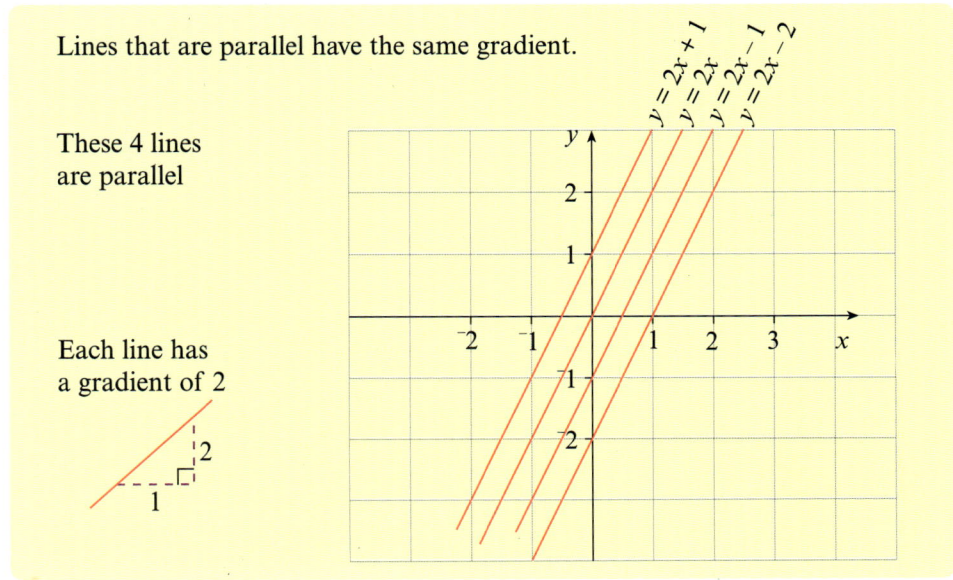

Lines that are parallel have the same gradient.

These 4 lines are parallel

Each line has a gradient of 2

Line A and line B are at right angles to each other.

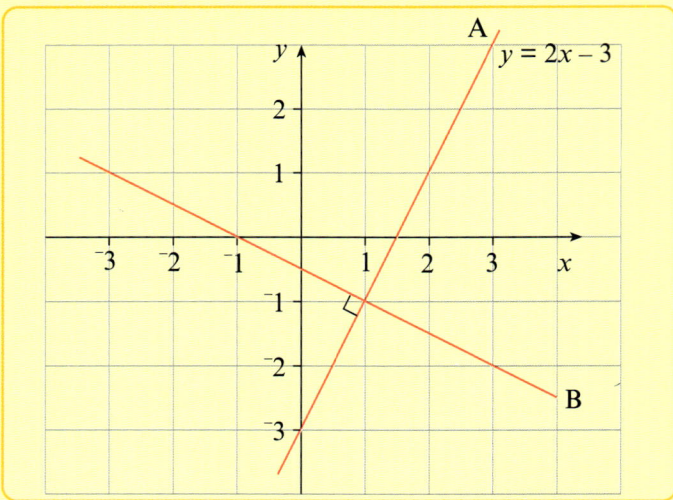

The gradient of line A is 2

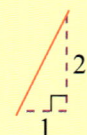

The gradient of line B is $-\frac{1}{2}$

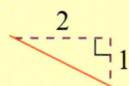

So there is a connection between the gradients of two lines that are at right angles to each other.

Here are some examples.

Lines A and B are at right angles to each other:

Gradient line A	Gradient line B
2	$-\frac{1}{2}$
3	$-\frac{1}{3}$
4	$-\frac{1}{4}$
5	$-\frac{1}{5}$
etc	etc

Gradient of one line × Gradient of the other line = –1 when the lines are at right angles

You can check if two lines are at right angles in this way:

Example

Line A has a gradient of 8.

Line B has a gradient of $-\frac{1}{8}$.

Is line A at right angles to line B?

Multiply the gradients : $-\frac{1}{8} \times 8 = -1$

So, the two lines must be at right angles.

Exercise 4.5
Parallel lines and lines
at right angles

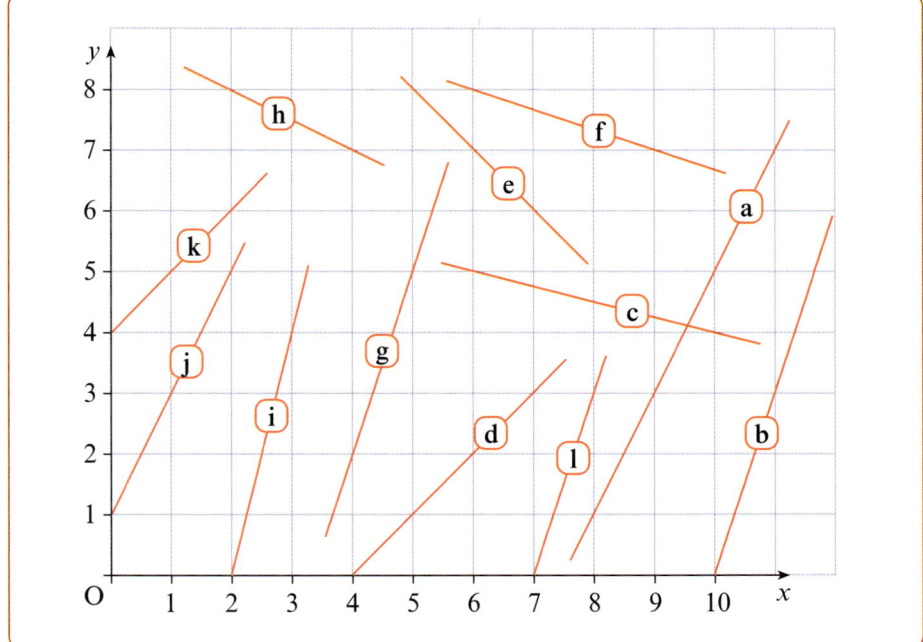

1   What is the gradient of line **a**?

2   Give the gradient of a line at right angles to line **a**.

3   Which of the lines on the grid is at right angles to **a**?

4   **a**   What is the gradient of line **l**?

    b   What is the gradient of line **f**?

    c   Multiplying the gradients of lines **l** and line **f**.

    d   Are lines **l** and **f** at right angles to each other?

5   Explain why line **c** must be at right angles to line **i**.

6   Which line is parallel to line **b**?

    Explain your answer.

7   Are lines **d** and **e** at right angles to each other?
    Explain your answer.

8   Which line is parallel to line **k**?

9   Which line is at right angles to line **k**?

10 Prove that line **g** and line **f** will meet at right angles.

Drawing linear graphs from a table of values

A linear graph has a rule that links the *x*-coordinates and the *y*-coordinates. We call this rule the **equation** of the line, and we can use it to find the coordinates of points on the line.

Example Find the coordinates of a point on the line $y = 2x + 3$.

To find the *y*-coordinate you multiply the *x*-coordinate by 2, then add 3
When $x = {}^-2$, $y = (2 \times {}^-2) + 3$, which gives $y = {}^-4 + 3$: so $y = {}^-1$

For any *x*-coordinate you choose, you can calculate the value of the *y*-coordinate that goes with it.

To draw the graph of $y = 2x + 3$:

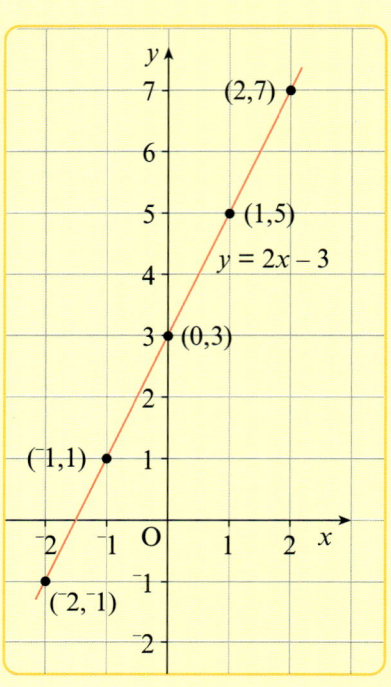

♦ choose values for *x*
  (say $^-2$, $^-1$, 0, 1 and 2)

♦ draw up a table and calculate values for *y*

x	$^-2$	$^-1$	0	1	2
y	$^-1$	1	3	5	7

♦ from the table, list points on the line
  $(^-2, {}^-1)$, $(^-1, 1)$, $(0, 3)$, $(1, 5)$, $(2, 7)$

♦ on a pair of axes, plot the points

♦ draw the line through the points for the graph of $y = 2x + 3$.

From the table of values you can decide how to scale your axes.

Here:

the *x*-axis shows from $^-2$ to 2

and

the *y*-axis shows from $^-1$ to 7

Exercise 4.6
Drawing graphs from tables of values

1   For the equation $y = 3x + 2$:

  a Explain in words how to calculate values for *y*.
  b Draw up a table of values with these values of *x*: $^-2$, $^-1$, 0, 1, 2 and 3.
       For each value of *x* calculate a value for *y*.
  c From the table list points on the line.
  d Draw a pair of axes. Plot and label each point with its coordinates.
  e Draw a line to show the graph of $y = 3x + 2$.

2   For the equation $y = 2x - 3$:

  a Explain in words how to calculate values for *y*.
  b Draw up a table of values. Use values of *x* from $^-1$ to 3.
  c Draw the graph of $y = 2x - 3$.

3   **a** Draw up a table of values for $y = 2x + 1$. Use values of *x* from $^-2$ to 3.
  b Draw the graph of $y = 2x + 1$.

4   Draw the graph of $y = 4x - 1$.
    Use values of *x* from $^-2$ to 2.

5   **a** On the same pair of axes draw graphs of $y = 3x - 3$ and $y = 2x - 1$.
  b Give the coordinates of the point where the two graphs cross.

Linear graphs and their equations

For equations to be the same they do not have to be written in the same order.

For example:

$y = 2x + 1$ is the same as:
$$y = 1 + 2x$$
$$2x + 1 = y$$
$$1 + 2x = y$$

and

$$y = 3x - 2 \text{ is the}$$
same as: $y = {}^-2 + 3x$

Three linear graphs are shown here, and their equations.

We can list the gradient of each line:
$y = 2x - 2$ has a gradient of 2
$y = 2x + 1$ has a gradient of 2
$y = 2x + 4$ has a gradient of 2.

We can list where each line crosses the y-axis:
$y = 2x - 2$ crosses the y-axis at $^-2$
$y = 2x + 1$ crosses the y-axis at $^+1$
$y = 2x + 4$ crosses the y-axis at $^+4$.

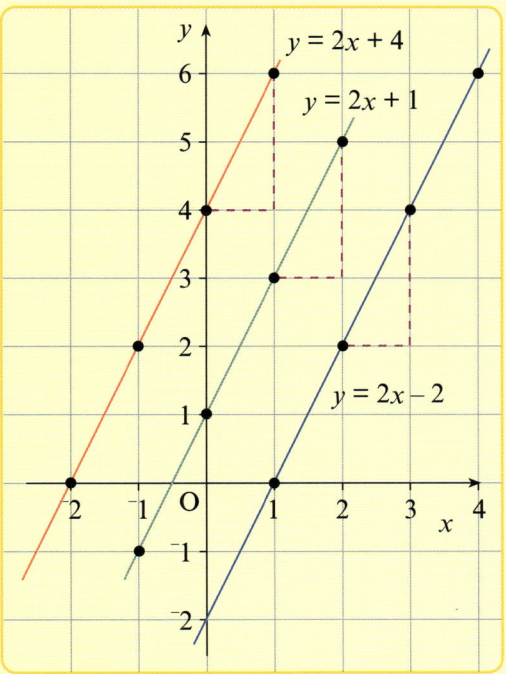

So we have:

$y = 2x - 2$ has a gradient of 2
it crosses the y-axis at $^-2$

$y = 2x + 1$ has a gradient of 2
it crosses the y-axis at $^+1$

$y = 2x + 4$ has a gradient of 2
it crosses the y-axis at $^+4$.

From the equation of a line we can tell:
♦ the gradient of the line
♦ where the line crosses the y-axis.

Exercise 4.7
Using equations of linear graphs

1   For the graph of $y = 2x - 5$
   a   give the gradient of the line   **b**   where does the line cross the y-axis?

2   For the line $y = 3x - 1$
   a   what is the gradient of the line?   **b**   where does the line cross the y-axis?

Where a line crosses the y-axis is called the y-intercept.

3   Which of these lines has a gradient of 4 and crosses the y-axis at $^+2$?
   a $y = 2x + 4$ **b** $y = 4x + 2$ **c** $y = 2 + 4x$

4   For each of these lines give the gradient and the y-intercept.
   a $y = 3x - 2$ **b** $y = x + 2$ **c** $y = 4x$ **d** $y = 5x - 4$
   e $y = 3 + 2x$ **f** $y = 3x - \frac{1}{2}$ **g** $5x + 1 = y$ **h** $\frac{1}{2}x = y$
   i $y = 5 - 5x$ **j** $y = x$ **k** $x - 2 = y$ **l** $y = {}^-4x$

5   Here is a description of five different linear graphs.
   Line K has a gradient of 3 and a y-intercept at $^+5$
   Line M has a gradient of $\frac{1}{2}$ and a y-intercept at $^-1$
   Line P has a y-intercept at 1.5 and a gradient of 2
   Line R has a y-intercept at $^+2$ and a gradient of $^-3$
   Line T has a gradient of 1 and a y-intercept at 0.
   Give an equation for each of these lines.

Drawing linear graphs from their equation

As the equation of a line tells us about:
♦ the gradient of the line
♦ the y-intercept
it is possible to draw a linear graph just by using the data in the equation.

Example

To draw a linear graph of the equation $y = 3x - 4$

We can tell that the gradient of the line is 3 and the y-intercept is ⁻4.

To draw the graph of $y = 3x - 4$:

♦ Draw a pair of axes.

♦ **Step 1**
Mark the
y-intercept, ⁻4.

♦ **Step 2**
From the y-intercept
draw in the gradient 3.

♦ **Step 3**
Join the points with
a line and label.

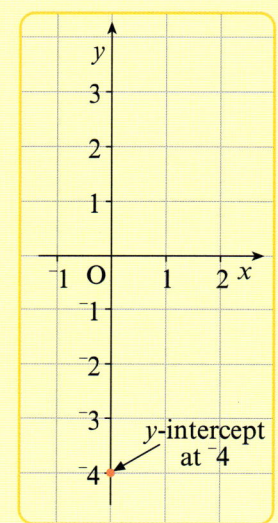

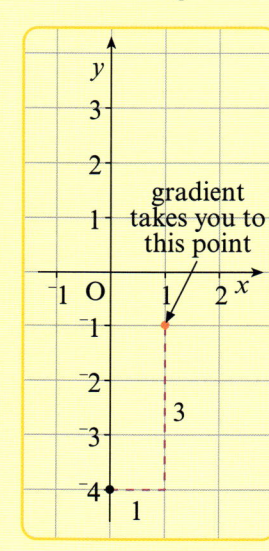

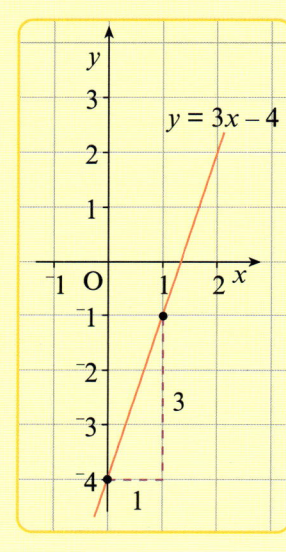

Exercise 4.8
Drawing linear graphs
from equations

1 **a** Draw a pair of axes. Use x from ⁻1 to 2, and y from ⁻5 to 5.
   b On your axes draw and label graphs of these lines:
   $y = 4x - 3$ $y = 2x + 1$ $y = x - 1$

2 **a** Draw a pair of axes. Use x from ⁻1 to 4, and y from ⁻4 to 6.
   b On your axes draw graphs of these lines:
   $y = 3x - 3$ $y = x + 2$ $y = 3x$
   c Which two of these lines go through the point (1, 3)?
   d Which of these lines go through the point (3, 5)?

3 **a** What is the gradient of the line $y = 2 - 3x$? What is the y-intercept?
   b Draw the graph of $y = 2 - 3x$.

Exercise 4.9
Building equations
from graphs

1

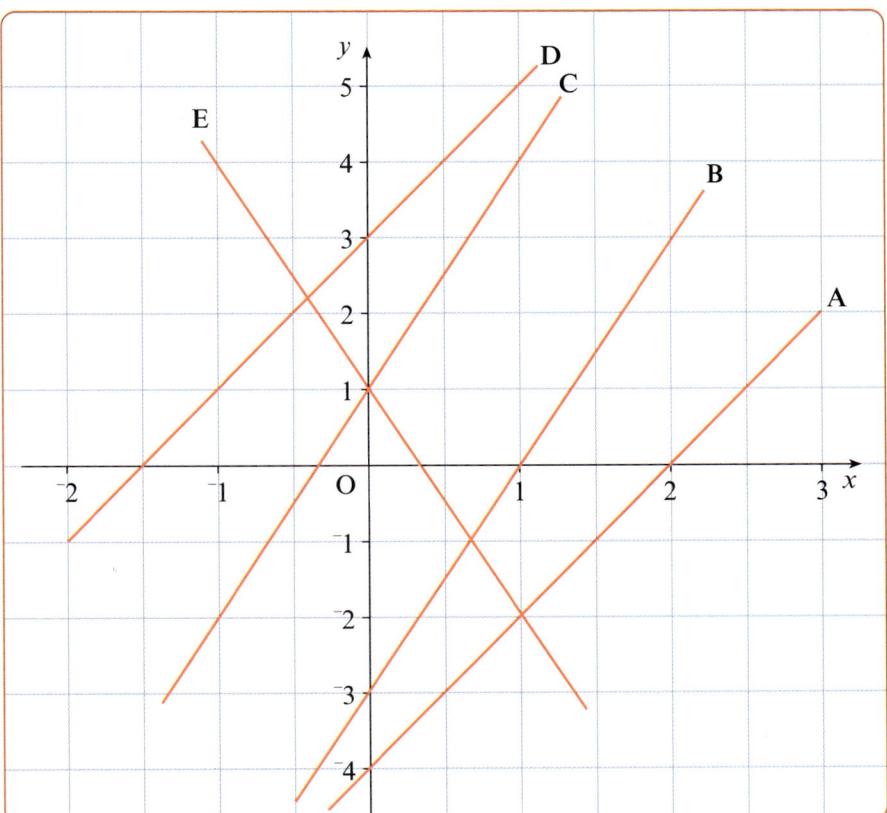

a What is the gradient of line A?
b What is the y-intercept for line A?
c Write an equation for line A.

2 a What is the y-intercept for line B?
   b Write an equation for line B.

3 Write an equation for each of the lines C, D, and E.

Writing equations in the form $y = mx + c$

'Rewrite an equation' is not the same as 'change an equation'.

An equation, and its rewritten form give the same data, we say they are equivalent.

When the equation of a linear graph starts $y = \ldots$, it is easy to draw a graph. Sometimes you will have to rewrite an equation to make it read $y = \ldots$
One way to rewrite an equation is to divide each term by the same number.

Examples

To rewrite the equation: $3y = 6x - 3$

divide each term by 3

$$y = 2x - 1$$

as $3y \div 3 = y$, $6x \div 3 = 2x$
and $^-3 \div 3 = ^-1$

To rewrite the equation: $5y = 3x + 10$

divide each term by 5

$$y = \tfrac{3}{5}x + 2$$

as $5y \div 5 = y$, $3x \div 5 = \tfrac{3}{5}x$
and $10 \div 5 = 2$

Exercise 4.10
Rewriting equations

1 a To rewrite the equation $4y = 8x - 20$, to make it read $y = \ldots$, what number will you divide each term by?
   b Rewrite $4y = 8x - 20$, to make it read $y = \ldots$.

2 **a** To rewrite $2y = 3x + 4$ in the form $y = \ldots$,
what will you divide each term by?

   b Rewrite $2y = 3x + 4$ in the form $y = \ldots$.

3 Rewrite each of these equations in the form $y = \ldots$.

   a $2y = 4x + 8$ **b** $3y = 4x - 9$ **c** $3y = 3x - 6$ **d** $5y = 3x$
   e $5y = 2x + 5$ **f** $2y = 6x - 4$ **g** $2y = 3 - 4x$ **h** $4y = x$
   i $4y = 4 - 4x$ **j** $6y = 12 + 12x$ **k** $5y = x + 1$ **l** $2x = 3y$

4 Which of these equations can be rewritten as $y = 2x - 3$?

   a $2y = 3x - 6$ **b** $4y = 8x - 16$ **c** $3y = 6x - 9$ **d** $2y = 4x$

5 Write three different equations that can be rewritten as $y = 3x - 1$.

Rewriting equations and drawing graphs

Example

Draw the graph of $3y = 4x - 6$
One way is with the equation in
the form $y = \ldots$.

- Rewrite the equation

  $3y = 4x - 6$

  Divide each term by 3

  $y = \frac{4}{3}x - 2$

- Draw the graph from the equation:

  gradient $\frac{4}{3}$

  y-intercept, $^-2$

- Label the graph with
  its equation $3y = 4x - 6$.

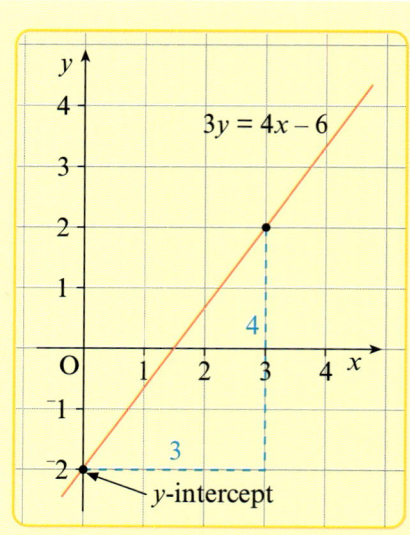

When you divide your
equation, the gradient can
work out as a fraction.

A fraction is easier to work
with for the gradient than a
decimal value.

**Exercise 4.11
Drawing graphs**

1 For the graph of $2y = 3x + 2$:

   a Rewrite the equation in the form $y = \ldots$.
   b What is the gradient of the line?
   c What is the y-intercept?
   d Draw a pair of axes: use x from $^-1$ to 3 and y from $^-1$ to 5.
   e Draw the graph of $2y = 3x + 2$.

2 For the graph of $3y = 2x - 3$:

   a What is the gradient of the line?
   b What is the y-intercept?
   c Draw a graph of $3y = 2x - 3$.

3 Draw a graph of $4y = 8x - 12$.

4 **a** Draw a pair of axes, use x from $^-1$ to 5 and y from $^-4$ to 4.
   b On your axes draw a graph for each of these:
      $2y = x$ $4y = x + 4$ $4y = 5x - 12$
   c Give the coordinates of a point where all three lines cross.

5 Draw a graph of $3y = 4x - 2$.

6 Is the point $(2, 3)$ on the graph of $2y = 5x - 4$?
Explain your answer.

Linear graphs that cross

Any linear graph is part of a straight line drawn through every point that fits its equation.

Where two graphs cross, the point must be on both lines.

For example, at A the graphs of:
$$y = 2x - 2$$
and $\quad 2y = x + 2$

both have a point of (2, 2)

We can say that at (2, 2) the equations must both be true for these values of x and y.

You can test this by using the values of the x- and y-coordinates in each equation.

At A: $x = 2$ and $y = 2$
$y = 2x - 2$ is true as:
$\quad y = 2$ and $2x - 2 = 2$
$2y = x + 2$ is true as
$\quad 2y = 4$ and $x + 2 = 4$

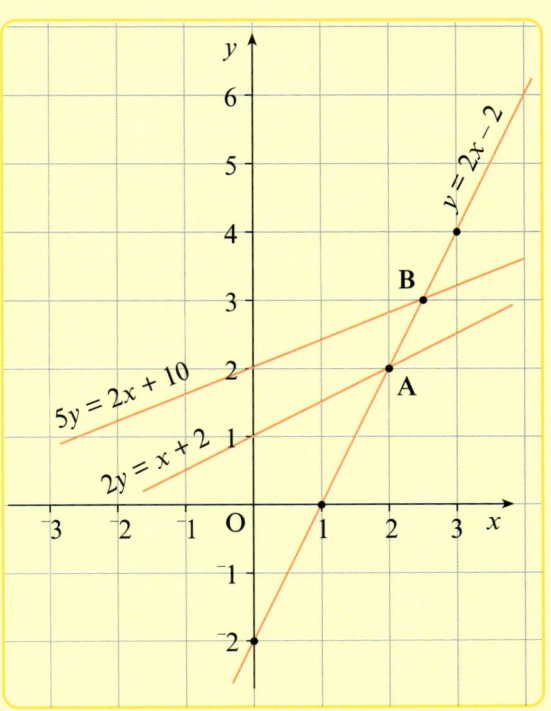

Exercise 4.12
Crossing points

1 a Give the coordinates of point B.
   b Give the equations of two lines that go through B.
   c Test that the values for x and y at B are true for the equations.

2 a On a pair of axes draw the graphs of:
      $y = x + 1$ and $y = 2x - 1$
   b Give the coordinates of a point that both lines are drawn through.
   c Test that the equations are true at this point.

3 As lines C and D cross, they must have one point that is the same.

   a Estimate the coordinates where lines C and D cross.
   b Test these values on each equation.
   c Try other estimates to see if you can find the exact coordinates.

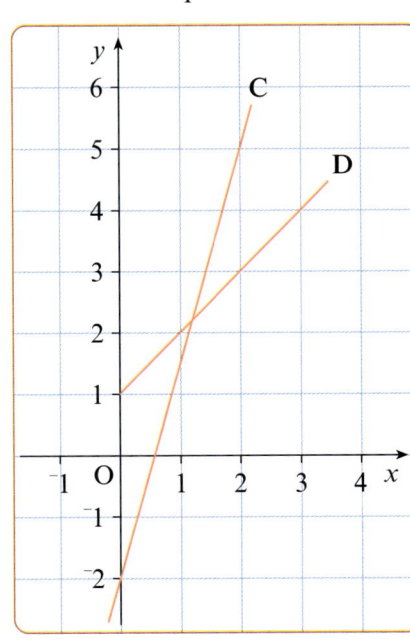

End points

You should be able to ...

... so try these questions

A Explain gradients of linear graphs

A1   **a**   In words, and with a diagram, describe a positive gradient.

     b   In words, and with a diagram, describe a negative gradient.

     c   What does the gradient of a line tell you:
- the length of the line
- the colour of the line
- something else?

       Explain your answer.

B Give the value of the gradient of a linear graph

B1   Give the gradient of line D.

B2   Give the gradient of line A
   a   as a fraction
   b   as a decimal.

B3   Give the gradients of lines B and C.

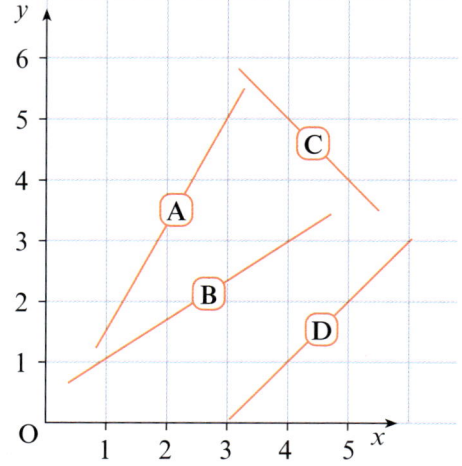

C Use a table of values

C1   Draw up a table of values for the graph of $y = 3x - 4$
      Use values of x from $^-1$ to 3.

C2   Draw the graph of $y = 3x - 4$.

D Rewrite equations to read $y =$

D1   Change the equation $4y = 3x + 6$ to read $y = \dots$.

E Use $y = mx + c$

E1   For this line, give the gradient and the y-intercept.
$$y = 2x - 5$$

E2   Draw the graph of $y = \frac{1}{2}x + 1$.

E3   Draw a graph of $3y = 5x - 6$.

F Test to see if a line goes through a point

F1   Which of these lines pass through the point (3, 4)?
$$3y = 2x + 6 \qquad y = 2x - 2 \qquad 3y = 4x + 1$$

Some points to remember

- $y = mx + c$ is the general form of a linear equation where, m is the gradient and c the y-intercept.
- Rewriting a linear equation is not the same as changing the equation.

Starting points
You need to know about ...

... so try these questions

A Types of data

- There are two main types of data:
  - data that is divided into **categories**, such as:
    make of car,
    colour of car.
  - data that is **numerical**, such as:
    number of people in car,
    length of car.

Car	Make of car	Number of people in car
	TRAFFIC SURVEY	
A	Peugeot	1
B	Ford	3
C	Vauxhall	1
D	Ford	2
E	Rover	2

B Presenting data in frequency tables

Make of car	Tally	Number of cars
Ford	⁴⁄ ////	9
Rover	⁴⁄ /	6
Vauxhall	//	2
Others	⁴⁄ //	7

Number of people in car	Frequency
1	3
2	10
3	6
4	3
5	2

C Diagrams that present one set of data

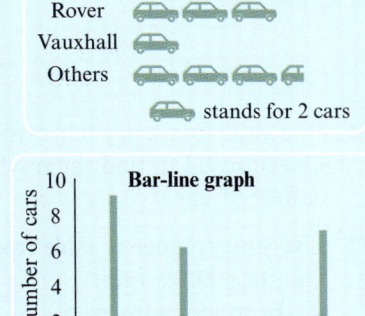

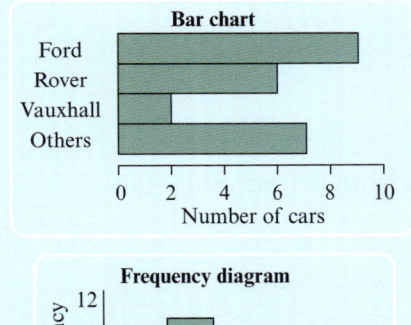

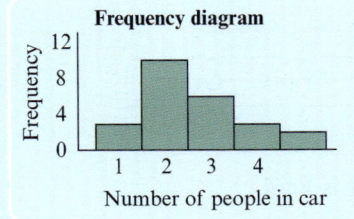

D Constructing a pie chart

- To calculate the percentage for each category.
  For Vauxhall, there are 2 cars out of 24 so
  $100 \div 24 \times 2 = 8.3\%$ (to 1 dp).

	No. of cars	%
Ford	9	37.5
Rover	6	25
Vauxhall	2	8.3
Others	7	29.2
Totals	24	100

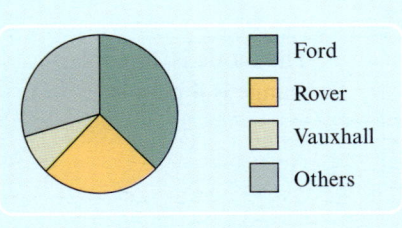

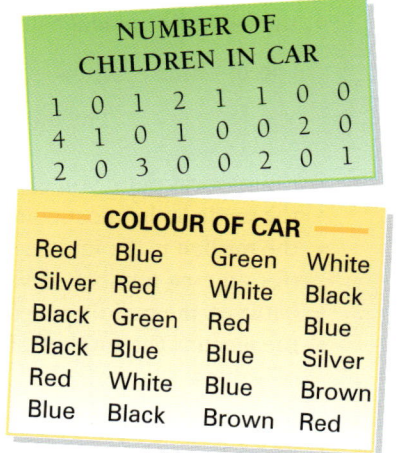

NUMBER OF CHILDREN IN CAR

1	0	1	2	1	1	0	0	
4	1	0	1	0	0	2	0	
2	0	3	0	0	2	0	1	

COLOUR OF CAR

Red	Blue	Green	White
Silver	Red	White	Black
Black	Green	Red	Blue
Black	Blue	Blue	Silver
Red	White	Blue	Brown
Blue	Black	Brown	Red

B1 Use a tally to present the number-of-children-in-car data in a frequency table.

B2 Present the colour-of-car data in a frequency table.

C1 Draw a bar-line graph to show the colour-of-car data.

C2 Draw a frequency diagram to show the number-of-children-in-car data.

C3 Draw a pictogram to show the colour-of-car data.

C4 Draw a bar chart to show the colour-of-car data.

D1 For the colour-of-car data, calculate the percentage for each colour.

D2 Draw a pie chart to show the colour-of-car data.

E Finding averages and the range

- For data in categories, you can only find one average:
  - ❖ the **mode** (the **modal** category is the most common)

  > Red   Blue   Yellow   Blue   Green   Black   Red   Blue

  The modal colour is **Blue**.

- For numerical data, you can find several averages:
  - ❖ the **mode** is the most common value (or values)
  - ❖ the **median** is the middle value when the data is in order (for an **even** number of values, take the median as halfway between the middle pair of values)
  - ❖ the **mean** is the total of all the values divided by the number of values.

- A measure of how spread out the data is can also be found:
  - ❖ the **range** is the difference between the highest and lowest values.

  > 46   27   82   46   102   27   60

  Mode = **27** and **46**
  Median = **46**   (27   27   46   46   60   82   102)
  Mean = **55.7** $\left(\dfrac{46 + 27 + 82 + 46 + 102 + 27 + 60}{7}\right)$
  (to 1 dp)
  Range = **75**   (102 − 27)

  > 87   43   101   56   87   67

  Mode = **87**
  Median = **77**   (43   56   67   87   87   101)
  Mean = **73.5**   (441 ÷ 6)
  Range = **58**   (101 − 43)

F Finding the mode and the range from a frequency table

- For data in categories:
  - ❖ the **mode** is the category with the highest frequency.

Make of car	Ford	Rover	Vaux.	Other
Number of cars	9	2	6	7

The modal make of car is **Ford**.

- For numerical data:
  - ❖ the **mode** is the value (or values) with the highest frequency
  - ❖ the **range** is the difference between the highest and lowest values.

Number of people in car	Frequency
1	3
2	10
3	6
4	3
5	2

The mode is **2** people.
The range is **4** people (5 − 1).

E1

> 15   42   33   37   84   42   50
> 81   29   26   67   15   19   55

For this set of data, find:
a the mode
b the median
c the mean
d the range.

E2

> 38   25   106   78   44   62
> 13   90   25   31   25

For this set of data, find:
a the mode
b the median
c the mean
d the range.

E3

> Mode 3 and 7
> Median 6
> Mean 6

These are averages for a set of data with 8 values.
List what the values might be.

F1 Use your frequency table from Question **B2** to find the modal colour of car.

F2 Use your frequency table from Question **B1** to find:
a the modal number of children in car
b the range of the number of children in car.

G Deciding which average to use

♦ The **mean** is the most widely used average.
It is not a sensible average to use for data with **extreme values**, values which are much smaller or much greater than the others.

G.T. Small & Son – Monthly Salaries (£)					
870	870	870	870	1050	1050
1050	1210	1210	1210	**2080**	**2330**

Mean = £1222.50
Median = £1050
Mode = £870

10 of the 12 salaries are smaller than the mean, so the **median** is the more sensible average to use.

♦ The **mode** can also be a poor choice of average.
For this data, it is a poor choice because it is the lowest salary.

H Diagrams that present two or more sets of data

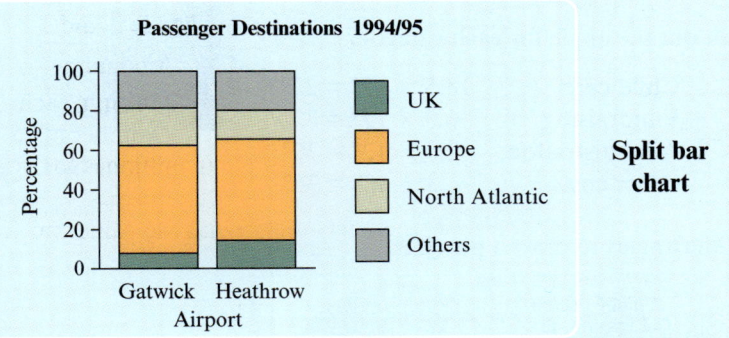

Split bar chart

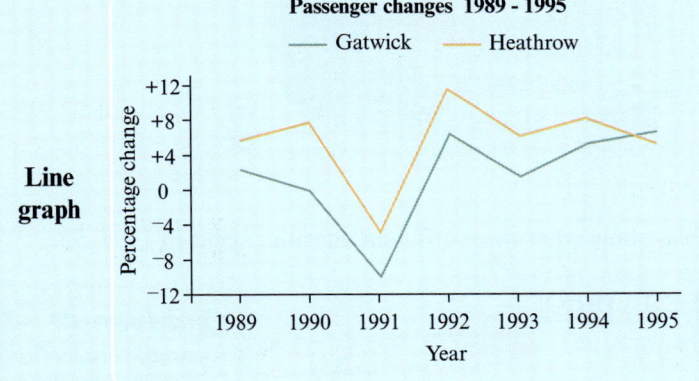

Line graph

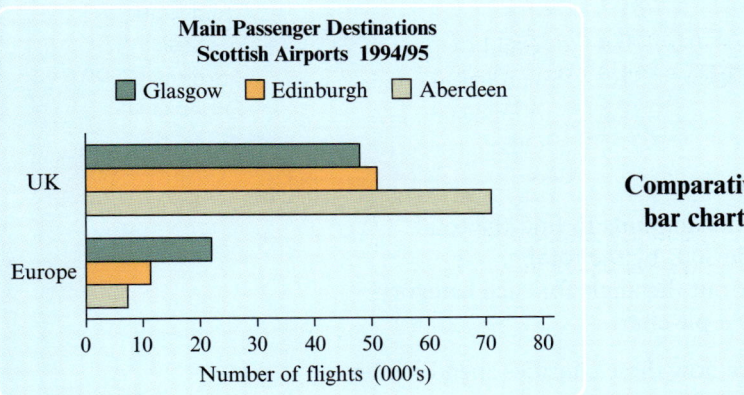

Comparative bar chart

G1

S. Fry & Partners – Weekly Wages (£)				
112	285	285	340	340
340	372	372	388	

Find
a the modal wage
b the median wage
c the mean wage
d For each average wage:
  i decide whether it is sensible to use or not
  ii if you think it is not sensible, explain why.

H1

Southampton to Channel Islands 1994/95		
	Number of passengers (000's)	Number of flights (000's)
Jersey	156	4.0
Guernsey	100	3.4
Alderney	31	2.9
Totals	287	10.3

Calculate the percentage of passengers going to:
a Jersey   **b** Guernsey
c Alderney

H2 Calculate the percentage of flights going to each of the three islands.

H3 Draw a split bar chart to show the two sets of data.

H4

% change in no. of passengers					
	1990	1991	1992	1993	1994
Glasgow	11.0	⁻3.1	12.4	7.4	8.8
Edinburgh	5.3	⁻6.1	8.4	7.2	10.3

Draw a line graph to show this data for Glasgow and Edinburgh.

H5

NUMBER OF FLIGHTS 1994/95 (000's)		
	Heathrow	Gatwick
UK	75	33
Europe	252	111
North Atlantic	40	19
Others	47	21

Draw a comparative bar chart to show this data.

Constructing a pie chart using degrees

£m stands for millions of pounds.

- A set of data in categories can be shown on a pie chart.

- To construct a pie chart using degrees:

CHILD CONCERN		£m		£m
Expenditure	Child care	18.5	Administration	2.5
(1994/95)	Fundraising	6.0	Other costs	3

❖ add the amounts to find the total

$$18.5 + 6.0 + 2.5 + 3 = \textbf{30}$$

£28.8m is shared between the 360° in a circle: each £1m is given 12°

There are 360° at the centre of a circle.

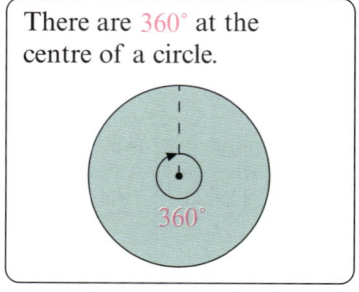

360°

❖ divide 360° by the total

$$360° \div 30 = \textbf{12}°$$

❖ work out the angle for each category

Child care	$18.5 \times 12° = 222°$
Fundraising	$6.0 \times 12° = 72°$
Administration	$2.5 \times 12° = 30°$
Other costs	$3 \times 12° = 36°$

Multiply each number of millions by 12°

❖ use the angles to draw a pie chart.

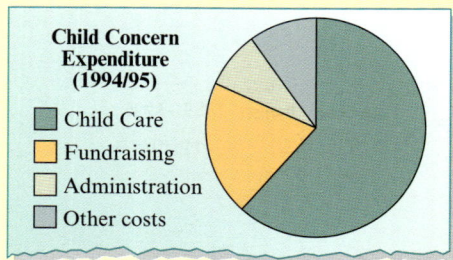

Child Concern Expenditure (1994/95)

- Child Care
- Fundraising
- Administration
- Other costs

Exercise 5.1
Presenting data divided into categories

Use the same method for data given in %:
divide 360° by 100.

1   These tables show what three different charities spent in 1994–95.

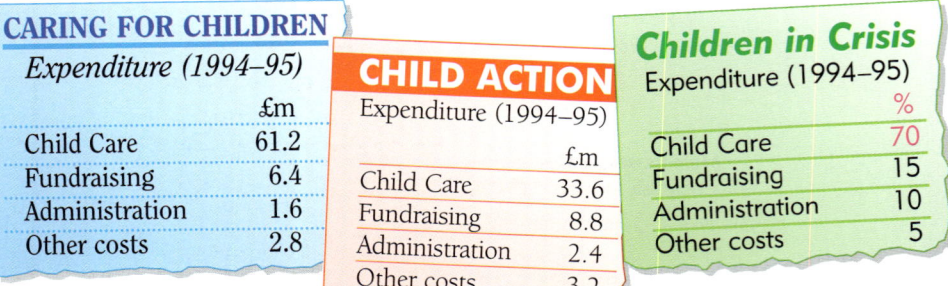

CARING FOR CHILDREN	
Expenditure (1994–95)	
	£m
Child Care	61.2
Fundraising	6.4
Administration	1.6
Other costs	2.8

CHILD ACTION	
Expenditure (1994–95)	
	£m
Child Care	33.6
Fundraising	8.8
Administration	2.4
Other costs	3.2

Children in Crisis	
Expenditure (1994–95)	
	%
Child Care	70
Fundraising	15
Administration	10
Other costs	5

For each charity:

a   add the amounts to find the total
b   divide 360° by the total
c   work out the angle for each category
d   draw a pie chart.

2   Compare how these charities spend money.

Thinking ahead to ...
finding the median

A   Archers use a 10-ring target.
Give the ring score for:

a   the outer red ring
b   the inner black ring.

Finding the median of a frequency distribution

10-ring target

Inner gold ring score: 10
Outer gold ring score: 9
Inner red ring score: 8
and so on.

The frequency here is
the number of arrows.

◆ A set of data presented in a frequency table is called a **frequency distribution**.

◆ To find the median by listing the data:

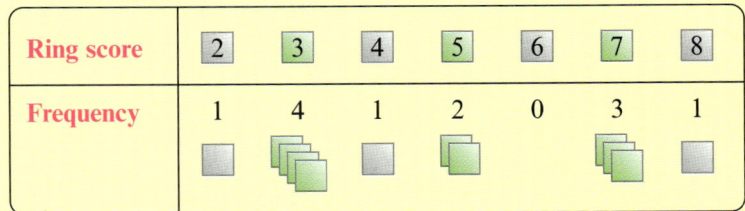

Ring score	2	3	4	5	6	7	8
Frequency	1	4	1	2	0	3	1

❖ list the data in order

2 3 3 3 3 4 5 5 7 7 7 8

❖ find the middle value.

2 3 3 3 3 4 5 5 7 7 7 8
Median score = **4.5**

Exercise 5.2
Finding the median
by listing the data

1

Ring score	5	6	7	8	9
Frequency	4	2	1	2	1

Ravi

a   How many of Ravi's arrows hit the 8 ring?
b   List Ravi's scores for his ten arrows.
c   Find Ravi's median score.

2

Ring score	5	6	7	8	9	10
Frequency	3	2	4	0	3	3

Sally

a   How many arrows did Sally fire in total?
b   List all Sally's scores.
c   Find Sally's median score.

3

Ring score	3	4	5	6	7	8	9
Frequency	1	5	6	2	1	1	2

Peta

Peta hit seven different rings: 3 4 5 6 7 8 9
The middle ring of these is the 6 ring.
a   Explain why 6 is not Peta's median score.
b   Find Peta's median score.

4   Karl fires 14 arrows: his lowest score is 2,
his highest score is 10,
his median score is 6.5
Write a possible frequency distribution for Karl's scores.

Thinking ahead to ...
calculating the mean of a
frequency distribution

A

Ring score	1	2	3	4	5	6	7
Frequency	1	0	1	3	5	4	3

Saul

What is the total score for Saul's arrows that hit the 6 ring?

Calculating the mean of a frequency distribution

♦ To calculate the mean of a frequency distribution:

Ring score	2	3	4	5	6	7	8
Frequency	1	4	1	2	0	3	1

❖ calculate the total of all the values

Ring score	Frequency	Total score in ring	
2	1	2 × 1	2
3	4	3 × 4	12
4	1	4 × 1	4
5	2	5 × 2	10
6	0	6 × 0	0
7	3	7 × 3	21
8	1	8 × 1	8
Totals	12		57

1 arrow hit the 2 ring: a total score of 2.

2 arrows hit the 5 ring: a total score of 10.

12 arrows scored 57 in total.

❖ divide the total of all the values by the **total frequency**.

Mean score = $\frac{57}{12}$ = **4.8** (to 1 dp)

Exercise 5.3
Calculating the mean

1

Ring score	3	4	5	6	7	8
Frequency	2	5	2	2	4	1

Jodie

Geeta

Ring score	Frequency
6	2
7	6
8	7
9	4
10	1

For each archer:
a calculate the total of all their scores
b calculate the total frequency
c calculate their mean score.

Give your answers to 1 dp.

2

Ring score	6	7	8	9	10
Frequency	4	2	1	2	1

Mean score = $\frac{6+7+8+9+10}{5}$ = $\frac{40}{5}$ = 8 ✗

The calculation at the side is wrong.
a Explain the mistakes.
b Calculate the correct mean score.

Comparing sets of data

A measure of spread measures how spread out data is.

The simplest measure of spread is the range.

◆ One way to compare sets of data is to compare two types of value:
   A – an average
   B – a measure of spread.

Example

Compare Kate's and Bob's scores using the median and the range.

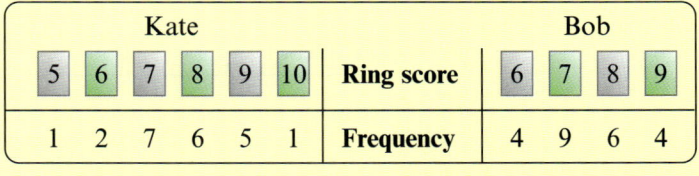

	Kate								Bob			
Ring score	5	6	7	8	9	10		6	7	8	9
Frequency	1	2	7	6	5	1		4	9	6	4

Kate   5 6 6 7 7 7 7 7 7 7 8 8 8 8 8 8 9 9 9 9 9 10

Median = **8**
Range = 10 – 5 = **5**

Bob   6 6 6 6 7 7 7 7 7 7 7 7 8 8 8 8 8 8 9 9 9 9

Median = **7**
Range = 9 – 6 = **3**

A – Kate's scores are higher on average.
B – Bob is more consistent because his scores are less spread out.

Exercise 5.4
Comparing sets of data

1

Javed

Ring score	3	4	5	6	7	8	9	Total
Frequency	3	2	6	5	5	3	1	25

Lisa

Ring score	4	5	6	7	Total
Frequency	4	8	9	3	24

a   Find the median score for each frequency distribution.
b   Calculate the range of the scores for each distribution.
c   Use the median and range to compare Javed's and Lisa's scores.

2

Amy

Ring score	4	5	6	7	8	Total
Frequency	3	5	4	3	2	17

Paul

Ring score	3	4	5	6	7	8	Total
Frequency	4	2	0	7	5	2	20

a   Calculate the mean and the range of each distribution.
b   Compare Amy's and Paul's scores using the mean and the range.

3

Imran

Ring score	1	2	3	4	5	6	7	8	9	Total
Frequency	1	1	1	2	3	2	1	2	2	15

Viv

Ring score	1	2	3	4	5	6	7	8	9	Total
Frequency	1	0	0	2	1	3	4	0	1	12

a   Compare Imran's and Viv's scores using the mode and the range.
b   Explain why the range is not a sensible measure of spread for Viv's scores.

Calculating the interquartile range

◆ The **interquartile range** of a set of data measures how spread out the middle 50% of the data is.

◆ To calculate the interquartile range:

Ring score	1	2	3	4	5	6	7	8	9
Frequency	1	0	0	2	1	3	4	0	1

❖ list the data in order

$$1 \quad 4 \quad 4 \quad 5 \quad 6 \quad 6 \quad 6 \quad 7 \quad 7 \quad 7 \quad 7 \quad 9$$

❖ divide the data into four quarters

$$1 \quad 4 \quad 4 \mid 5 \quad 6 \quad 6 \mid 6 \quad 7 \quad 7 \mid 7 \quad 7 \quad 9$$

❖ find the end values of the middle 50% of the data

$$1 \quad 4 \quad 4 \mid 5 \quad 6 \quad 6 \mid 6 \quad 7 \quad 7 \mid 7 \quad 7 \quad 9$$

4.5 **7**

Lower quartile **Upper quartile**

❖ calculate the difference between the upper and lower quartiles.

Interquartile range = 7 − 4.5
= 2.5

> You can divide the data into four quarters by first dividing it into two halves.
>
> The end values are the medians of the two halves.

Exercise 5.5
Calculating the interquartile range

1

William

Ring score	2	3	4	5	6	7	8	Total
Frequency	1	0	3	2	5	3	2	16

Bryony

Ring score	4	5	6	7	8	9	Total
Frequency	2	4	3	3	0	2	14

Daniel

Ring score	2	3	4	5	6	7	8	9	10	Total
Frequency	1	1	2	3	5	3	0	3	2	20

For each archer:
a list the data
b find the upper and lower quartiles
c calculate the interquartile range of their scores.

2

Sheera

Ring score	1	2	3	4	5	6	7	8	9	Total
Frequency	1	2	5	4	2	1	0	2	1	18

Dave

Ring score	1	2	3	4	5	6	7	Total
Frequency	2	4	4	3	4	5	2	24

Is Sheera or Dave the more consistent archer?
Give reasons for your answer.

> Calculate some figures to support your explanation.

Thinking ahead to ...
constructing a cumulative
frequency table

A

Ring score	2	3	4	5	6	7	8	Total
Frequency	1	4	1	2	0	3	1	12

Calculate how many arrows in total scored:

a 2 or 3
b less than or equal to 5
c less than or equal to 6.

Constructing a cumulative frequency table

◆ The total of frequencies up to a particular value in a set of data is called the **cumulative frequency**.

◆ To construct a cumulative frequency table from a frequency table.

Frequency table

Ring score	2	3	4	5	6	7	8	Total
Frequency	1	4	1	2	0	3	1	12

≤ stands for
'is less than or equal to'

Cumulative frequency table

Ring score	Cumulative Frequency
≤2	1
≤3	5
≤4	6
≤5	8
≤6	8
≤7	11
≤8	12

6 arrows hit the
2, 3 or 4 rings:
1 + 4 + 1

11 arrows hit the
2, 3, 4, 5, 6 or 7 rings:
1 + 4 + 1 + 2 + 0 + 3

◆ You can find the cumulative frequency for each score by adding its frequency to the cumulative frequency for the previous score.

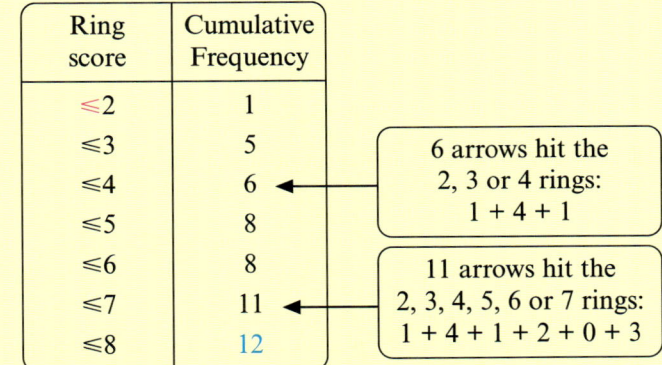

Ring score	2	3	4	5	6	7	8
Frequency	1	4	1	2	0	3	1
Cumulative frequency	1	5	6	8	8	11	12

1

Ring score	5	6	7	8	9	Total
Frequency	2	5	12	7	5	31

André

Ring score	Cumulative Frequency
≤5	
≤6	
≤7	
≤8	
≤9	

a Copy this cumulative frequency table.
b Calculate the cumulative frequencies for André's scores.

2

Ring score	2	3	4	5	6	7	8	
Frequency	3	7	12	19	13	9	4	Simone
Cumulative frequency								

   a Copy this table.

   b Calculate the cumulative frequencies for Simone's scores.

3 This distribution shows the ages of archers at a junior event.

Age	10	11	12	13	14	15	16	Total
Number of archers	8	7	10	11	8	9	11	64

  How many archers can enter these age group competitions?

   a 11 and under   **b** 13 and under   **c** 16 and under.

4 **a** Copy this cumulative frequency table for the data in question 3.

     b Calculate the cumulative frequency for each age group.

Age	Cumulative frequency
10 and under	
11 and under	
12 and under	
13 and under	
14 and under	
15 and under	
16 and under	

5 Dani fired 20 arrows at the target from each of seven different distances.

Distance to target in metres	10	15	20	25	30	35	40
Number of arrows hitting target	20	19	17	16	15	13	11

  Construct a cumulative frequency table for this distribution.

6 This table shows Jade's scores in the April, May and June competitions.

	Ring score										Total
	1	2	3	4	5	6	7	8	9	10	
April	4	7	7	8	5	6	2	3	5	1	48
May	6	5	4	9	5	3	8	2	2	4	48
June	3	6	6	4	5	4	8	4	2	6	48

  a Construct a cumulative frequency table like this.

  b Do you think Jade is improving as an archer or getting worse? Explain why.

	Ring score		
	≤1	≤2	≤3	≤4
April	4	11	18	
May	6	11	1	
June	3	9		

Thinking ahead to ...
finding the median

A

Ring score	5	6	7	8	9	Total
Frequency	7	8	13	18	14	60

Emily

a   List all Emily's scores.
b   Find Emily's median score.

Using cumulative frequencies to find the median of a frequency distribution

You can find the median for a small set of data by listing it.

This method is quicker for a distribution with a large total frequency.

♦ The median can be found without listing all the data.

♦ To find the median using cumulative frequencies:

❖ construct a cumulative frequency table
❖ use the total frequency to decide where the median is
❖ find the median.

Ring score	Frequency	Cumulative Frequency
5	7	7
6	8	15
7	13	28
8	18	46
9	14	60

The total frequency is 60, so the median score is halfway between the 30th largest and the 31st largest.

The 28th largest score is 7.

All the scores between the 29th and 46th largest are 8.

The 30th and 31st largest scores are both 8, so
Median score = **8**

Exercise 5.7
Finding the median using cumulative frequencies

1

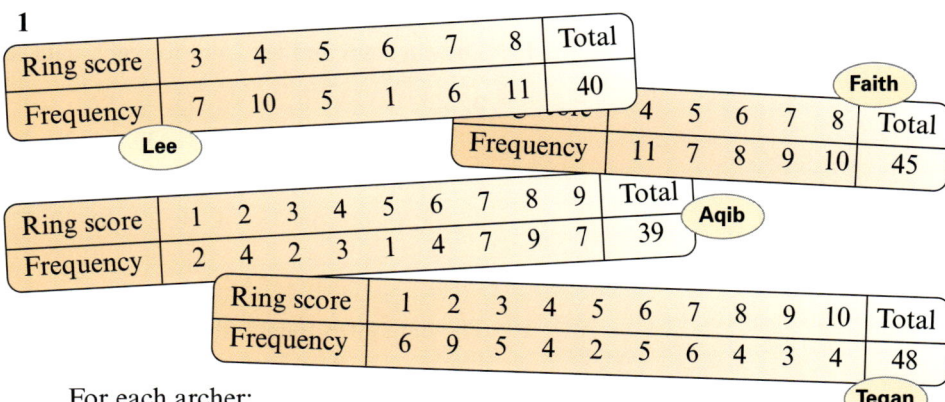

Ring score	3	4	5	6	7	8	Total
Frequency	7	10	5	1	6	11	40

Lee

Faith

	4	5	6	7	8	Total
Frequency	11	7	8	9	10	45

Ring score	1	2	3	4	5	6	7	8	9	Total
Frequency	2	4	2	3	1	4	7	9	7	39

Aqib

Ring score	1	2	3	4	5	6	7	8	9	10	Total
Frequency	6	9	5	4	2	5	6	4	3	4	48

Tegan

For each archer:
a   construct a cumulative frequency table
b   decide where the median is
d   find the median score.

2

Ring score	4	5	6	7	8	9	10
Cumulative Frequency	7	18	31	40	48	62	80

Jake

a   Explain why Jake's median score is 7.5.
b   Explain why the interquartile range of this distribution is 3.

Drawing and using a stem and leaf plot

◆ The stem and leaf plot is a frequency diagram.

◆ The actual data is displayed together with its frequency.

◆ Part of the value of each piece of data is used to fix the data class ie the **stem**.

Part of the value is listed in the diagram ie the **leaves**

Example

Show this set of data with a stem and leaf plot

6, 13, 23, 42, 30, 32, 36, 27, 24, 32, 40, 45, 8

As the data is listed and shown in order on a stem and leaf plot the median value can be easily found. Here the median value is 6.

```
        4 | 0  2  5
        3 | 0  2  2  .6
stem    2 | 3  4  7         leaves
        1 | 3
        0 | 6  8
```

Exercise 5.8
Stem and leaf plots

1 Draw a stem and leaf plot for each data set.

   a 23, 12, 14, 26, 38, 46, 33, 32, 19, 7, 24, 12, 40, 20

   b 15, 22, 24, 20, 10, 12, 31, 30, 24, 26, 26, 24, 32, 40

   c 42, 16, 9, 7, 6, 18, 25, 6, 24, 30, 36, 25, 30, 41, 16, 44

   d 38, 14, 22, 56, 18, 24, 28, 37, 40, 30, 32, 28, 46, 40, 28.

2 The data gives the number of minutes waited by people to fly on the 'London Eye'.

   44, 56, 50, 52, 42, 40, 38, 44, 52, 54, 53, 38, 30
   41, 40, 51, 59, 62, 34, 40, 47, 50, 54, 50, 57

   a Draw a stem and leaf plot to show this data.
   b Find the median value for the data set.

Thinking ahead to ...
box-and-whisker plots

A A survey of three fast food restaurants recorded the number of chips in a serving. A sample of 80 servings was taken from each restaurant.

Restaurant	Number of chips													Total
	32	33	34	35	36	37	38	39	40	41	42	43	44	
X	1	2	3	5	9	13	14	12	8	5	4	3	1	80
Y	4	6	11	14	13	11	8	6	4	3	0	0	0	80
Z	1	1	2	6	9	13	15	13	10	6	2	1	1	80

a **i** Which restaurant do you think gives more chips in a serving?
   ii Explain your answer.

b **i** Find the median number of chips for each restaurant.
   ii Calculate the interquartile range for each distribution.

Drawing box-and-whisker plots

◆ A **box-and-whisker** plot shows a frequency distribution by using:

❖ the lowest and highest values
❖ the median
❖ the lower and upper quartiles.

Example

Draw a box-and-whisker plot for these chips from restaurant W.

Number of chips	34	35	36	37	38	39	40	41	42	43
Frequency	1	2	4	7	12	14	16	15	7	2
Cumulative frequency	1	3	7	14	26	40	56	71	78	80

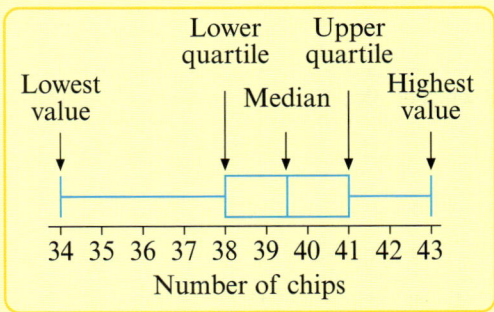

Exercise 5.9
Drawing box-and-whisker plots

1 Copy this box-and-whisker diagram on squared paper. Use the data from Question **A** to draw a box-and-whisker plot for each restaurant.

2 Write a short report that compares the number of chips per serving from the four fast food restaurants.

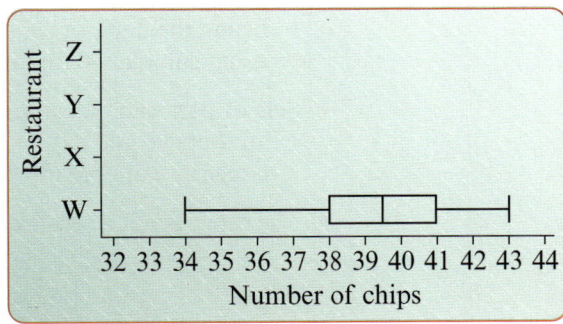

65

End points

You should be able to ...   ... so try these questions.

A	Construct a pie chart using degrees

A1

SAVING CHILDREN		£m		£m
Expenditure (1994/95)	Child care	9.2	Administration	1.0
	Fundraising	3.4	Other costs	0.8

Draw a pie chart to show how Saving Children spend their money.

Manoj

Ring score	2	3	4	5	6	7	Total
Frequency	1	0	3	5	7	2	18

Pat

Ring score	1	2	3	4	5	6	7	8	Total
Frequency	2	1	5	8	7	5	3	2	33

B	Find the median of a frequency distribution

B1 **a** Find Manoj's median score by listing the data.
     b **i** Construct a cumulative frequency table for Pat's scores.
         ii Find Pat's median score.

C	Calculate the mean of a frequency distribution

C1 Calculate the mean of:
     a Manoj's scores   **b** Pat's scores
Give your answers to 1 dp.

D	Compare sets of data

D1 Compare Manoj's and Pat's scores using the mean and the range.

Kim

Ring score	2	3	4	5	6	7	8	Total
Frequency	3	2	1	3	4	5	2	20

Stuart

Ring score	3	4	5	6	7	8	9	10	Total
Frequency	2	0	3	6	5	2	3	1	22

D2 Compare Kim's and Stuart's scores using the mode and the range.

E	Draw a box-and-whisker plot and a stem and leaf plot

E1 Draw box-and-whisker plots on the same diagram to show
     a Kim's scores   **b** Stuart's scores

Some points to remember

- The median of a frequency distribution can be found by:
    - listing the data, or
    - using cumulative frequencies.
- Sets of data can be compared using:
    - an average, and
    - a measure of spread.
- The range is not a sensible measure of spread to use when there are extreme values in the data.

Starting points

You need to know about ...

... so try these questions

A Perimeter and circumference

The perimeter of a shape is the distance around the edges of the shape.

Example

Find the perimeter of ABCDE.

The perimeter (*p*) of ABCDE
is given by:

$p = 6 + 8.5 + 6 + 2.5 + 4.5$
$= 27.5$

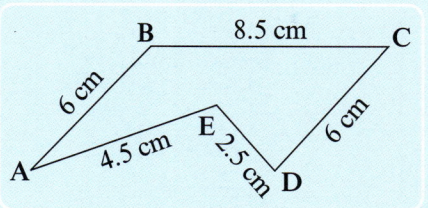

Perimeter of ABCDE is 27.5 cm.

The perimeter of a circle is called the circumference.
The formula for the circumference (*c*) of a circle is:

$c = \pi D$ (where *D* is the diameter of the circle)

Example

The circumference (*c*) of this circle
is given by:

$c = \pi \times 4.5$
$c = 14.1$ (to 1 dp)

**The circumference of a circle of
diameter 4.5 cm is 14.1 cm (1 dp).**

B Area

The area of a rectangle is given by: Base × Height

The area of a parallelogram
is given by: Base × Height

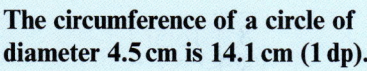

The area of a triangle
is given by: 0.5 × Base × Height

or $\dfrac{\text{Base} \times \text{Height}}{2}$

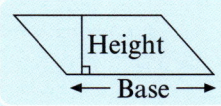

The area of a circle
is given by: πr^2 (where *r* is the radius)

Example

The area (*A*) of this circle
is given by:

$A = \pi \times 5.4^2$ ($\pi \times 5.4 \times 5.4$)
$A = 91.6$ (to 1 dp)

The area of a circle of radius 5.4 cm is 91.6 cm² (1 dp).

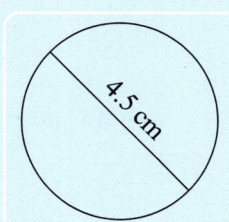

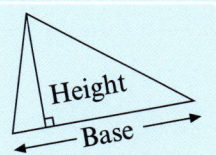

A1 Calculate the perimeter of a
square of side 5.8 cm.

A2 Find the perimeter of RSTVW.

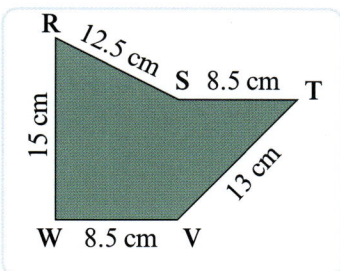

A3 A rectangle has:
long sides of 12.4 cm
and short sides of 9.8 cm
Calculate its perimeter.

A4 Calculate the circumference of
each of these circles (to 1 dp):
  a circle with diameter 5.2 cm
  b circle with diameter 0.7 cm
  c circle with radius 2.4 cm.

B1 A rectangle has:
long sides of 15.2 cm
and short sides of 8 cm.
Calculate its area.

B2 Calculate the area of CDEF.

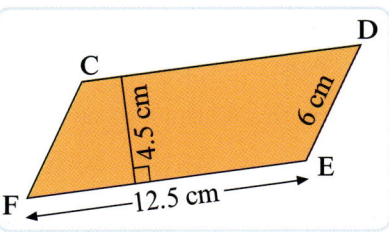

B3 Find the area of triangle JKL.

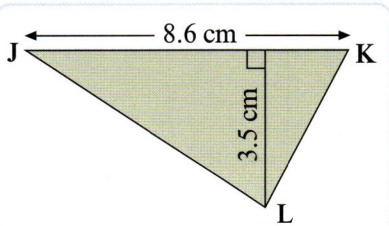

B4 Find the area of a circle of
radius:
  a 4.3 cm   **b** 7.5 cm   **c** 12.4 cm

C Splitting up a shape to find its area

Often you can find the area of a complicated shape by splitting it into simple shapes that you can calculate the area for.

Example

Find the area of the whole shape. One way to split the shape is to make two rectangles and one triangle as shown.

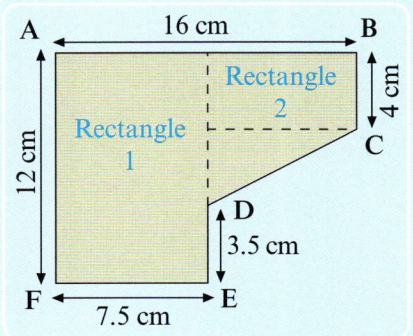

The area (A) of ABCDEF is given by:

A = Area of rectangle 1 + Area of rectangle 2 + Area of triangle

Area Rectangle 1 given by:	Area Rectangle 2 given by:	Area Triangle given by:
7.5×12 $= 90$	8.5×4 $= 36$	$0.5 \times 8.5 \times 4.5$ $= 19.125$

So: $A = 90 + 36 + 19.125 = 145.125$

Area of whole shape = 145.1 cm² (1 dp)

D Pythagoras' rule

Pythagoras' rule is:
In any **right-angled triangle**, the area of the square on the hypotenuse is equal to the sum of the areas of the squares on the other two sides.

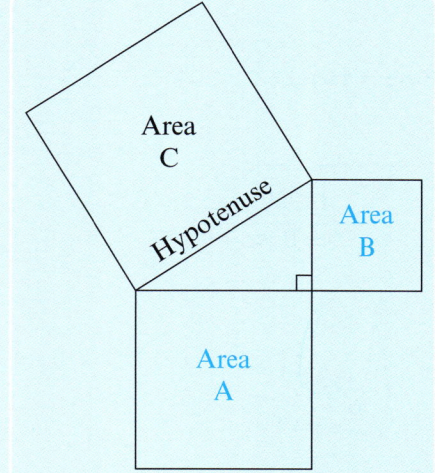

For this triangle:

Area C = Area A + Area B

From this we can also say:

Area A = Area C – Area B

and

Area B = Area C – Area A

For RST: to find the length of RS

$RS^2 = RT^2 - ST^2$
$RS^2 = 6.4^2 - 3.7^2$
$RS^2 = 40.96 - 13.69 = 27.27$
$RS = 5.2$ (1 dp)

RS is 5.2 cm long (1 dp).

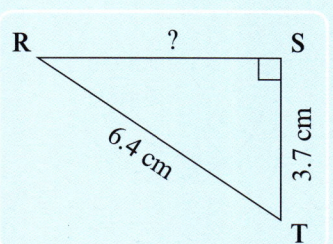

C1 Calculate the area of each shape.

a

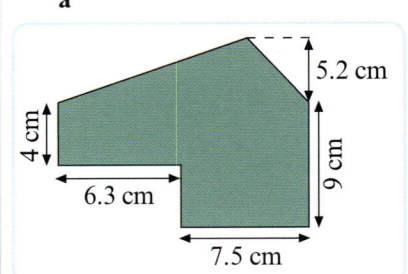

b

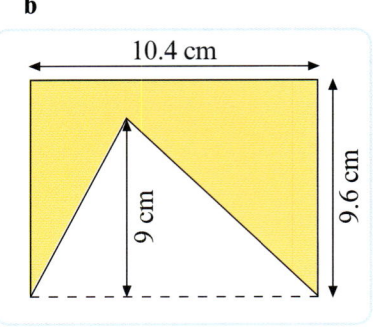

D1 Calculate the length marked x in each triangle.

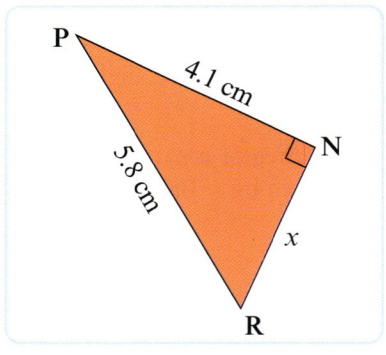

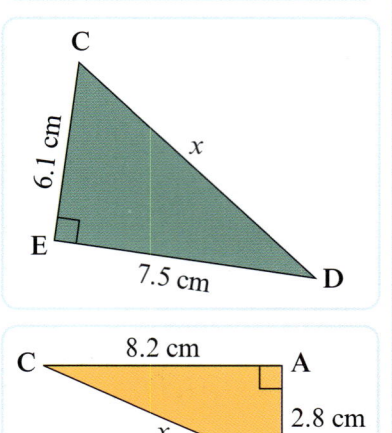

The area of a trapezium

> A quadrilateral is a shape with four straight edges.

A trapezium is a quadrilateral with one pair of opposite sides that are parallel.

ABCD is a trapezium as:
- ◆ it is a quadrilateral
- ◆ AB and CD are parallel.

One way to find the area of ABCD is to split it into two triangles and a rectangle, then calculate the total area.

Another way to find the area is to use the formula for the area of a trapezium.

Start with ABCD, then add to it a rotation of ABCD to make a parallelogram.

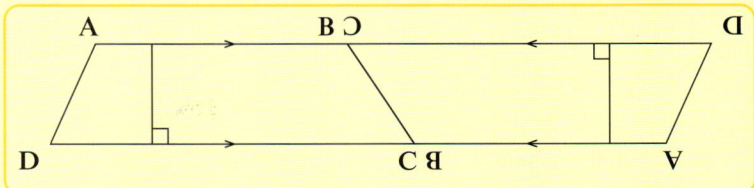

The formula for the area (A) of a parallelogram is: A = Base × Height

The area of this parallelogram is: (DC + BA) × Height
                                      Base

The area of the trapezium ABCD is half the area of the parallelogram
So, the area (A) of trapezium ABCD is given by:

$$A = 0.5 \times (AB + CD) \times \text{Height}$$

In general, the formula for the area (A) of a trapezium is:

$$A = 0.5 \times (\text{sum of the parallel sides}) \times \text{Height}$$

> The sum is the result of adding.
> For example:
> the sum of 3 and 5 is 8.

> In calculations with decimals you should either:
> ◆ give your answer to the degree of accuracy asked for e.g. 1 dp, or 2 sf
> or
> ◆ give your answer to the same degree of accuracy as is used in the question.

Example

Calculate the area of STVW.

Area (A) given by
$A = 0.5 \times (ST + VW) \times 6.6$
$ = 0.5 \times (12.4 + 7.5) \times 6.6$
$ = 0.5 \times 19.9 \times 6.6$
$ = 65.7$ (to 1 dp)

Area of STVW is 65.7 cm².

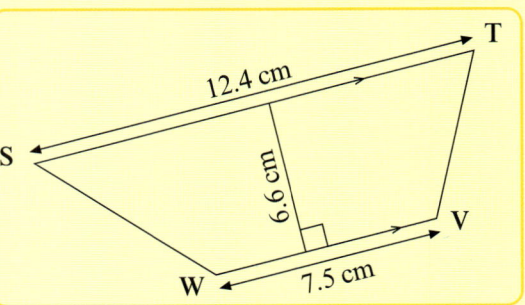

Exercise 6.1
Using the formula for the area of a trapezium

1 Calculate the area of STVW:
   a   when ST = 15.2 cm, VW = 9.5 cm, and the height is 8.3 cm
   b   when the height is 1.7 cm, VW = 2.1 cm, and ST = 4.5 cm.

2 Calculate the area of each shape:

a

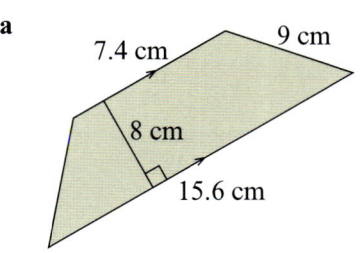

7.4 cm 9 cm
8 cm
15.6 cm

b

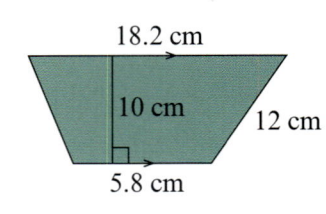

18.2 cm
10 cm 12 cm
5.8 cm

c

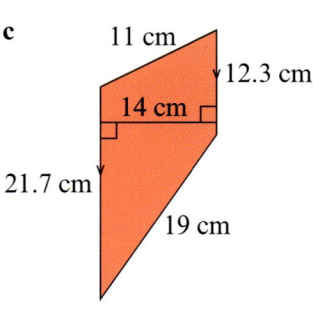

11 cm
12.3 cm
14 cm
21.7 cm
19 cm

d

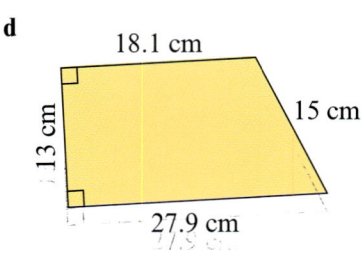

18.1 cm
15 cm
13 cm
27.9 cm

e

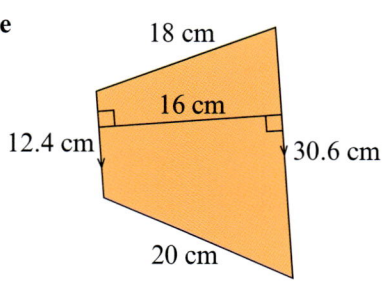

18 cm
16 cm
12.4 cm 30.6 cm
20 cm

f

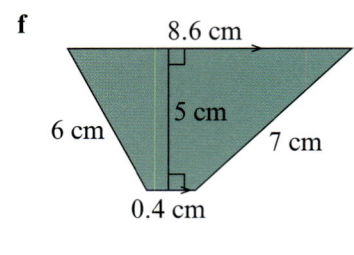

8.6 cm
6 cm 5 cm 7 cm
0.4 cm

g

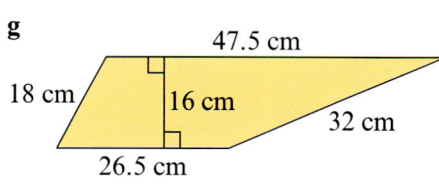

47.5 cm
18 cm 16 cm
32 cm
26.5 cm

h

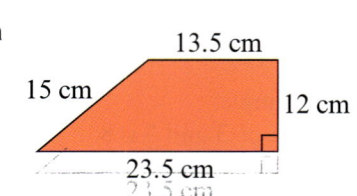

13.5 cm
15 cm 12 cm
23.5 cm

Calculate the shaded area of these shapes:

i

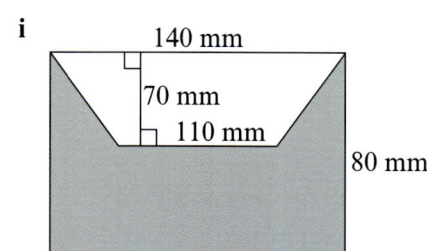

140 mm
70 mm
110 mm
80 mm

j

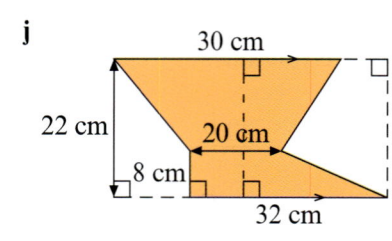

30 cm
22 cm 20 cm
8 cm
32 cm

3   This logo uses the same shape trapezium three times.

   a   Calculate the area of CDEF.

   b   What is the area of the complete logo?

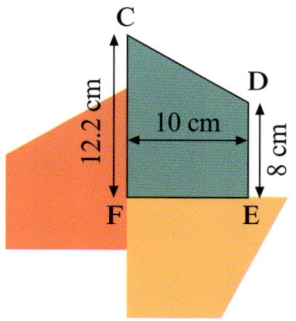

4   The diagram shows the door of a cupboard under some stairs.
The door is 76.4 cm wide.

   a   Draw a diagram of the door and show
all the distances you know.

   b   What shape is the front of the door?

   c   Calculate the area of the front
of the door.

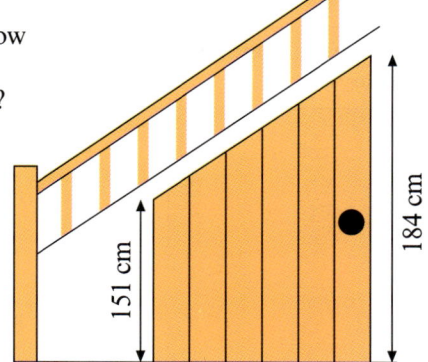

5   The diagram shows the baffles on the
front of a spotlight.
All four baffles are the same size.

   a   Draw a diagram of one baffle,
and label all the dimensions
you know.

   b   Calculate the area of
one baffle.

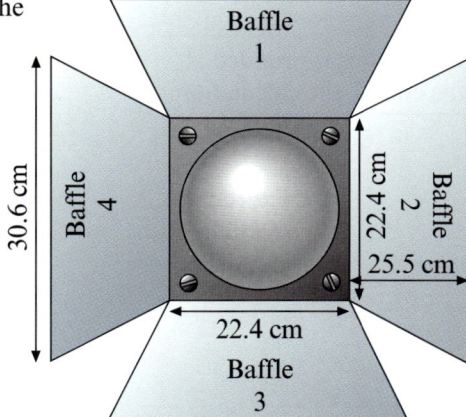

6   The diagram shows the vertical end of a
feed bin.

Calculate the area of the end of
the bin.

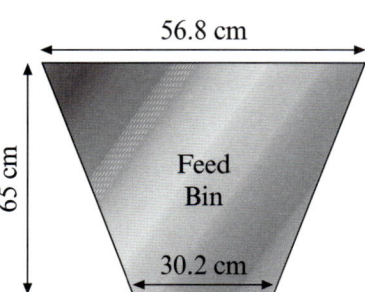

71

Composite shapes

Composite shapes are made up from more than one shape.
Sometimes you can find the area by adding areas, at other times it is easier to subtract.

Example 1

Find the area of ABCDEF.

The area (A) of ABCDEF can be given by:
$$A = (12.4 \times 4.5) + (4.6 \times 4.8)$$
$$= 55.8 + 22.08$$
$$= 77.9 \text{ (1 dp)}$$

The area of ABCDEF is 77.9 cm²

or

The area (A) of ABCDEF is given by:
$$A = \text{Area of large rectangle} - \text{Area cut out for L-shape}$$
$$A = (12.4 \times 9.3) - (4.8 \times 7.8)$$
$$A = 115.32 - 37.44$$
$$\mathbf{A = 77.9 \text{ (1 dp)}}$$

> It can be cut out in this way.

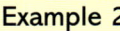

Example 2

Find the shaded area.

The area of the shaded part (A) is given by:
$$A = \text{Area of square} - \text{Area of circle}$$
$$A = (8.6 \times 8.6) - (\pi \times 4.3^2)$$
$$A = 73.96 - 58.09\ldots$$
$$\mathbf{A = 15.9 \text{ (1 dp)}}$$

> The answer is rounded at the end of the calculation.

The shaded area is 15.9 cm²

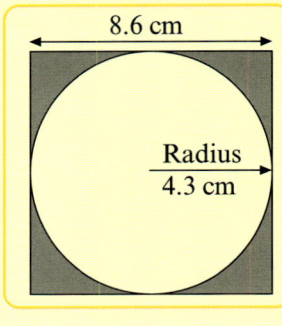

Exercise 6.2
Working with composite shapes

1 **a** Calculate the area of a circle of radius 7.2 cm.
  b A circle has a diameter of 12.4 cm. Find its area.
  c A circle has a diameter of 3 cm.
      Show that its area can be written as 2.25π.
  d The radius of a circle is given as $\sqrt{8}$.
      Write an expression for the area of the circle in terms of π.
  e The area of a circle is given as 53π.
      What is the radius of the circle?

2 This shows how milk bottle tops are cut from a strip of foil.

  a What is the radius of a foil bottle top?
  b Calculate the area of a foil bottle top.
  c How long is this strip of foil?
  d Calculate the area of foil wasted by cutting out these five tops.

3 Tops are cut from a 7.2 m strip of foil. Calculate the area of wasted foil.

4   A rectangle of card 7.4 cm by 8.3 cm had a piece cut from it to leave shape R.

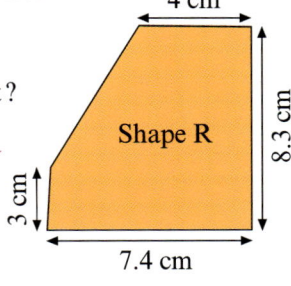

   a   What shape was cut from the rectangle?
   b   What was the area of the rectangle before the cut?
   c   Calculate the area of shape R.

5   This card is used to frame photographs.
   A circle of radius 4 cm is cut from the card.
   Gold lines are printed 1 cm from the edge of
   the circle and the card as shown.

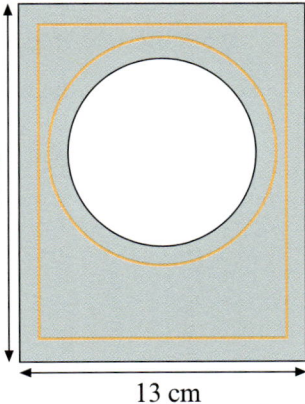

   a   Calculate the area of card after the cut out.
   b   Is the area of photograph showing more or less
      than $\frac{1}{4}$ of the blue card?

      Explain your answer.

   c   Calculate the length of gold line on the card.

6   This diagram shows the net of
   a box for playing cards.
   The net is cut from a
   rectangle of card.

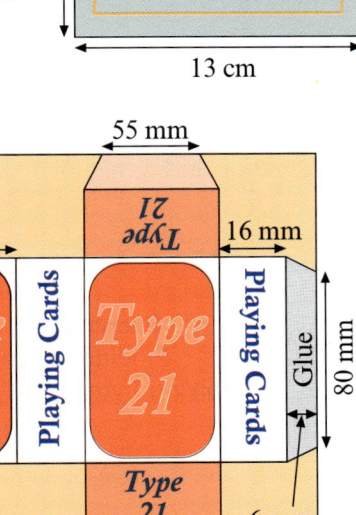

   For the rectangle of card:

   a   what is its length?
   b   how wide is it?
   c   Calculate the area of the
      glue flap.
   d   The box opens at either end.

   Find the area of an opening end.

   e   Calculate the total area of
      the net.
   f   Calculate the area of waste card.
   g   Roughly, what fraction of the card is wasted? Explain how you decided.

7   This logo design uses two parts of a circle in a red square.

   a   What is the radius of the circle?
   b   What fraction of the circle stands for
      the visor (the blue part)?
   c   Calculate the area of the visor in the
      logo.
   d   Find the area of the helmet in the logo
      (the yellow part).
   e   Calculate the area of the logo that is
      red.

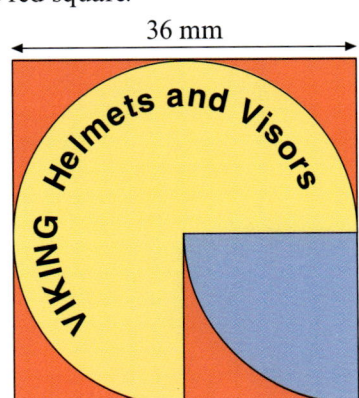

Involving Pythagoras

Sometimes to find the area of a shape you have to calculate a distance before you can calculate the area.

Example

Calculate the area of ABC.

The area of ABC is given by:
$0.5 \times \text{Base} \times \text{Height}$

Think of AC as the base
and
DB as the height.

The area of ABC (A) can be given by:

$$A = 0.5 \times AC \times DB$$

but we first have to calculate the length of AC:

$$AC = AD + DC$$

To find AD. In triangle ADB:

$$AD^2 = AB^2 - DB^2$$
$$AD^2 = 3.5^2 - 2.8^2 = 12.25 - 7.84 = 4.41$$
$$AD = \sqrt{4.41} = 2.1$$

To find DC. In triangle DBC:

$$DC^2 = BC^2 - DB^2$$
$$DC^2 = 5.5^2 - 2.8^2 = 30.25 - 7.84 = 22.41$$
$$DC = \sqrt{22.41} = 4.73 \dots$$

The answer is rounded at
the end of the calculation.

So $AC = 2.1 + 4.73 \dots = 6.83 \dots$

Area of ABC (A) = $0.5 \times AC \times DB = 0.5 \times 6.83 \dots \times 2.8 = 9.6$ (1 dp)

The area of ABC is 9.6 cm^2

Exercise 6.3
Finding areas
involving Pythagoras

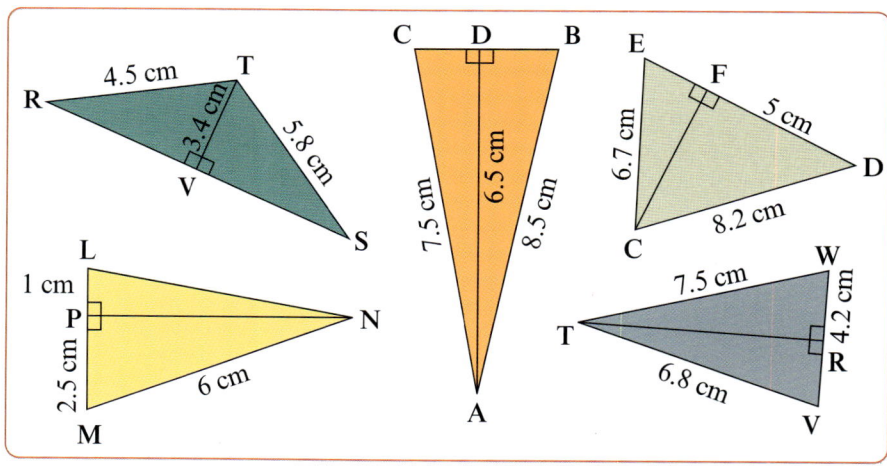

1   In triangle RST:
   a   Calculate the length of RV.
   b   Calculate the length of VS.
   c   Calculate the area of triangle RST.

2   Calculate the area of triangle ABC.

3   **a**   Calculate the height of triangle CDE.
   b   Calculate the area of triangle CDE.

4   **a**   Calculate the area of LMN.
   b   Calculate the area of TVW.

Each face of this box is a triangle. This 3-D shape is a **tetrahedron**.

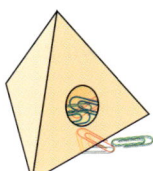

5   The diagram shows the net for this paperclip box.

Calculate the length marked *h*.

6   Find the area of the bottom of the box.

7   Calculate the area of a glue tab.

8   Find the area of the face with the circle cut out.

9   Calculate the total area of the net.

10   Find the area wasted, when the net is cut from a 26 cm square of card.

11   Calculate the perimeter of the net (including the glue flaps).

Using circumference

Exercise 6.4
Using circumference

1   **a**   Find the circumference of a circle of radius 9.5 cm.
  b   A circle has a diameter of 24.5 cm. Find its circumference.
  c   Show that the circumference of a circle of radius 4.8 cm can be written as 9.6π.
  d   The circumference of a circle is given as 15π. What is the radius of the circle?
  e   The semicircle has a diameter of 21.5 cm. Find the perimeter of the shape.

Remember

Diameter = 2 × radius.

2   The larger of these crop circles has a diameter of 31.5 metres, and the smaller a radius of 5.25 metres.

Calculate the circumference of:

  a   the larger circle
  b   the smaller circle.

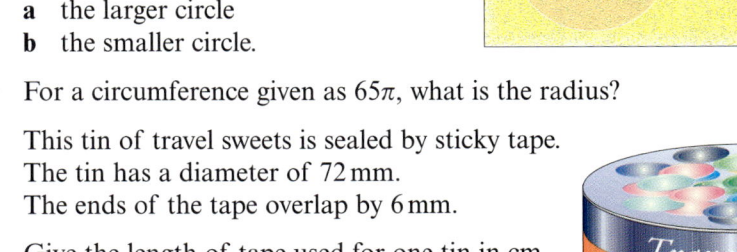

3   For a circumference given as 65π, what is the radius?

4   This tin of travel sweets is sealed by sticky tape.
The tin has a diameter of 72 mm.
The ends of the tape overlap by 6 mm.

Give the length of tape used for one tin in cm.

5   Rolls of tape are 1450 metres long, and 1 metre is wasted.

How many tins can be sealed with one roll of tape?

2D representations of 3D shapes

3D shapes can be shown in 2D in several ways.

For example:

♦ A drawing or sketch.

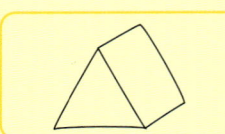

An isometric grid is like this:

this
way
up

♦ A diagram on isometric grid paper.

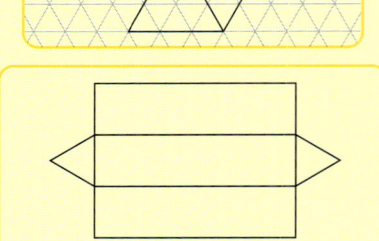

♦ A drawing of a net of the shape.

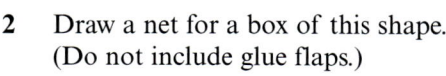

Exercise 6.5
2D/3D representation

1 This is a drawing of a box for a double CD.

Draw a diagram of a box like this on isometric paper.

2 Draw a net for a box of this shape. (Do not include glue flaps.)

3 This drawing is of a box for sweets. What shape is:

a an end of the box
b a side of the box
c the bottom of the box?

An accurate drawing has angles and lengths drawn using measuring instruments.

4 Make an accurate drawing of one end of the box.

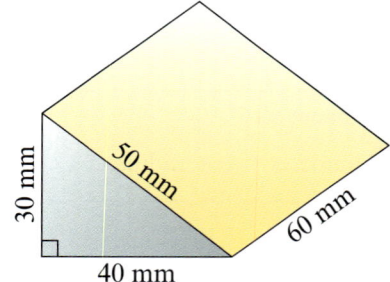

5 Calculate the total area of all five faces of the box.

6 Draw an accurate net for the box.

7 This a sketch of a stack of blank dice.

a Draw this stack on isometric paper.
b How many dice are in this stack?

8 A stack of fifteen 3 cm dice is packed in a box.

a Sketch a possible box you think can be used.
b Draw an accurate net for your box.

End points
You should be able to ...   ... so try these questions

A Use the formula for the area of a trapezium

A1

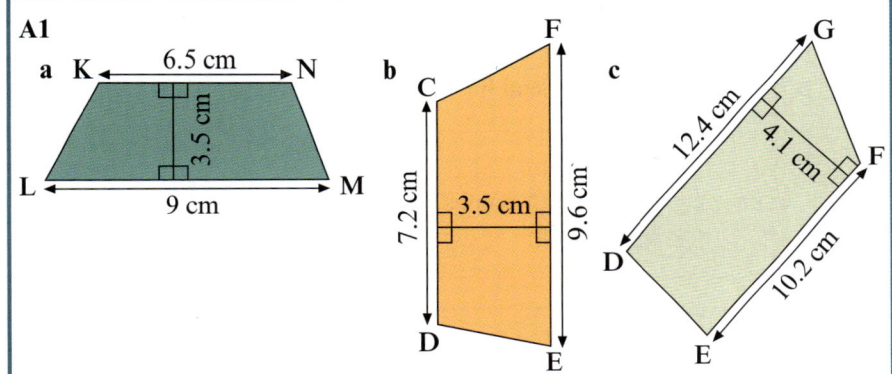

a K ⟷ 6.5 cm ⟶ N
L ⟷ 9 cm ⟶ M
3.5 cm

b C
7.2 cm
3.5 cm
F
9.6 cm
D
E

c G
12.4 cm
4.1 cm
F
10.2 cm
D
E

Calculate the area of each shape.

B Work with composite shapes

B1 Explain, with a diagram, what is meant by a composite shape.

B2 a Calculate the area of shape P.

Shape P is cut from a 16 cm square of card.

b Calculate the area of waste card.

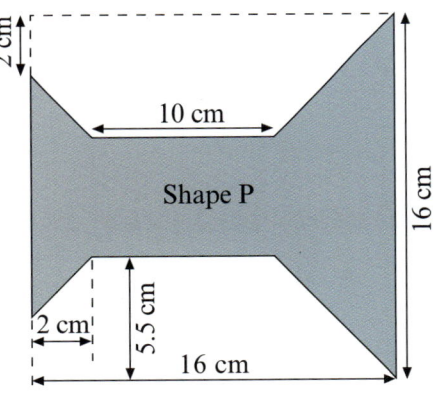

2 cm
10 cm
Shape P
16 cm
2 cm
5.5 cm
16 cm

B3 Calculate the area of RST.

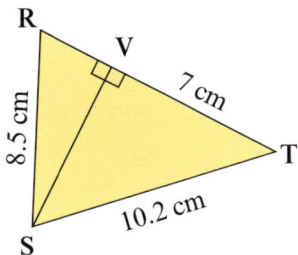

R
V
8.5 cm
7 cm
T
10.2 cm
S

C Calculate perimeters

C1 This hoax crop circle was made by two people with a piece of rope 3.2 metres long.
They simply made two half-circles.

Calculate the perimeter of the crop circle.

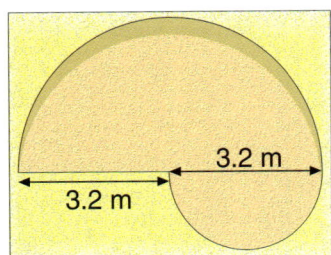

3.2 m
3.2 m

D Work with 2D representations of 3D objects

D1 a Make a sketch of the two parts of a matchbox.
b Draw the two parts of a matchbox on isometric grid paper.

7

About the Cornish Seal Sanctuary

It was founded in 1957.

In 1975, it moved to a site of about 40 acres in Gweek.

It has a fully equipped seal hospital and 10 outdoor pools.

Cafe prices		
Coffee		60p
Tea		40p
Cola		50p
Rolls:	Bacon	£1.50
	Sausage	£1.25
	Egg	£1.10
Chips		70p
Beans		40p
Salad		55p

Opening hours
Open every day except Christmas
09 00 to 18 30

Feeding times
11 00 • 13 30 • 16 00

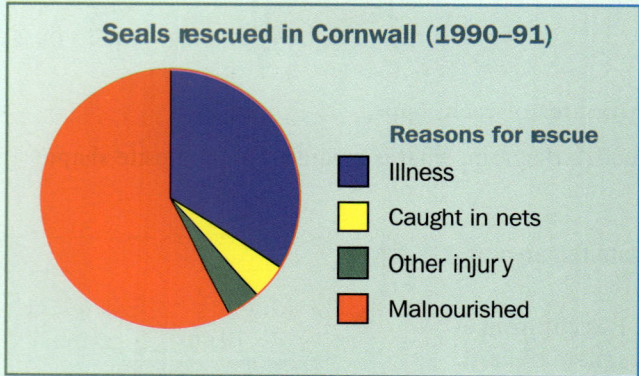

Seals rescued in Cornwall (1990–91)

Reasons for rescue
- ■ Illness
- ■ Caught in nets
- ■ Other injury
- ■ Malnourished

Ratio of male: female seals rescued between Land's End and Porthleven	
Year	Ratio
1992–93	1:1
1993–94	1:2
1994–95	2:1

About the grey seals

Weights of pups rescued in December 1992		
Name	Weight at rescue (kg)	Weight at release (kg)
Bill	23.4	65.0
Ben	16.8	56.0
Mandy	12.3	66.0
Tony	24.0	98.0
Rory	18.5	85.0

Seals are usually released when they weigh about 60 kg.

Seals at the sanctuary eat over a tonne of fish per week.

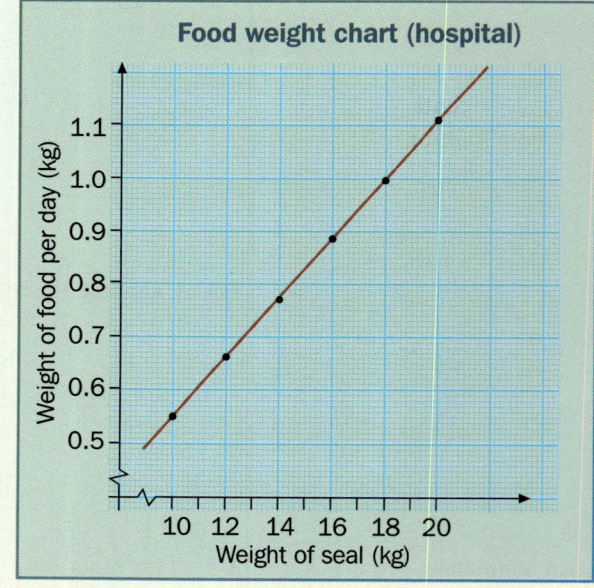

Food weight chart (hospital)

Weight of food per day (kg) vs Weight of seal (kg)

Major breeding areas for grey seals			
Location	Pups born (1989)	Pups born (1990)	Total population in 1990 (to nearest 100)
Inner Hebrides	2051	2256	7800
Outer Hebrides	9537	9823	34000
Orkney	7038	7319	25400
Isle of May	933	1185	4100
Farne Islands	892	1004	3500

The world population of grey seals is estimated at 120000.

About two thirds of them live around the British coastline.

In water, seals can reach speeds of up to 20 km per hour.

1   How many years is it since the Seal Sanctuary was founded?

2   How long is it open each day?

3   In the cafe, Pritpal orders a cola, bacon roll and chips.
How much does this cost?

4   What was the most common reason for a seal being rescued in Cornwall in 1990–91?

5   What was the weight of the lightest pup rescued in December 1992?

6   About how many kilograms of fish do the seals at the Sanctuary eat each week?

7   What weight of food will be given to a 10 kg seal pup during a day in hospital?

8   What percentage of seals rescued between Land's End and Porthleven in 1992–93 were male?

9   How many pups were born in Orkney in 1990?

10 How long has the Seal Sanctuary been at Gweek?

11 Pritpal gets there at 8.25 am. How many minutes is it till the Sanctuary opens?

12 About what fraction of the seals rescued in 1990–91 were ill?

13 About what percentage of the seals rescued in 1990–91 were malnourished?

14 24 seals were rescued in Cornwall in 1990–91. About how many of them were ill?

15 What fraction of the seals rescued between Land's End and Porthleven in 1994–95 were female?

16 About what percentage of the seals rescued in Cornwall in 1990–91 were caught in nets?

17 4 seals were rescued between Land's End and Porthleven in 1992–93.
How many were male?

18 3 seals were rescued between Land's End and Porthleven in 1993–94.
How many were female?

19 In 24 hours in hospital, what weight of food is given to a 15 kg seal pup?

20 In a day, what weight of food in grams will be given to a pup that weighs 14 kg?

21 In a day, a pup in hospital is given 800 grams of food. Give the weight of this pup in kilograms.

22 Which of the pups rescued in December 1992 gained the most weight before being released?

23 Write Tony's rescue weight as a percentage of his release weight.

24 For the rescue weights, find:
a   the mean   b   the median
c   the range

25 Write each rescue weight rounded to the nearest kilogram.

26 Write each release weight in grams.

27 About how many kilograms of fish do the seals at the Sanctuary eat each day?

28 Out of the major breeding areas, which location had the highest total population of grey seals in 1990?

29 For each location in 1989, write down how many pups were born, to the nearest hundred.

30 For each location in 1990, write down how many pups were born, to the nearest thousand.

31 Draw a graph to show the number of pups born in 1989 and 1990 for these locations.

32 About how many grey seals live around the British coastline?

33 At 20 km per hour:
a   How far could a seal travel in 2 hours?
b   How long would it take to travel 35 km?
c   How long would it take to travel 1 km?
d   How far could a seal travel in 1 minute?

34 In June about 800 people visit the Seal Sanctuary each day.
In total, about how many people visit in June?

35 Susan arrives at the Sanctuary at 3.15 pm.
How long has she to wait for feeding time?

36 In the cafe, Susan asks her aunt for a hot drink and a roll.
If her aunt chooses a hot drink and roll at random, what is the probability of her choosing:
a   a coffee and an egg roll?
b   a tea and a bacon roll?

37 If cafe prices are increased by 15%, what is the new price of:
a   a tea?   b   a sausage roll?
c   chips?

Dale Valley Railway Spring 1997

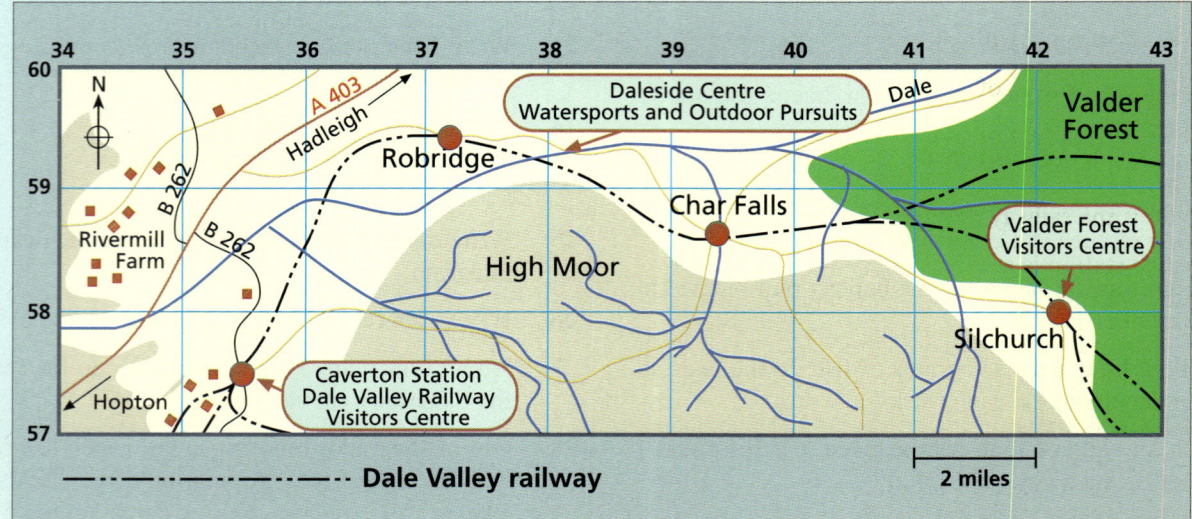

————·—··—·—··— **Dale Valley railway**

2 miles

Dale Valley Railway

Timetable 1 May to 30 September

	Depart	Depart	Depart
Caverton Station	1000	1200	1430
Robridge	1012	1212	1445
Char Falls	1025	1225	1505
Silchurch	a 1040	a 1240	a 1520

	Depart	Depart	Depart
Silchurch	1105	1330	1610
Char Falls	1120	1345	1628
Robridge	1135	1400	1645
Caverton Station	a 1145	a 1410	a 1700

a – arrival time

** Note – in May and September there is no 1200 departure

Fares

Adult	£5.50
Child (under 14 years)	£3.00
Senior Citizen	£3.50

* Note all fares are return
no single tickets are for sale

Festival Special

Adult	£12.50
Child (under 14 years)	£6.50
Senior Citizen	£9.50

* Note all fares include lunch
at the Railway Arms in Silchurch

Information card

Weights and Measures

Locomotive	55 tons
Type A coach	27 tons
Type B coach	37 tons
Coal	1.5 tons
Water	500 gallons

1 gallon of water weighs 10 lb
1 ton = 2240 lb

Tally for Festival Special

Adult	/////////////////////////////// ////////////////////////
Child	/////////////////
Senior Citizen	///////////////////// ///////////////////////////// //////////////////////

1 Estimate how far it is by rail from Caverton Station to Silchurch.

2 What area is shown by one square on the map?

3 Estimate the area of High Moor on the map.

4 Estimate the area of Valder Forest on the map.

5 When you travel from Robridge to Char Falls by road, is the railway on your left or right?

6 Which station is at the eastern end of the Dale Valley railway?

7 **a** Which station is roughly north-east of Caverton?
   b Give the bearing of Robridge from Silchurch.

8 Give a 6-figure grid reference for:
   a the Daleside Centre
   b Rivermill Farm
   c places where the railway crosses the River Dale
   d where the B262 crosses the River Dale
   e the Valder Forest Visitors Centre.

9 As a very good guide:
   100 miles is equivalent to 160 kilometres.
   Draw a conversion graph for miles to kilometres.

10 Estimate these distances in kilometres:
   a 30 miles   **b** 55 miles   **c** 80 miles
   d 25 miles   **e** 130 miles   **f** 285 miles

11 Estimate these distances in miles:
   a 55 km   **b** 96 km   **c** 40 km
   d 120 km   **e** 235 km   **f** 375 km

12 1 km is roughly what fraction of a mile?
   Explain your answer.

13 In kilometres, roughly how far is it by rail from Char Falls to Caverton Station?

14 What is the mean journey time for a journey between Caverton and Silchurch stations.
   Give your answer to the nearest minute.

15 For a return journey, the same day, what is:
   a the shortest time between departure and return?
   b the longest time?

16 Between 10 00 and 17 00, how long in total does the train spend in Silchurch?

17 Calculate the shortest journey time between Robridge and Char Falls.

18 **a** For how many days does the timetable run?
   b How many return trips will the train make during the season?

Next year the Dale Valley Railway estimate that they will carry 25 000 passengers.
They expect the ratio of adults to children to be 5 : 3.

19 **a** How many adults do they expect to carry?
   b How many children?

20 Will £120 000 be a good estimate of ticket sales? Explain your answer.

21 Calculate the mean number of passengers expected per day (1 May – 30 Sept).

22 DVR expect about 35% of the adults to be Senior Citizens.

   How many Senior Citizens are expected?

23 The yearly running costs of the railway are estimated at £7000 per mile between Caverton and Silchurch.

   Do you expect the railway to make a profit next year? Explain your answer.

24 To build a cafe, Dale Valley Railway plan to increase the ticket charges by 6%, then to round the new charge to the nearest 10 pence.

   Give the new charge for each type of ticket.

25 The Festival Special ran on 1 May 1992.
   Draw up a frequency table to show the number of tickets sold.

26 What was the total amount taken in ticket sales for the Festival Special?

27 To the nearest penny, what was the mean price of a ticket on the Festival Special?

28 Draw a pie chart to show the numbers of tickets sold for the Festival Special.

29 What is the weight of 500 gallons of water?

30 Which is heavier:
   500 gallons of water or 1.5 tons of coal?

31 At the start of the day, the locomotive is loaded with 500 gallons of water and 1.5 tons of coal.
   What is the total weight of the locomotive?

32 An eight coach train is made up of:
   p type A coaches, and n type B coaches.
   Write an equation with p and n for the number of coaches in the train.

33 **a** Write an expression in p and n, for the weight of the coaches in the train.
   b Write an expression for the weight of coaches and loco.

Starting points
You need to know about ...

... so try these questions

A Finding the value of linear expressions

- Examples of linear expressions are: $2m + 4$ $6p - 4t$
- The value of an expression depends on the value of the letters.

When $p = 2$ and $t = 1$, $6p - 4t = (6 \times p) - (4 \times t)$
$$= (6 \times 2) - (4 \times 1)$$
$$= 12 - 4 = 8$$

A1 Find the value of these expressions when $t = 5$.
  a $5t - 8$ **b** $11 - t$
  c $2(t + 7)$ **d** $3(t - 6)$

A2 Evaluate these expressions when $a = 2$ and $b = 5$.
  a $4a + b$ **b** $3b - 4a$
  c $7(a + b)$ **d** $2a - 3b$

B Adding like terms

- In the expression $2x + 4y + 3x + 5y$:

  - $2x$, $4y$, $3x$ and $5y$ are called **terms**.
  - $2x$ and $3x$ are called **like terms** as both give the number of xs.

- The expression can be simplified by adding like terms:

$$2x + 4y + 3x + 5y$$
$$= 2x + 3x + 4y + 5y$$
$$= 5x + 9y$$

$2x + 4y + 3x + 5y$ and $5x + 9y$ are **equivalent expressions**.

B1 Which of these is equivalent to $5x + 2y + 3x + y$?
  A $15x + 2y$ B $8x + 2y$
  C $8x + 3y$ D $15x + 3y$

B2 Simplify each of these by adding like terms.
  a $5t + 3t + t$
  b $7s + 3s + 5k + 4k$
  c $2c + 6b + c + 2b$
  d $6x + 2 + 5x$
  e $5v + 6 + 2v + 10$

C Solving linear equations

- Solving an equation is finding the possible values for each letter.

- With the value of one letter to find, add, subtract, multiply or divide **both** sides of the equation by equal amounts.

Example

Solve $6n - 2 = 2n + 8$

$+2$ ⟨ $6n - 2 = 2n + 8$ ⟩ $+2$
$-2n$ ⟨ $6n = 2n + 10$ ⟩ $-2n$
$\div 4$ ⟨ $4n = 10$ ⟩ $\div 4$
    $n = 2.5$

The **solution** of this equation is $n = 2.5$.

C1 Which of these is the solution of:
$4x - 5 = 2x + 7$?
  A $x = 1$ B $x = 2$ C $x = 6$

C2 Solve:
  a $6z + 1 = 10$
  b $5y - 2 = 13$
  c $3x + 14 = 8$
  d $2w + 5 = 3w + 1$
  e $5 + v = 2v - 3$
  f $6t + 8 = t + 3$
  g $5s - 1 = 7s - 5$

D Linear graphs

- An example of an equation of a straight line is $y = 3x + 5$.

- For any x-coordinate, you can calculate the y-coordinate.

When $x = {^-}1$, $y = (3 \times {^-}1) + 5$
$$= {^-}3 + 5 = 2$$

So the line $y = 3x + 5$ goes through the point $({^-}1, 2)$.

D1 Which of these points is on the line $y = 5x - 1$?
  A $(1, 4)$ B $(2, 11)$ C $({^-}1, {^-}6)$

D2

x	$^-2$	$^-1$	0	1	2
y					

Copy and fill this table for:
  a $y = 2x + 5$ **b** $y + x = 6$

Thinking ahead to ...
simplifying linear
expressions

P $10 + (1 \times 0.6)$ Q $10 + (2 \times 0.6) - (3 \times 0.6)$ R $10 - (5 \times 0.6)$

S $10 - (1 \times 0.6)$ T $10 - (2 \times 0.6) - (3 \times 0.6)$

U $10 - (2 \times 0.6) + (3 \times 0.6)$

A Sort these calculations into pairs with the same value.

Simplifying linear expressions

To simplify an expression   ◆ reorder the terms if you need to
◆ add or subtract like terms

Examples

| $6y - x - 3x$ $= 6y - 4x$ | $5p + 6 - 3p - 1$ $= 5p - 3p + 6 - 1$ $= 2p + 5$ | $7a - 3b - 4a + 2b - a$ $= 7a - 4a - a - 3b + 2b$ $= 2a - b$ |

Exercise 7.1
Adding and subtracting
like terms

1 Simplify each of these expressions:

a	$4t + 7t - 3t$	b	$6x - 5 - x + 5$	c	$6 + 4h - 3 + h$
d	$3a - 2b + 2b - 2a$	e	$2 - y + 6 - 4y - 7$	f	$10 - 5f - 4 + 2f$
g	$3x + 4y - 5 - 2x$	h	$3x - 5y - x$	i	$5x - 4 - 3x + 2y$
j	$2ax + 3x + 4ax$	k	$xy + 2x + y + xy$	l	$4x - 3 - 2y - 7$
m	$3x + 4y + x + y - 3$	n	$5y - x - 2y - 3x$	o	$4 - 5y - x - 2y$
p	$12 - 7y - 5x - 8y$	q	$7x - 3 - 8x + 2y$	r	$ax + ay + 3ax + y$

2 Some expressions are arranged in a square.

$3x - y$	$x - 2y$	$2x$
x	$2x - y$	$3x - 2y$
$2x - 2y$	$3x$	$x - y$

In a magic square, the
numbers in each row, each
column and each diagonal
add to give the same total.

a Find the value of each expression in the
square when $x = 5$ and $y = 2$.
b Draw the square with these values.
Is it a magic square?
c For each row, column and diagonal, find the total of the
three expressions.
d Explain why your totals show that any values for x and y will give a
magic square.

Thinking ahead to ...
using brackets

A Sort these calculations into pairs with the same value.

(A) $200 - 30$ (B) $5 \times (100 + 6)$ (C) $5 \times (100 + 30)$

(D) $5 \times (100 - 30)$ (E) $500 + 30$ (F) $2 \times (100 - 15)$

(G) $(5 \times 100) + (5 \times 30)$ (H) $(5 \times 100) - (5 \times 30)$

Using brackets in linear expressions

To multiply out brackets:

♦ Multiply **every** term inside the brackets.

Examples

$$2(n + 3) = 2 \times (n + 3)$$
$$= (2 \times n) + (2 \times 3)$$
$$= 2n + 6$$

$2(n + 3)$ and $2n + 6$ are
equivalent expressions

$$5(3n - 2) = 5 \times (3n - 2)$$
$$= (5 \times 3n) - (5 \times 2)$$
$$= 15n - 10$$

$5(3n - 2)$ and $15n - 10$ are
equivalent expressions

Exercise 7.2
Using brackets

1 For $n = 5$, find the value of $2(n + 3)$ and $2n + 6$.

2 For $n = 2$, find the value of $5(n - 2)$ and $5n - 10$.

3 Sort these into pairs of equivalent expressions.

$2(a + 4)$ $8a - 12$ $2(a + 2)$ $2a + 8$

$4(2a - 3)$ $2a - 12$ $2a + 4$ $2(a - 6)$

4 A student has tried to multiply out the brackets from $3(4 + x)$.

$3(4 + x) = 4 + 3x$ **✗**

a Explain the mistake you think she has made.
b Multiply out the brackets from $3(4 + x)$.

5 Two students find expressions for the nth term in a sequence.

Sue: nth term $= 5(2n - 1)$ **Ahmet:** nth term $= 10n - 5$

Show that the two expressions are equivalent.

6 Multiply out the brackets from:

a $4(n + 1)$	**b** $5(m - 3)$	**c** $6(c - 9)$	**d** $2(3p + 2)$
e $8(2s - 5)$	**f** $4(3 + t)$	**g** $3(5 - k)$	**h** $2(7n - 5)$
i $30(2f - 13)$	**j** $180(5 + 3h)$	**k** $10(3 - 7q)$	**l** $5(2y + 3z)$

7 Show that $2(n + 2) + 2n$ and $4n + 4$ are equivalent expressions.

8 Expand and simplify:

a $5(p + 2) + 3p$	**b** $4(2q - 1) - 3q$	**c** $3(x + 2) - 4(x - 5)$
d $4r + 10(3 + 5r)$	**e** $3(2s - t) + 5t$	**f** $2(x - 3) + 5(x + 1)$
g $3(x - 4) - 5(x - 2)$	**h** $2(x - 3) + 3(x - 1)$	**i** $5(2x - 3) - 4(3x + 2)$

Solving linear equations

For linear equations with brackets you can start by multiplying out the brackets.

Examples

- Solve $2(p + 4) = 8p - 1$

$$2(p + 4) = 8p - 1$$
$$2p + 8 = 8p - 1$$
$$8 = 6p - 1 \quad \substack{-2p}$$
$$9 = 6p \quad \substack{+1}$$
$$1.5 = p \quad \substack{\div 6}$$

$-2p$ $+1$ $\div 6$

You can check by finding the value of both sides of the equation for your solution. The values should be equal.

- Solve $3(4 - t) + 5 = 13 - 5t$

$$3(4 - t) + 5 = 13 - 5t$$
$$12 - 3t + 5 = 13 - 5t$$
$$17 - 3t = 13 - 5t$$
$$17 + 2t = 13 \quad \substack{+5t}$$
$$2t = {}^-4 \quad \substack{-17}$$
$$t = {}^-2 \quad \substack{\div 2}$$

$+5t$ -17 $\div 2$

Exercise 7.3
Solving linear equations

1 Solve these equations.

a $3x + 5 = 17$	**b** $4y - 3 = 21$	**c** $5x + 8 = 43$
d $2(w + 8) = 52$	**e** $3(2v - 1) = 21$	**f** $6(3 + 5u) = 54$
g $4(t + 1) = 33$	**h** $4(3s - 1) = 4s + 39$	**i** $5(2r + 3) = 7r + 42$
j $2(q - 3) + 3q = 3$	**k** $10 - p = 4p$	**l** $14 - 4n = 2(1 + 2n)$
m $3m - 5 = 11 - m$	**n** $2(5l - 3) = 48 - 5l$	**o** $5(2k - 3) = 15 - 2k$
p $4j - 7 = 3(18 - j) + 9$	**q** $17 - 2h = 22 - 3h$	**r** $2(9 - 2g) = 25 - 6g$

Each equation has a negative number as its solution.

2 Solve these equations.

a $2z + 12 = 3z + 17$	**b** $4(y + 6) = 20$	**c** $5(2x + 7) = 15$
d $5(w + 5) = 2(w + 8)$	**e** $v + 17 = 9 - v$	**f** $3u + 35 = 5 - 2u$
g $2(3 - t) = 4t + 9$	**h** $8 - 5s = 2 - 10s$	**i** $2r + 29 = 1 - 6r$

$\dfrac{2x + 1}{4}$ stands for $(2x + 1) \div 4$

3 Solve these equations.

a $4z + 1 = 10z - 1$ **b** $2(3y - 1) = 12y - 7$ **c** $\dfrac{2x + 1}{4} = 2$

d $\frac{1}{4}(3w + 1) = 4$ **e** $\frac{1}{3}(2v - 1) = v - 1$ **f** $\dfrac{5u + 6}{2} = 4$

Forming and solving linear equations

For some problems, you can form a linear equation and then solve it.

Example

The lengths of the sides of a triangle are y cm, $2y$ cm and $(y + 3)$ cm.

If the perimeter of the triangle is 63 cm, what is the length of each side?

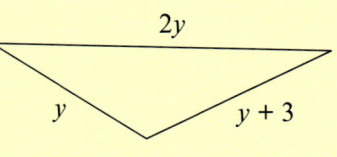

The perimeter is the total distance around the outside edge of a shape.

- The perimeter, in terms of y, is: $y + 2y + y + 3 = 4y + 3$
- The perimeter is 63 cm so: $4y + 3 = 63$
- Find the value of y: $4y = 60$
                                     $y = 15$
- Find the value of $2y$ and $y + 3$: $2y = 30$ and $y + 3 = 18$
- So the lengths of the sides are: **15 cm, 30 cm and 18 cm.**

Exercise 7.4
Forming and solving
linear equations

1 The lengths of the sides of a triangle are x cm, $2x$ cm and $(3x - 4)$ cm.

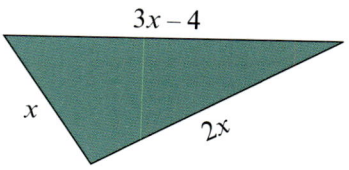

> An expression in terms of x does not include any letters other than x.

a List the lengths of the sides when $x = 5$.
b What is the perimeter when $x = 4$?
c What is the perimeter of the triangle in terms of x?
d If the perimeter is 104 cm, form an equation in x and solve it to find the length of each side.

2 A square has sides of length $2p$ metres.
A rectangle has width p metres and length $(p + 5)$ metres.

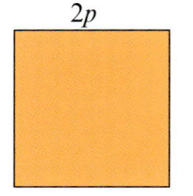 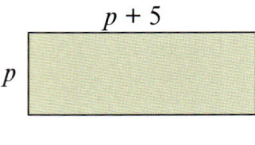

a What is the perimeter of each shape when $p = 10$?
b Find the perimeter of the square in terms of p.
c Find the perimeter of the rectangle in terms of p.
d If the perimeter of the rectangle is 90 m, how long is each of its sides?
e Find a value of p so that the perimeters of the square and rectangle are equal.

3 An equilateral triangle has sides of length $(t + 1)$ cm.
A rectangle has width t cm and length $2t$ cm.

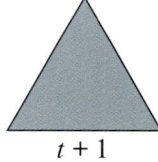 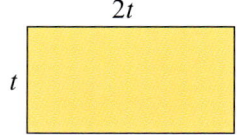

Find a value of t so that the perimeter of the triangle and rectangle are equal.

4 These expressions are arranged in a magic square.

$2x$	$x - 2$	$3x - 1$
$3x - 2$	$2x - 1$	x
$x - 1$	$3x$	$2x - 2$

> The magic total is the total of the numbers in each row, column or diagonal.

a What value for x gives 21 at the centre of the magic square?
b Draw the complete magic square.
c Find a value for x that gives a magic square with a magic total of 117.
d **i** Make a magic square where the four numbers in the corner squares add to give 100.
ii Explain how you decided on a value for x.
e With these expressions, explain why the total of the numbers in the corner squares will always be four times the number in the centre square.

Some number puzzles can be solved using equations.

Example

I think of a number, add 1, and then double. I get the same answer if I subtract my number from 14. What is my number?

> The left-hand side of the *equation* is for 'add 1 and then double' $2(n + 1)$ is equivalent to $(n + 1) \times 2$.
>
> The right-hand side is for 'subtract ... from 14' $14 - n$ is **not** equivalent to $n - 14$.

- ◆ Choose a letter to stand for the number: n is the number
- ◆ Write an *equation* for the puzzle: $2(n + 1) = 14 - n$
- ◆ Solve the equation to find the number: $2(n + 1) = 14 - n$

$$
\begin{array}{ll}
+n \left(\begin{array}{l} 2n + 2 = 14 - n \end{array} \right) +n \\
-2 \left(\begin{array}{l} 3n + 2 = 14 \end{array} \right) -2 \\
\div 3 \left(\begin{array}{l} 3n = 12 \end{array} \right) \div 3 \\
\qquad\qquad n = 4
\end{array}
$$

- ◆ So the number is **4**.

Exercise 7.5
Solving number puzzles

1 I think of a number, subtract 2 and then double. I get the same answer if I multiply my number by 3 and subtract from 21. What is my number?

Which of these equations fits the number puzzle?

A $2(n - 2) = 3n - 21$ **B** $2n - 2 = 21 - 3n$ **C** $2n - 2 = 21 - 3n$

D $2(n - 2) = 21 - 3n$ **E** $n - 4 = 21 - 3n$

> Use n to stand for the number each time.

2 For each puzzle A to E, write an equation and solve it to find the number.

A I think of a number, multiply it by 3, and add 5. I get the same answer if I subtract my number from 13. What is my number?

B I think of a number, subtract 2, and multiply by 4. I get the same answer if I subtract 3 and multiply by 5. What is my number?

C I think of a number, subtract 1, and multiply by 3. I get the same answer if I subtract my number from 8 and multiply by 4. What is my number?

D I think of a number, multiply it by 6, and subtract from 10. I get the same answer if I multiply my number by 3 and subtract from 7. What is my number?

E I think of a number, double it, and subtract 11. I get the same answer if I double my number and subtract from 15. What is my number?

3 **a** Write a number puzzle for the equation $3(n - 5) = 20 - 2n$
    b Solve the equation to find the value of n.

4 Make up some number puzzles for someone else to solve.

Using graphs to solve problems with two values to find

For some problems, you can form two equations and use graphs to solve them.

Example

Two families visit a funfair.
One family buys 1 adult ticket and 2 child tickets. The total is £10.
The other family buys 2 adult tickets and 1 child ticket. The total is £14.

How much is each type of ticket?

♦ Choose letters to stand for each type of ticket:
  Use a to stand for the cost in pounds of an adult ticket.
  Use c to stand for the cost in pounds of a child ticket.

♦ Write an equation for each family: $a + 2c = 10$
  $2a + c = 14$

♦ Draw up a table of values for each equation:

> Negative values for a and c are not included as the cost of a ticket cannot be negative.

$a + 2c = 10$

a	0	1	2	3	4	5
c	5	4.5	4	3.5	3	2.5

$2a + c = 14$

a	0	1	2	3	4	5
c	14	12	10	8	6	4

♦ Draw a graph for each equation, so that the lines cross at a point:

♦ Where the lines cross, the values of a and c fit both equations, so they give the cost of each ticket:

The lines cross when $a = 6$ and $c = 2$,

So an adult ticket costs £6 and a child ticket costs £2

♦ Check your solution:

When $a = 6$ and $c = 2$,

$a + 2c = 6 + (2 \times 2)$
$= 10$

$2a + c = (2 \times 6) + 2$
$= 14$

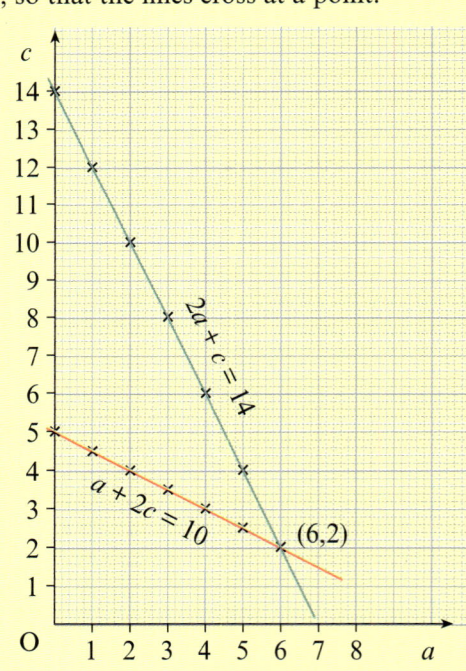

♦ These are the correct values so the solution is correct.

Exercise 7.6
Using equations and graphs

> £1.30 = 130p

1 Two people visit a cafe.
Rita pays 70p for 1 tea and 2 biscuits.
Simon pays £1.30 for 3 teas and 1 biscuit.

   a Write an equation for each person, using t for the cost of a tea in pence and b for the cost of a biscuit in pence.

   b Draw up a table of values like this for each equation for values of t from 0 to 40.

t	0	10	20	30
b				

   c Draw a graph for each equation so that the lines cross.

   d From your graphs, how much does each item cost?

In Questions **2**, **3** and **4** the letters can stand for any type of number, including decimals and negative numbers.

2 Equation 1 $x + y = 6$

Equation 2 $3x + y = 9$

a   For each equation, what is the value of y when $x = 4$?
b   Draw up a table of values for each equation with these values of x: 0, 1, 2, 3, 4, 5.
c   On one set of axes, draw a graph for each equation.
d   What are the values of x and y at the point where the lines cross?
e   Check that these values fit Equations 1 and 2.

3 Equation 1 $y = x + 4$

Equation 2 $y = 2x + 5$

a   For each equation, what is the value of y when $x = {}^-2$?
b   Draw up a table of values for each equation with these values of x: ${}^-4$, ${}^-2$, 0, 2, 4.
c   On one set of axes, draw a graph for each equation.
d   Use your graphs to find the values of x and y that fit both equations.

4   A $y = x - 5$ B $y = 2x + 6$ C $y = 2x$
            $y = 3x - 11$ $y = 4x + 11$ $y = 7x - 9$

For each pair of equations:

a   Draw up a table of values for each equation with these values of x: ${}^-4$, ${}^-2$, 0, 2, 4.
b   On one set of axes, draw a graph for each equation.
c   Use your graphs to find the values of x and y that fit both equations.

Solving problems without using graphs

Exercise 7.7
Solving problems
without using graphs

1   In a cafe, 2 teas and 4 coffees cost £4.60.
From this information, which of these can you find the cost of?

A   4 teas and 8 coffees   B   1 tea and 4 coffees

C   1 tea and 2 coffees   D   1 coffee   E   6 teas and 8 coffees

2   In a shop:

1 cola and 3 bags of crisps cost £1.52
2 colas and 4 bags of crisps cost £2.48

Find the cost of:

a   2 colas and 6 bags of crisps   b   1 cola and 2 bags of crisps
c   3 colas and 7 bags of crisps   d   1 cola and 1 bag of crisps
e   1 bag of crisps   f   1 cola

3   A woman has twins.
The twins are the same height.
Adding the height of the woman to the height of one twin gives 280 cm.
The total height of the woman and her twins is 400 cm.

a   What is the height of each person?
b   Make up a problem like this for someone else to solve.

Thinking ahead to ...
using algebra

A   A value for p and a value for q fit both these equations.

$$3p + 4q = 23$$

$$p + 2q = 11$$

a   Copy and complete these equations for p and q.
i $3p + 6q = \square$ **ii** $2p + \square = 22$ **iii** $4p + 6q = \square$
iv $2p + 2q = \square$ **v** $p + q = \square$

b   Find the values of p and q.

Using algebra to solve problems with two values to find

Values that fit two equations can be found using algebra.

Example

Solve these equations to find the values of a and b.

$$2a + b = 19$$
$$3a + 4b = 26$$

◆ Label the equations (1) and (2): $2a + b = 19 \ldots$ (1)
                                                                        $3a + 4b = 26 \ldots$ (2)

◆ Multiply both sides of
   equation (1) by 4 to give   (1) × 4 ... $8a$ + $4b$ = 76 ... (3)
   two equations with '+ $4b$': $3a$ + $4b$ = 26 ... (2)

◆ Subtract: (3) − (2) ... $(8a - 3a) + (4b - 4b) = 76 - 26$
                                                          $5a = 50$

◆ Find the value of a: $a = 10$

◆ Substitute the value of a in one $2a + b = 19$ ... (1)
   equation to find the value of b: $(2 × 10) + b = 19$
                                                          $20 + b = 19$
                                                          $b = {}^-1$

◆ **So the solution is $a = 10$, $b = {}^-1$.**

Subtract to remove '+ $4b$'
from each equation:
$4b - 4b = 0$

Equation (2) could also be
used to find the value of b:

$3a + 4b = 26$
$(3 × 10) + 4b = 26$
$30 + 4b = 26$
$4b = {}^-4$
$b = {}^-1$

Exercise 7.8
Using algebra

1   For each pair of equations, use algebra to find the values of x and y.
a $x + 4y = 42$ **b** $11x + 3y = 91$ **c** $5x + 7y = 32$
       $2x + 5y = 57$ $3x + y = 25$ $x + 3y = 12$

2 For which pair of equations is it true that $s = 5$ and $t = 2$?

A
$$s + 3t = 11$$
$$5s + t = 32$$

B
$$3s + t = 17$$
$$4s + 5t = 30$$

C
$$3s + 2t = 20$$
$$s + 2t = 9$$

3 A value for m and a value for n fit both these equations.

$$2m + 3n = 28 \ ...(1)$$
$$3m + 4n = 37 \ ...(2)$$

> Multiply **both** equations to give two equations with '$6m$'.

a Multiply equation (1) by 3.
b Multiply equation (2) by 2.
c Subtract to find the value of n that fits both equations.
d Substitute in one of the equations to find the value of m.

4 For each pair of equations, use algebra to find the values of v and w.

a $2v + 3w = 40$
$5v + 2w = 34$

b $3v + 2w = 3$
$6v + 10w = 24$

c $4v + 2w = 9$
$3v + 7w = 4$

Sometimes it is simpler to **add** the equations.

Example

Solve these equations to find the values of x and y.

$$6x - 2y = 18$$
$$5x + 3y = 1$$

◆ Label the equations (1) and (2):

$$6x - 2y = 18 \quad (1)$$
$$5x + 3y = 1 \quad (2)$$

> This is one way to use algebra to solve this problem. There are other ways.

◆ Multiply equation (1) by 3 and equation (2) by 2 to give '$-6y$' and '$+6y$':

$(1) \times 3$
$(2) \times 2$

$$18x \quad - \quad 6y \quad = 54 \quad (3)$$
$$10x \quad + \quad 6y \quad = 2 \quad (4)$$

> Add to remove
> '$-6y$' and '$+6y$':
> $-6y + 6y = 0$

◆ Add:

$(3) + (4)$
$(18x + 10x) + (-6y + 6y) = 54 + 2$
$$28x = 56$$

◆ Find the value of x:
$$x = 2$$

◆ Substitute the value of x in one equation to find the value of y:

$$5x + 3y = 1 \quad (2)$$
$$(5 \times 2) + 3y = 1$$
$$10 + 3y = 1$$
$$3y = {}^-9$$
$$y = {}^-3$$

◆ **So the solution is $x = 2$, $y = {}^-3$.**

**Exercise 7.9
Using algebra**

1 For each pair of equations, use algebra to find the values of a and b.

a $a - b = 8$
$4a + b = 42$

b $5b + 2a = 29$
$b - 2a = 1$

c $3a - b = 15$
$4a + 2b = 25$

d $b + 3a = 2$
$3b - a = 26$

e $5a + 2b = 17$
$2a - 3b = 3$

f $7a + 5b = 27$
$3a - 2b = 24$

2 Two numbers m and n fit both these equations.

$$5m - n = 15$$
$$3m - n = 5$$

> Subtract to remove '$-n$'
> from each equation:
> $(-n) - (-n) = 0$

a Subtract to find the value of m that fits both equations.
b Substitute in one of the equations to find the value of n.

3   Find the values of p and q that fit: $6p - 2q = 16$
and $p - 2q = 1$

4   Two numbers m and n fit both these equations:

$5m - 2n = 28$ (1)
$7m - 5n = 37$ (2)

> Here we multiply **both** equations to give two equations with '$-10n$'.

a   Multiply equation (1) by 5.
b   Multiply equation (2) by 2.
c   Subtract to find the value of m that fits both equations.
d   Substitute in one of the equations to find the value of n.

5   For each pair of equations, use algebra to find the values of x and y.

a $5x - 2y = 16$ b $4y - x = 17$ c $5x - 3y = 7$
$2x - 3y = 2$ $3y - 4x = 3$ $2x - 4y = 0$

d $5y + 3x = 15$ e $4x + 3y = 17$ f $3x + 2y = 13$
$5y + 7x = 25$ $5x - 7y = 32$ $5x - 6y = 59$

Using algebra to solve word problems

Example

A father's age and his son's age add to give 54.
The father is 30 years older than his son.

How old is his son?

♦ Use f to stand for the father's age and
s to stand for the son's age:

♦ The ages add to give 54, so: $f + s = 54$ ... (1)
♦ The father is 30 years older than the son, so: $f - s = 30$ ... (2)

♦ Add: (1) + (2) ... $(f + f) + (s + - s) = 54 + 30$

$2f = 84$

♦ Find the value of f: $f = 42$

♦ Substitute the value of f in one $f + s = 54$ ... (1)
equation to find the value of s: $42 + s = 54$
$s = 12$

♦ **So the son is 12 years old.**

Exercise 7.10
Solving word problems

1   Susan's age and her brother's age add to give 73.
Susan is 3 years older than her brother.
Find the ages of Susan and her brother.

2   A bag contains a mixture of large and small marbles.
Each small marble weighs 2 g.
Each large marble weighs 5 g.
The total weight of the marbles in the bag is 256 g.
Altogether there were 89 marbles in the bag.

Use x to stand for the number of small marbles.
Use y to stand for the number of large marbles.

a   Use this information to write two equations in x and y.
b   Solve these equations to find the number of each type of marble.

End points

You should be able to ...

A Add and subtract like terms in linear expressions

B Form and solve linear equations

C Solve problems using graphs

D Solve problems using algebra

... so try these questions

A1 Simplify:

a $6k + 8m - 4k + m$ b $5p - 1 - 3p + 9$

B1 Solve:

a $2z - 1 = 12$ b $4y + 1 = 2y + 7$
c $6(x - 1) = 3$ d $15 - 3w = w + 11$
e $3(2v + 7) = v + 6$ f $2(t + 3) = 18 - 6t$

B2 The lengths of the sides of a triangle are x cm, $(x + 8)$ cm and $(x - 6)$ cm.

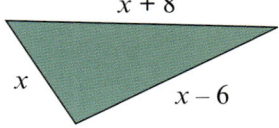

a What is the perimeter of the triangle in terms of x?
b The triangle has a perimeter of 56 cm.
Write down an equation in x, and solve it to find the length of each side of the triangle.

B3

I think of a number, double it and subtract 1. I get the same answer if I multiply my number by 4 and subtract from 14. What is my number?

Write an equation for this number puzzle and solve it to find the number.

C1 Equation 1 $y = 2x - 3$

Equation 2 $y = 4x - 4$

a Draw up a table of values for each equation with these values of x: ⁻4, ⁻2, 0, 2, 4.
b On one set of axes, draw a graph for each equation.
c Use your graphs to find the values of x and y that fit both equations.

D1 For each pair of equations, use algebra to find the values of m and n.

a $3m + 2n = 5$ b $2m + 3n = 5$ c $2m - n = 22$
  $5m + 2n = 8$ $3m + 4n = 6$ $3m + 2n = 40$

D2 A slot machine takes only 20p and 50p coins.
It contains a total of 140 coins.
The value of the coins is £45.10.

Use x to stand for the number of 20p coins.
Use y to stand for the number of 50p coins.
a Show that $20x + 50y = 4510$.
b Write down a different equation in x and y.
c Use your equations to find the number of each type of coin.

Some points to remember

- When solving an equation, add, subtract, multiply or divide **both** sides by equal amounts.
- Problems where two values have to be found can be solved using graphs or algebra.
  The most accurate answer can be found using algebra.

Starting points

You need to know about ...

... so try these questions

A Rounding numbers

Rounding is a way to approximate numbers when an exact value is not needed.

Whole numbers are usually rounded to the nearest 10, 100, 1000 and so on.

Example

3837.6 rounded	to the nearest whole number is 3838
	to the nearest 10 is 3840
	to the nearest 100 is 3800
	to the nearest 1000 is 4000
	to the nearest 10 000 is 0

Decimals are rounded to a given number of decimal places (dp).

Example

3.6748 rounded	to 1 dp is 3.7
	to 2 dp is 3.67
	to 3 dp is 3.675

Halfway numbers can be rounded either way but are usually rounded up.

Example

735 to the nearest ten is 740
56.75 to 1 dp is 56.8

A1 Round 2175.6 to the nearest:
  a thousand **b** hundred
  c ten   **d** whole number

A2 Round 45.638 to:
  a 2 dp   **b** 1 dp

A3 Round these numbers.
  a 34.597 to 2 dp
  b 2.501 to the nearest whole number
  c 38.45 to 1 dp
  d 3496 to the nearest ten

A4 Which numbers are not 2.56 when rounded to 2 dp?
  a 2.5555   **b** 2.5648
  c 2.5651   **d** 2.55099
  e 2.5500   **f** 2.5666

B Adding and subtracting decimals

When you add or subtract decimal numbers you may find it easier to arrange the digits in columns.

Example 1

45.346 + 8.6 + 237

1000	100	10	U		$\frac{1}{10}$	$\frac{1}{100}$	$\frac{1}{100}$
		4	5	.	3	4	6
			8	.	6		
	2	3	7	.			
	2	9	0	.	9	4	6

+

Example 2

34.6 – 2.784

```
   3   4   .   6   0   0
               2   .   7   8   4   −
   3   1   .   8   1   6
```

B1 Add these decimal numbers:
  a 45.73 + 8.423 + 123.6
  b 14 + 0.563 + 28.9
  c 0.004 + 0.03 + 0.95 + 3
  d 17.8 + 3425 + 0.0895

B2 Subtract these numbers:
  a 34.78 – 6.8
  b 14 – 2.83
  c 154.36 – 4.5
  d 256 – 23.764

B3 What mistake has been made here?
  34.5 + 2.34 = 57.9

B4 Explain the mistake made in this calculation.
  34.6 – 2.278 = 32.478

Rounding up or down?

People make estimates every day.
They often base an estimate on a calculation they do in their head.

Sometimes it is best to **overestimate** so they **round up** their answer.
At other times it is best to **underestimate** and **round down**.

We should be able to cycle about 78 miles each day so how far apart do we want the hostels to be?

In this case it might be better to **round down** the 78 to say 60 miles a day just in case they felt tired or had an accident.

I've worked out that we need 4756 bricks for the extension. How many should I order?

Here, it would be better to **round up** the number of bricks. An order of say 5000 bricks would allow for breakage or error. Bricks ordered later might not be exactly the same colour.

Exercise 8.1
Rounding up or down?

1   In each of these situations do you think it is better to round up, round down, or not to round at all. Explain why.

   a   You have to draw out some money from the bank.
           You work out that you need £8.35 for your trip.

   b   The seat number on your concert ticket is 213.
           You must decide where to sit.

   c   You calculate that you need 11 rolls of wallpaper for your room.
           You go into the shop to buy the paper.

   d   Don lives at 26 Hayward Road. You decide to pay him a call.

   e   You think you may earn £360 from your holiday job.
           You look at hi-fis you think you will be able to afford.

2   Describe a new situation for each of these:

   a   when it would be a good idea to round up

   b   when it would be best to round down

   c   when it would be silly to do any rounding.

3   For each of these situations decide if it is better to round up or down.
    Say what number you would use, and explain why.

   a   You calculate that the gap for a desk is 98 cm wide.
           You have to decide what width of desk to ask for.

   b   The milometer in your sister's car reads 67 673 miles.
           You advertise the car with the mileage it has done.

   c   You expect 74 people for the school-leavers meal.
           You have to hire some glasses.

   d   You calculate that you need 15.3 metres of wood for some shelves.
           You have to buy the wood.

Significant figures

Exercise 8.2
Rounding

1 In this extract some numbers are given to a greater accuracy than they need to be.

Practical Green Keeping

March edition

Crew measure up while the games are on

THE MAINTENANCE CREW arrived at the stadium at 10:43 while the athletics events were taking place. The 71 934 crowd was already seated when measuring up started. The perimeter railings were measured as 63 479.6 cm long and the supports as 19.6 cm thick. At one point the crowd rose to its feet as Mary Taylor set a new European record of 10.84 sec for the 100 metres sprint. From their calculations the crew estimated the area of grass which needed re-seeding was 1452.56 metres². The measuring was completed in about 56 minutes with little disruption to the 492 or so competitors.

The maintenance contract with the sports committee expires in 2003

> 'Appropriate' means 'sensible'.

a List the numbers which you think are more accurate than is appropriate. Write what you think each one should be rounded to.

b Which numbers should not be rounded? Explain why.

c In both 63 479.6 and in 19.6 the last digit stands for $\frac{6}{10}$.
In which of these numbers do you think this 6 is more significant? Explain why.

> 'Significant' in this case means 'important'.

2 Draw a line on your page.
Measure its length as accurately as you can.
How many digits are in the number you have written?

3 Estimate how far it is from Land's End to John O'Groats.
How many of the digits in your answer are not zero?

4 The digit 4 is in both of these numbers: **24.6   10243**

a In which number does the 4 have the greater actual value?
b In which number do you think the 4 is more significant?

5 To what accuracy do you think a 100 metre running track must be measured when it is marked out?

Significant figures means the most important digits in a number.

In 6351.2 the 6 and 3 are the two most significant figures because they show the largest numbers 6000 and 300.

Significant figures can be written as **sf**.

Ways of approximating include rounding a number to the nearest ten or to a set number of decimal places.

Another way is to round a number is to a set number of significant figures.

63 479.6   rounded to **2 significant figures** is 63 000

The three zeros are added to keep the value of the number about the same.

63 479.6   rounded to **3 significant figures** is 63 500

Note how the 4 has rounded up to 5 because the next digit 7 is above halfway.

63 479.6   rounded to **4 sf** is   63 480

63 479.6   rounded to **5 sf** is   63 480

Exercise 8.3
Rounding using significant figures

1   Round each of these numbers to 3 sf:

   a 1452.56   **b** 21 675   **c** 142.51   **d** 2 134 518.4   **e** 149 625

2   Round 71 934 to:

   a 1 sf   **b** 2 sf   **c** 3 sf   **d** 4 sf

3   When 63 479.6 is rounded to 4 sf or 5 sf the answer is the same.
   Why do you think this is?

4   A number, rounded to 2 sf, is 3200.
   Give three numbers it could be.

5   Copy and complete this table.

Number	45 287	2395	302 604.32	14.823
to 2 sf				
to 3 sf				
to 4 sf				

6   The grass that needed re-seeding in the stadium was a rectangle this size:

54.2 m

Area = 1452.56 m²

26.8 m

In calculations you should either:

◆ give your answer to the degree of accuracy asked for e.g. 1 dp or 2 sf, or

◆ give your answer to the same degree of accuracy as is used in the question.

Why do you think it is not sensible to use all the digits for the area?
What would you round the area to?

7   A number, when rounded to 2 sf or to 3 sf is 420 000.
   Give an example of what the number might be.

8   To how many significant figures could a number be rounded to give 164 000?

Numbers less than 1 can also be rounded using significant figures.

0.0753   rounded to **1 significant figure** is   0.08

> The first significant figure is this 7. The zeros at the start do not count as significant.

> The 7 rounds up to 8 because the next digits are above halfway.

0.004 03   rounded to **2 significant figures** is   0.0040

> This zero is significant because it is between two other significant figures.

> This zero must stay to make it clear there are 2 significant figures.

Exercise 8.4
Significant figures

1   The width of a human hair found at the scene of a crime was 0.007 64 cm. Round this number to 1 sf.

2   When 0.000 524 6 is rounded to 2 sf it becomes 0.000 52. 0.000 520 0 is not really correct. Explain why.

3   For these conversions round the blue numbers.

Metric/Imperial conversions

	to 1 sf	to 2 sf	to 3 sf
1 inch = 0.0254 metres	0.03	0.025	0.0254
1 yard = 0.009 144 kilometres			
1 millimetre = 0.0394 inches			
1 kilometre = 0.6214 miles			
1 millilitre = 0.001 76 pints			
1 centimetre^3 = 0.061 023 74 inches^3			
1 foot^3 = 0.0283 metres^3			
1 pound = 0.004 535 9237 tonnes			

4   The table shows land areas of some countries.
Which country has a land area of four hundred thousand km^2 when rounded to 1 sf?

5   Why might it not be helpful for a book to give all the areas to 1 sf?

6   Give the following land areas:

  a   United Kingdom to 5 sf
  b   Andorra to 2 sf
  c   Japan to 3 sf
  d   China to 4 sf
  e   Monaco to 1 sf.

7   For two countries the area stays the same when rounded to 3 sf, 4 sf or 5 sf. Which countries are these?

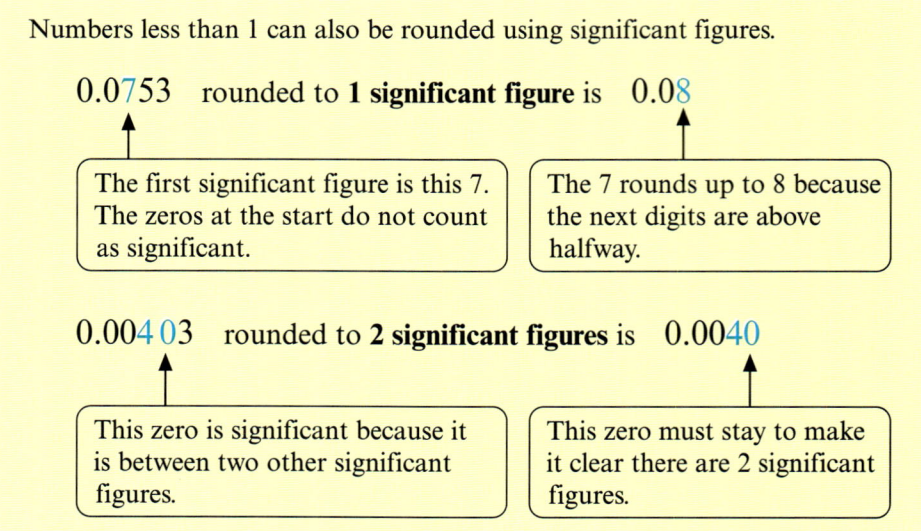

COUNTRY	AREA (km²)
Andorra	464
Argentina	2 758 829
Bangladesh	143 998
Cambodia	181 035
Cameroon	475 499
China	9 560 948
Congo	348 999
Denmark	43 030
Japan	369 698
Monaco	1.6
United Kingdom	244 019

Using significant figures to estimate answers

One way to check if a calculation gives an answer of about the right size is to round each of the numbers to 1 sf.

For example, to estimate an answer:

Here are some answers given by four students when they had to calculate the value of **43 183.5 × 184.23** without using a calculator.

a 795 569.6   b 79 556 962   c 7 955 696   d 79 557

Which answer is likely to be most accurate?

> A value which is of the correct order of magnitude is about the right size.

To 1 sf these numbers become   40 000 × 200

which is   40 000 × 2 × 100

= 80 000 × 100

= 8 000 000

Answer **c**. 7 955 696 is about the same order of magnitude as 8 000 000 so it is most likely to be most accurate.

Exercise 8.5
Estimating answers

1 Work out estimates of the answers to each of these.
Show all the stages you use.

a 31.2 × 41.45   b 677 × 3.764
c 856 × 83.42   d 542 × 52
e 56 234 ÷ 82.5   f 62 381.23 ÷ 478.23
g 452 ÷ 2.34   h 28 536 ÷ 0.9623

2 A theatre sells 562 tickets at £28.50 each.

a Roughly what is their income from ticket sales?
b Why does rounding both numbers to 1 sf give too large an estimate?

> Population density is the average (mean) number of people to each square kilometre.

3 France has a land area of 549 619 km².
In 1990 the population was 56 304 000.
Estimate the population density in people per km².

4 When the numbers in 341.2 × 14.25 are rounded to 1 sf and then multiplied the estimate is much smaller than the true answer.
For the problem 156 ÷ 34.7 the estimate is much larger.
Explain why.

5 For each of these problems, say if rounding all numbers to 1 sf makes estimates too large, too small, or about the right size.

a 56.5 × 1763.2   b 184 ÷ 19.6
c 491.432 × 2061.4   d 445 × 84 632
e 2265 ÷ 27.7   f 6834 ÷ 14.23
g 453 782 + 242 565   h 7452.3 − 2837.324

6 For the problem 342 561 + 453, why is rounding to 1 sf not helpful?

Thinking ahead to ...
working with numbers
less than 1

Width is 2.7 cm
So the area of the tape = 34.2 x 0.027
Area = 923.4 metres²

2.7 cm

The answer in the book says 9.234, but this must be wrong because the numbers get bigger when you multiply.

A Work out 342×0.027 with a calculator.
Is the answer larger or smaller than 342?

B Work out $342 \div 0.027$. Is this answer larger or smaller than 342?

> You may need to do some more calculations to decide this.

C What can you say about the answer when you:
  a multiply by a number less than 1
  b divide by a number less than 1?

Working with numbers less than 1

Example 1   Estimate the value of 342×0.052.

> Approximating to 1 sf, this becomes 300×0.05.
>
> The answer to this estimate will be smaller than 300.
> One way to work out the value is to look for patterns.
>
> $300 \times 5 = 1500$
>
> $300 \times 0.5 = 150$
>
> $300 \times 0.05 = 15$ **So the estimate is 15.**

Example 2   Estimate for the value of $26 \div 0.0056$.

> To 1 sf this is $30 \div 0.006$.
>
> $30 \div 6 = 5$
>
> $30 \div 0.6 = 50$
>
> $30 \div 0.06 = 500$
>
> $30 \div 0.006 = 5000$ **So the estimate is 5000.**

Exercise 8.6
Calculating and
estimating answers

1 a Calculate 342×52 without a calculator.
  b Use the example above to help decide what 342×0.052 is.

2 Estimate the value of 45×0.0023.
Show all the stages you use to find the estimate.
Calculate the exact answer and check it with your estimate.

3 Estimate, then calculate, the exact values of these.
  a $346.3 \div 0.04$ **b** 26.23×0.6 **c** 876.2×0.0002
  d $2.448 \div 0.0018$ **e** $35\,465 \times 0.000\,12$ **f** $53\,546.2 \div 0.045$

4

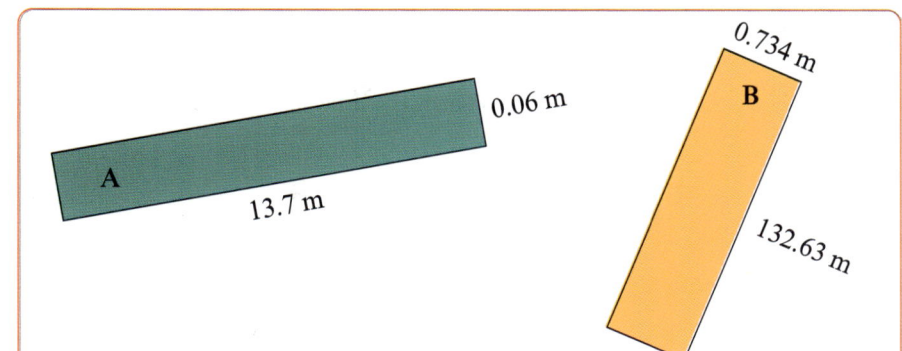

a Estimate the areas of shapes A and B.
Show the stages in your working of the estimate.

b Calculate the areas without using a calculator.

5

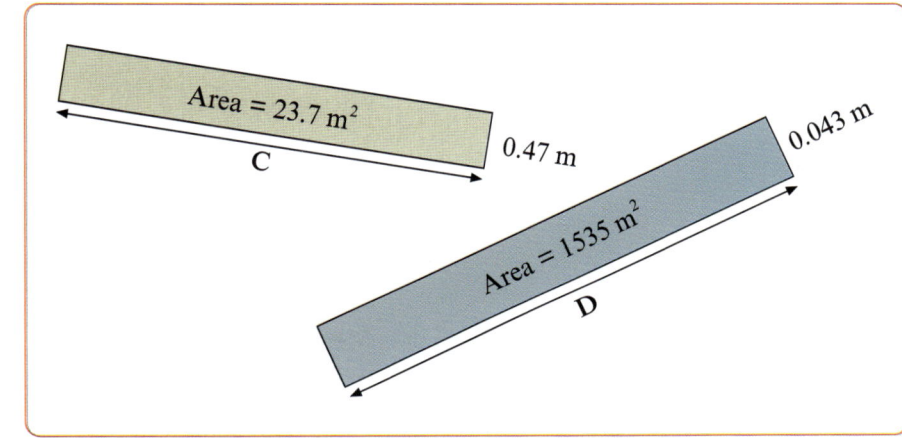

Estimate the dimensions C and D.
Show the stages in your working of the estimate.

6

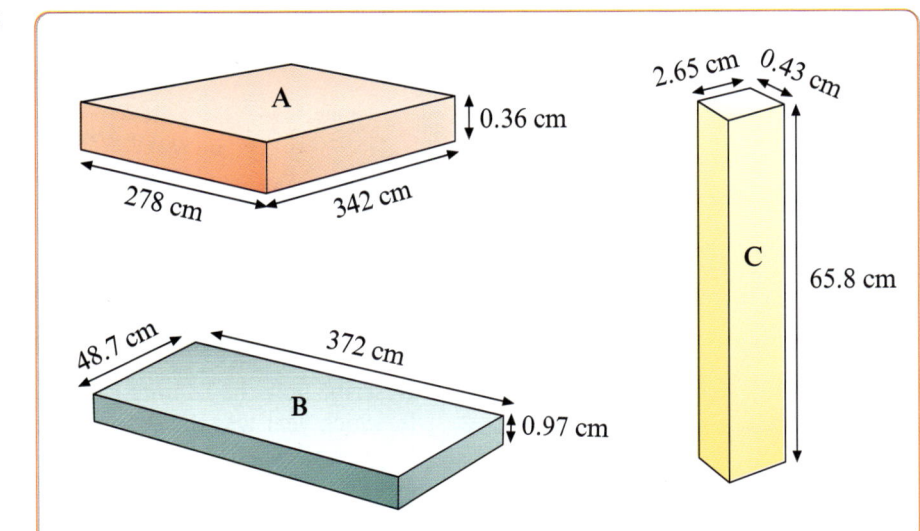

Estimate the volumes of cuboids A to C.
Explain your working.

7 A cuboid is 113.6 cm long, 0.42 cm deep and has a volume of 18.42 cm³.
Estimate its width.

When should you round?

What is the area of the £20 note? Round your answer to the nearest cm².

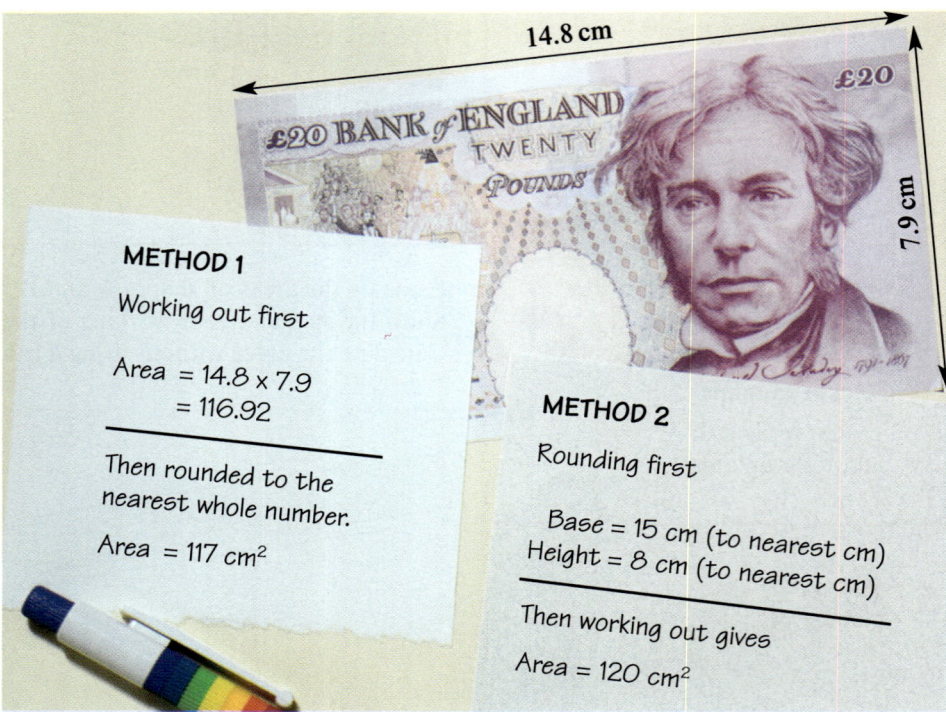

METHOD 1

Working out first

Area = 14.8 x 7.9
     = 116.92

Then rounded to the nearest whole number.

Area = 117 cm²

METHOD 2

Rounding first

Base = 15 cm (to nearest cm)
Height = 8 cm (to nearest cm)

Then working out gives

Area = 120 cm²

14.8 cm

7.9 cm

Alison uses Method 1. She calculates then rounds the answer.
Mark uses Method 2. He rounds all the numbers, then calculates.

Exercise 8.7
The effect of rounding

1 The £5 note has changed in size since 1900.

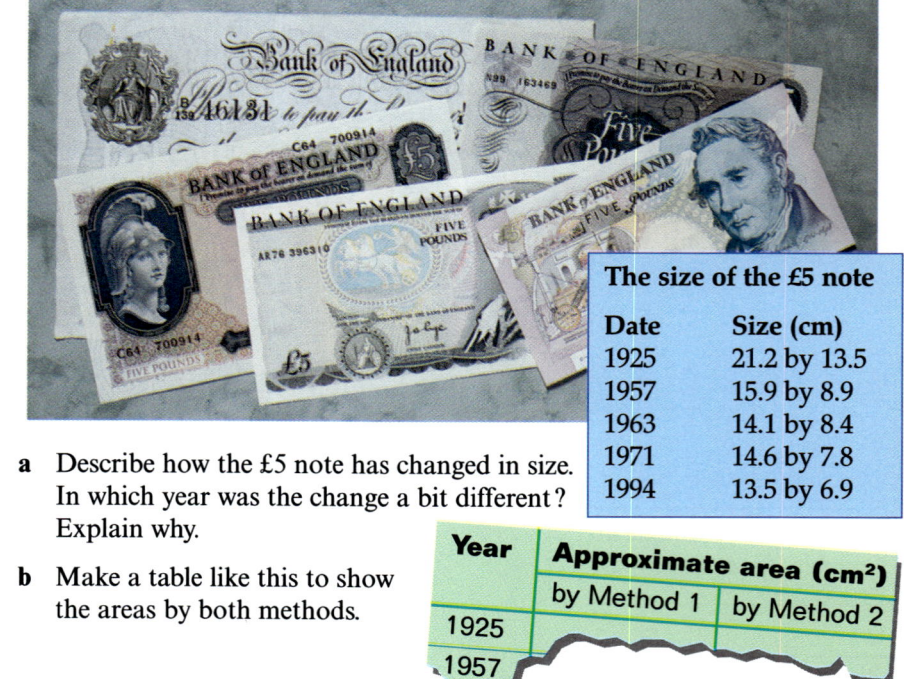

The size of the £5 note

Date	Size (cm)
1925	21.2 by 13.5
1957	15.9 by 8.9
1963	14.1 by 8.4
1971	14.6 by 7.8
1994	13.5 by 6.9

a Describe how the £5 note has changed in size. In which year was the change a bit different? Explain why.

b Make a table like this to show the areas by both methods.

Year	Approximate area (cm²)	
	by Method 1	by Method 2
1925		
1957		

c Which method do you find easier to use?

d Which method gives the more accurate answer? Explain why.

End points

You should be able to ...

... so try these questions

A Decide what rounding is appropriate for the situation

A1 Alez tuned into her favourite station, Channel 162 on the infrawave. She knew the slot lasted for about 91 minutes so she would need a compulsory meal before the end. She dined on 27 of her favourite food pills with 785 ml of ice-cold isophoric delight. Jeq materialised in 11 minutes raving about some antique maths book with about 416 pages that he'd found and dated as 1997.

Rewrite this extract and round numbers where it is appropriate.

A2 For each of these situations would you round up, down or not at all? Explain why.

a You find the wall area of your bedroom is 330 square feet.
You go to buy paint.

b The manual says your car can pull a trailer with a maximum weight of 470 kg.
You are loading your camping gear.

B Round a number to a given number of significant figures

B1 Round 345.683 to:

a 3 sf   **b** 5 sf   **c** 1 sf

B2 Round each number to 3 sf.

a 34.673   **b** 1.974   **c** 194 638.2
d 0.003 186   **e** 143.5   **f** 6.987 36

C Estimate then calculate the answers to problems and know when to round

C1 For the cuboid:

a Estimate its volume.
b Calculate the volume and give your answer correct to 2 dp.

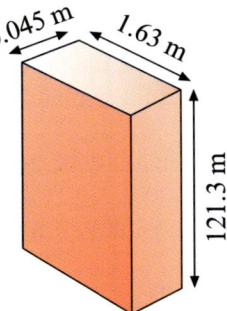

0.045 m   1.63 m   121.3 m

Some points to remember

- Give your final answers to an appropriate degree of accuracy for the situation.
- In exams, round your answers when asked to do so, and give any units.
- One way to estimate an answer is to round all the numbers to 1 sf.
- It is a good idea to estimate the order of magnitude of your answers before you calculate.
- Do not round at the start of, or during, a calculation if you want an accurate answer.
- When you multiply a number by a number less than 1, the answer is smaller.
- When you divide a number by a number less than 1, the answer is larger.

Starting points
You need to know about...

...so try these questions

A Probability from equally likely outcomes

If outcomes are equally likely, then you can calculate the probability that something will happen by counting the outcomes.

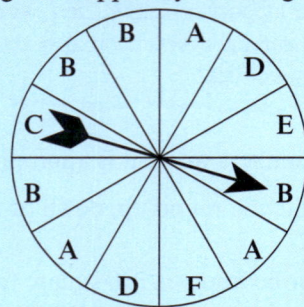

For example: There are twelve sections of equal size.
Four sections have B on them.
So the probability that the spinner stops on B is $\frac{4}{12} = \frac{1}{3}$.

Three sections have A on them and four sections have B.
So the probability that it stops on either A or B is $\frac{7}{12}$.

Probabilities can be shown on a probability scale from 0 to 1.

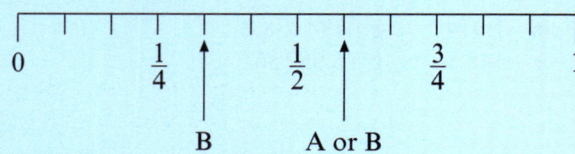

B The probability of a non-event

If you know the probability of something happening, then you can also calculate the probability of it **not** happening.

The probability of **getting B** on the spinner above is $\frac{1}{3}$.

The probability of **not getting B** is $1 - \frac{1}{3} = \frac{2}{3}$.

C Multiplication of fractions

In probability, sometimes you need to multiply fractions.

You can think of $\frac{3}{4} \times \frac{2}{5}$ like this:

$3 \times 2 = 6$

$\frac{3}{4} \times \frac{2}{5} = \frac{6}{20} = \frac{3}{10}$

$4 \times 5 = 20$

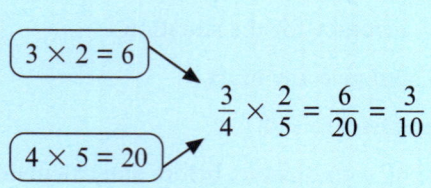

- ◆ Multiply the numerators.
- ◆ Multiply the denominators.
- ◆ Then reduce to the simplest terms if you need to.

A1 What is the probability that the wheel will stop on:
a A   b E   c C
d D   e F?

A2 What is the probability that the wheel will stop on:
a either B or F
b either E or A
c either A, B or C
d a letter after D in the alphabet
e a letter of the alphabet
f the letter N?

A3 Draw a probability scale and show the probabilities of the wheel stopping on each of A, B, C, D, E and F.

A4 For a 1 to 6 dice what is the probability that for one roll you will get:
a the number 5
b an even number
c a number less than 3?

B1 The probability of getting a red colour on a spinner is $\frac{4}{5}$.

What is the probability of not getting red?

C1 Multiply these fractions.
a $\frac{3}{4} \times \frac{1}{4}$
b $\frac{5}{8} \times \frac{1}{2}$
c $\frac{2}{3} \times \frac{3}{7}$
d $\frac{1}{8} \times \frac{3}{4}$

D Sample space diagrams

A sample space diagram can be used to show the outcomes from two events which are not linked (independent events).

For example, when a coin and a dice are thrown there are twelve different pairs of outcomes.

Outcome of coin						
H	H1	H2	H3	H4	H5	H6
T	T1	T2	T3	T4	T5	T6
	1	2	3	4	5	6

Outcome of dice

These diagrams can be used to show the probability of two things happening. For example, to find the probability of a tail on the coin and a number more than 3 on the dice.

The three pairs which match have been circled in red
so the probability is $\frac{3}{12} = \frac{1}{4}$.

E Tree diagrams

A tree diagram has branches showing different events. The probabilities of different events can be shown on the branches.
This is a tree diagram for the spin of a coin, then Spinner A from the top of the page.

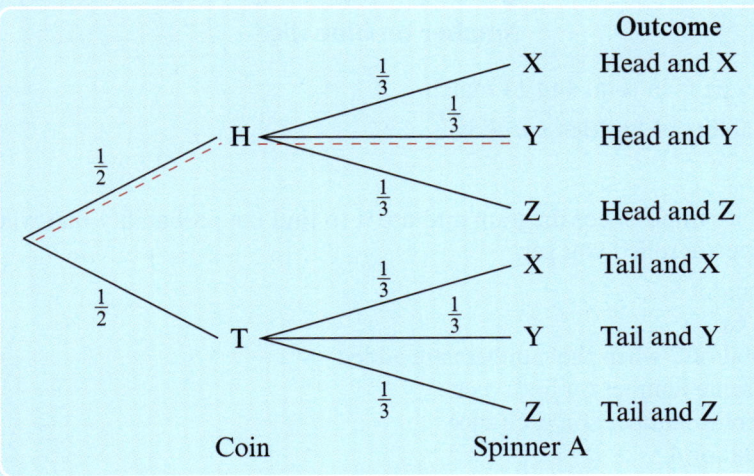

Outcome

- H — X — Head and X
- H — Y — Head and Y
- H — Z — Head and Z
- T — X — Tail and X
- T — Y — Tail and Y
- T — Z — Tail and Z

Coin   Spinner A

To find the probability of a particular outcome you can multiply the probabilities on the branches that lead to it.

For example, to find the probability of a head and Y you multiply the probabilities along the dotted branches.

Probability of a head and Y is $\frac{1}{2} \times \frac{1}{3} = \frac{1}{6}$.

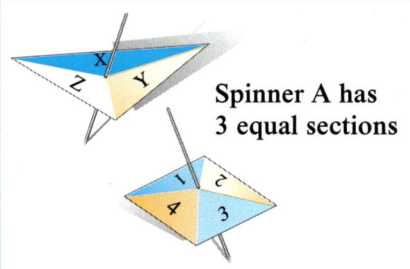

**Spinner A has
3 equal sections**

**Spinner B has
4 equal sections**

D1 Draw a sample space diagram to show the outcomes of letters and numbers for the two spinners, A and B.

D2 From your diagram calculate the probability that with a spin of each you get:
  a a 4 and a Y
  b a number less than 3 and an X.

E1 Draw a tree diagram to show the spinning of spinner B then a coin.

E2 From your tree diagram calculate the probability of a 3 on the spinner and a head on the coin.

Counting outcomes

The outcomes from rolling two dice can be shown by a sample space diagram. Each pair of numbers is equally likely.

Sample space diagram

Number on red dice						
6	6, 1	6, 2	6, 3	6, 4	6, 5	6, 6
5	5, 1	5, 2	5, 3	5, 4	5, 5	5, 6
4	4, 1	4, 2	4, 3	4, 4	4, 5	4, 6
3	3, 1	3, 2	3, 3	3, 4	3, 5	3, 6
2	2, 1	2, 2	2, 3	2, 4	2, 5	2, 6
1	1, 1	1, 2	1, 3	1, 4	1, 5	1, 6
	1	**2**	**3**	**4**	**5**	**6**

Number on blue dice

To calculate the probability that something will happen you can count those pairs that match the question.

For example, to find the probability that at least one dice shows a 4.

On the sample space diagram you can find all the outcomes which show 'at least one 4'.

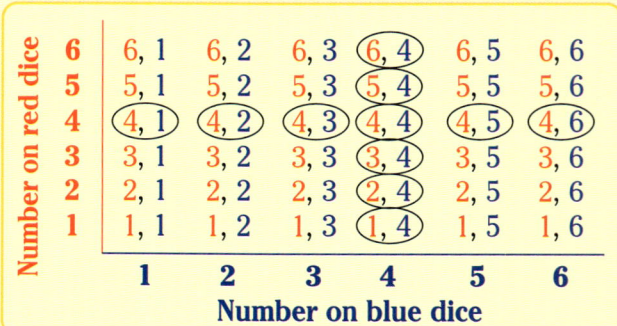

There are 36 pairs in total and 11 match.

So the probability of 'at least one 4' is $\frac{11}{36}$.

Exercise 9.1
Counting outcomes

1   Copy the sample space diagram and use it to find the probability that when both dice are rolled you get:

  a   a 1 and a 5
  b   two 6s
  c   a total of 7 when the numbers are added
  d   the same number on both dice
  e   different numbers on each dice
  f   a total of 4
  g   a total less than 6
  h   two prime numbers
  i   a total which is a prime number.

2   Draw a sample space diagram for a red 0 to 9 dice and a blue 0 to 9 dice.

A 0 to 9 dice has the numbers 0, 1, 2, 3, 4, 5, 6, 7, 8, 9.

3   Use your sample space diagram to find the probability of getting:

  a   two numbers the same
  b   a total greater than 13
  c   two even numbers.

More than two events

A sample space diagram can be used for two rolls of a dice but it is not suitable for three rolls. Here a tree diagram is better.

Example A six-sided dice has 2 red faces, 2 blue faces and 2 yellow faces. Each colour, therefore, is equally likely.
This dice is rolled three times.
What is the probability of rolling three faces of the same colour?

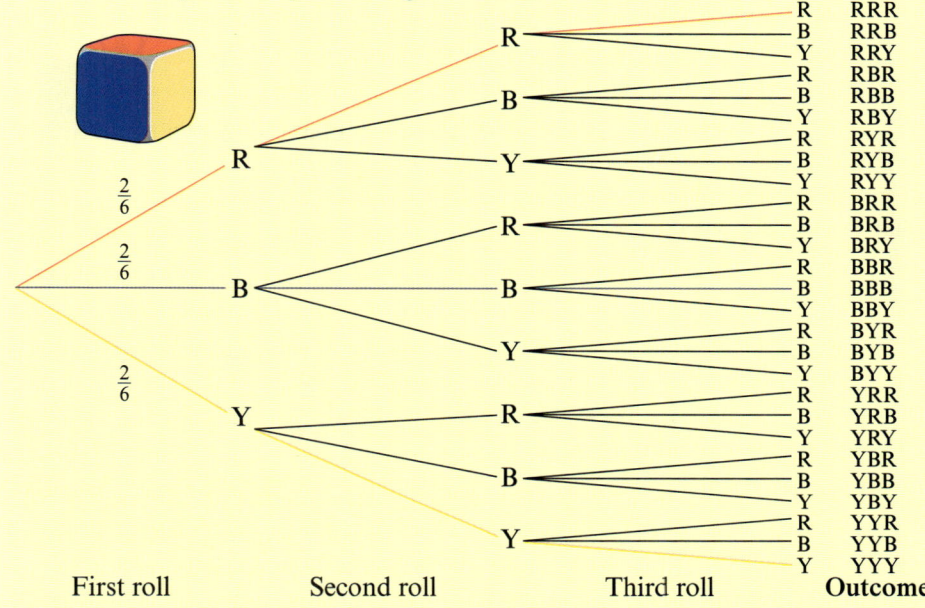

First roll	Second roll	Third roll	Outcome

The easiest way to calculate probabilities here is to look at the set of outcomes. There are 27 outcomes and 3 of these have the same colour (RRR, BBB and YYY).

So the probability is $\frac{3}{27} = \frac{1}{9}$.

Exercise 9.2
Three events

1 For three rolls of the dice above what is the probability of:

   a exactly two blues

   b at least two reds

   c all the colours different

   d only two colours

   e at least one yellow

   f all three blues

   g two blues and a red

   h exactly two colours the same?

2 An eight-sided dice has four red faces and four blue faces.
It is rolled three times.
Draw a tree diagram to show the different outcomes.

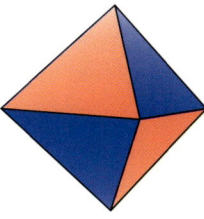

3 For three rolls of the eight-sided dice what is the probability that:

   a all three are blue

   b all three are the same colour

   c at least one is blue

   d exactly two have the same colour

   e all three are different colours

   f there are two blues and one red?

Ways of arranging things

Exercise 9.3
Arrangements

There is a saying that, given an enormously long time and a typewriter, a tribe of monkeys would type the complete works of Shakespeare just by hitting the keys at random.

What if ... a typewriter has just 4 keys?

E H N W

What is the probability of typing 'WHEN' by hitting the keys at random if each letter can be typed only once?

> Typing at letter at random gives each letter an equal chance of being typed.

When you look at the different arrangements of these four letters you get:

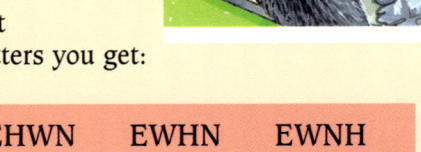

EHNW   EHWN   EWHN   EWNH
EN   HEWN

1 **a** List all the arrangements which use all the letters E, H, N and W, once only.
   b How can you be sure that you have found all the arrangements?
   c What is the probability of typing WHEN just by chance if each letter can only be typed once?

2 Suppose the typewriter has just the keys N, O and T.
   List all the different arrangements possible which use each letter once.

3 How many different ways are there to arrange two letters?

4 **a** Copy this table and fill in the arrangements for up to 4 letters.

No. of letters	1	2	3	4	5	6
No. of arrangements	1					

Multiply by a Multiply by b Multiply by c Multiply by d Multiply by e

   b What are the values of a, b, and c for this table?
      What pattern can you find in these numbers?
   c Use your pattern to decide how many arrangements there are for 5 different letters and 6 different letters.

5 On a real typewriter it is not actually true that each letter can only be typed once. If you have only the keys N, O and T, you can still type arrangements such as OOT and TNN.

   a With these three letters list the different arrangements of three letters it is possible to make.
   b How many arrangements are there?
   c How could a tree diagram have helped you decide how many arrangements are possible?

Relative frequency

A fair dice is one where each of the numbers 1 to 6 is equally likely.

A biased dice is made so that one number comes up more often. It is how some people cheat when gambling.

When a fair dice is rolled or a coin is spun you can calculate the probability of something happening because outcomes are equally likely. When outcomes are not equally likely you must use other methods to find the probability.

To estimate the probability that Julie will win a tennis match you can look at data to see how well she played in other matches.

Julie Matthews	
Matches played	105
Wins	68

An estimate of the probability that Julie will win her next match is given by:

Probability = $\dfrac{68}{105}$ = 0.65

Of her next 12 matches Julie is likely to win about 8 (because 0.65 × 12 = 7.8).

To estimate the chance that a biased dice will give a six you can do an experiment.

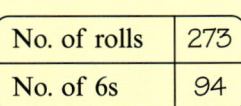

No. of rolls	273
No. of 6s	94

An estimate of the probability of a six is given by:

Probability = $\dfrac{94}{273}$ = 0.34

This estimate of probability, based on collected data is also called the relative frequency. It is usually given as a decimal.

You do not usually need to give the relative frequency to more than 2 decimal places.

Exercise 9.4
Relative frequency

1   Copy and complete this table for the biased dice shown above.

Rolls of dice	Number of sixes	Relative frequency
20		0.34
50		
70		

2   Of the 241 trains that arrived at Wayhurst station, 56 were late.
   a   What is the relative frequency of a train being late?
   b   For the next 9 trains, how many would you expect to be late?
   c   What is the relative frequency of trains arriving on time?

Gregor Mendel was an Austrian monk who did experiments on heredity in about 1856. His work led to a breakthrough in our knowledge of how young inherit the characteristics of their parents.

3   Mendel crossed plants from peas with smooth skins with those from peas with wrinkled skins, and planted the seeds.
   He found that out of 7324 new plants 1850 had wrinkled peas.
   a   What is the relative frequency of a plant giving wrinkled peas?
   b   What is the relative frequency of smooth skins?
   c   When 200 of these peas are planted how many would you expect to grow into plants with smooth peas?

4   Mendel also crossed plants with long stems with plants with short stems.
   He found that he got 787 plants with long stems and 277 with short stems.
   What is the relative frequency of getting a plant with a short stem?

Note that the answer is not 0.35.

5   When the relative frequency of male births is 0.51 what is the relative frequency of female births?

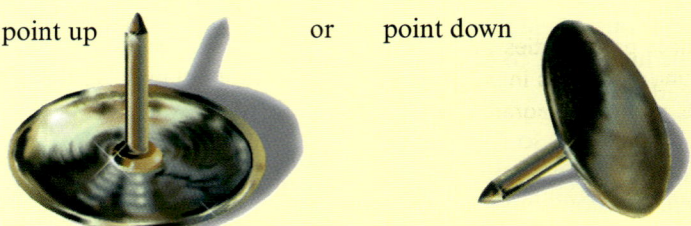

When a drawing pin is dropped it can come to rest either:

point up   or   point down

If you want to find an estimate of the probability that it will come to rest point up, you can experiment to find the relative frequency. The more drops you make the more likely it is that your relative frequency is reliable.

Exercise 9.5
Drawing pin experiment

1   Draw up a frequency table like this.

Position of pin	Tally	Frequency
Point up		
Point down		
	Total	

> You must decide how many times to drop the pin to give a reliable result.

2   Drop a drawing pin a large number of times and record in the table how it lands, point up, or point down.

3   From your results calculate the relative frequency that a pin will land
    a   point up   b   point down.

4   In twenty more drops predict how many times the pin will land point up.
    Check your prediction with twenty more drops.
    Why might the answer not match your prediction?

5
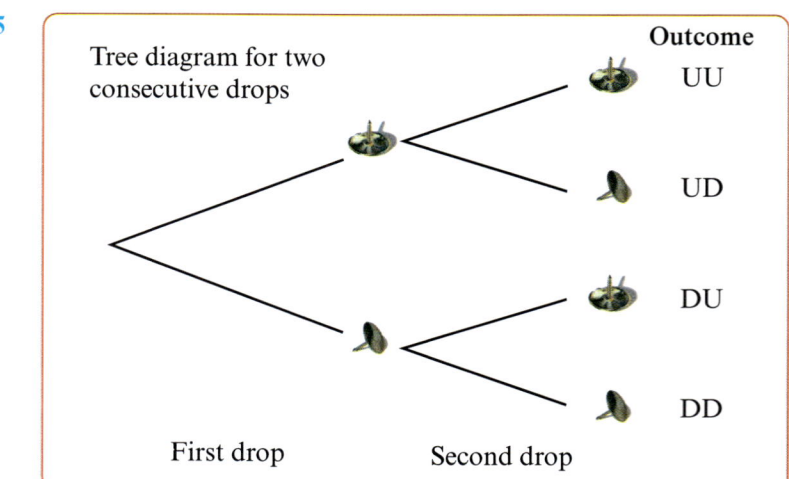

Tree diagram for two consecutive drops

Outcome

UU

UD

DU

DD

First drop   Second drop

a   Copy the tree diagram for two drops of a pin.
    Fill in the probability (relative frequency) on each branch.

b   Calculate the probability that a pin will land point up in two consecutive drops.

End points
You should be able to ...

       ... so try these questions

A Calculate probabilities by counting outcomes in sample space, and tree diagrams

A1 Draw a sample space diagram to show a spin of spinner A then a spin of spinner B.

Spinner A has 3 equal sections

A2 From your sample space diagram find the probability that for a spin of each spinner you get:
    a two of the same colour
    b a blue and a pink
    c two yellows
    d two pinks.

Spinner B has 4 equal sections

A3 Draw a tree diagram to show three consecutive spins of a coin.

A4 Use your tree diagram find the probability that in three spins you get:
    a three heads
    b exactly two coins showing the same
    c all three coins showing the same
    d two heads and one tail.

These four cards are put down in random order.

B Find how many ways there are to arrange a number of items

B1 **a** How many different ways of doing this are there?
      b What is the probability that the red and yellow cards would be next to each other?

C Calculate and use relative frequency

C1 Give the relative frequency of a plane arriving late.

Plane arrivals – Fridays	
Late	On time
183	327

Some points to remember

♦ In most cases when you are asked a question about probability from equally likely outcomes it is best to list all the outcomes and count the ones which match.

♦ You may be asked to find the number of ways to arrange items or to pick from a set of items. It is often a good idea to try it with fewer items and look for patterns.

♦ When you calculate relative frequency make sure you use the total number as the denominator.
(For example, in Question **D1**, add together 183 + 327 for the denominator.)

Starting points
You need to know about ...

A Multiplying out brackets

- To multiply out brackets, multiply every term inside the bracket by the term outside.

  Example
  $$2(a - 8) = 2 \times (a - 8)$$
  $$= (2 \times a) - (2 \times 8)$$
  $$= 2a - 16$$

B Collecting like terms

- In the expression $7a + 4b - 3a + 6b - 2a$
  - ❖ $7a$, $3a$ and $2a$ are like terms as they all give the number of as
  - ❖ $4b$ and $6b$ are like terms as they all give the number of bs.
- The expression can be simplified by collecting like terms.

  $$7a + 4b - 3a + 6b - 2a$$
  $$= 7a - 3a - 2a + 4b + 6b$$
  $$= 2a + 10b$$

C Solving linear equations

- To **solve an equation** find the possible values for each letter.

  Example Solve $5(2a - 1) = 2(3 + 4a)$

  To solve this equation
  - ❖ simplify each expression
  - ❖ add, subtract, multiply or divide both sides of the equation by equal amounts.

  $$5(2a - 1) = 2(3 + 4a)$$
  $$+5 \left(\; 10a - 5 = 6 + 8a \;\right) +5$$
  $$-8a \left(\; 10a = 11 + 8a \;\right) -8a$$
  $$\div 2 \left(\; 2a = 11 \;\right) \div 2$$
  $$a = 5.5$$

  The **solution** to this equation is $a = 5.5$.

D Indices

- Indices are used as shorthand for multiplication
n^2 stands for	$n \times n$		
$2n^2$ stands for	$2 \times n^2$	or	$2 \times n \times n$
$2mn^2$ stands for	$2 \times m \times n^2$	or	$2 \times m \times n \times n$

E Evaluating an expression

- The value of an expression depends on the value of each letter.

  Example Evaluate $2a^2 + 3ab + 8$ when $a = 6.4$ and $b = 2.1$

  $$2a^2 + 3ab + 8 = (2 \times 6.4^2) + (3 \times 6.4 \times 2.1) + 8$$
  $$= (2 \times 40.96) + (3 \times 6.4 \times 2.1) + 8$$
  $$= 81.92 + 40.32 + 8$$
  $$= 130.24$$

... so try these questions

A1 Multiply out the brackets from:
- **a** $3(a + 4)$ **b** $2(b - 6)$
- **c** $3(2c + 5)$ **d** $4(2d - 2)$
- **e** $4(8 - 2e)$ **f** $5(3 + 6f)$

B1 Simplify each of these.
- **a** $8a - 3a + 4b - 6$
- **b** $7p + 2q - 3p + 4q - 10$

B2 Simplify these.
- **a** $2(a - 4) + 3(2a + 6)$
- **b** $4(2a + 3) + 2(3a - 1)$

C1 Solve these equations.
- **a** $2p + 5 = 3p + 2$
- **b** $16 + 4q = 6q - 4$
- **c** $10 - 5t = 3t + 4$
- **d** $5(2x - 6) = 6x + 45$
- **e** $2(x - 1) = 2(3x + 8)$

D1 Find 3 pairs of equivalent terms.

A	B	C
ab^2	a^2	a^2b

D	E
$a \times b \times a$	$a \times b \times b$

F	G
$a \times a$	$a \times a \times b \times b$

E1 Evaluate these expressions when $p = 3.4$ and $q = 1.8$
- **a** $2p + 4pq$
- **b** $5p + 2q - 8$
- **c** $pq - 2$
- **d** $8p - 3q + pq$
- **e** $p^2 + 3p + 8$
- **f** $2pq + 2p^2 + 3q^2$

Calculating missing dimensions

Exercise 10.1
Missing dimensions

1   This is a plan of a garden. There is a fence round the garden and edging between the lawn and the flower beds.

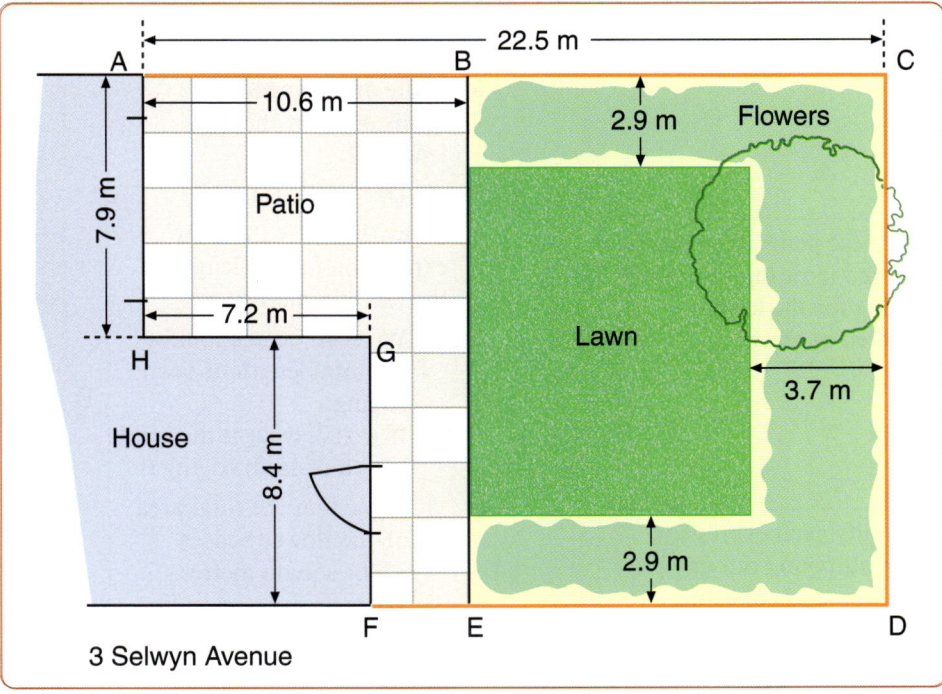

3 Selwyn Avenue

It may help to make a sketch of the garden and mark the lengths on it.

The area of the garden includes the patio.

a   Calculate the length of:
  i BE   **ii** EF   **iii** DE
b   The fence is shown in orange on the plan.
  Calculate the total length of the fence.
c   Calculate the dimensions of the lawn.
d   The edging for the lawn is shown in green.
  What is the total length of edging used for the lawn?
e   What is the total area of the flower bed?
f   Calculate the area of the whole garden.

2   The fence and edging for the lawn are shown in the same way on this plan.

  a   Calculate the total length of the fence.
  b   What length of edging is used for the lawn?
  c   What is the total area of the flower bed?
  d   Calculate the area of the whole garden.

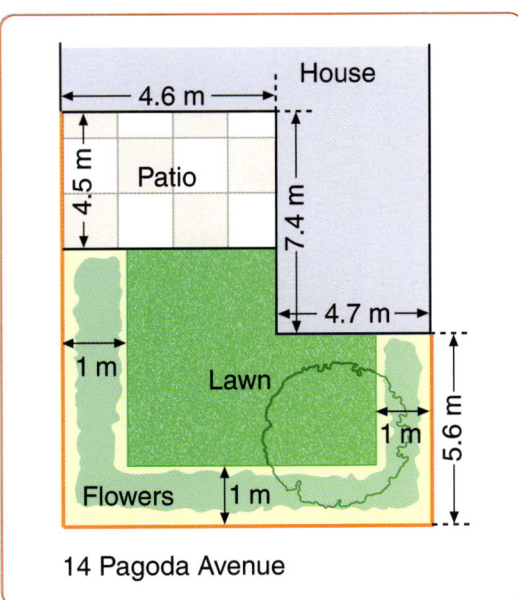

14 Pagoda Avenue

Writing expressions

Exercise 10.2
Writing expressions

On these plans the fence is marked in orange and the edging for the lawn is marked in green.

Simplify each expression by multiplying out any brackets and collecting like terms.

1 On this plan the flower beds are *w* metres wide.

a Explain why 8.1 + *w* is an expression for the length of BC in metres.

b Write an expression in terms of *w* for the length of:
   i DF **ii** HI

c Show that 43 + 2*w* is an expression for the length of the fence in metres.

d **i** Write an expression for the total length of lawn edging.
   ii If *w* = 2, what is the total length of lawn edging?

e **i** Show that the total area of the flower beds is 32.4*w* square metres.
   ii Write an expression for the area of the lawn.

f What is the value of *w* when the area of the lawn is the same as the total area of the flower bed?

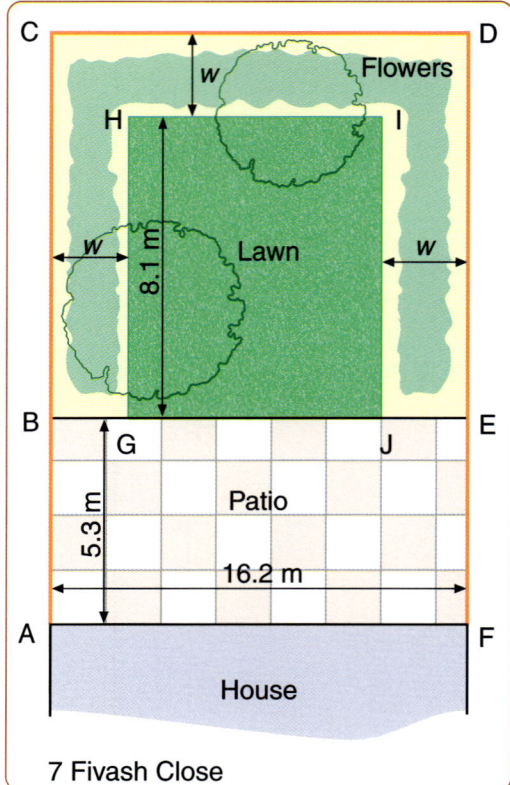

7 Fivash Close

2 On this plan the flower beds are *p* metres wide.

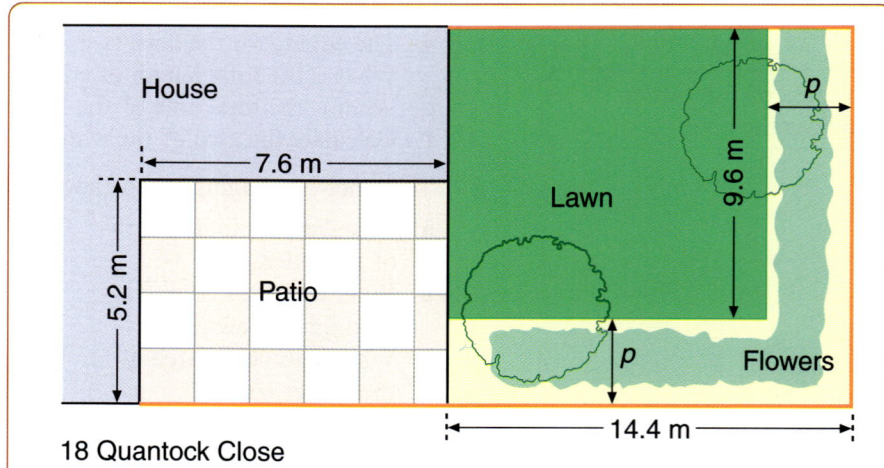

18 Quantock Close

a Write an expression in terms of *p* for:
   i the total length of the fence
   ii the total length of the lawn edging
   iii the area of the lawn
   iv the total area of the flower bed.

b Calculate the total length of the fence if *p* = 2.

c What value of *p* makes the area of the lawn twice the total area of the flower bed?

Using brackets

◆ You can use brackets to write an expression for the shaded area in each of these rectangles.

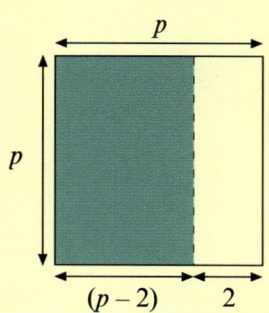

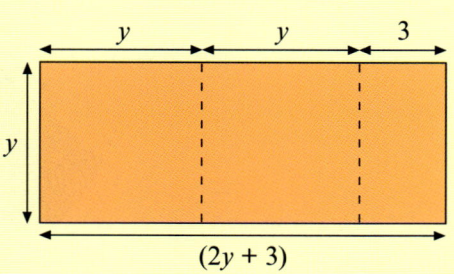

Shaded area = $p(p - 2)$

Shaded area = $y(2y + 3)$

◆ To multiply out a bracket it may help to use a table.

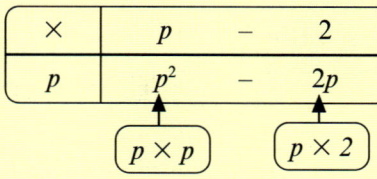

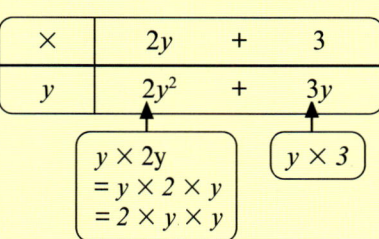

So $p(p - 2) = p^2 - 2p$

So $y(2y + 3) = 2y^2 + 3y$

Exercise 10.3
Using brackets

1 Multiply out each of these.

a $2(a + 4)$ b $3(b - 5)$ c $4(2c + 3)$ d $d(d + 7)$
e $e(6 - e)$ f $f(4 + f)$ g $p(2p + 9)$ h $r(4r - 2)$

2 For each of these rectangles write an expression for the shaded area:

a with brackets   b without brackets.

For some of the shaded rectangles you will need to find an expression for the length or the width.

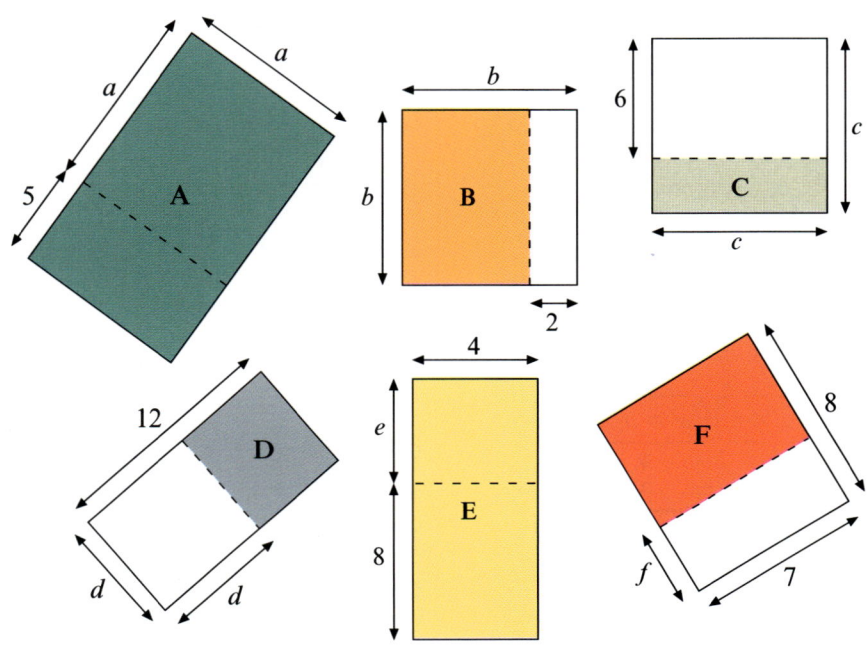

3 For each of these rectangles P, Q and R write an expression for the shaded area:

   a with brackets   **b** without brackets.

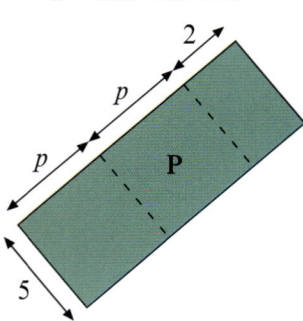
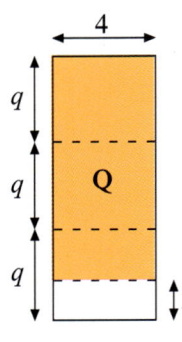
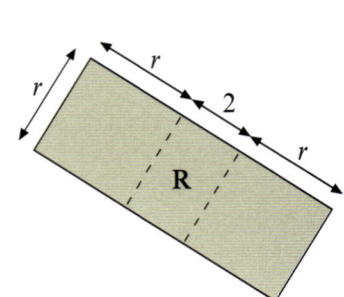

4 Write down the widths of rectangles A, B and C.

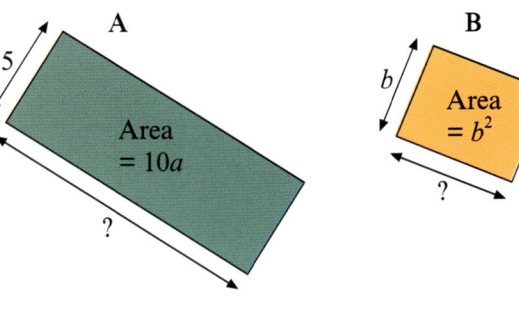
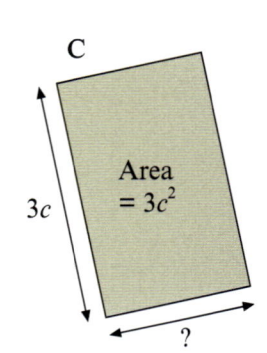

5 Write an expression for the widths of rectangles D to G.

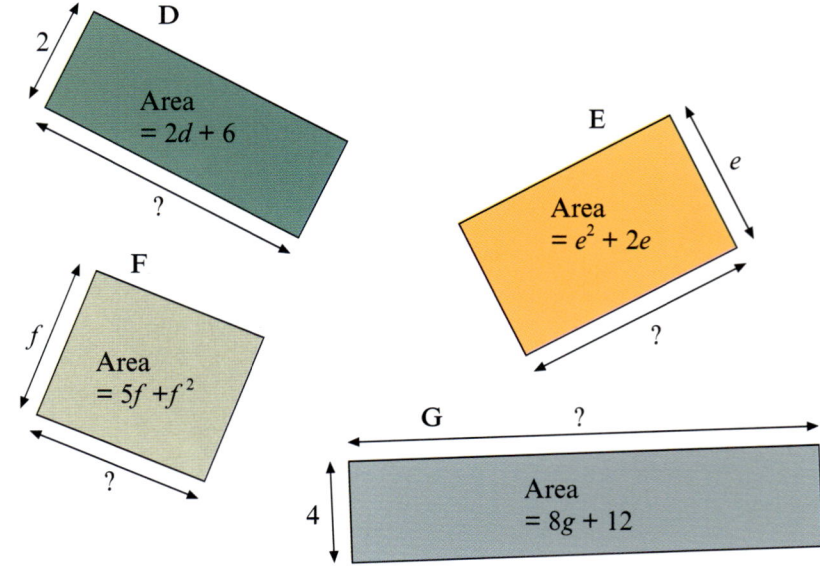

When you complete an expression using brackets check that the two expressions are equivalent.

For example:

$4a - a^2 = a(\square - \square)$

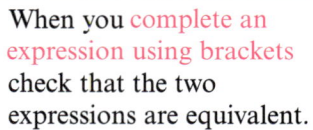

$4a = a \times 4$ $a^2 = a \times a$

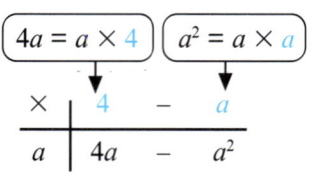

×	4	−	a
a	4a	−	a²

So $4a - a^2 = a(4 - a)$

6 Copy and complete these expressions using brackets.

  a $4x + 12 = 4(\square + \square)$ **b** $6p - 4 = 2(\square - \square)$

  c $2a + 8 = \square(a + \square)$ **d** $10 - 5q = \square(2 - \square)$

  e $b^2 - 6b = b(\square - \square)$ **f** $y^2 + 4y = y(\square + \square)$

  g $2r^2 + 3r = r(\square + \square)$ **h** $3d^2 - 5d = d(\square - \square)$

  i $3t^2 + 4t = \square(\square + \square)$ **j** $6s^2 + 7s = \square(\square + \square)$

Writing expressions to solve problems written in words

Example

The length of a rectangle is 4 m greater than the width.
The perimeter is 36 m. What is the area of the rectangle?

To find the length of this rectangle:

- ◆ Choose a letter to stand for the width.
  Let the width be w metres.

- ◆ Write the length in terms of this letter.
  Length = $w + 4$

- ◆ Draw and label a diagram.

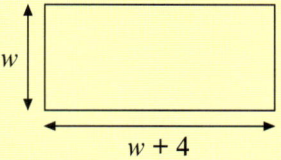

- ◆ Write and solve an equation for the information given.

Perimeter = $2w + 2(w + 4)$	Perimeter = 36 metres
$= 2w + 2w + 8$	So $4w + 8 = 36$
$= 4w + 8$	$4w = 28$
	$w = 7$

So Width = 7 metres
Length = 7 + 4 metres
= 11 metres

- ◆ Answer the problem.
  Area = 7×11 m^2
  So the area of the rectangle is 77 m^2.

Exercise 10.4
Solving problems
written in words

1 The length of a rectangle is 8 centimetres more than its width.

 a If the width is a centimetres, write an expression
   for the length of the rectangle.
 b Write an expression for the perimeter of
   the rectangle in terms of a.
 c If the perimeter of the rectangle is 60 centimetres
   what is the area of the rectangle?

2 In this triangle AB is twice the
  length of BC.

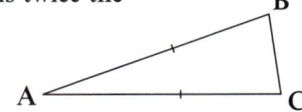

 a If BC is d centimetres, write an expression for the
   perimeter of the triangle in terms of d.
 b If the perimeter is 58 cm, what is the length of AB?

3 The perimeter of a square is 60 metres.
  What is the area of the square?

4 The length of a rectangle is twice its width.
  The perimeter is 39 centimetres.
  What is the area?

Multiplying terms and simplifying

In any term the letters are usually written in alphabetical order.

For example:

$2ba$ is usually written as $2ab$

$4n^2m$ is usually written as $4mn^2$.

♦ You can **multiply any terms** by grouping together the numbers and each of the letters.
For example:

$$
\begin{aligned}
&2m \times 3n \\
&= 2 \times m \times 3 \times n \\
&= 2 \times 3 \times m \times n \\
&= 6mn
\end{aligned}
\qquad
\begin{aligned}
&2ab \times a \\
&= 2 \times a \times b \times a \\
&= 2 \times a \times a \times b \\
&= 2a^2b
\end{aligned}
\qquad
\begin{aligned}
&2p \times 3p^2 \\
&= 2 \times p \times 3 \times p \times p \\
&= 2 \times 3 \times p \times p \times p \\
&= 6p^3
\end{aligned}
$$

The letters in each term are usually written in alphabetical order.

♦ **Like terms** must have exactly the same letters in them.
For example:

$2p^2q = 2 \times p \times p \times q$
$8p^2q = 8 \times p \times p \times q$
So $2p^2q$ and $8p^2q$ are like terms.

$3pq^2 = 3 \times p \times q \times q$
$2p^2q = 2 \times p \times p \times q$
So $2p^2q$ and $3pq^2$ are not like terms.

♦ To **simplify an expression** collect together any **like terms**.
For example:

These expressions **can be simplified** by collecting together like terms.
$2a^2b + 3ab^2 + 4a^2b = 6a^2b + 3ab^2$
$2x^2 + 2x + 3x^2 - x + 4 = 5x^2 + x + 4$

These expressions **cannot be simplified** because there are no like terms.
$2a^2b + 3ab^2$
$2x^2 + 4x + 3$

**Exercise 10.5
Multiplying terms
and simplifying**

1 Multiply these terms.

a $3a \times 2b$ b $p \times 3q$ c $4y \times 5x$
d $5q \times 6p$ e $x \times 2x$ f $ab \times a$
g $2xy \times y$ h $2ab \times 3a$ i $2p^2 \times 3q$
j $a^3 \times a^2$ k $2b^2 \times 3b$ l $5c^3 \times 2b$

2 Find four pairs of equivalent terms.

A $(2b^2)^3$ B $6b^5$ C $3b^2 \times 2b^4$
D $5b^5$ E $8b^6$ F $3b^2 \times 2b^3$
G $6(b^4)^2$ H $6b^6$ I $6b^8$

To multiply out each bracket multiply each pair of terms.

$\times$	m	$+$	$3n$
$2n$	$2mn$	$+$	$6n^2$

$2n \times m$
$= 2 \times n \times m$

$2n \times 3n$
$= 2 \times n \times 3 \times n$
$= 2 \times 3 \times n \times n$

So $2n(m + 3n) = 2mn + 6n^2$

3 Multiply out these brackets.

a $a(b + 4)$ b $x(y + z)$ c $m(2n + 3p)$
d $2x(3y + 2z)$ e $c(a + c)$ f $p(p - q)$
g $3b(c + b)$ h $4a(a - b)$ i $p(3p - 4)$
j $a(2b - 4c)$ k $2a(3a + 4b)$ l $4p(2q - 3p)$
m $2pq(3p + 2q)$ n $4xy(x - 2y)$ o $3ab(x^2 - y^2)$

4 Simplify these where possible.

a $5a - 3b + 4a + 25b$ b $4x^2 + x - 2x^2$
c $5a + 2ab - a + 3ab$ d $x^2 + x^3 - 2x$
e $4a - 3b + 7a + 5b$ f $7x^2 + xy - x^2 + 3xy$
g $4p^2 - pq + 6q + pq$ h $ab + 2a - ab + 4a$

5 Multiply out these brackets and simplify.

a $2(2a + 3b) + 5(a + 4b)$ **b** $2(2x + 4y) + 3(2x - y)$
c $x(x - 3) + x(x + 4)$ **d** $2x(3x + 2y) + x(4x - y)$
e $ab(a + b) + ab(a - b)$ **f** $3a(ab + b) + 2b(ab - b)$

6 Simplify these expressions.

a $2a(3a - b) + 4b(2a + 3b)$ **b** $5xy(2x + 4y) + 3x(2xy - 2y)$
c $pq(2p - 3q) + 2p(3pq - 2q^2)$ **d** $2mn(3m - 4n) + 4m^2(2n - 3m)$

Factorising

When you look for **common factors** in algebra the factor might be numerical (a number) or algebraic (a letter).

Example

Factorise this expression: $6a + 3bx - 9y$

There is a common factor of 3 in each term:

$$2 \times 3 \quad 3 \quad 3 \times 3$$
$$6a + 3bx - 9y$$

So we can take the factor of 3 outside a bracket.
So $6a + 3bx - 9y$ **factorised** is $3(2a + bx - 3y)$

Example

Factorise this expression: $3ax + 6ay - 9ac$

There is a common factor of $3a$ in each term so we can take the $3a$ outside a bracket so $3ax + 6ay - 9ac$ factorised is $3a(x + 2y - 3c)$

> You can check if you have factorised correctly by multiplying out the bracket. You should get the expression you started with, for example
> $3(2a + bx - 3y)$
> $= 6a + 3bx - 9y$

Exercise 10.6
Common factors

1 Factorise each expression.

a $5ax + 10y$ **b** $6ay - 15bc$
c $9xy - 12a$ **d** $4xy + 10x$
e $21xy + 35c$ **f** $36ab - 45y$
g $6ax + 12y + 15x$ **h** $8ac + 20cy - 24ax$
i $9x + 15ay - 21xy$ **j** $30kx + 18y + 42ac$
k $28xy + 56y - 42c$ **l** $144ac - 60y + 108x$

2   Factorise each expression.

a	$3ax - 5ay$		**b**	$6xy + 7ax$
c	$9xy - 5ay$		**d**	$6ax - 17xy$
e	$3abc + 5ax$		**f**	$9ab + 4b$
g	$7xy - 9y$		**h**	$6ax + 7axy$
i	$7abc - 8c$		**j**	$3ax + 5xy - 7cx$
k	$5xy - 7ay + 6y$		**l**	$12xy - 19ay + 15y$
m	$6abc + 5bc + 7c$		**n**	$8x + 7axy + 9bxy$
o	$7abc + ab$		**p**	$16xy + 25y - 13axy$
q	$21abx - 15cbx$			

You may be able to factorise by finding more than one common factor.

Example   Factorise fully $6axy - 9abc$

> Each term has a common factor of 3.
> (6 can be written as 2×3 and 9 as 3×3)
> Each term has a common factor of a.
> So we can take $3a$ outside a bracket.
> So $6axy - 9abc$ fully factorised is $3a(2xy - 3bc)$

Check the answer.
$3a(2xy - 3bc)$
gives
$6axy - 9abc$

3   Factorise these fully.

a	$2ax - 8ay$		**b**	$25xy + 15ax$
c	$16ay + 28by$		**d**	$32xy - 24ax$
e	$42abc - 35bx$		**f**	$56ab + 42by$
g	$6ab - 9ay + 12a$		**h**	$4xy + 6ay - 10cy$
i	$21xy - 15x + 12ax$		**j**	$15kxy + 25axy$
k	$18cx + 12axy + 15x$		**l**	$21abc + 56acx$
m	$15axy + 18ax - 30xy$		**n**	$abc + 5bc - 15bcx$
o	$24xy - 16x^2 + 8ax$		**p**	$2x^2 + 4x - 6xy$
q	$5x - 10xy - 25x^2$		**r**	$6ab - 9a^2b + 3abc$

End points
You should be able to ...

... so try these questions

| **A** Calculate missing dimensions |

A1 This is the plan of a paved area.

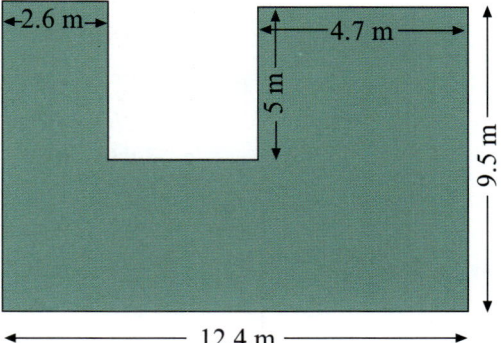

Calculate the total perimeter of the paved area.

A2 This is the plan of a play area.

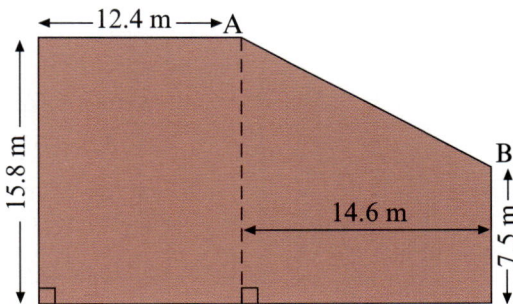

The total perimeter of the play area is 79.4 metres.
Calculate the length of AB.

A3 This is the plan of a circular pond.

The distance around the edge of the pond is 100 m.
a Find the radius of the pond to 1 decimal place.
b Calculate the area of the pond to 3 significant figures.

End points
You should be able to ...   ... so try these questions

B Simplify expressions

B1 Multiply out the brackets from:

a $4(5f - 4)$ b $m(m + n)$
c $2p(r - p)$ d $ab(a + b)$
e $mn(2m + 3n)$ f $2xy(3y - 5x)$

B2 Simplify these expressions.

a $2(2b - 4) + 6(4 + 3b)$ b $4(2b - 4) + 6(5 + 3b)$
c $3ab(b - a) + 7ab(b + a)$ d $3xy(x - 3) + 7xy(y + 4)$

C Write expressions to solve a problem

C1 The shape ABCDEF is cut from a square. The square is p centimetres wide.

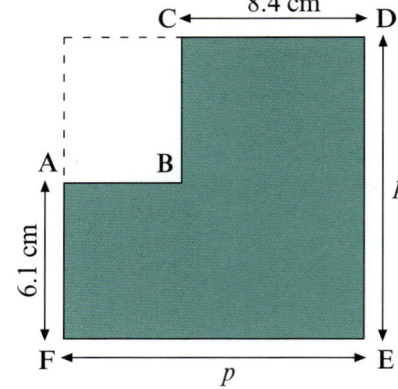

a Write an expression for the length of:

   i AB   **ii** BC

b **i** Write an expression for the perimeter of ABCDEF.
   ii What is the value of p if the perimeter is 56 cm?

C2 A rectangle is a centimetres wide.
The length is 5 centimetres more than the width.

a Write an expression for the length of the rectangle in centimetres.
b Write an expression for the perimeter of the rectangle in centimetres.
c Show that $a^2 + 5a$ is an expression for the area of the rectangle in square centimetres.
d What value of a makes the perimeter 56 cm?
e If $a = 5.1$, what is the area of the rectangle?

D Factorise expressions

D1 Factorise these fully.

a $8p + 4q$ b $6a - 12b$ c $a^2 + ab$
d $7m + 3mn$ e $8xy + 10y$ f $4ab + 6a^2$
g $3xy^2 - 5x^2y$ h $4pq - 6p^2q$ i $2g^2h - 5h^2$
j $30xy + 24x^2 - 16xy^2$ k $25a^2b - 15b^2a$
l $14axy - 21x^2 + 35x^2y$ m $100xy + 50ay^2 - 250x^2y$

Some points to remember

- Like terms must have exactly the same letters in.
- When simplifying an expression, add and subtract like terms.
- When factorising an expression check your answer by multiplying out the bracket.

Starting points
You need to know about ...

... so try these questions

A Some mathematical terms

Lines AB and CD are **perpendicular** to line XY because they would meet XY at right angles.

XY is also **perpendicular** to AB and CD.

AB and CD are **parallel**.

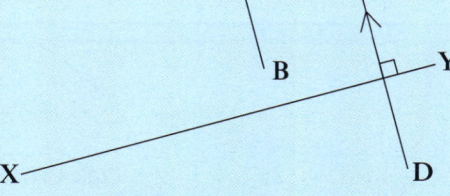

These terms are for parts of a circle.

Arc
Chord
Diameter
Radius
Tangent

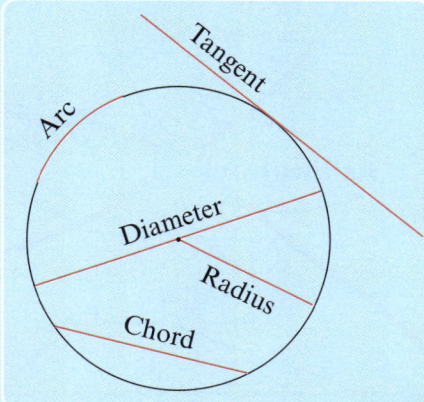

These terms are for different types of triangle.

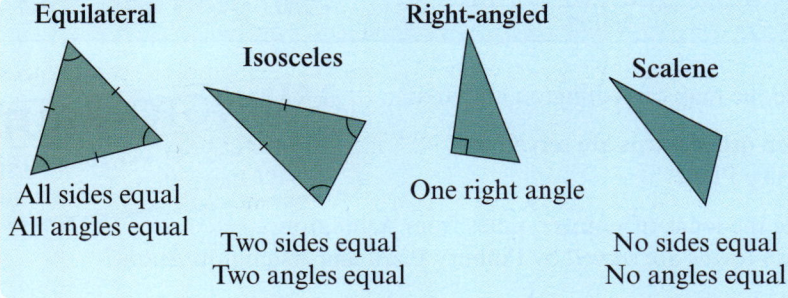

Equilateral
All sides equal
All angles equal

Isosceles
Two sides equal
Two angles equal

Right-angled
One right angle

Scalene
No sides equal
No angles equal

B Congruent triangles

Triangles are said to be **congruent** if they have the same shape or size or if they are reflections of each other.

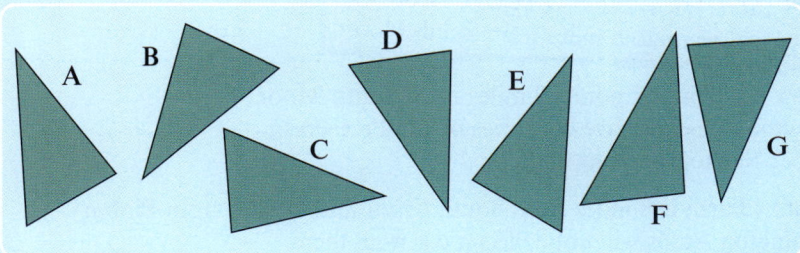

Triangles A and B are **congruent** to each other.

A1 You will need to measure for some of these questions.

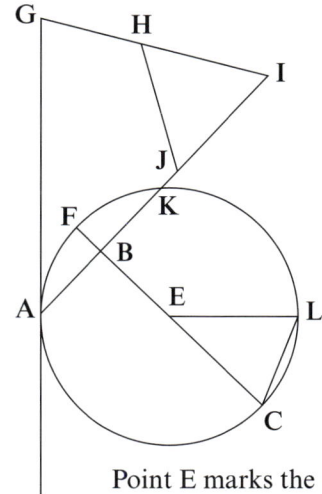

Point E marks the centre of the circle

a Which line is perpendicular to line FC?
b Give a line which is perpendicular to GD.
c Give a line which is a radius of the circle.
d Which triangle is isosceles but not equilateral?
e Which triangle is scalene?
f Which triangle is equilateral?
g Which line is a chord?
h Which line is a tangent?
i Which line is a diameter?

B1 Which of the triangles B to G is not congruent to triangle A?

B2 Draw a different triangle which is congruent to triangle A.

Regions and points on maps

A locus shows where a set of points satisfy a given condition.

The plural of locus is **loci**.

Deliveries are often made up to a certain distance from a town centre.

The locus of all points 6 miles from Pinbury is shown by the circle.

PINBURY PIZZA
Cold pizza? Not any more !
We now deliver up to 6 miles from Pinbury.
Ring 544364 for pizza to your door.

Exercise 11.1
Regions and points on maps

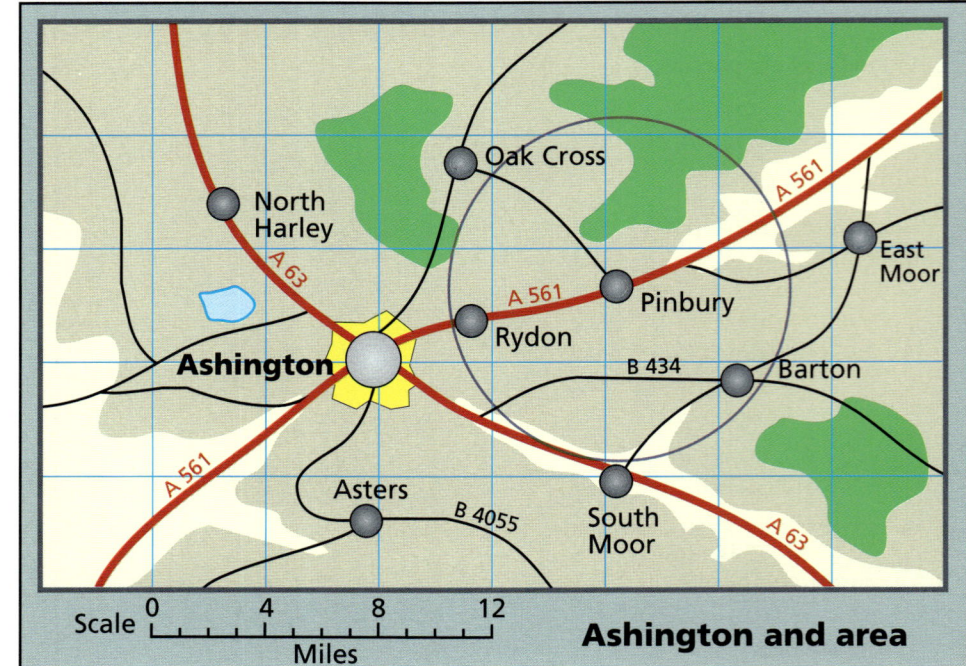

Ashington and area

You will need tracing paper and a pair of compasses for Exercise 11.1.

Scale: 0, 4, 8, 12 Miles

All distances are given 'as the crow flies' from the town centres.

1 Trace the map of Ashington and area.

2 Which other towns are served by Pinbury Pizza?

ASHINGTON AUTOS
We collect your car for you if you are not more than 5 miles away.

3 Draw the locus of points 5 miles from Ashington.
   Which towns are served by Pinbury Pizza **and** Ashington Autos?

4 Shade in the region which is up to 6 miles from Pinbury **and** up to 5 miles from Ashington.

5 The village of Newton is 6 miles from Pinbury and 5 miles from Ashington.
   Mark an X on your map for any place where Newton could be.

6 ## = South Moor Free Press =
   We deliver up to 8 miles from South Moor

   Show the locus of points 8 miles from South Moor.
   Newton does not have deliveries of the Free Press.
   Label Newton on your map.

7 Brooks Farm is 8 miles from South Moor and 6 miles from Pinbury.
   Ashington Autos will not collect a car from there.
   How far is Brooks Farm from North Harley?

Constructing triangles

Loci can also be used to construct triangles, when you know all three sides.

Example

Construct a triangle with sides of 4.4 cm, 3.8 cm and 3.2 cm.

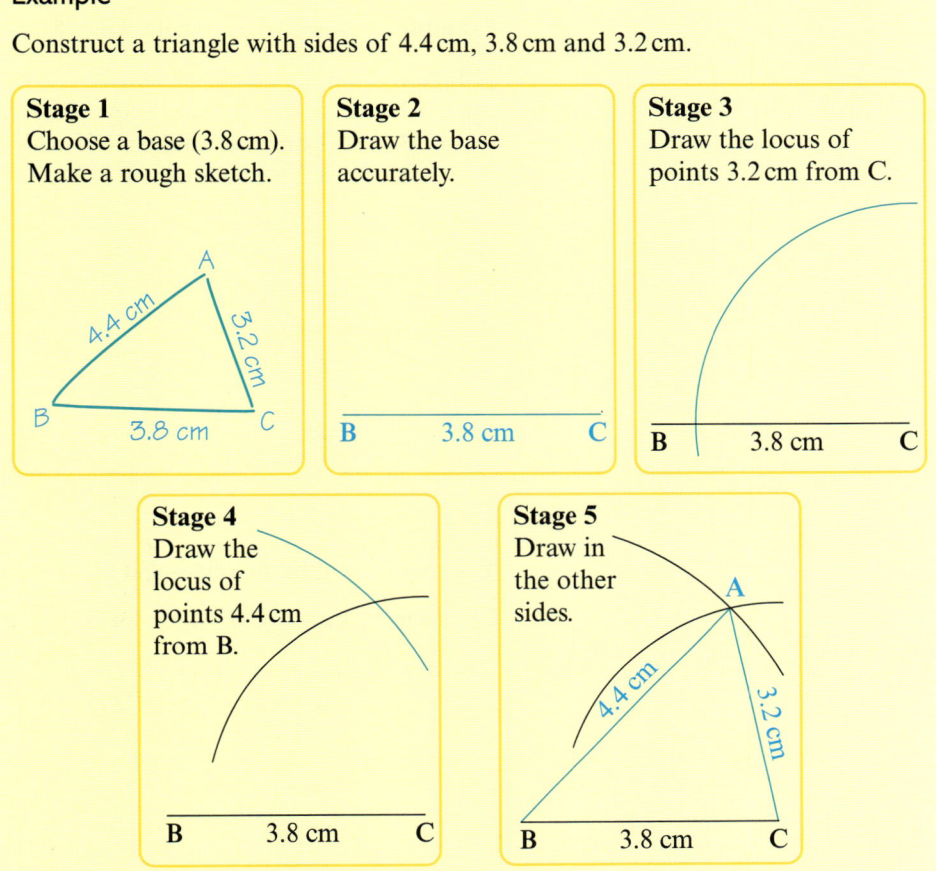

Stage 1
Choose a base (3.8 cm).
Make a rough sketch.

Stage 2
Draw the base
accurately.

Stage 3
Draw the locus of
points 3.2 cm from C.

Stage 4
Draw the
locus of
points 4.4 cm
from B.

Stage 5
Draw in
the other
sides.

You will need tracing paper
and a pair of compasses
for Exercise 11.2.

Exercise 11.2
Constructing triangles

1 Choose one of the other sides for a base of triangle ABC.
Construct the triangle on this base.
Check that both triangles are congruent.

2 Construct a triangle with sides of 7 cm, 3 cm and 7 cm.
What type of triangle is this?

3 Construct triangles with these dimensions.
In each case state the type of triangle.

   a 4.4 cm, 3.3 cm and 5.5 cm   **b** 5.2 cm, 4 cm and 8 cm
   c 7.5 cm, 7.5 cm and 7.5 cm   **d** 4.3 cm, 9.2 cm and 12.8 cm

4 Construct a triangle with sides of 10 cm, 6 cm and 3 cm.
What problems did you find? Explain why.

A scale of 1:10 000
means that 1 cm stands
for 10 000 cm.

So 1 cm stands for
100 metres.

5 A triangular field is 350 metres by 840 metres by 760 metres.
Use a scale of 1:10 000 to construct a scale drawing of the field.

6 Four ships R, S, T and U have the following distances between them.
R to S is 60 miles; R to T is 62 miles; S to T is 45 miles; S to U is 76 miles;
T to U is 43 miles.
Use a scale of 1 cm to 10 miles to construct a scale drawing of
their positions.
From your drawing find the distance between R and U in miles.

Constructing triangles from other data

You can construct a triangle when you are not given the length of all three sides. You might be given **one side and two angles**.

Example Draw △ABC where AB = 5 cm, ∠B = 55°, and ∠A = 43°.

> **Stage 1** Make a rough sketch.
> **2** Make the side you know (AB) the base and draw it.
> **3** At A draw an angle of 43° with a protractor.
> **4** At B draw an angle of 55°.

△ABC means Triangle ABC.

∠B means angle B.

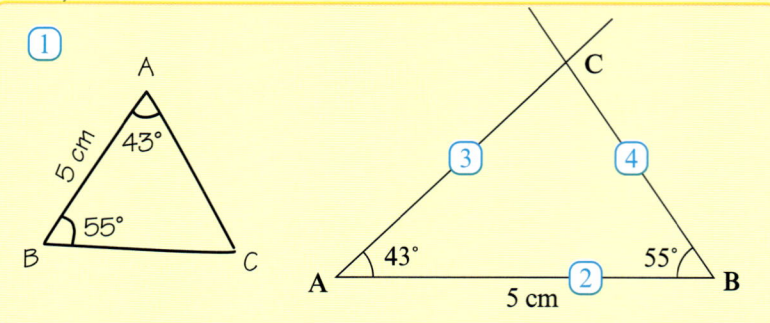

You might be given **two sides and one angle**.

Example Draw △RST where RT = 6.5 cm, RS = 4.5 cm and ∠R = 46°.

> **Stage 1** Make a rough sketch.
> **2** Make the long side (RT) the base, and draw it.
> **3** At R draw an angle of 46° and mark S, 4.5 cm from R.
> **4** Draw the last side TS.

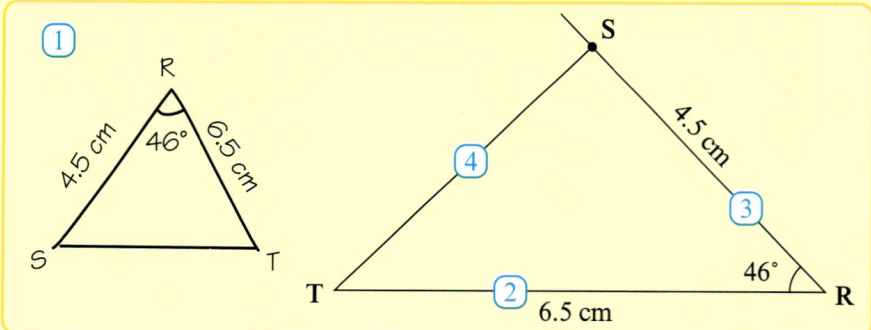

Exercise 11.3
Constructing triangles

1 Construct these triangles.
  a △DEF, where DF = 6 cm, EF = 5 cm and ∠F = 69°
  b △GHI, where GI = 7.2 cm, ∠G = 53° and ∠I = 42°
  c △JKL, where JK = 8 cm, ∠J = 25° and ∠K = 125°
  d △MNP, where NP = 6.3 cm, MP = 5.2 cm and ∠P = 131°

2 Construct △QRS, where RQ = 8.3 cm, ∠R = 35° and ∠S = 68°. You will need to calculate another angle first.

3 Construct these triangles to decide which two look identical.
  a △IJK, where JK = 5.5 cm, IJ = 8 cm and ∠J = 50°
  b △LMN, where LM = 8 cm, ∠L = 60° and ∠M = 50°
  c △PQR, where PR = 6.5 cm, QR = 7.5 cm and ∠R = 70°

Constructing a perpendicular from a point to a line

A line segment is a measurable portion of a straight line. It is finite in length. Straight lines can be infinite.

To construct a perpendicular from point A to the line BC:

- ♦ With centre A draw an arc that it cuts the line BC twice [X and Y]

- ♦ With centre X open the compass over half the length of the line segment XY and make an arc.

- ♦ With the same radius and centre Y make another arc crossing the arc made from X.

- ♦ Join the crossing point of the arcs and A, extending the line through to BC.

- ♦ The line is perpendicular to BC.

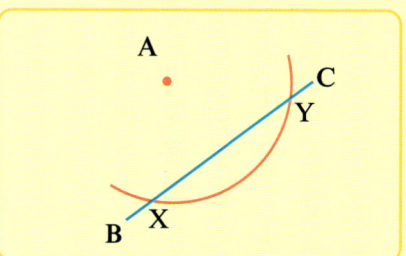

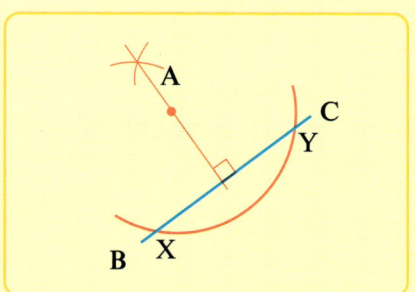

Exercise 11.4
A perpendicular from a point to a line

1 a Draw a straight line CD with points A and B either side of the line as in the diagram.

   b Construct perpendiculars from A and B to CD.

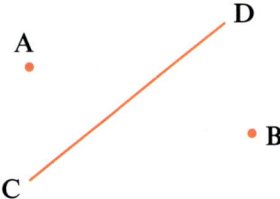

2 a Construct this triangle.

   b Construct a perpendicular from C to AB.

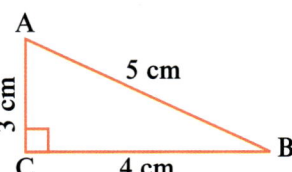

127

Thinking ahead to ...
perpendicular bisectors

A   Draw a straight line and
     mark two points, A and B,
     6 cm apart.

B   Draw a circle at A and
     another with the same
     radius at B.

> When two lines cross they
> are said to intersect.

C   Draw larger circles with
     equal radii at A and B.
     If they intersect, then mark
     the points of intersection.

D   Continue by drawing
     larger circles and marking
     the points of intersection.

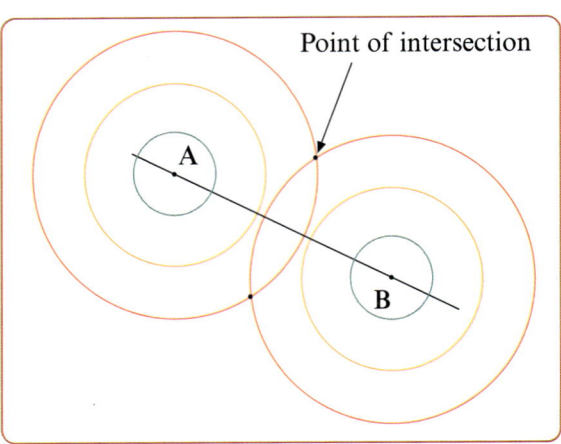

E   a   Draw the locus of all points of intersection of equal circles.
     b   Describe the link between this locus and the line AB.

Perpendicular bisectors

A line which cuts a straight line exactly in half at right angles is called a
perpendicular bisector.

To construct the perpendicular
bisector of the line EF.

♦ Draw the line EF

♦ With centre E draw an arc
  with a radius greater than
  half of EF.

♦ With centre F draw another
  arc with the same radius.

♦ Join the two points of
  intersection.

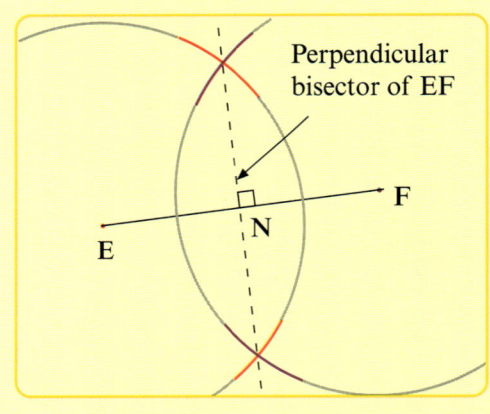

N marks the midpoint of EF.

Exercise 11.5
Perpendicular bisectors

1   a   Draw a line AB, 6 cm long. Construct the perpendicular bisector of AB.
     b   Mark any point on the bisector and measure its distance to A and to B.
     c   What can you say about the distance of any point on the
          bisector from A and from B?

2   Draw the perpendicular bisectors of lines with these lengths.
     Check that both sides are equal in length and that you have right angles.

     a   10 cm   b   7.7 cm   c   4.6 cm

3   a   Draw this triangle.
     b   Construct the perpendicular
          bisector of each side.
     c   Where do all three bisectors intersect?
     d   Does this happen for other triangles?

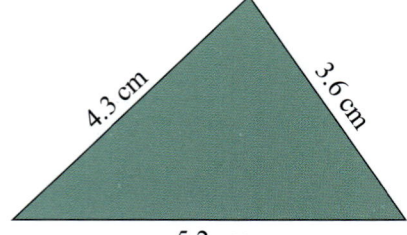

Thinking ahead to ...
bisecting an angle

A pedestrian area is edged by two buildings which meet at an angle of 36°.
The plans say trees must be planted an equal distance from both buildings.

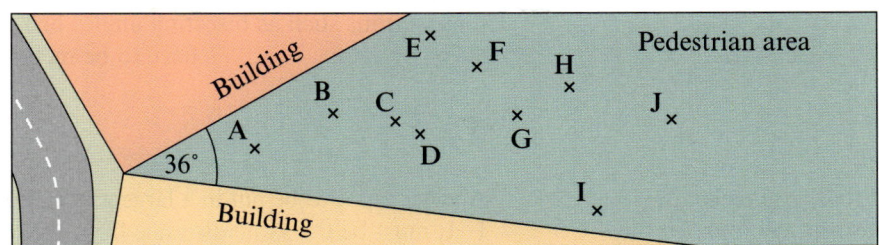

> 'Equidistant from' means 'the same distance from'.

A Which crosses on the diagram mark where trees could be planted?

B Draw two lines which meet at an angle of 36°.
Mark the locus of all points which are equidistant from both lines.

Bisecting an angle

> To bisect an angle means to draw a line which cuts it in two equal parts from its vertex.

It is useful to be able to bisect an angle without measuring it.
You can do this using a pair of compasses.

Stage 1
Draw an angle ABC.

Stage 2
With centre B draw an arc so it cuts AB and BC at D and E.

Stage 3
With centre D, draw an arc.
With centre E, draw an arc.

Stage 4
Draw the bisector BF.

Exercise 11.6
Bisecting an angle

1 Use a protractor to draw each angle then bisect it using compasses.
   a 44°   **b** 100°   **c** 146°   **d** 90°

2 Use compasses to draw an equilateral triangle with sides of 8 cm.
What size is each angle?
Bisect one of the angles. What angle have you made?

Meeting conditions

Rules that have to be met such as 'the tap must be the same distance from A and B' are known as conditions.

Constructions such as bisecting angles or lines can be used for making scale drawings where conditions have to be met.

Example

A water tap is to be put in a large garden but it must meet these conditions:
1   It must be the same distance from the two greenhouses, A and B.
2   It must be the same distance from the grape vine wires as from the hedge.

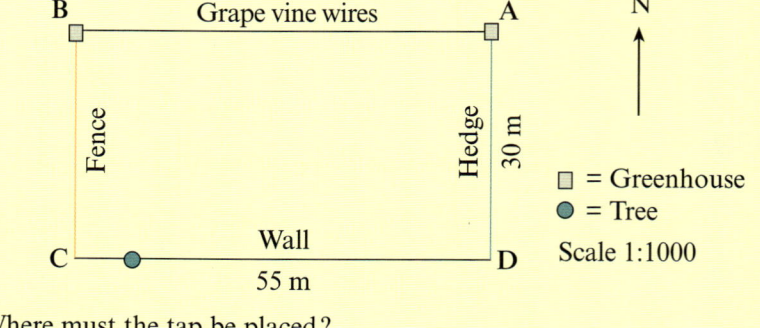

Where must the tap be placed?

To meet condition 1 you draw the perpendicular bisector of BA.
This line is the locus of all points equidistant from B and A.

To meet condition 2 you bisect angle BAD.
This line is the locus of all points equidistant from line BA and line AD.

Where the two loci intersect both conditions are met – so the tap must be at this point.

Scale 1:1000

Exercise 11.7
Meeting conditions

1   a   On the scale diagram above measure the distance in centimetres between the tap and the tree.
    b   What is this actual distance in the garden?
    c   Make a scale drawing to show where the tap will be if:
        condition 1 stays the same
        condition 2 says the tap must be equidistant from the wall and the hedge.

2 An Olympic javelin field has lines which make an angle of 29° to each other. A thrower aims the javelin so that it flies equidistant from both lines. The thrower hopes to reach the club record of 88 metres.

> This diagram only approximates to how a true javelin field is marked out. The throwing point actually lies on an arc about 2 metres wide which comes at the end of a 36 metre run-up.

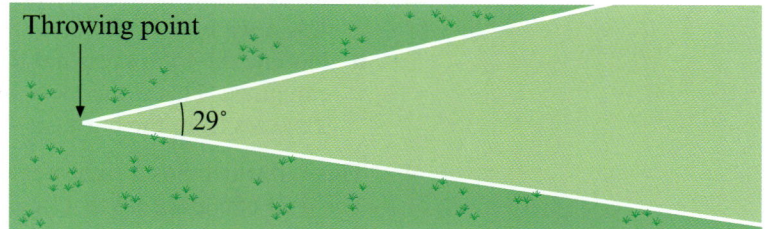

Throwing point

29°

a Make a scale drawing of the field for throws up to 100 metres. Use a scale of 1:1000.
b Mark the locus of all points 88 metres from the throwing point.
c Construct the locus of points equidistant from the sidelines.
d Mark where the thrower hopes the javelin will land.

3 Two lighthouses are 3.6 miles apart on a straight coastline. A ferry sails into port by keeping the same distance from both lighthouses. A fishing boat sails so that it is always 3 miles from the coast.

a Make a scale drawing of the coast to show the position of the lighthouses. Use a scale of 1 cm to 0.5 miles.
b Show and label the course taken by the fishing boat.
c Construct and label the course taken by the ferry.
d Mark the point where there is the greatest risk of a collision.

Loci and regions

Exercise 11.8
Loci and regions

Goats eat almost every type of plant.
They are often tied by a rope to limit their grazing.

Their rope can be fixed to a ring which can slide along a rail.

For example, a goat is tied in this way to a rail 8 metres long. The rope allows the goat to graze 2 metres from the rail.

Ring

Goat

Rail

When the ring is halfway along the rail the goat can graze the area inside the circle shown.

Scale 1:100

Rope

Rail

1 Make a scale drawing of the grazing area shown above.

2 On the same diagram draw the grazing areas for other positions of the ring along the rail.
Outline in red the shape of the total area the goat can graze.

3 A farmer puts a special rail around a barn to allow the goat's ring to travel right round. The goat can graze 2 metres from the rail.

  a Make a scale drawing of the barn. Use a scale of 1:100.
  b Outline the total area that the goat can graze. This is the locus of points 2 metres from the barn.

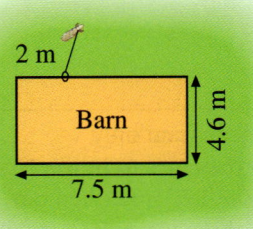

4 This rail which has a right-angled bend in it. The goat can graze up to 11 metres from the rail.

  a Draw the rail to a scale of 1:1000.
  b Outline the total area the goat can graze.
  c Are all the corners of this locus smooth curves? If not, which ones are not?

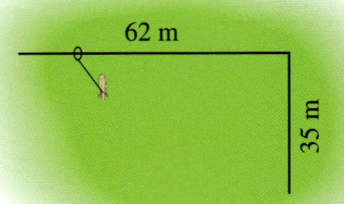

5 Goat A can graze 3 metres from barn A and goat B can graze 2.5 metres from barn B.

  a Make a scale drawing of the barns.
  b Draw the grazing area for each goat.
  c Shade in the area of grass that both goats can graze.

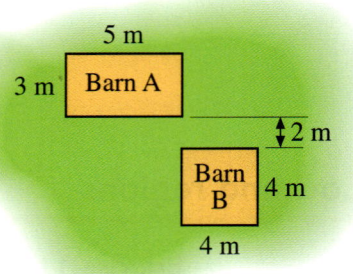

6 For each of these shapes, copy the shape and draw the locus of points 1 cm from the shape.

Copy one shape and draw the locus before you draw the next shape.

End points

You should be able to ...   ... so try these questions

A Construct and draw triangles when you are given their sides or angles

A1 Construct Δ ABC where AB = 7 cm, AC = 6 cm and BC = 6 cm.

A2 Draw Δ DEF, where ∠EDF = 92°, ∠DFE = 47° and DF = 4.5 cm.

A3 Draw Δ PQR, where PQ = 7.4 cm, QR = 4.3 cm and ∠Q = 123°.

B Construct and use perpendicular bisectors

B1

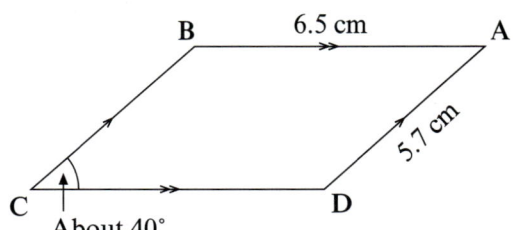

Draw the parallelogram ABCD with BĈD about 40°.
Construct perpendicular bisectors for the sides BC and AD.
What can you say about the gradients of the two bisectors?

B2

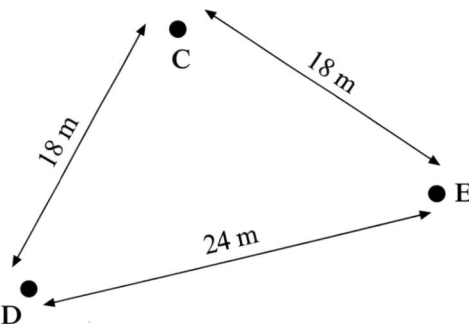

Two dogs are tied at points D and E, 24 metres apart.
A cat at point C wishes to pass between the dogs so that it is always the same distance from each one.
Make a scale drawing and construct the path the cat must take.
Use a scale of 1 cm to 3 metres.

C Construct and use the bisectors of an angle

C1 Draw an angle of 110° with a protractor.
Construct the bisector of the angle.

C2 Construct Δ ABC where AB = 4.3 cm, BC = 6.8 cm and AC = 5.9 cm.
Bisect each of the angles.
What do you notice about where the bisectors intersect?

D Construct a perpendicular from a point to a line

D1 Trace this diagram.
Construct perpendiculars from K and L to BA.

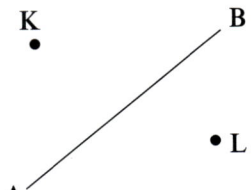

E Draw loci from some conditions given

E1 This garden is the shape of a trapezium with the dimensions given.

Trees are at the midpoints of sides AD and DC.

A new bush is to be planted:
- ◆ equidistant from each fence
- ◆ equidistant from each tree.

Make a scale drawing of the garden and show by construction the position of the bush. Use a scale of 1 : 100.

Some points to remember

- ◆ To construct an angle of 90° draw a line and construct its perpendicular bisector.

- ◆ When asked to construct something do not rub out the construction lines when you finish.

- ◆ To construct an angle of 60° construct an equilateral triangle.

- ◆ The perpendicular bisector of any chord will pass through the centre of the circle.

- ◆ All angles in a semicircle which are made by the diameter and a point on the circumference are right angles.

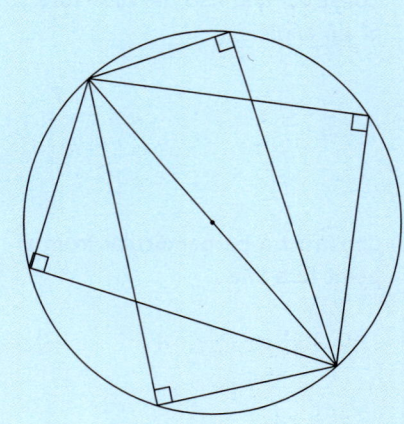

Starting points
You need to know about ...

... so try these questions.

A Writing ratios

- A **ratio** compares the size of two or more quantities.

$$3:4 \qquad 1:4:2 \qquad 3\tfrac{1}{2}:\tfrac{1}{2}$$

- In the ratio $3:4$, the **number of parts** are 3 and 4.

B Equivalent ratios

- The ratios $3:4$ and $6:8$ are **equivalent ratios**.

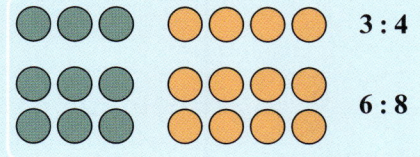

$3:4$

$6:8$

- To write an equivalent ratio:
  - multiply or divide each number of parts by the same number.

$$\overset{\div 2}{\underset{3:4}{6:8}}\div 2 \qquad \overset{\times 5}{\underset{10:15:35}{2:3:7}}\times 5$$

- A ratio in its **simplest terms**, or **lowest terms**, is written with the smallest whole numbers possible.

$$\overset{\times 4}{\underset{4:1:12}{1:\tfrac{1}{4}:3}}\times 4 \qquad \overset{\div 3}{\underset{5:8}{15:24}}\div 3$$

C Writing fractions as decimals and percentages

- To write a fraction as a decimal or a percentage:
  - divide the numerator by the denominator.

$$\frac{3}{8} = 3 \div 8 = \mathbf{0.375} = \mathbf{37.5\%} \qquad \frac{5}{6} = 5 \div 6 = \mathbf{0.8\dot{3}} = \mathbf{83.\dot{3}\%}$$

$$\frac{7}{4} = 7 \div 4 = \mathbf{1.75} = \mathbf{175\%} \qquad \frac{5}{3} = 5 \div 3 = \mathbf{1.\dot{6}} = \mathbf{166.\dot{6}\%}$$

D Fibonacci sequences

- A sequence of numbers is called a **Fibonacci sequence** when:
  - the first two terms are any numbers
  - all other terms are the total of the two terms before.

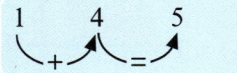 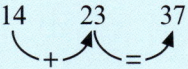
$$1 \quad 4 \quad 5 \quad 9 \quad 14 \quad 23 \quad 37 \quad 60 \ldots$$

- The most common Fibonacci sequence is:

$$1 \quad 1 \quad 2 \quad 3 \quad 5 \quad 8 \quad 13 \quad 21 \quad 34 \quad 55 \quad 89 \ldots$$

A1 Roughcast is used to cover walls. It is a mix of 3 parts mortar to 1 part gravel.
Write this mix as a ratio of:
  a mortar to gravel
  b gravel to mortar.

B1 Copy and complete these sets of equivalent ratios.
  a $4:1$ **b** $2:5:4$ **c** $9:12$
  $12:?$ $1:?:2$ $?:4$
                              $6:?$

B2
$2:5$	$1:3:2$	$10:20$
$10:25$	$2\tfrac{1}{2}:5$	$1:2:3$
$3:9:6$	$2:1$	$1:2\tfrac{1}{2}$

List the sets of equivalent ratios.

B3 Write each of these ratios in its simplest terms.
  a $9:6$ **b** $14:21$
  c $4:8:2$ **d** $5:\tfrac{1}{2}$
  e $2:1\tfrac{1}{2}:3$

C1 Write each of these fractions as a decimal.
  a $\frac{5}{8}$ **b** $\frac{9}{5}$ **c** $\frac{7}{12}$ **d** $\frac{15}{11}$

C2 A test has 72 marks.
Write each of these test scores as a percentage.
  a $\frac{44}{72}$ **b** $\frac{36}{72}$ **c** $\frac{51}{72}$ **d** $\frac{63}{72}$

D1 Copy and complete the first eight terms of each of these Fibonacci sequences.
  a $1, 3, \_, \_, \_, \_, \_, \_, \ldots$
  b $2, 2, \_, \_, \_, \_, \_, \_, \ldots$
  c $\_, 5, \_, 11, 17, \_, \_, \_, \ldots$
  d $3, \_, \_, 15, \_, 39, \_, \_, \ldots$

E Naming shapes, sides and angles

- A **vertex** is a point where the edges of a shape meet.
  The plural of vertex is **vertices**.

- If the vertices of a shape are labelled with letters, then
  the shape, the sides and the angles can be named.
  For example:

  - this shape is PQRS

  - the sides of PQRS are
    PQ, QR, SR, PS.

    PQ can also be used for
    the **length** of side PQ.

  - the angles inside PQRS are:
    ∠ QRS (or QR̂S or R̂)
    ∠ PSR (or PŜR or Ŝ)
    ∠ SPQ (or SP̂Q or P̂)
    ∠ PQR (or PQ̂R or Q̂).

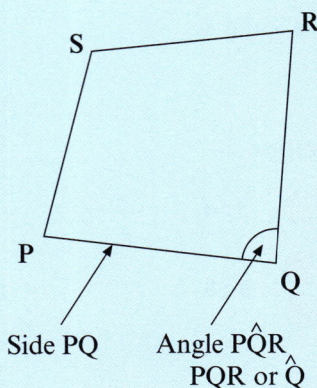

Side PQ   Angle PQ̂R
           PQR or Q̂

F Enlargement and similar shapes

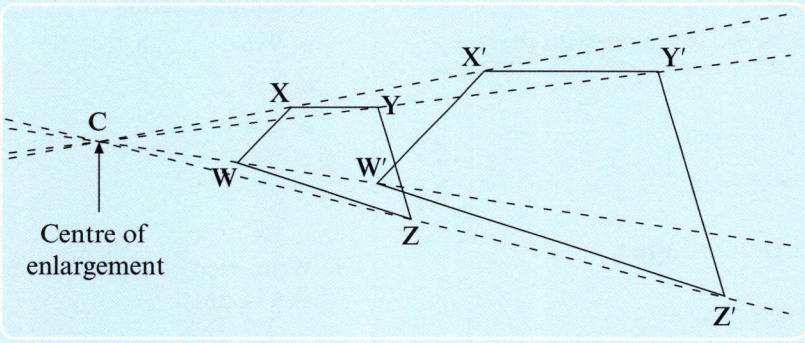

Centre of
enlargement

- Shape W'X'Y'Z' is an **enlargement** of shape WXYZ so:

  - WXYZ is the **object** and W'X'Y'Z' is the **image**

  - all **corresponding angles** are equal
    $$\hat{W} = \hat{W}', \ \hat{X} = \hat{X}', \ \hat{Y} = \hat{Y}', \ \hat{Z} = \hat{Z}'$$

  - any length on WXYZ can be multiplied by the **scale factor** of
    the enlargement to give the corresponding length on W'X'Y'Z'

    2 is the scale factor of this enlargement so,
    for **corresponding sides**,
    WX × 2 = W'X'   XY × 2 = X'Y'
    YZ × 2 = Y'Z'   ZW × 2 = Z'W'

- If the scale factor of an enlargement is **less than 1**, then
  the image is **smaller** than the object.

- If one shape is an enlargement of another, then the
  two shapes are **similar**.

E1

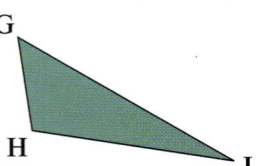

In triangle GHI:
a which side is the longest?
b which angle is the largest?
c which angles are acute?
d which is the shortest side?

F1

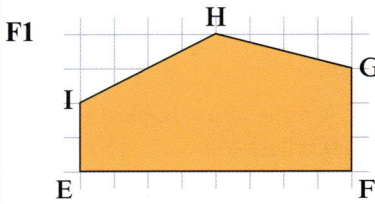

a Copy shape EFGHI on to
square grid paper.
b Mark a centre of
enlargement, C.
c Draw an enlargement of
EFGHI, scale factor 1.5.

F2 Repeat Question **F1** using a
scale factor of 0.5.

F3

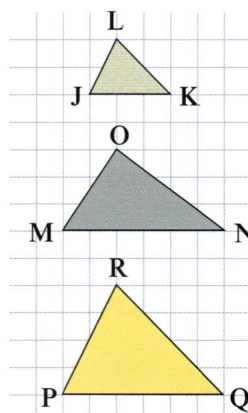

Which of these triangles are
similar?

Multiplying parts of a ratio

◆ The number of parts in a ratio can be multiplied to solve problems.

Example

An orange dye is a mix of red and yellow in the ratio 2:3.
a What amount of red is mixed with 480 ml of yellow?
b What is the total amount of orange dye produced?

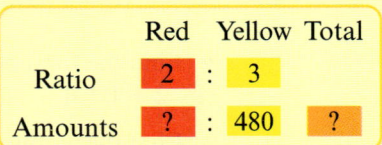

	Red	Yellow	Total
Ratio	2 :	3	
Amounts	? :	480	?

◆ Find the amount for one part:
480 ml ÷ 3 = 160 ml.

◆ Multiply each number of parts by the amount for one part.

◆ Find the total of the amounts.

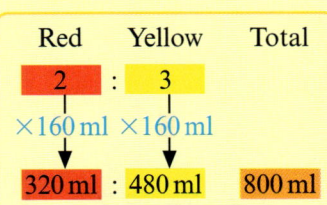

	Red	Yellow	Total
	2 :	3	
	×160 ml	×160 ml	
	320 ml :	480 ml	800 ml

Exercise 12.1
Multiplying parts of a ratio

For Questions **1** and **2** refer to the panel above.

1 Calculate:
a the amount of red dye to mix with 720 ml of yellow dye
b the total amount of orange dye produced.

2 Calculate:
a the amount of yellow dye to mix with 210 ml of red dye
b the total amount of orange dye produced.

3 Tropical Fruit Juice is a mix of pineapple and grapefruit in the ratio 5:2.
Calculate:
a the amount of grapefruit juice to mix with 125 ml of pineapple juice
b the total amount of Tropical Fruit Juice produced.

4 Walls can be covered with mortar to protect them from the weather.
a For a clay wall how much cement is mixed with 1500 kg of sand?
b For a concrete wall:
 i write the mortar mix as a ratio in its simplest terms
 ii calculate how much concrete is produced with 400 kg of cement.

Wall surface	Mortar mix Cement : Lime : Sand
Clay	1 : 1 : 6
Concrete	$1 : \frac{1}{2} : 4\frac{1}{2}$

5 A plan is drawn using a scale of 1:200.
a What actual distance does 4 cm stand for on the plan?
b What length on the plan stands for a distance of 18 m?

6 For a plan using a scale of 1:30, calculate:
a the length on the plan which stands for a distance of 4.5 m
b the actual distance shown on the plan as 7.5 cm.

7 On a plan, 4.8 cm stands for a distance of 240 m.
a What length on the plan stands for a distance of 315 m?
b What scale is used on the plan?

Ratio and proportion

◆ An amount can be shared in a given ratio.

Example Share £140 in the ratio 2 : 1 : 4.

			Total
Ratio	2 : 1 : 4		
Amounts	? : ? : ?	£140	

◆ Find the total number of parts:
  2 + 1 + 4 = 7

◆ Find the amount for one part:
  £140 ÷ 7 = £20

◆ Multiply each number of parts by the amount for one part.

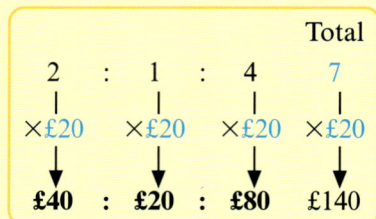

			Total
2 :	1 :	4	7
×£20	×£20	×£20	×£20
£40 :	**£20** :	**£80**	£140

Exercise 12.2
Sharing in a
given ratio

‘in proportion to’
means
‘in the same ratio as’

1 Sarah and Denzil share 35 conkers in the ratio 4 : 3.
Calculate how many conkers each gets.

2 Tim spends his money on sweets, comics and videos in the ratio 1 : 2 : 3.
Calculate how much he spends on each item from £18.

3 Amy, Ben and Zoe are left £150 by their grandmother.
The money is shared in proportion to their ages: 5, 3 and 2.
Calculate how much each child gets.

4 Liz is training for the 100 m hurdles.
She splits her time between speed work and technique in the ratio 3 : 5.
Calculate how long:

a she spends on technique in two hours of training

b she spends on speed work in four hours of training.

◆ Some problems need amounts to be kept in proportion.

Example This recipe for Swiss Hot Chocolate serves 2 people.
How much of each ingredient is needed for 5 people?

Swiss Hot Chocolate
• 600 ml milk
• 140 g drinking chocolate
• 80 ml whipped cream

Milk	Chocolate	Cream	Serves
600 ml :	140 g :	80 ml	2
? :	? :	?	5

◆ Find the **multiplier**:
  5 ÷ 2 = 2.5

◆ Multiply each amount by the multiplier.

Milk	Chocolate	Cream	Serves
600 ml :	140 g :	80 ml	2
×2.5	×2.5	×2.5	×2.5
1500 ml :	350 g :	200 ml	5

Exercise 12.3
Keeping in proportion

1 This recipe serves 6 people.
How much of each ingredient do you need for 9 people?

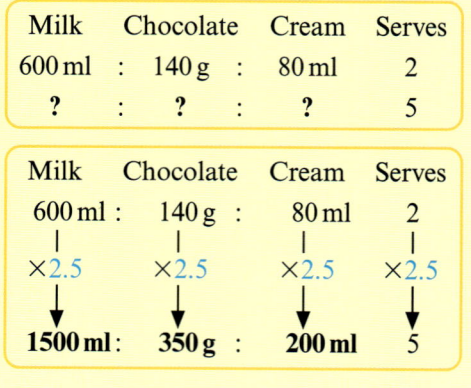

Chilled Chocolate Drink
230 g caster sugar • 300 ml water
50 g cocoa • 1200 ml chilled milk

> The multipliers here are less than 1.

2 This recipe for chocolate fudge makes 36 pieces.

  a How much cocoa is needed to make 72 pieces?

  b How much milk is needed to make 18 pieces?

  c Calculate how much of each ingredient you need to make 24 pieces.

Chocolate Fudge
- *450 g white sugar*
- *150 ml milk*
- *150 ml water*
- *75 g butter*
- *30 g cocoa*

3 This recipe makes 24 sweets.

  a Calculate how many drops of peppermint essence are needed to make 42 sweets.

  b How many egg whites do you need to make these 42 sweets?

Chocolate Peppermint Creams
- *230 g icing sugar* • *1 egg white*
- *4 drops peppermint essence*
- *100 g plain chocolate*

> Discuss your answer.

Ratios as fractions

♦ A ratio that compares two quantities, $p:q$, can be given in the form $\frac{p}{q}$.

Example

The ratio of **width to height** is $4:3$, i.e.

$$\frac{\text{Width}}{\text{Height}} = \frac{4}{3}$$

so the width is $\frac{4}{3}$ of the height.

♦ In some ratios, you can use the **total** number of parts to give other fractions.

Example The ratio of boys to girls in a class is $1:2$

Boys	:	Girls	Total
1	:	2	3

Ratio of **boys to girls** is $1:2$

$$\frac{\text{Boys}}{\text{Girls}} = \frac{1}{2}$$

Ratio of **boys to total** is $1:3$

$$\frac{\text{Boys}}{\text{Total}} = \frac{1}{3}$$

The number of boys is $\frac{1}{2}$ of the number of girls and $\frac{1}{3}$ of the total.

Exercise 12.4
Ratios as fractions

For Questions **1** and **2** refer to the panel above.

1 Give the ratio of girls to boys in the form:

  a Girls : Boys   **b** $\dfrac{\text{Girls}}{\text{Boys}}$

2 Give the ratio of girls to total in the forms $p:q$ and $\frac{p}{q}$.

3 For this photograph, give each of these as a fraction:

  a the ratio of width to height

  b the ratio of height to width.

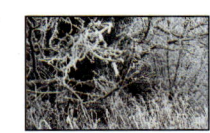

4 The ratio of women to men in a fitness class is $1:4$. Give the fraction of the class which are:

  a women   **b** men.

5 The ratio of height to diameter for this tin is $2:3$.

  a What fraction of the diameter is the height?

  b Explain why the fraction $\frac{2}{5}$ has no meaning for this tin.

Ratios and enlargement

> The ratio is calculated by dividing a length in the **image** by the corresponding length in the **object**.

♦ The scale factor of an enlargement is the ratio of corresponding lengths.

Example

Calculate the scale factor of this enlargement, and find the height K'L'.

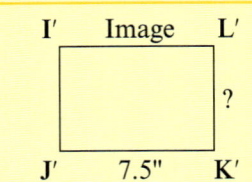 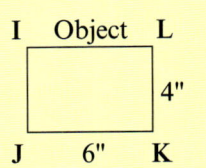

♦ Divide width J'K' by the corresponding width JK.

$$7.5 : 6 = \frac{7.5}{6} = 1.25$$

♦ Multiply the corresponding height by the scale factor.

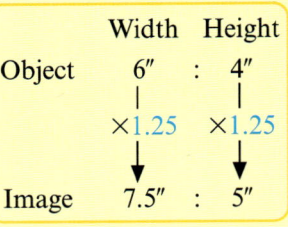

Exercise 12.5
Enlargement

1

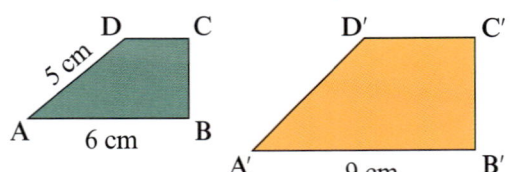

Trapezium A'B'C'D' is an enlargement of trapezium ABCD.

> If the image is larger than the object, the scale factor is greater than 1.

a Calculate the scale factor of the enlargement.
b Use your scale factor to find the length A'D'.

2

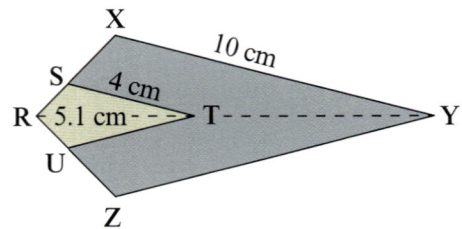

Kite RXYZ is an enlargement of kite RSTU.

a Calculate the scale factor of the enlargement.
b Find the length RY.

3

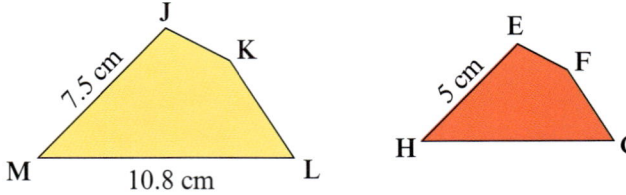

Shape EGFH is an enlargement of shape JKLM.

> If the image is smaller than the object, the scale factor is less than 1.

a Calculate the scale factor of the enlargement.
b Find the length HG.

End points

You should be able to ...

A Multiply the number of parts in a ratio

B Share an amount in a given ratio

C Keep amounts in the same proportion

D Write ratios as fractions

... so try these questions

A1 Tropical Fruit Juice is a mix of pineapple and grapefruit in the ratio 5:2. Calculate:

    a the amount of pineapple juice to mix with 180 ml of grapefruit juice

    b the total amount of Tropical Fruit Juice produced.

B1 Amy, Ben and Zoe are left £400 by their grandfather.
The money is shared in proportion to their ages: 10, 8 and 7.
Calculate how much each child gets.

C1 This ice cream recipe serves 8 people.

    a Calculate how much of each of these ingredients you need to serve 20 people:

        i caster sugar

        ii boiling water.

    b How many eggs are needed to serve 6 people?

> **Chocolate Ice Cream**
> • 150 g caster sugar
> • 20 g cocoa • 4 eggs
> • 410 g can evaporated milk
> • 4 tablespoons boiling water

D1 For this photograph, give the ratio of height to width:

    a in its simplest terms

    b as a fraction

6"

8"

D2 The ratio of girls to boys in a class is 2:3.
What fraction of the class are girls?

Some points to remember

♦ Write in headings to help you get the ratio the right way round.
For example, a ratio of boys to girls of 1:3 can be written as:

        Boys : Girls
          1 : 3

♦ The scale factor of an enlargement is the ratio of corresponding lengths.

SUMMER

NINE ELMS GARDEN CENTRE

Watering cans - plastic

5 litre	8 litre	10 litre	12 litre
99p	£1.29	£1.69	£1.99

Watering cans - traditional

1 gallon	1.5 gallon	2 gallon	2.5 gallon
£4. 99	£6.49	£7.99	£9.99

Special offers for July

Nine Elms Mulchbags

Large	£1.49
Super	£1.99
Major	£2.99
Professional	£4.99

15% OFF THESE PRICES

Mower Madness !!!
Can you believe this offer ?

- 20% deposit
- 0% interest
- 6 equal monthly payments

Super Hover £179.99	Mowhog £199.99	Mastermow £249.99

Rolls of Plastic Sheet

The easy way to stop weeds
15 metres long
and
1650 mm wide

Only £3.75
per roll
while stocks last

LIQUIDGRO
Add one scoop
(15 grams) per gallon

£3.99

Superbraid
50 metre rolls £12.99
30 metre rolls £7.49
Any length in bulk: just 28p per metre

Connectors 75p each

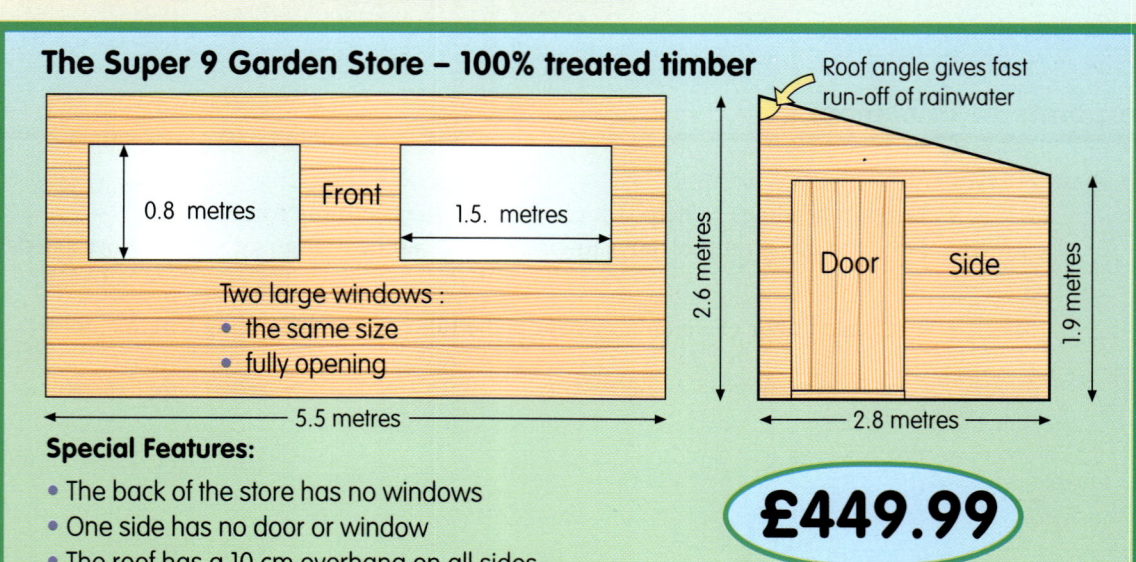

The Super 9 Garden Store – 100% treated timber

Roof angle gives fast run-off of rainwater

0.8 metres   Front   1.5. metres

Two large windows :
- the same size
- fully opening

Door   Side

2.6 metres

1.9 metres

5.5 metres

2.8 metres

Special Features:
- The back of the store has no windows
- One side has no door or window
- The roof has a 10 cm overhang on all sides

£449.99

1   List the discount price of each size Mulchbag, to the nearest penny.

2   Which plastic watering can do you think is best value for money?
    Explain your answer.

3   Which watering can holds most water: the 12 litre or the 2.5 gallon?
    Explain your answer.

4   Darren says that a plastic watering can is only about 20% of the cost of a traditional one.
    Do you agree? Explain your answer.

5   Which traditional watering can do you think is best value for money?
    Explain your answer.

6   A 5 litre plastic watering can weighs 105 grams.
    A gallon of water weighs 10 lb.
    What do you expect a 5 litre plastic watering can to weigh when it is full?

7   An empty 1.5 gallon traditional can weighs 0.7 lb.
    a   Give the weight of this can in kilograms.
    b   Do you expect a full 1.5 gallon traditional can to weigh more, or less, than 7.5 kg?
        Explain your answer.

8   a   To the nearest gram, how much Liquidgro is needed for 10 litres of water?
    b   To be fairly accurate how many scoops would you put in the 10 litres of water?
        Explain your answer.

9   If you measured the Liquidgro accurately:
    a   How many gallons will the 1 kg pack treat?
    b   How much spare liquidgro is in the pack?
    c   What is this spare Liquidgro as:
        i   a fraction of the pack?
        ii   a decimal of the pack?
        iii a percentage of the pack?

10 The manufacturers of Liquidgro say that a 1 kg pack will treat 300 litres of water.
    Is this a fair claim? Explain your answer.

11 Liquidgro costs £285 per tonne to produce.
    Packaging costs are 6.8 pence per pack.
    Distribution costs are 13.5 pence per pack.

    How much profit do the manufacturers make on one tonne of Liquidgro?

12 a   What is the area of plastic sheet on a roll?
    b   To the nearest penny, what is the price per square metre?

13 The sheet is rolled on a tube of radius 4.5 cm.
    The tube is made from a rectangle of card, with a 1.5 cm overlap allowed for gluing.

    What area of card is used for the tube?

14 Which do you think is better value for money: the 50 metre or the 30 metre roll of hose pipe?
    Explain your answer.

15 A hockey club needs a hose 180 metres long.
    a   List three different ways to buy 180 metres of hose in the July special offers.
    b   What is the cheapest way for the hockey club to buy the hose they need?
        Explain your answer.

16 Calculate the total area of glass used for the windows of the Super 9 Garden Store.

17 a   What shape is the back of the Garden Store?
    b   What are the dimensions of the back?

18 Calculate the area of one side of the store (including the door).

19 The front, back and both sides of the store have to be sprayed with timber preservative.
    Calculate the total area to be sprayed.

20 Preserver is sprayed at 250 ml per m².
    Will 5 litres of preserver cover one store?
    Explain your answer.

21 Calculate the size of the roof angle.

22 a   What shape is the roof of the store?

    One dimension of the roof is 5.7 metres (including overhangs).
    b   What is the other dimension of the roof?
    c   Calculate the area of the roof.

    The roofing used weighs 4.4 kg per m².
    d   What is the weight of the roof?

23 Labour charges are 40% of the price of the store which takes $12\frac{1}{2}$ hours to make.

    a   What is the labour charge for a store?
    b   Calculate the labour charge per hour.

    The door takes 45 minutes to make.
    c   What is the labour charge for a door?

24 For each mower, give the monthly payment.

Toujours Paris

Ile de la Cité

This boat-shaped island in the River Seine is where Paris was first inhabited by Celtic tribes over 2000 years ago. It is where Notre Dame is situated. This cathedral is a superb example of French medieval architecture and is particularly known for its wonderful stained glass rose windows.

The width of this South Window is 13 metres.

In the Ile de la Cité you will also find Point Zéro. This is a geometer's mark from which all distances in France are measured.

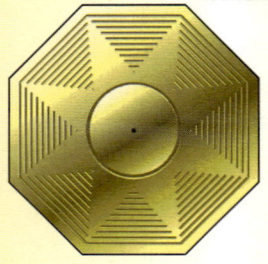

It measures 29 cm across and each side is 12 cm long.

Paris au Quotidien

Arrondissements: Il faut le savoir, Paris est divisé en 20 arrondissements se déroulant en spiraleà partir du 1er (le quartier du Louvre).

Banques: Ouvertes en général du lundi au vendredi de 9h à 16h30, quelques (rares) agences le samedi. Les Caisses d'Epargne ouvrent plus souvent le samedi et ferment le lundi.

Change: On ne peut pas tout prévoir à l'avance; vous pourrez changer vos devises dans les gares, les aéroports, les grandes agences de banque, les points change (ouverts tard le soir), ainsi qu'à notre bureau d'accueil des Champs-Elysées.

Daily Life in Paris

Districts: You should know that Paris is divided into 20 districts numbered in a circular direction, starting with the 1st district (the Louvre area).

Banks: They are generally open from Monday to Friday from 9 am to 4.30 pm, some (rare) branches on Saturday. The savings banks are more often open on Saturday and closed on Monday.

Exchange: You cannot foresee everything; you can therefore change your foreign currency in railway stations, airports, major bank branches, exchange offices (open late in the evening), as well as in the visitors office on the Champs-Elysées.

Eiffel Tower Factfile — 1996

- Built in 1889 for the Universal Exhibition
- Built by Gustave Eiffel (1832–1923)
- Built from pig iron girders
- Total height 320 metres
- The world's tallest building until 1931
- Height to 3rd level is 899 feet
- There are 1652 steps to the third level
- Two and a half million rivets were used
- The tower is 15 cm higher on a hot day.
- Its total weight is 10100 tonnes
- 40 tons of paint are used every 4 years
- On a clear day it is possible to see Chartres Cathedral, 72 km away to the South West.
- The tower is visited by about $5\frac{1}{2}$ million people every year

Admission charge   56 Francs

DAY TRIPS BY EUROSTAR
Waterloo Station (London) to Paris
Celebrate that special occasion in style with a day trip to Paris!
ADULT £
CHILD (4–11 YRS) £

Entry fees in Paris – 1996

Eiffel Tower	56 FF
Louvre	45 FF
Pompidou Centre	35 FF
Picasso Museum	28 FF
Museum of Modern Art	27 FF
Versailles Palace	45 FF
Parc de la Villette	45 FF
Cluny Museum	28 FF

Paris Lucky dip – Superb value !

In the hat are four tickets to the: Eiffel Tower, Picasso Museum, Versailles Palace and the Museum of Modern Art. Pick two tickets at random from the four. Entry fee – only 83 Francs.

83 FF

Datafile
Population of Paris (in 1982)   2 188 918
In 1996, £1 sterling was equivalent to 7.54 French Francs.
1 metre = 3.281 feet
1 kilometre = 0.62 miles

1 How many lines of symmetry has the South Window of Notre Dame?

2 What order of rotational symmetry has the South Window?

3 What is the circumference of the South Window?

4 Calculate the area of the South Window.

5 For Point Zéro in the Ile de la Cité, the outer polygon is regular.
   a What is the name of this polygon?
   b Calculate the size of an exterior angle.
   c Calculate the size of an interior angle.

6 Calculate the total area of Point Zéro.

7 How many planes of symmetry do you think the Eiffel Tower has?

8 How many tons of paint will have been used on the tower from when it was built up to the year 2000?

9 How much higher than the third level is the total height of the tower?
Give your answer to the nearest metre.

10 What is the mean height of a tower step in cm?

11 The ratio of steps to the third level, to steps to the first level is about 9 to 2.
About how many steps are there to the first level?

12 Use standard form to give:
   a the weight of the tower in tonnes
   b the number of rivets used.

13 Round the number of steps to the third level to:
   a the nearest ten
   b the nearest hundred
   c the nearest thousand.

14 Which one of the following gives the approximate percentage that the tower grows on a hot day?
   a 5%   **b** 0.5%   **c** 0.05%   **d** 0.005%

15 What is the approximate bearing of:
   a Chartres from Paris
   b Paris from Chartres?

16 In 1996, what was the Eiffel Tower entry fee equivalent to in £ Sterling?

17 **a** Approximately how much money in Francs was made from entry fees to the Eiffel Tower in 1996?
   b What was this equivalent to in £ Sterling?

18 Give the population of Paris in 1982 to:
   a 2 sf   **b** 3 sf   **c** 4 sf   **d** 5 sf.

19 Compare the French and English texts in the extract on Daily Life in Paris.
What is the relative frequency of a vowel in each language? (Vowels are a, e, i, o, and u.)

20 Compare the French and English extracts. Which language uses the longest words? Describe how you decided.

21 For the Paris Lucky Dip use the following shorthand:
   E – Eiffel Tower
   P – Picasso Museum
   V – Versailles Palace
   M – Museum of Modern Art

   a List each pair of tickets that could be picked at random.
   b What is the probability that a pair of tickets is picked which includes the Versailles Palace?
   c What is the probability of picking P and M?
   d Give the probability of picking a pair of tickets
     i worth more than the entry fee
     ii worth less than the entry fee.
   e Is the seller likely to make a profit or loss on every hundred entries? Give your reasons.

22 These two bills were for Eurostar day trips to Paris.

Enclosed are tickets for:
2 adults and 3 children
Total charge £465

Enclosed are tickets for:
3 adults and 1 child
Total charge £386

Calculate:
   a the total cost for these 5 adults and 4 children
   b the cost of an adult's ticket
   c a child's ticket
   d the total cost for 1 adult and 2 children.

Starting points
You need to know about ...

... so try these questions

A Units for measuring distance

♦ Kilometres (km), metres (m), centimetres (cm) and millimetres (mm) are **metric** units.

♦ Miles, yards, feet and inches are **imperial** units.

Metric	Imperial
1 km = 1000 m	1 mile = 1760 yards
1 m = 100 cm	1 yard = 3 feet
1 cm = 10 mm	1 foot = 12 inches

Some approximate conversions

Think of: 5 miles as 8 km
$1\frac{3}{4}$ pints as 1 litre
30.5 cm as 1 foot
2.2 lb as 1 kg

B Interpreting graphs

♦ Graphs that help you change from one unit to another are called **conversion graphs**.

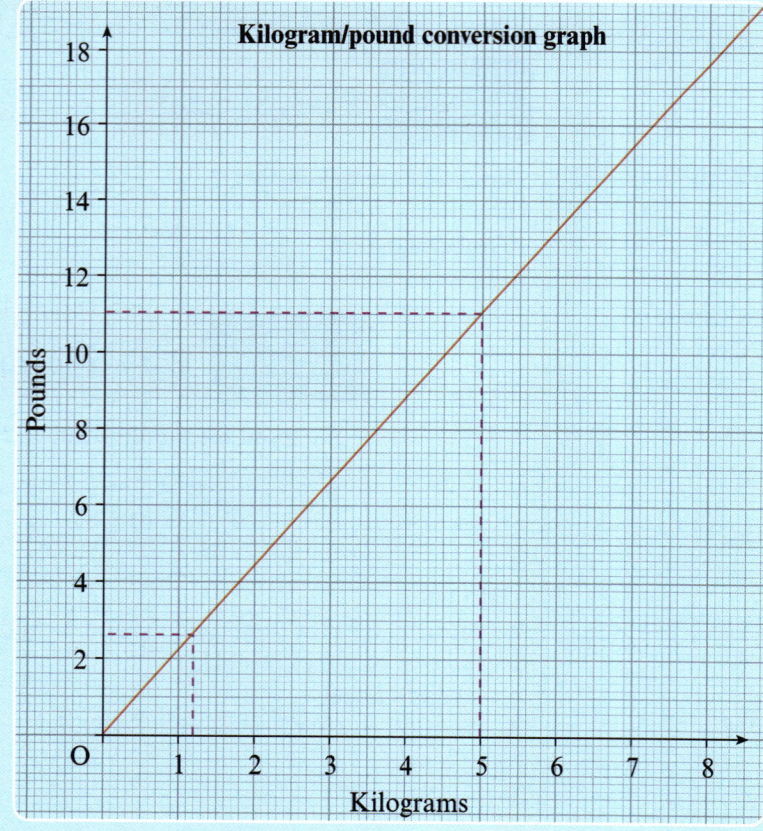

Examples   5 kilograms ≈ 11 pounds
2.6 pounds ≈ 1.2 kilograms

A1 Write these in metres.
 a 5 km   **b** 340 cm

A2 Write these in metres. Choose an appropriate degree of accuracy for each answer.
 a 3 miles   **b** 5 feet
 c 2 yards

A3 Write these in centimetres, correct to the nearest cm.
 a 40 mm   **b** 4 km
 c 3.9 m   **d** 7 inches

A4 Write these in kilometres, correct to 2 dp.
 a 5130 metres   **b** 7.1 miles

A5 Write 20 km in miles, correct to 1 dp.

B1 Use the graph to give an estimate of these in pounds, to 1 dp.
 a 7 kilograms
 b 3.8 kilograms

B2 Use the graph to give an estimate of these in kilograms, to 1 dp.
 a 16 pounds   **b** 5 pounds
 c 7.8 pounds   **d** 4.5 pounds

B3 Use the graph to estimate the number of pounds in 1 kilogram, to 1 dp.

B4 1 gallon ≈ 4.55 litres.
 a Use this approximation to copy and complete this table up to 8 gallons.

Gallons	1	2	3
Litres	4.55	9.1	

 b Make a conversion graph for gallons/litres up to 8 gallons.
 c Use your graph to give an estimate of:
   i 2.5 gallons in litres
   ii 15 litres in gallons.

C Calculating with time

- Times can be written:
  - in 12-hour time using am or pm
  - in 24-hour time.

  Examples   6:20 am is 06:20 in 24-hour time.
  6:20 pm is 18:20 in 24-hour time.

- Some units for measuring time are hours (h),
  minutes (min) and seconds (s):

  > 1 day   = 24 hours
  > 1 hour   = 60 minutes
  > 1 minute = 60 seconds

- Times can be written in different ways.

  Examples   250 minutes is 4 hours and 10 minutes.
  3 minutes and 12 seconds is 192 seconds.

D Timetables and time intervals

This shows part of a bus timetable from Paignton to Heathrow.

PAIGNTON	0625	0825	0955	1145	1345	1700
TORQUAY	0640	0840	1010	1205	1400	1715
Newton Abbot	0700	0900	1030	1225	1425	1735
EXETER	0730	0930	1100	1300	1500	1805
Taunton						1855
Calcot Coachway	↓	↓	1345	↓	1745	↓
HEATHROW AIRPORT	1105	1310	1440	1635	1840	2145

- Times for each different bus are shown in vertical columns.

  Example   The 0825 bus from Paignton stops at
  0840 in Torquay,
  0900 in Newton Abbot, …

- The arrows show that the bus does not stop.

  Example   The 0825 bus from Paignton does not stop at
  Taunton or Calcot Coachway.

This distance table shows distances in miles between four cities.

London

117	Birmingham		
159	76	Sheffield	
397	292	248	Glasgow

- To find the distance
  between two cities, read
  down and across.

Example

The distance from London
to Sheffield is 159 miles.

C1 Write these times using
am or pm.
a 09:30   **b** 14:21

C2 Write these in 24-hour time.
a 2:23 am   **b** 5:25 pm

C3 How many minutes are in
1 h and 15 min?

C4 Write 400 minutes in hours
and minutes.

D1 What is the latest time you
can catch a bus from Exeter
to reach Heathrow before
6:00 pm?

D2 In hours and minutes, calculate
how long each bus takes to go
from Paignton to Heathrow.

D3 Which bus takes the longest
time to travel from Torquay
to Exeter?

D4 After 3:00 pm, what is the time
of the first bus from Torquay
to Exeter?

D5 It takes David 8 hours to drive
from London to Glasgow.
About how long do you think
it would take him to drive
from London to Sheffield?

D6 Suki drives from Glasgow to
London via Birmingham.
She leaves at 2.15 pm and
arrives in Birmingham at
7.15 pm. She leaves
Birmingham at 7.45 pm.
About what time do you think
she will arrive in London?

Distance, time and constant speed

A **constant speed** is a steady speed.

An object travelling at a constant speed does not slow down or get faster.

At a constant speed of 9 metres per second, how far would a car travel in 20 seconds?

♦ In 1 second the car travels 9 metres.

♦ So in 20 seconds the car would travel 9 × 20 = 180 metres.

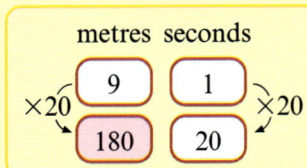

Distance = Speed × Time

To find the formula for speed:

Distance = Speed × Time

Now divide both sides by Time:

$$\frac{\text{Distance}}{\text{Time}} = \text{Speed}$$

At a constant speed, a car travels 8 kilometres in 5 minutes. How fast is it travelling?

♦ In 5 minutes the car travels 8 kilometres.
♦ In 1 minute the car travels 8 ÷ 5 = 1.6 kilometres.
♦ So the speed of the car is 1.6 kilometres per minute.

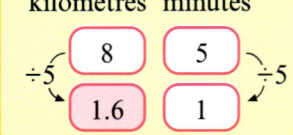

Speed = Distance ÷ Time

To find the formula for time:

Distance = Speed × Time

Now divide both sides by Speed:

$$\frac{\text{Distance}}{\text{Speed}} = \text{Time}$$

At a constant speed of 20 miles per hour, how long would it take for a car to travel 50 miles?

♦ To travel 20 miles takes 1 hour.

♦ So to travel 50 miles takes 50 ÷ 20 = 2.5 hours.

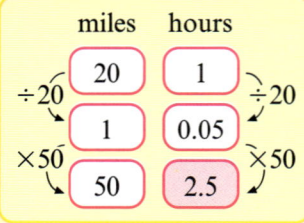

Time = Distance ÷ Speed

**Exercise 13.1
Calculating with
constant speeds**

Accuracy
In this exercise, give distances in metres, to 1 dp, and times to the nearest second.

1 At the Cairngorm Ski Area, you can go up and down the ski slopes by chairlift. The chairs on the White Lady Chairlift travel at a constant speed of 2.55 metres per second.

a How far would you travel on this chairlift in:
i 2 seconds   **ii** 10 seconds   **iii** 4.5 seconds?

b At this speed, how many metres would a chair travel in:
i 1 minute   **ii** 1 hour?

Chairs on the White Lady Chairlift travel 1054 metres to the top of the slope.

c How long does it take a chair to travel to the top of the slope:
i in seconds   **ii** in minutes and seconds?

2   An escalator is a set of stairs that moves at constant speed.

   a   Why do you think the speed of an escalator is constant?

This table shows data on some escalators in London Underground stations.

> m/s stands for metres per second.

Station	Normal speed (m/s)	Length (m)	No. of steps
Alperton	0.46	13.7	102
The Angel	0.75	60.0	318
Chancery Lane	0.60	9.1	84
Kentish Town	0.66	44.1	237

   b   Calculate how long it takes to travel up each escalator.

   c   At The Angel, 250 people travel up the escalator.
Each person stands on the step just below the person in front.
If the first person steps onto the escalator at 2.00 pm, when will the last person reach the top of the escalator?

> Assume that each person stands still on one step of the escalator.

For problems with hours and minutes, it is often easier to work in minutes.

Example

A plane travels 670 miles at a constant speed in 1 hour and 42 minutes.
Calculate its speed in miles per hour.

miles   minutes

÷102 670 102 ÷102
   6.56 ... 1
×60 394.1 ... 60 ×60

- Time taken is 60 + 42 = 102 minutes.
- Speed = Distance ÷ Time
  = 670 ÷ 102
  = 6.5686 ... miles per minute
- In 60 minutes (1 hour) the plane travels
  6.5686 ... × 60
  = 394.1176 ... miles
- So the speed is about 394.1 miles per hour.

**Exercise 13.2
Calculating with hours and minutes**

1   A plane travels 1300 miles at a constant speed in 2 hours and 30 minutes.
Calculate its speed in miles per hour. Do not use a calculator.

2   A student has tried to solve a problem that involves hours and minutes.

> **Accuracy**
> In this exercise, give all answers correct to 1 dp unless stated otherwise.

A plane travels 1115 miles at a constant speed in 1 hour and 50 minutes.
Calculate its speed in miles per hour.

Speed = distance ÷ time
= 1115 ÷ 1.50
≈ 743.3 miles per hour. ✗

   a   Explain the mistake you think she has made.

   b   Solve the student's problem to find the speed of the plane.

3   A Douglas DC-10 jet cruises at a constant speed of 620 miles per hour.

   a   Find this speed in miles per minute, correct to 2 dp.

   b   At this speed, how many miles would it travel in 10 minutes?

4   A Boeing 757 jet cruises at a constant speed of 403 mph.

   a   Find this speed in miles per minute, correct to 2 dp.

   b   How many minutes would it take to travel 120 miles at this speed?

> mph stands for miles per hour.

5   Concorde cruises at a constant speed of 1350 mph.

   a   At this speed, how long would it take to travel 800 miles?

   b   How far would it travel in 2 hours and 35 minutes?

Constant speeds and graphs

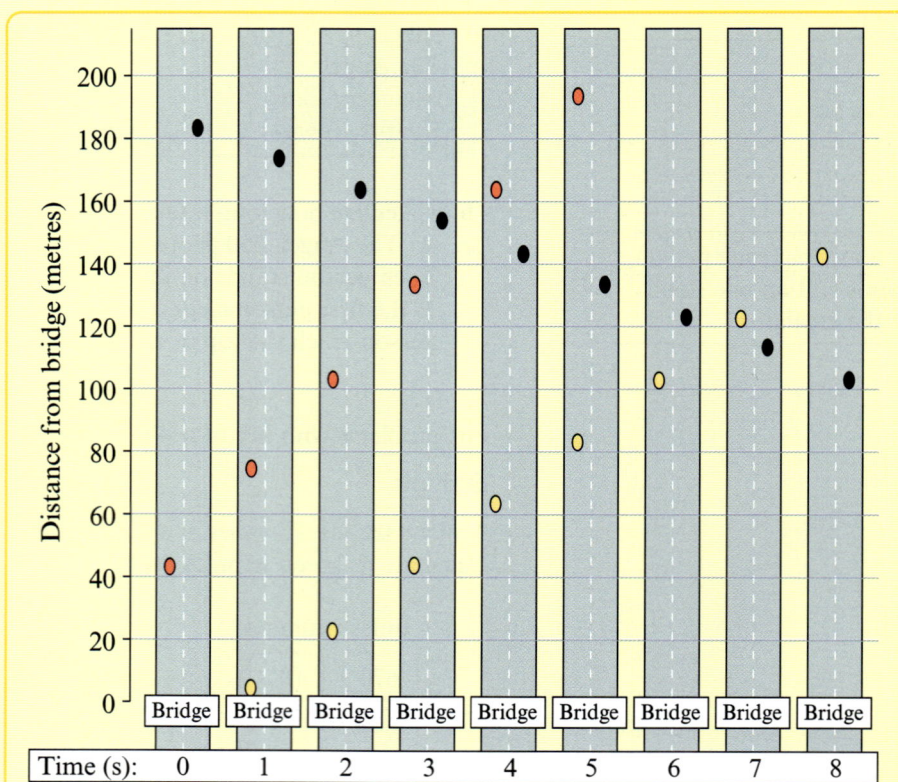

These diagrams show the positions of three cars on a road at one-second intervals. Each car is travelling at a constant speed.

The scale shows the distance along the road from the bridge in metres

This is part of a **distance–time graph** for the red car.

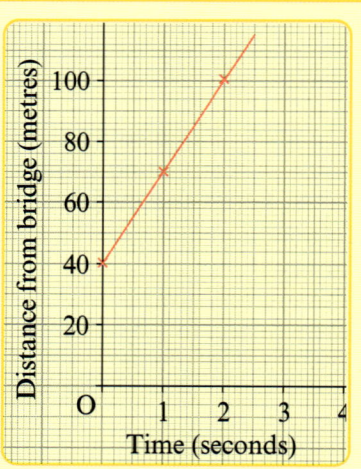

♦ The shortest distance is measured from each car to the bridge.

♦ The red car travels 30 metres every second.

It has a constant speed of 30 metres per second or 30 m/s.

Exercise 13.3
Distance-time graphs

Title your graph 'Distance–time graph for cars'.

For Question **1** refer to the panel above.

1 a i Draw a set of axes from:
  ♦ 0 to 8 seconds on the horizontal axis
  ♦ 0 to 200 metres on the vertical axis.
  ii Draw the complete distance–time graph for the red car.

b Use your graph to estimate the distance of the red car from the bridge after 4.5 seconds.

c After how many seconds was the red car 120 metres from the bridge?

d How far was the black car from the bridge after 6 seconds?

e Draw the graphs for the yellow car and the black car on your axes.
   f **i** Find the speeds of the yellow car and black car in metres per second.
      ii Which was travelling fastest: the red, yellow or black car?
      iii How can you tell this from your graph?
   g **i** After how many seconds did the red car pass the black car?
      ii Explain how you found your answer.

2 This distance–time graph shows part of the journeys of four cars on a motorway.

> The distances are measured from a speed camera.

> The colour of each line shows the colour of the car.

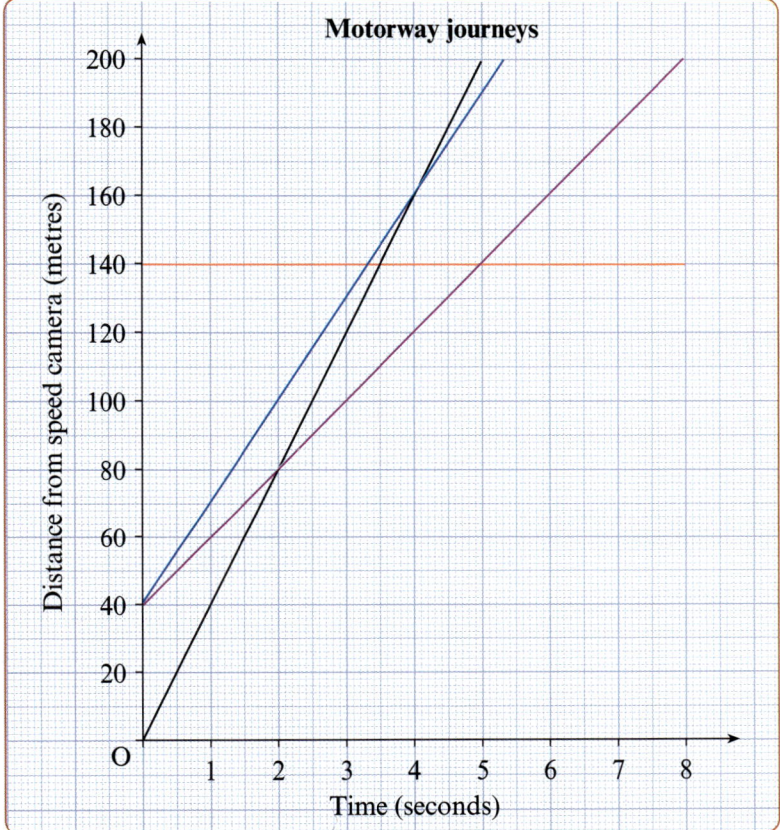

> Use tracing paper over the graph if you need to make marks to help you.

   a Estimate how far each car was from the speed camera after 2.5 seconds.
   b Which car was travelling fastest?
   c After how many seconds did:
      i the black car pass the blue car
      ii the blue car pass the red car?
   d Find the speed of each car in m/s.
   e What do you think happened to the red car?

3 This diagram shows the position of three cars and their speeds in m/s.
   Each car is travelling in the direction of the arrow at a constant speed.

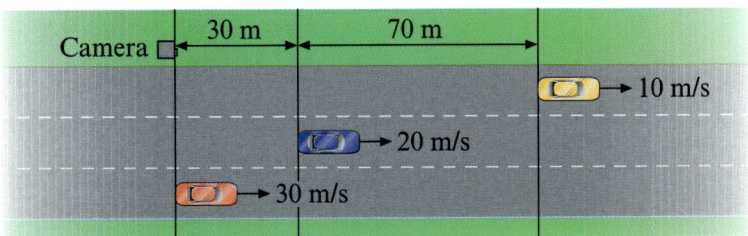

> Use a set of axes from:
> ◆ 0 to 10 seconds on the horizontal axis
> ◆ 0 to 300 metres on the vertical axis.

   a Draw a distance–time graph for the next 10 seconds.
   b How far is the red car from the camera when it passes the blue car?

Interpreting graphs

Exercise 13.4
Interpreting graphs

1 This is a sketch graph of Geeta's car journey to her workplace one morning.

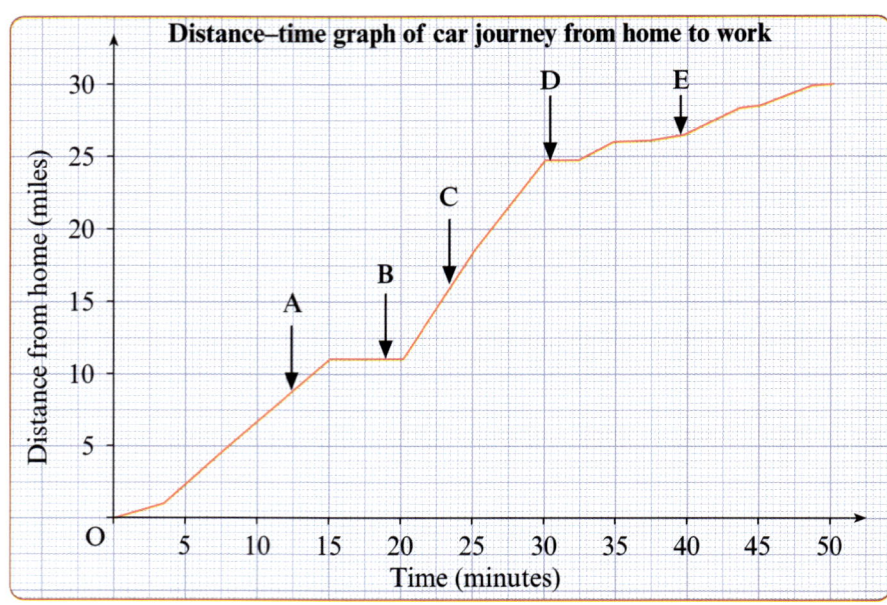

Distance–time graph of car journey from home to work

Use tracing paper over the graph if you need to make marks to help you.

a How far from Geeta's home is her workplace?
b How long did it take her to get to work?
c How far had Geeta travelled after 25 minutes?
d How long did it take her to travel:
   i the first 6 miles of her journey
   ii the last 6 miles of her journey?
e She stopped to buy some petrol.
   About how far from home do you think she was when she did this?
f Some points are labelled A to E on the graph.
   The statements 1 to 5 describe Geeta at different times on her journey.

① She is driving in a busy part of town.

② She is paying for her petrol.

③ She is travelling at a constant speed and sees her petrol is rather low.

④ She is travelling at the fastest speed of her whole journey.

⑤ She is waiting at traffic lights.

Match each statement to a labelled point.

g Sanjay describes points A to E on Geeta's journey.

At points A and C, Geeta was driving uphill. At points B and D, she was driving on level roads. At point E, she was driving on a bumpy road.

Explain what is wrong with this description.

2 Geeta described her journey to the gym one evening.

The gym is 10 miles away in a town. For the first 5 miles, I cycled at a fairly constant speed along country roads. I stopped for 5 minutes to buy some water and then cycled quickly for 3 miles to the edge of the town. I went on through the town to the gym. The journey took me 50 minutes.

Compare your sketch graph with someone else's.

Draw a sketch graph that could be for Geeta's journey to the gym.

Average speed

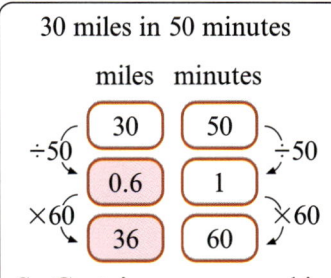

30 miles in 50 minutes

So Geeta's average speed is 36 miles per hour.

For a journey or part of a journey: **Average speed = Distance ÷ Time**

Example

Find Geeta's average speed for her car journey to work.

- She travelled 30 miles in 50 minutes.

- Average speed: Distance ÷ Time = 30 ÷ 50
  = 0.6 miles per minute.

- Her **average speed** for this journey is 0.6 miles per minute.
  In miles per hour this is: $0.6 \times 60 = 36$ mph.

Exercise 13.5
Average speed

1 Geeta cycles 10 miles to the gym in 50 minutes.
   Calculate her average speed for this journey:

   a in miles per minute   **b** in miles per hour.

2 Jason leaves home at 08.15 and drives 35 miles to his workplace.
   He arrives at 08.55.
   Calculate his average speed for this journey:

   a in miles per minute   **b** in miles per hour.

3 Geeta drives 373 miles from Bristol to Glasgow in 6 hours and 20 minutes.
   What is her average speed for this journey in mph, correct to 1 dp?

4 Jason cycles 50 miles from Perth to Braemar at an average speed of 12 mph.
   He leaves Perth at 9.30 am.
   When does he arrive at Braemar?

5 On long journeys, Geeta cycles at an average speed of 9 mph.
   She begins a long journey at 10.15 am.
   To the nearest mile, estimate how far she will have travelled by 12.00 noon.

6 This is part of a bus timetable.

Accuracy
In Question **6**, give all average speeds correct to the nearest whole number.

Taunton	0755	1000	1040	1205
Bridgwater	0815	1020	1100	1225
Burnham-on-Sea........	0845	↓		
Weston-super-Mare ..	0910	1050		↓
Bristol	1000	1140		1320

   a Each of these buses travels 11 miles from Taunton to Bridgwater.
      i How long do these buses take to travel from Taunton to Bridgwater?
      ii Calculate a bus's average speed between Taunton and Bridgwater.
   b From Taunton to Bristol:
      - the 0755 bus travels 53 miles
      - the 1000 bus travels 49 miles.
      Find the average speed in mph of each of these two buses.
   c The average speed of the 1205 bus between
      Taunton and Bristol is 36 mph.
      How far does it travel between Taunton and Bristol?
   d The 1040 bus goes to Heathrow Airport by a route 155 miles long.
      Between Taunton and Heathrow Airport its average speed is 50 mph.
      When does it arrive at Heathrow Airport?

Distance-time graphs and speed

This is a sketch graph for Jay's journey to college one morning.

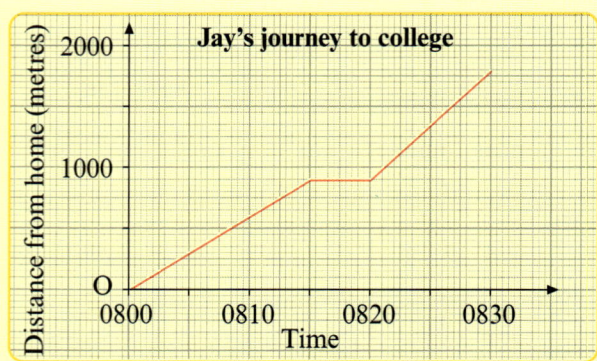

For different parts of Jay's journey, her average speed could be different.

For instance:

- She travels 900 metres in the first 15 minutes.
  So her average speed for the first 10 minutes is 900 ÷ 15 = 60 m/min.

- In total, she travels 1800 metres in 30 minutes.
  So her average speed for the whole journey is 900 ÷ 15 = 60 m/min.

> m/min stands for metres per minute.

Exercise 13.6
Graphs and average speed

1 Four students live in a block of flats and go to the same college.
The sketch graph shows each student's journey to college one morning.

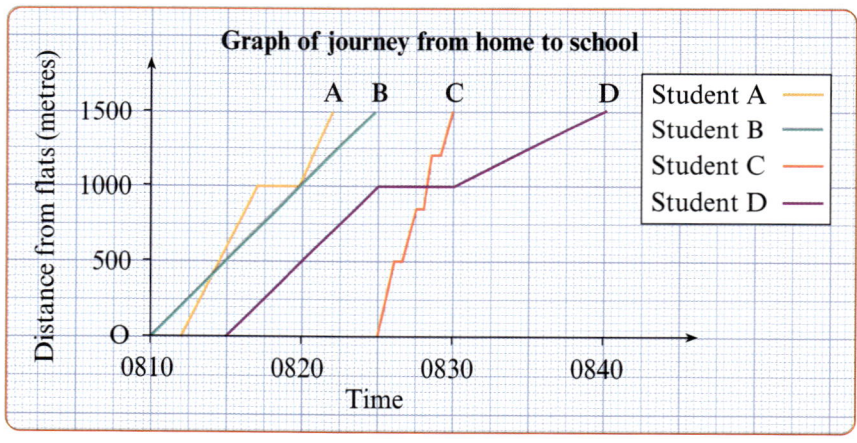

a How far is the journey from the flats to the college?
b Which student arrived at college first?
c How long did student A take to travel to college?
d The fire alarm was set off at 0822.
What distance had each student travelled at 0822?
e Find the average speed for student C's journey.
f The names of the four students are Jane, Lorna, Jason and Sharon.
Lorna was buying crisps in the corner shop when she saw Jason go by in a car. Jane walked to college. Sharon cycled past Jane on the way.
i Match each student (A, B, C and D) to Jane, Lorna, Sharon and Jason.
Explain how you decided.
ii How far is the corner shop from the college?
iii How do you think Lorna travelled to college?
Explain your answer.
g Calculate each student's average speed for the whole journey in:
i m/min ii m/h iii km/h.

> m/h stands for metres per hour.
>
> km/h stands for kilometres per hour.

Accuracy
In Question **2**, give all average speeds correct to the nearest whole number.

2 This sketch graph shows the journey of a car and a motorbike.
The car travelled from Leeds to Doncaster.
The motorbike took the same route to Doncaster but returned to Leeds.

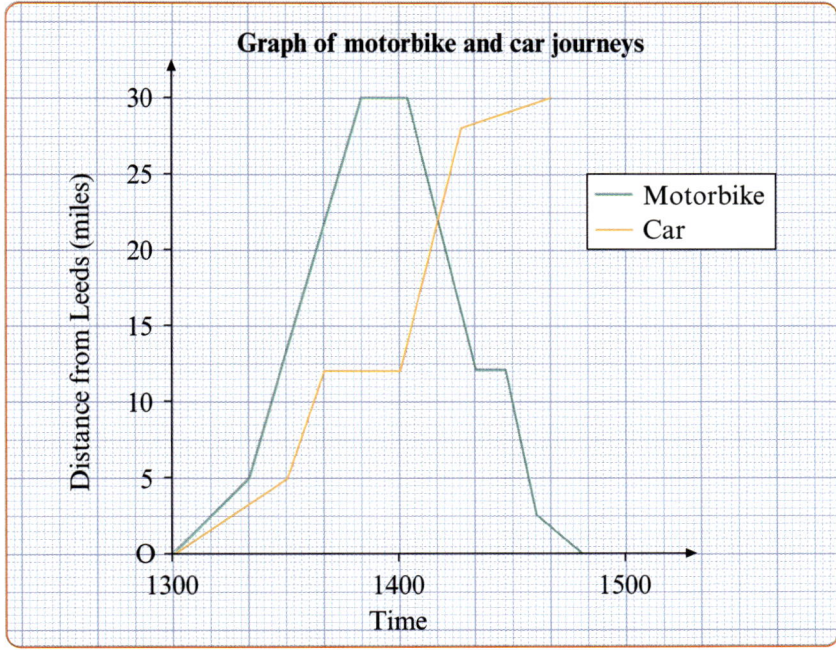

a **i** When did the car stop?
 ii For how many minutes did it stop?
b **i** Between what times was the car travelling fastest?
 ii What was its speed between these times?
c Describe what happened at 1410.
d Describe the motorbike's whole journey as fully as you can.
e **i** What was the total distance travelled by each vehicle on its journey?
 ii Calculate each vehicle's average speed for its whole journey in mph.

Exercise 13.7
Drawing and interpreting graphs

1 Ken and his sister Amy go on a cycling holiday.
The graph shows part of Ken's journey for the first day.

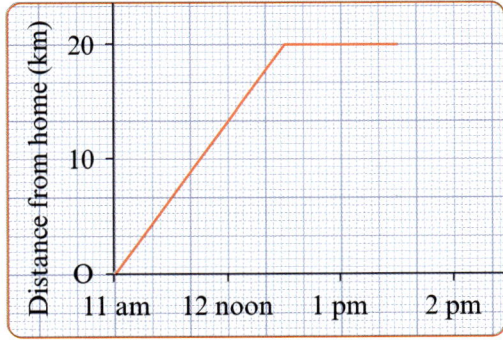

Accuracy
In this exercise, give all average speeds to 1 dp.

a After 1.30 pm he cycles at 15 km/h to reach a camp-site that is 50 km from home.
 i Copy and complete the graph for Ken's first day.
 ii Find Ken's average speed for his whole journey.
b Amy leaves their home at 1 pm and cycles along the same route at 22 km/h.
 i Draw a line on your graph to show Amy's journey.
 ii When did she reach the camp-site?
 iii At about what time did Amy pass Ken on her way to the camp-site?

2 Alice lives in Cupar and Emily lives 10 miles away in St Andrews.
One day Alice cycles to St Andrews.

This graph shows part of her journey.

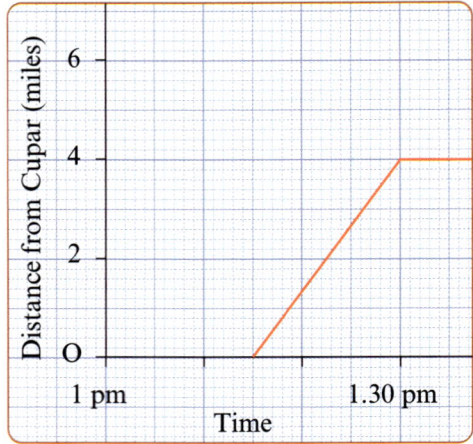

a At 1.30 pm, she stops for 10 minutes to buy cakes and then cycles
at a constant speed to reach Emily's house at 2.20 pm.

  i Copy and complete the graph for Alice's journey.
  ii What is her speed in mph after her stop to buy cakes?
  iii Calculate Alice's average speed in mph for her whole journey.

b On the same day, Emily decides to visit Alice.
She leaves at 1 pm and cycles at a speed of 12 mph for 4 miles.
She stops for five minutes and then carries on at a constant speed
to reach Alice's house at 1.47 pm.

  i How far is Emily from Cupar at the start of her journey?
  ii Show her journey on your graph.
  iii Find Emily's average speed in mph for her journey to Cupar.
  iv Explain why you think Emily did not see Alice on her journey.

c After ringing Alice's door bell for 3 minutes, Emily decides to cycle
home. She reaches St Andrews at the same time as Alice does.

  i Show Emily's journey home to St Andrews on your graph.
  ii How fast does she cycle home?

3 Sharon gave this information about a car journey.

> ● I travelled on motorways.
> ● My average speed for the whole journey was 50 mph.
> ● I set off at 9 am and arrived at 1.45 pm.
> ● I drove to my destination and did not make
>   a return journey.
> ● I made two stops: 30 minutes for coffee and
>   40 minutes for lunch.
> ● I did not break the motorway speed limit of 70 mph.

a Draw a travel graph that fits this car journey.
b Compare your graph with someone else's graph for this journey.

Compound measures

The rate at which a vehicle uses fuel is called the **fuel consumption**.
It is often given in **miles per gallon** (mpg).

Example

What is the fuel consumption of a car that uses
2.5 gallons of petrol to travel 118 miles?

118 ÷ 2.5 = 47.2
So fuel consumption is 47.2 mpg (miles per gallon).

Exercise 13.8
Compound measures

Accuracy
In this exercise, choose an
appropriate degree of
accuracy for each answer.

1

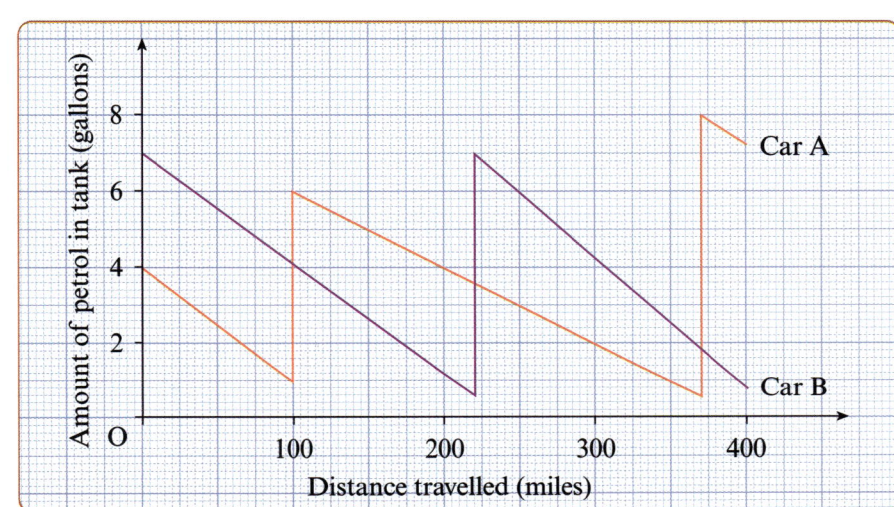

a   How much petrol was in each car's tank after 120 miles?
b   How far did car B travel on its first gallon of petrol?
c   How many times did car A stop for petrol?
d   How much petrol did car B buy on the journey?
e   How much petrol did each car use in the first 100 miles?
f   i   How much petrol did each car use over the whole journey?
    ii What was the fuel consumption of each car on this journey?

2   Here is some information about Britain in 1993.

Number of households	22 500 000
Number of cars	20 102 000
Total length of motorways in km	3141
Total length of all roads in km	364 477

For these figures:

a   What was the average number of cars per household?
b   If all the cars were on the road at the same time, how many cars would
    there be per kilometre of road?
c   Would it be possible to fit all the cars on the motorways
    at the same time? Explain your answer.

A car is about 4 m long.
Most motorways have
6 lanes.

3   In 1963 there were 7 479 000 cars and 16 700 000 households.

a   What was the average number of cars per household in 1963?
b   Describe how car ownership changed between 1963 and 1993.

End points

You should be able to ...

... so try these questions

A Calculate with distance, speed and time

A1 At the Cairngorm Ski Area, chairs on the Car Park Chairlift travel at a constant speed of 2.65 metres per second.
How far would you travel on the chairlift in 1 minute and 25 seconds?

A2 On the Car Park Chairlift, a chair travels 766 metres to the top of the ski-slope.
How long does it take a chair to travel to the top of the slope in minutes and seconds?

A3 Jane drives 76 miles from Birmingham to Sheffield in 1 hour and 40 minutes.
What is her average speed for this journey in mph, correct to 1 dp?

A4 Morag drives 159 miles from London to Sheffield at an average speed of 51 mph.
She leaves London at 13.25.
When does she arrive in Sheffield?

B Interpret distance–time graphs

B1 This sketch graph shows the journey of a car and a cycle.
The cyclist travelled from Leeds to York.
The car driver took the same route to York but returned to Leeds.

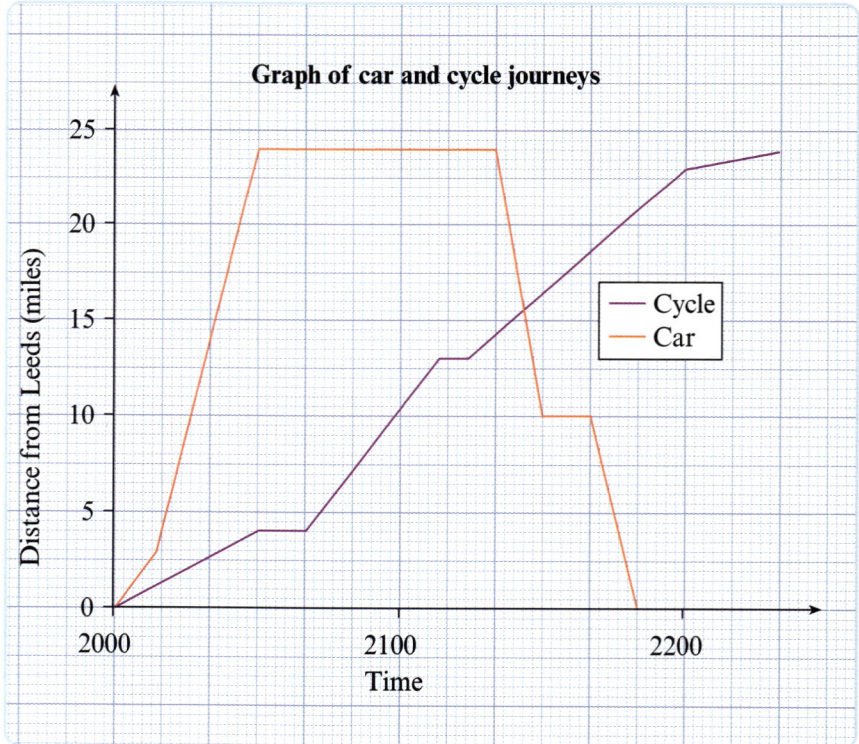

a How far did the cyclist travel in the first 20 minutes?
b How long did it take the car to travel the first 20 miles?
c How many times did the cyclist stop on her journey to York?
d What was the cyclist's average speed for this journey in mph correct to 1 dp?

e i Between which two times was the cyclist travelling fastest?
   ii Explain how you can tell from the graph.
f How long did the car driver spend in York?
g What was the speed of the car for the last 10 miles of its return journey?
h i At what time did the car and cycle pass each other?
   ii How far were they from Leeds when this happened?

C Draw distance–time graphs

C1 Liz and Pete cycle 17 miles from Keswick to Ambleside.
The graph shows part of Liz's journey.

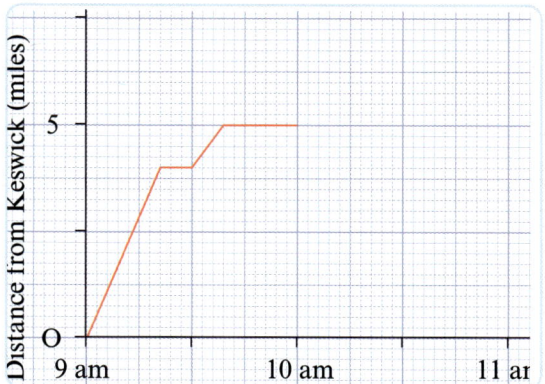

a After 10 am, she cycles at a constant speed of 10 mph to reach Ambleside.
Copy and complete the graph for Liz's journey.

b Find Liz's average speed for her whole journey in mph, correct to 1 dp.
c Pete leaves Keswick at 10 am. Apart from a 10 minute stop at 11 am, he cycles at a speed of 14 mph for the whole journey.
Show Pete's journey on your graph.

D Calculate with compound measures

D1 The White Lady Chairlift can carry 600 people per hour up the ski-slope.
At this rate, how many people would it carry up the slope between 9:30 am and 2:30 pm?

D2 Water flows out of a pipe at a constant rate.
In 30 minutes, 924 litres flow out of the pipe.
In litres per second, correct to 2 dp, at what rate is water flowing out of the pipe?

D3 In 1993, people in the UK spent a total of £113 500 000 on suncare products. The total population was 58 205 000.
Find the amount spent per person on suncare products in 1993.

Some points to remember

- Some rules for working with distance, speed and time:

  Distance = Speed × Time
  Time = Distance ÷ Speed
  Speed = Distance ÷ Time

  This triangle might help:

- Average speed = Total distance ÷ Total time

- For problems involving hours and minutes, it is often easier to work in minutes.

Starting points
You need to know about ...

... so try these questions

A Naming shapes, sides and angles

◆ The vertices of a shape can be labelled with letters.
The sides and angles can be named using these letters.

Example

❖ This is ΔPQR (triangle PQR).

❖ The **sides** of PQR are PQ, QR and PR.
PQ is also used for the length
of the line PQ.

❖ The **angles** in PQR are:
PQ̂R (or ∠ PQR or Q̂),
QR̂P (or ∠ QRP or R̂), and
RP̂Q (or ∠ RPQ or P̂).

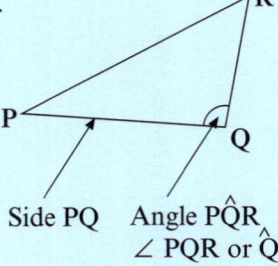

Side PQ   Angle PQ̂R
          ∠ PQR or Q̂

A1

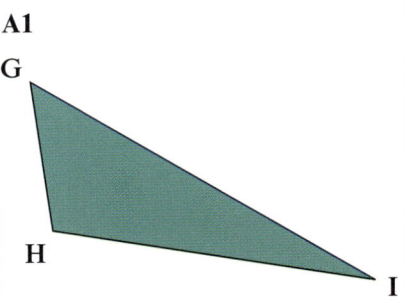

In triangle GHI:
a which side is the longest?
b which angle is the largest?
c which angles are acute?
d which is the shortest side?

B Similar triangles

◆ When two shapes are similar:
  ❖ the corresponding angles are equal
  ❖ the lengths of the corresponding sides
    are in the same ratio.

Two triangles are similar if either:

  ❖ corresponding angles are equal, or
  ❖ the lengths of corresponding sides
    are in the same ratio.

So to identify similar triangles you only need to check
one of these properties.

◆ This is one way to identify **corresponding sides** in a
pair of similar triangles.

In triangles ABC and PQR
AC and PR are
corresponding sides
because they are
opposite equal angles.

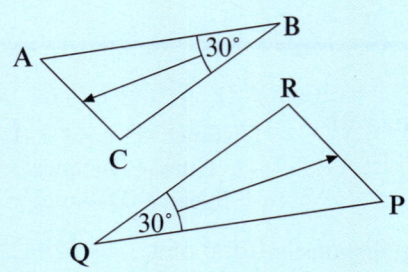

B1

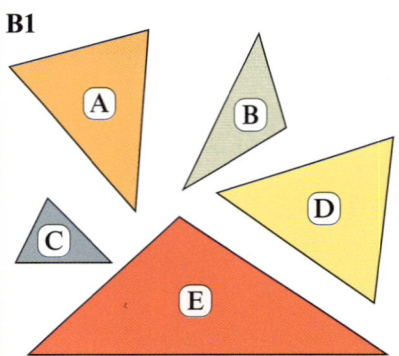

Using triangles A to E find
two pairs of similar triangles.

B2 ΔEFG and ΔXYZ are similar.

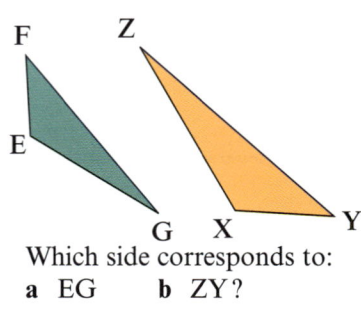

Which side corresponds to:
a EG   **b** ZY?

Thinking ahead to ...
similar triangles

A   Triangles ABC and PQR are similar.

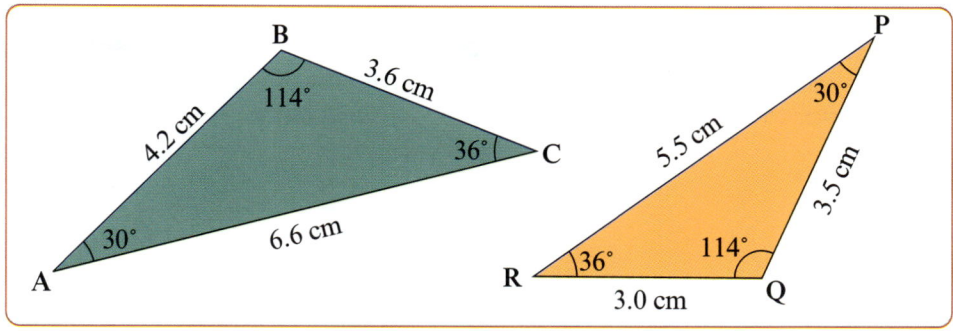

a   In triangle ABC, the shortest side is 3.6 cm and the longest side is 6.6 cm.

Check that the ratio $\dfrac{\text{Shortest side}}{\text{Longest side}}$ equals 0.55 (to 2 dp).

b   For triangle PQR calculate the ratio $\dfrac{\text{Shortest side}}{\text{Longest side}}$ (to 2 dp).

c   What do you notice?

B   **a**   Calculate the ratio $\dfrac{\text{Longest side}}{\text{Shortest side}}$ (to 2 dp) for triangles ABC and PQR.

b   What do you notice?

Similar triangles

> Triangles will only be
> similar if:
> ◆ all corresponding angles
>   are equal
> ◆ any pair of corresponding
>   ratios are equal

♦ In any triangle you can compare the length of two sides using a ratio.

Example

In triangle KLM:

$$\frac{KL}{ML} = \frac{1.9}{2.0} = 0.85$$

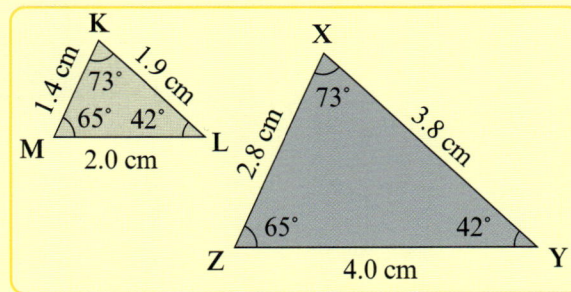

In triangles KLM and XYZ:

KM corresponds to XZ and ML corresponds to ZY.

So $\dfrac{KM}{ML}$ and $\dfrac{XZ}{ZY}$ can be called **corresponding ratios**.

$$\frac{KM}{ML} = \frac{1.4}{2.0} = 0.7 \quad \text{and} \quad \frac{XZ}{ZY} = \frac{2.8}{4.0} = 0.7 \qquad \text{So} \quad \frac{KM}{ML} = \frac{XZ}{ZY}$$

♦ For two similar triangles, **any** pair of corresponding ratios are equal.

So in triangles KLM and XYZ:

$\dfrac{ML}{KM}$ and $\dfrac{ZY}{XZ}$ are corresponding ratios, so $\dfrac{ML}{KM} = \dfrac{ZY}{XZ}$

$\dfrac{KL}{KM}$ and $\dfrac{XY}{XZ}$ are corresponding ratios, so $\dfrac{KL}{KM} = \dfrac{XY}{XZ}$

Exercise 14.1
Ratios of sides in
similar triangles

Accuracy
For this exercise round
each answer to 2 dp.

It may help if you redraw
the triangles in the same
orientation.

1   Triangles ABC, FDE and IHG are similar.

a   i   List the ratios that are equal
        to $\dfrac{GH}{GI}$

    ii Check by calculating the value
        of these ratios.

b   i   List the ratios that are equal
        to $\dfrac{GI}{GH}$

    ii Check by calculating the value
        of these ratios.

c   List other sets of corresponding ratios
    in triangles ABC, FDE and IHG.
    Find the value for each set.

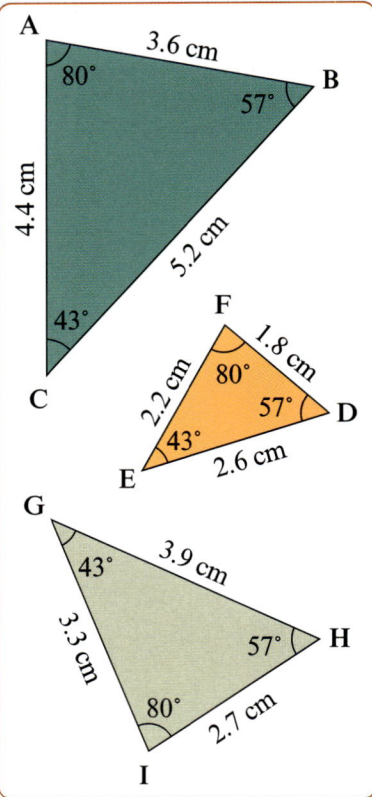

2

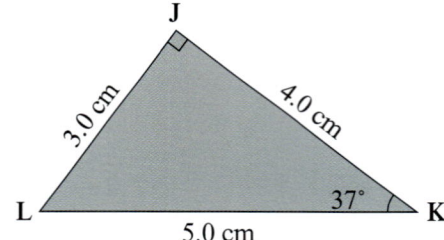

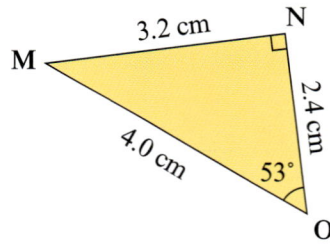

a   Explain why triangles JKL and NMO are similar.
b   Find the value of all the corresponding ratios in triangles JKL and NMO.

3   For two similar triangles, how many pairs of corresponding ratios are there?

Right-angled triangles

♦ The sides of a right-angled triangle can be labelled like this:

   the hypotenuse (the longest side)   hyp
   the side opposite x opp
   the side adjacent to (next to) x adj

   The **hypotenuse** is the side opposite
   the right angle.

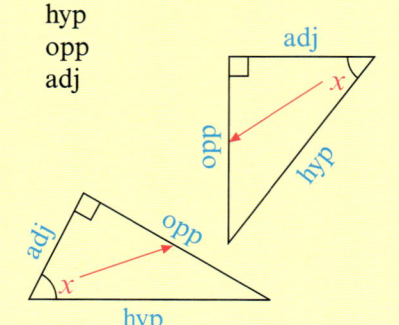

Exercise 14.2
Ratios of sides in
right-angled triangles

1 Triangle A is a 30° right-angled triangle.

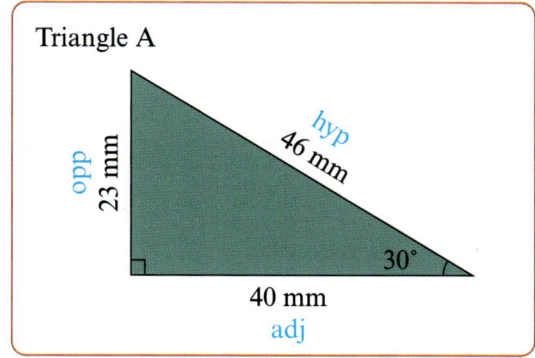

Triangle A

opp
23 mm

hyp
46 mm

40 mm
adj

30°

Make a ratio table like this and fill in the values for triangle A.

Sides labelled from an angle of	30°								
Length of sides (mm)			Ratios (to 2 dp)					
adj	opp	hyp	$\dfrac{\text{adj}}{\text{hyp}}$	$\dfrac{\text{adj}}{\text{opp}}$	$\dfrac{\text{opp}}{\text{hyp}}$	$\dfrac{\text{opp}}{\text{adj}}$	$\dfrac{\text{hyp}}{\text{adj}}$	$\dfrac{\text{hyp}}{\text{opp}}$
A 40	23	46	0.87					

2 Draw some other 30° right-angled triangles, and for each triangle:
  a label the sides from the 30° angle as hyp, opp or adj
  b measure the length of each side to the nearest millimetre
  c calculate the ratios for each pair of sides and fill in your ratio table.

Draw your triangles on
squared paper.

3 Describe anything you notice about your ratios.

4 **a** What do you think will happen to the ratios for
        other right-angled triangles?
  b Test your ideas on other right-angled triangles.

Trigonometric ratios

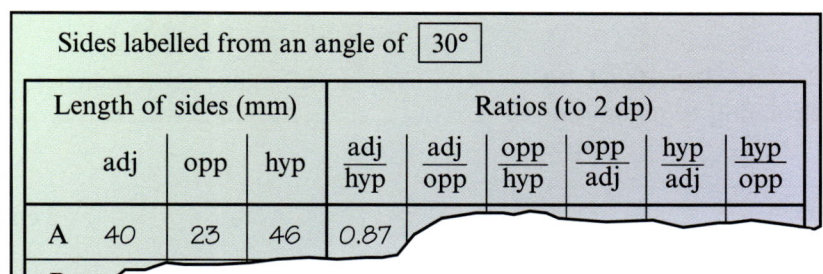

◆ The ratios of the sides in a right-angled triangle are
called **trigonometric ratios**.

In triangle ABC,

sine: sin 30°   = 0.50
cosine: cos 30°   = 0.87
tangent: tan 30°   = 0.58
               (to 2 dp)

B

opp   hyp

A   adj   30°   C

Exercise 14.3
Trigonometric ratios

Your ratios may not match
the values of sin 30°,
cos 30° and tan 30° exactly.

1 **a** Find these keys on your calculator: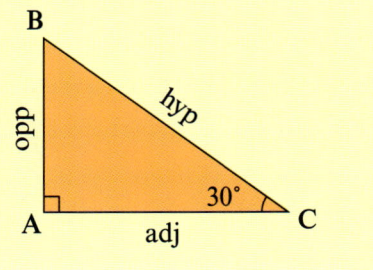
  b Use them to find the values of sin 30°, cos 30° and tan 30°.
  c Check the values match those given above.
  d Match each value to one of your ratios for the 30° triangles
        that you have drawn.

2 **a** How can you calculate the sine, cosine and tangent of an angle?
  b Check your ideas on the triangles you drew in Exercise 14.2.

Sine, cosine and tangent of an angle

A mnemonic (pronounced *ne-mon-ik*) is a memory aid.

This is a mnemonic used to remember the definitions for the sine, cosine and tangent of an angle:

Skive off homework
Cheat at homework
Telling off after

♦ These are definitions of three trigonometric ratios (or 'trig' ratios):

sine: $\sin x = \dfrac{\text{opp}}{\text{hyp}}$

cosine: $\cos x = \dfrac{\text{adj}}{\text{hyp}}$

tangent: $\tan x = \dfrac{\text{opp}}{\text{adj}}$

These trig ratios can be used to calculate any side or any angle in **right-angled triangles**.

Exercise 14.4
Calculating trigonometric ratios

Accuracy
For this exercise round each answers to 2 dp.

1 These are all right-angled triangles.

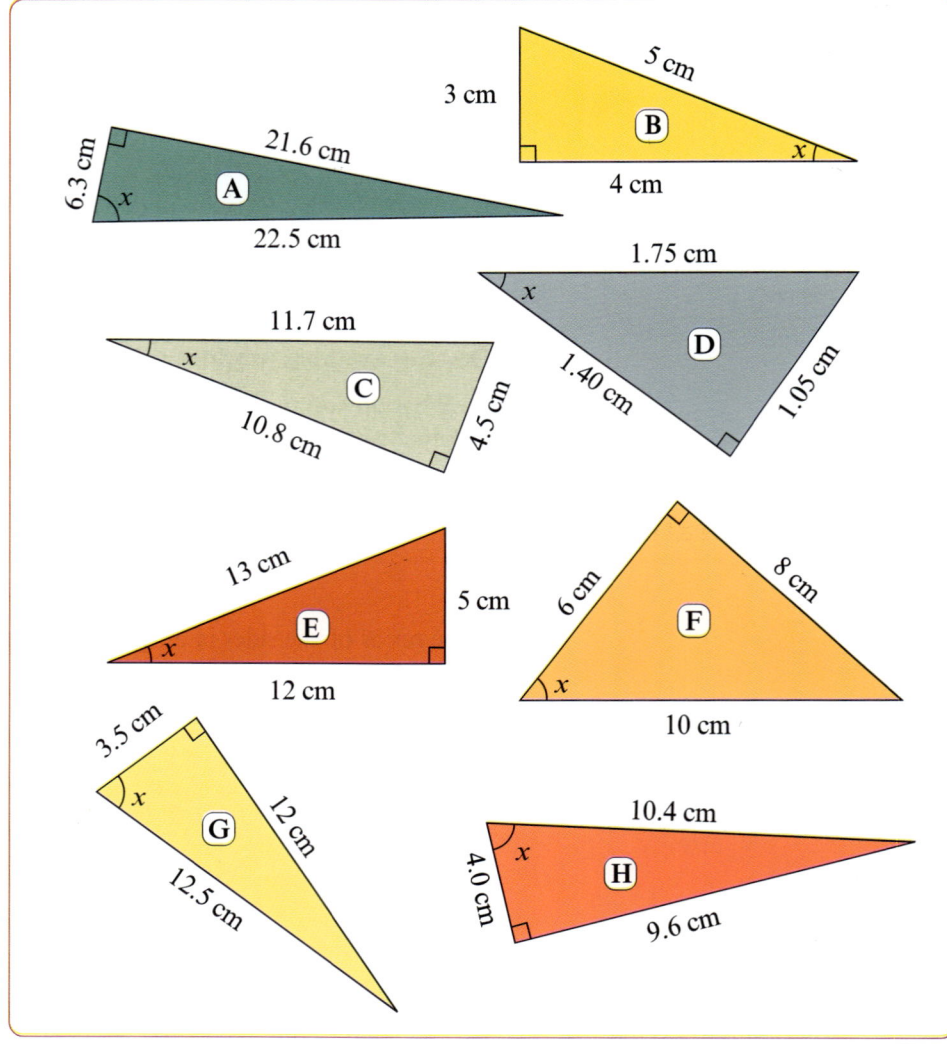

a Sketch each of these triangles and label the sides: opp, adj, hyp.

b For each triangle, find the value of sin x, cos x and tan x.
 i Give each value as a fraction
 ii Give each value as a decimal.

Finding a side

♦ Each trig ratio can be written in different ways.

Starting with the ratio $\sin x = \dfrac{\text{opp}}{\text{hyp}}$

we can write: $\dfrac{\text{opp}}{\text{hyp}} = \sin x$

Multiply both sides by hyp: $\times$ hyp $\qquad\qquad$ $\times$ hyp

$$\text{opp} = \text{hyp} \times \sin x$$

So $\sin x = \dfrac{\text{opp}}{\text{hyp}}$ gives: $\text{opp} = \text{hyp} \times \sin x$

In the same way, $\cos x = \dfrac{\text{adj}}{\text{hyp}}$ gives: $\text{adj} = \text{hyp} \times \cos x$

and $\tan x = \dfrac{\text{opp}}{\text{adj}}$ gives: $\text{opp} = \text{adj} \times \tan x$

♦ The length of any side in a right-angled triangle can be calculated using trigonometry.

Example In ∆ABC calculate the length of AC.

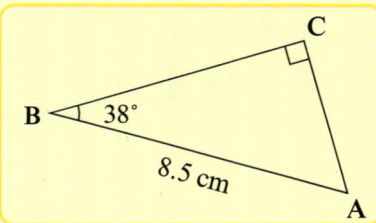

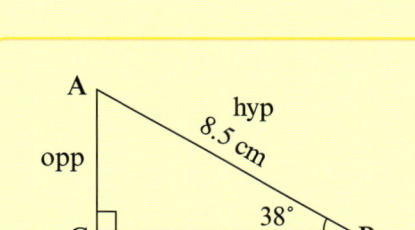

♦ **know:** AB (hyp)
        angle
♦ **find:** AC (opp)
♦ **use:** sine

$\text{opp} = \text{hyp} \times \sin x$
$\text{opp} = 8.5 \times \sin 38°$
$\text{opp} = 5.2$ (to 1 dp)

So the length of AC is 5.2 cm (to 1dp)

Use the full calculator value for sin 38° and round at the end of the calculation.

On some calculators this set of key presses can be used to calculate $8.5 \times \sin 38°$

Exercise 14.5
Finding lengths in right-angled triangles

Accuracy
For this exercise round each answer to 1 dp.

1 List the key presses you use to calculate $8.5 \times \sin 38°$ on **your** calculator.

2 Sketch each of these triangles and calculate the length of the blue side.

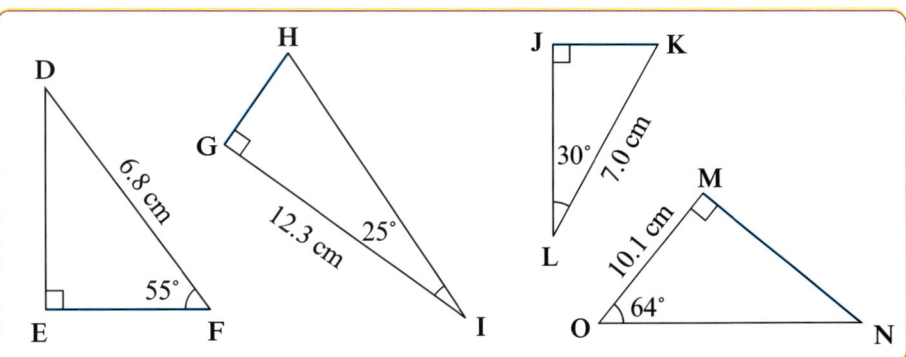

3 This is part of a calculation to find BC in △ABC.

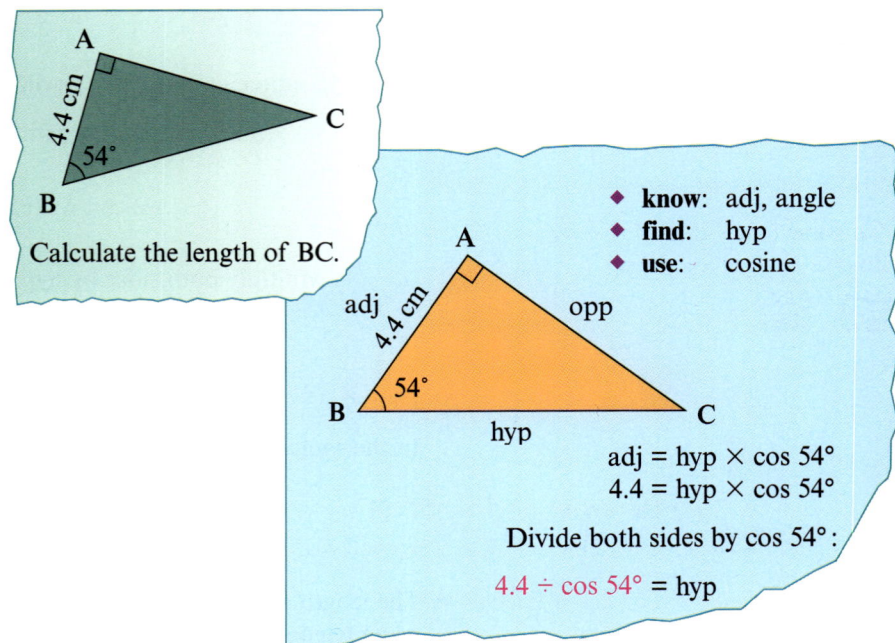

Calculate the length of BC.

♦ **know**: adj, angle
♦ **find**: hyp
♦ **use**: cosine

adj = hyp × cos 54°
4.4 = hyp × cos 54°

Divide both sides by cos 54°:

4.4 ÷ cos 54° = hyp

Use the full calculator value for cos 54° and round at the end of the calculation.

On some calculators this set of key presses can be used to calculate 4.4 ÷ cos 54°

a For **your** calculator, list the key presses to calculate 4.4 ÷ cos 54°
b What is the length of BC?

4

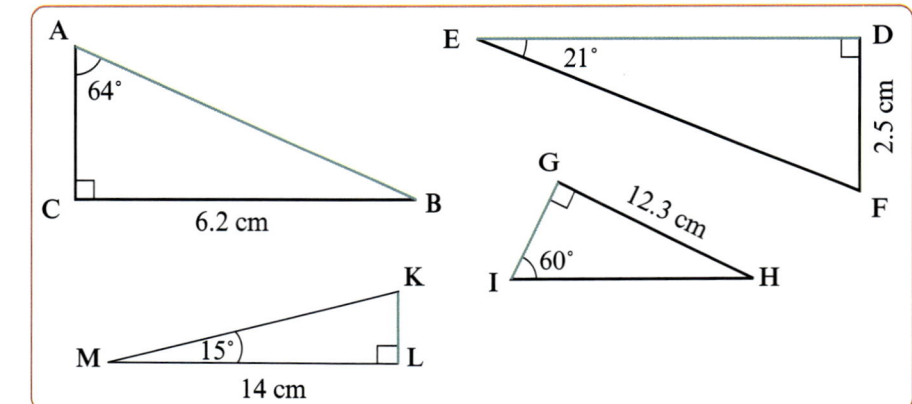

Calculate the length of AB, DE, KL and GI in these triangles.

5

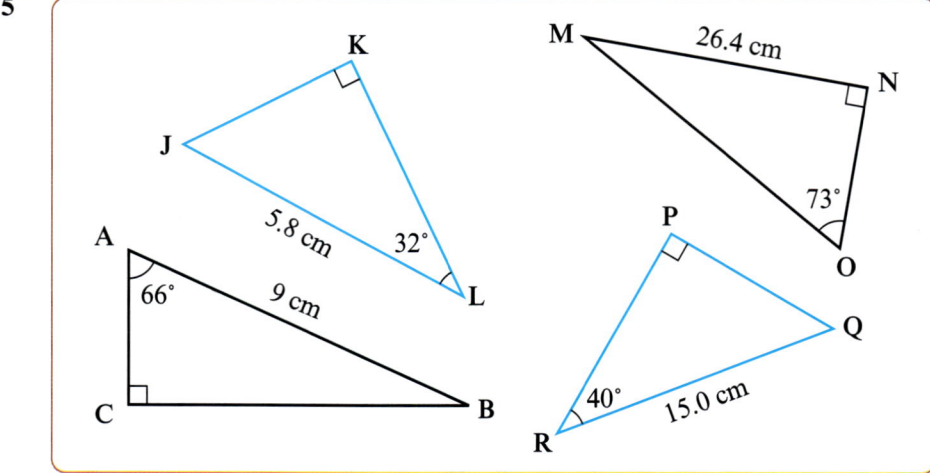

Calculate the length of JK, NO, AC and PR in these triangles.

Finding an angle

These are some pairs of
inverse operations:

to add / to subtract
to square / to square-root

On some calculators the
inverse button is marked

  or

♦ When you know the value of one trigonometric ratio for an angle, you
can use the inverse function on a calculator to find the size of the angle.

Example

Find x when $\sin x = 0.375$
On some calculators, these are the key presses.

$x = \mathbf{22°}$ (to the nearest degree)

Exercise 14.6
Calculating angles

Accuracy
For this exercise round each
angle to the nearest degree.

1 For your calculator, list the key presses to find x when $\sin x = 0.375$

2 Find the angle x when:
  a $\sin x = 0.3584$ **b** $\tan x = 1.0256$ **c** $\cos x = 0.351$

3 Find the angle x when:
  a $\tan x = \dfrac{5}{13}$ **b** $\cos x = \dfrac{3.5}{8.16}$ **c** $\tan x = \dfrac{8.9}{5.4}$

Do not round the value before using the inverse operation.

When you use
trigonometric ratios to
solve problems it is useful
to write down:

♦ what you know
♦ what you are trying to find
♦ what ratio you need to use.

4 This is part of a calculation to find x in triangle ABC.

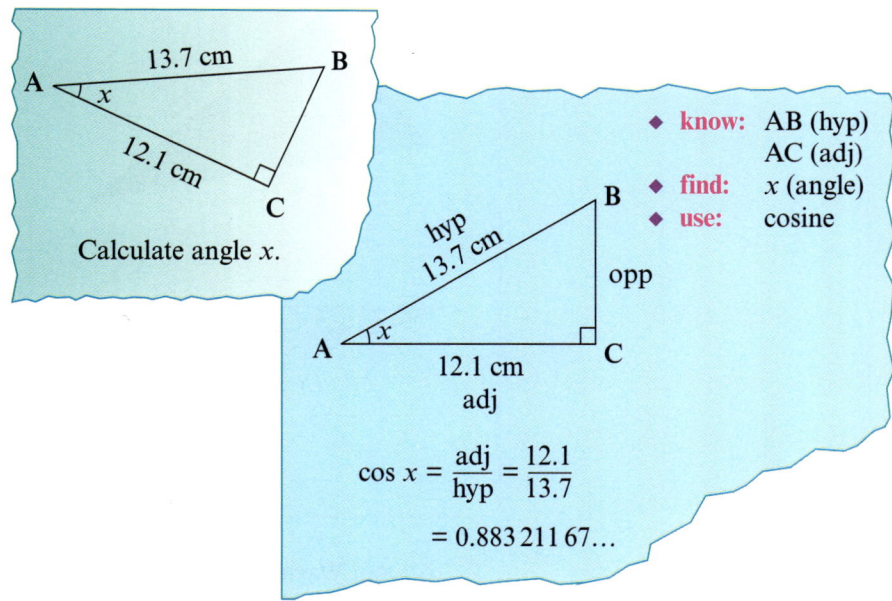

Calculate angle x.

♦ **know:** AB (hyp)
　　　　　AC (adj)
♦ **find:** x (angle)
♦ **use:** cosine

$$\cos x = \frac{\text{adj}}{\text{hyp}} = \frac{12.1}{13.7}$$

$$= 0.883\,211\,67\ldots$$

What is the size of angle x?

5

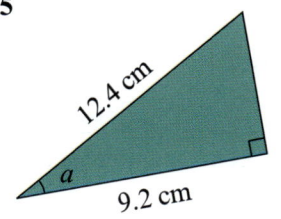

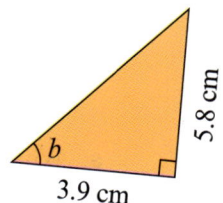

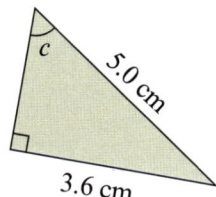

Calculate the size of angles a, b and c.

6 For each triangle, calculate the size of the lettered angle.

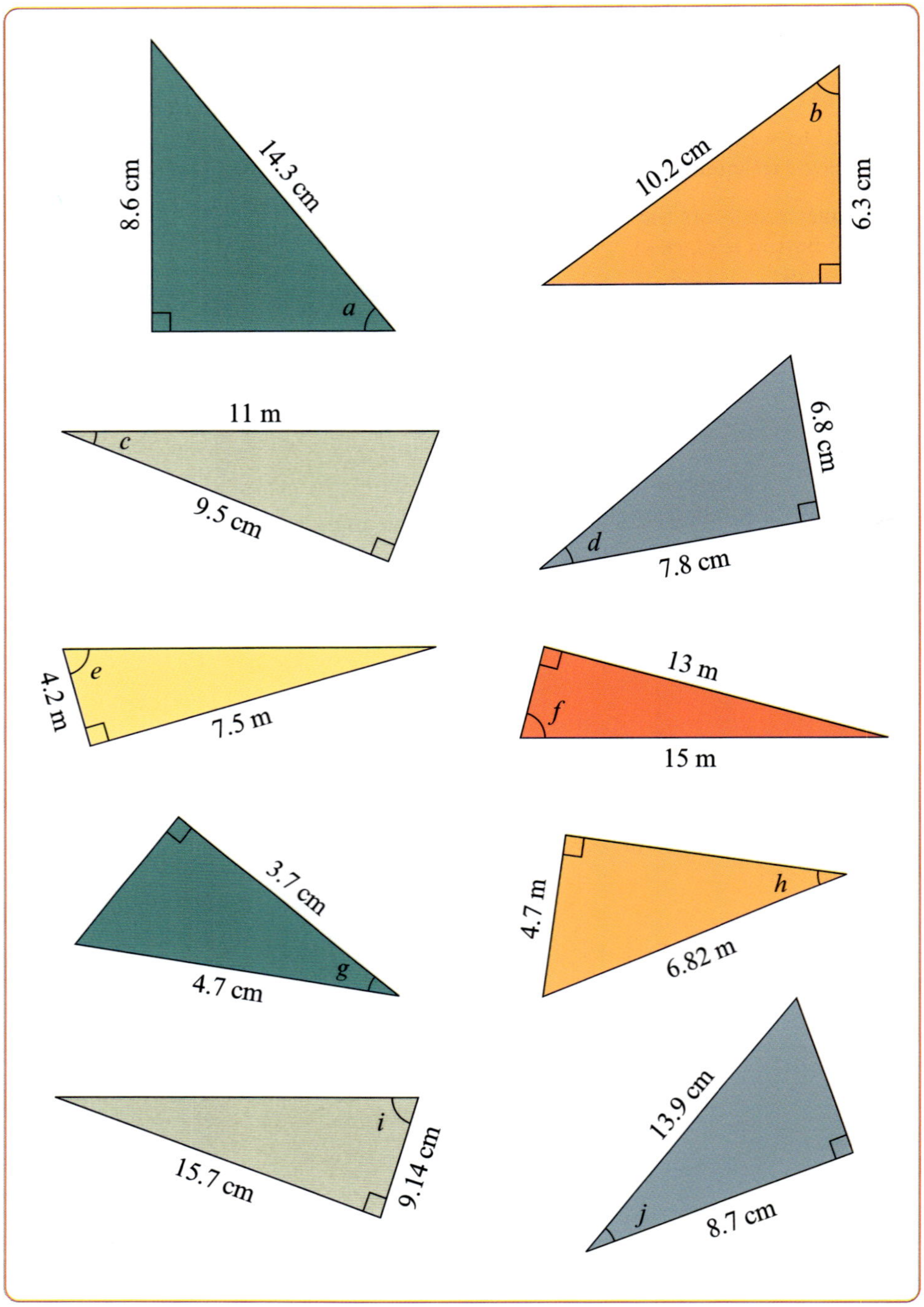

8.6 cm
14.3 cm
a

b
10.2 cm
6.3 cm

11 m
c
9.5 cm

6.8 cm
d
7.8 cm

e
4.2 m
7.5 m

13 m
f
15 m

3.7 cm
4.7 cm
g

4.7 m
h
6.82 m

i
15.7 cm
9.14 cm

13.9 cm
j
8.7 cm

7 Ali and Wayne calculate cos *x* for a triangle.
These are their answers:

Ali

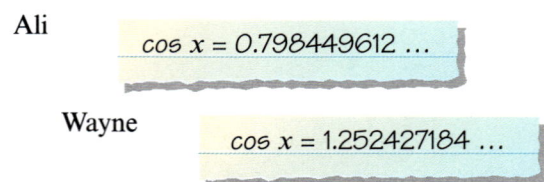

$\cos x = 0.798449612 \ldots$

Wayne

$\cos x = 1.252427184 \ldots$

a Which value for cos *x* must be incorrect?
b Explain why.

Thinking ahead to ...
finding sides and angles

A In triangle RST:
   a Is RS greater than or less than 5.6 cm?
   b Explain your answer

B Estimate the length of RS.

C Calculate the length of RS to 2 dp.

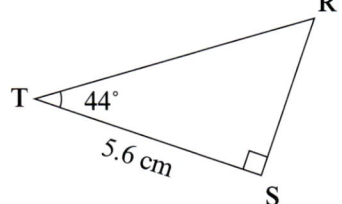

Finding sides and angles

Exercise 14.7
Estimating and calculating
sides and angles

Accuracy
For this exercise round
each answer to 3 sf.

1 The lengths and angles marked in blue on these triangles are incorrect.

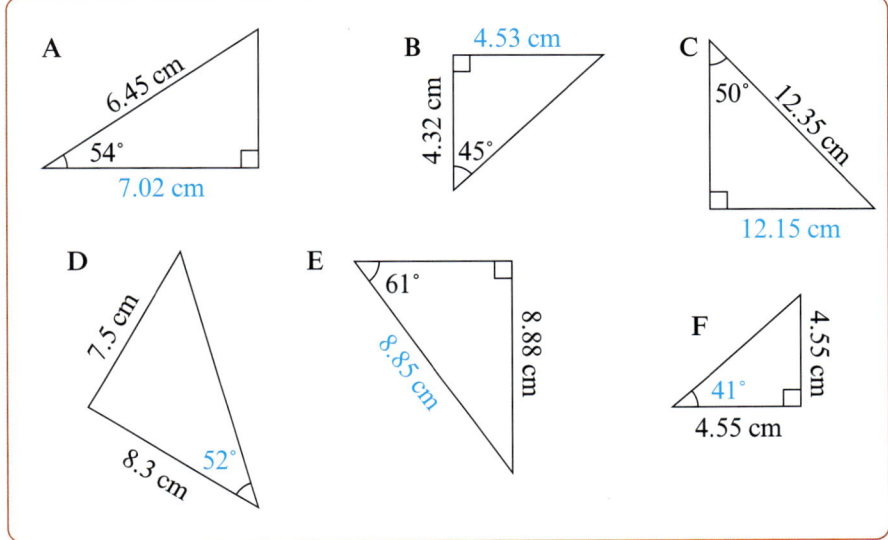

  a For each triangle, without calculating, explain why the length or angle
    marked in blue is wrong.
  b Calculate the correct length or angle for each triangle.

2 Sketch each triangle and calculate the length or angle marked with a letter.

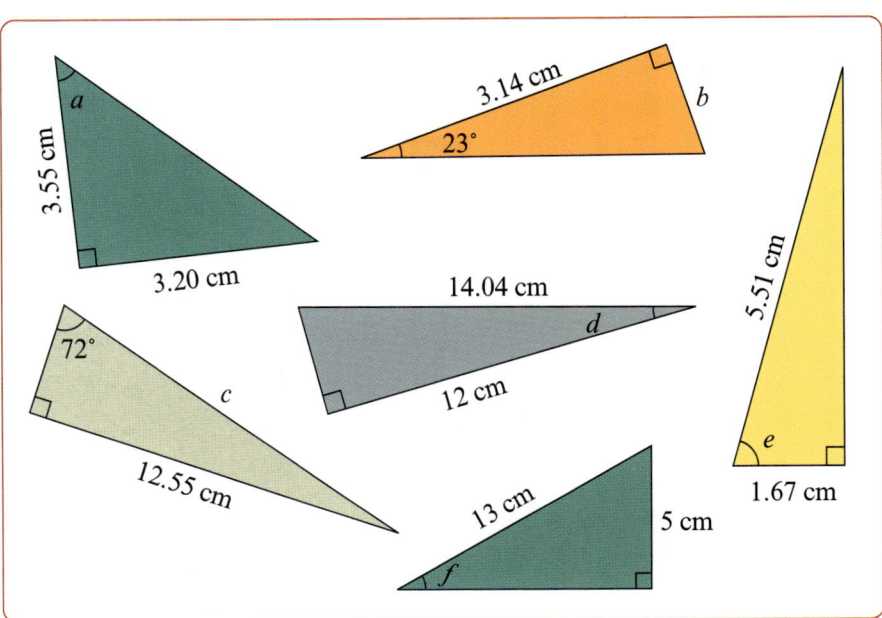

Trigonometry – solving problems

Exercise 14.8
Solving problems

> **Accuracy**
> For this exercise:
> ◆ round each angle to 2 dp
> ◆ round each distance to
>   the nearest cm.

> To write a length written
> in metres to the nearest
> centimetre you can round
> it to 2 dp.
>
> **Example**
>
> 5.3167 m
> = 5 metres 31.67 cm
> = 5 metres 32 cm
>   to the nearest cm
> = 5.32 m (to 2 dp)
>
> 25.12906 m
> = 25.13 m
>   to the nearest cm

1   This ladder is 5 metres long.
    The foot of the ladder is 1.25 metres
    from the wall.

    a The ladder is at an angle x to the
    horizontal. Calculate the angle x.

    b The ladder reaches a height
    of h metres up the wall.
    Calculate the height h to the
    nearest centimetre.

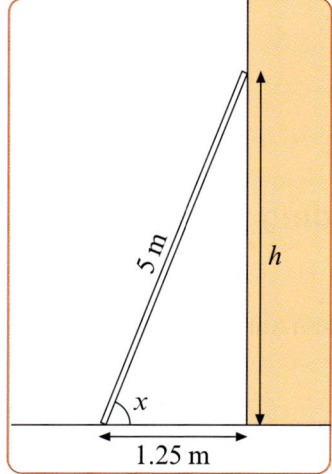

2   The firm Light Ladders recommends that the angle x, between the ladder and
    the horizontal, should be between 72° and 76°.

    The length of this ladder is 4.5 metres.
    The distance from the foot of the ladder
    to the wall is d metres.

    a Find the distance d when the
    angle x is 72°.

    b Find d when x is 76°.

    c When x is 72° how far up the wall
    does the ladder reach?

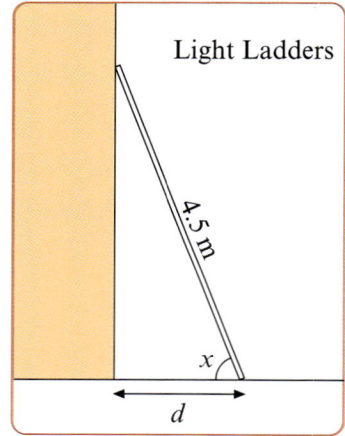

3   This is part of the safety label Light Ladders
    put on their extending ladder.

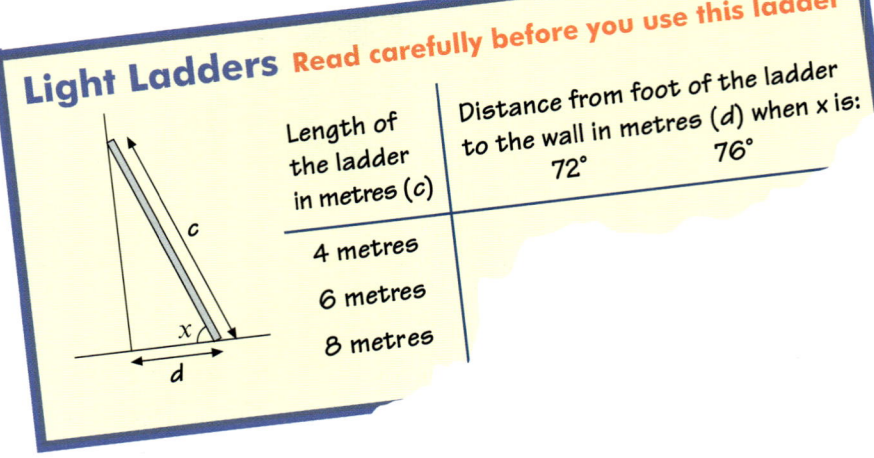

Make the safety label for this extending ladder.

4 This diagram shows some steps and a ramp outside a building.

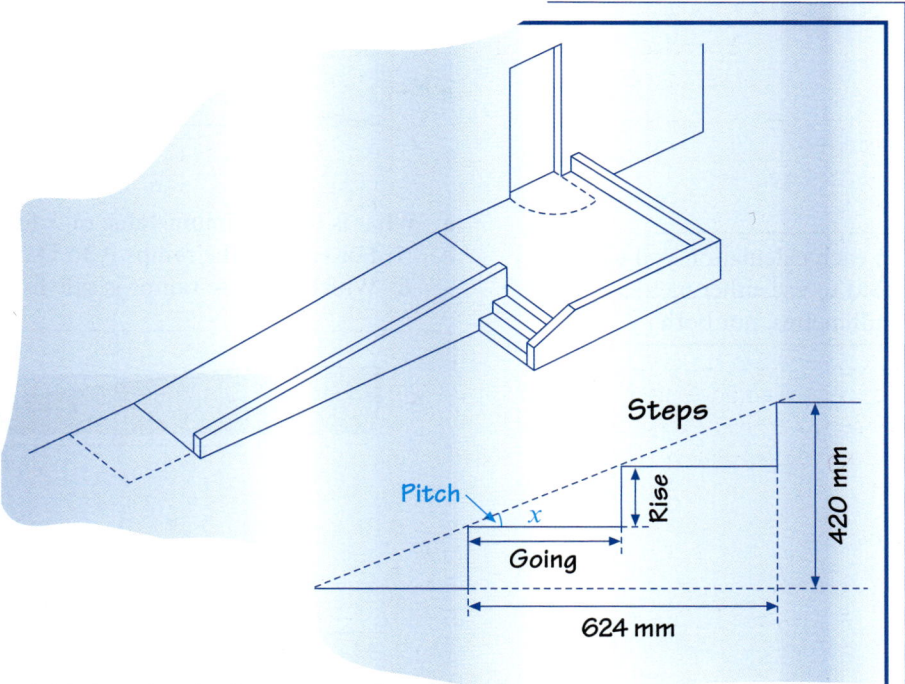

This is a sketch of a cross-section of the steps.
All the steps are the same size.

a For one step what is the length of the going?
b For each step what is the length of the rise?
c Complete this diagram for one step.
d Calculate the angle x for these stairs.

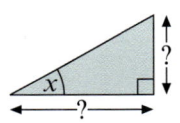

> Angle x is known as the pitch of the stairs.

5 This is a cross-section of the ramp.

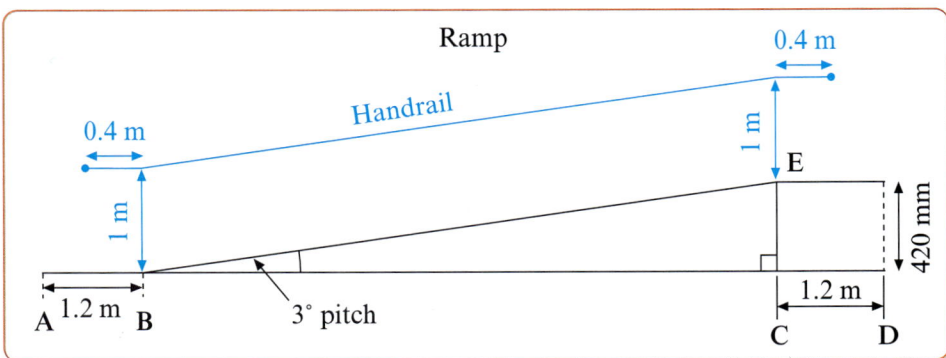

The pitch of the ramp is 3° and the height is 420 millimetres.

a Give the height of the ramp in metres.

b Calculate the length of the ramp, BE, in metres.

c A handrail is fixed to the wall 1 metre above the ramp.
It extends 0.4 metres beyond each end of the ramp.
What is the total length of the handrail?

d Calculate the horizontal distance BC.

e At each end of the ramp there must be a landing at least 1.2 metres long.
Calculate the horizontal distance AD.

> In each calculation you will need to use either metres or millimetres, not both.

6 For wheelchairs the maximum gradient of a ramp should be 1 in 20.

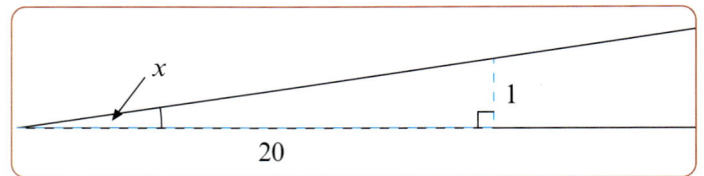

a What is the maximum value of x for a wheelchair ramp?
b **i** For each of the ramps A to D calculate the angle x.
    ii Which of these ramps is safe for wheelchair users?

> In each calculation you will need to use either metres or millimetres, not both.

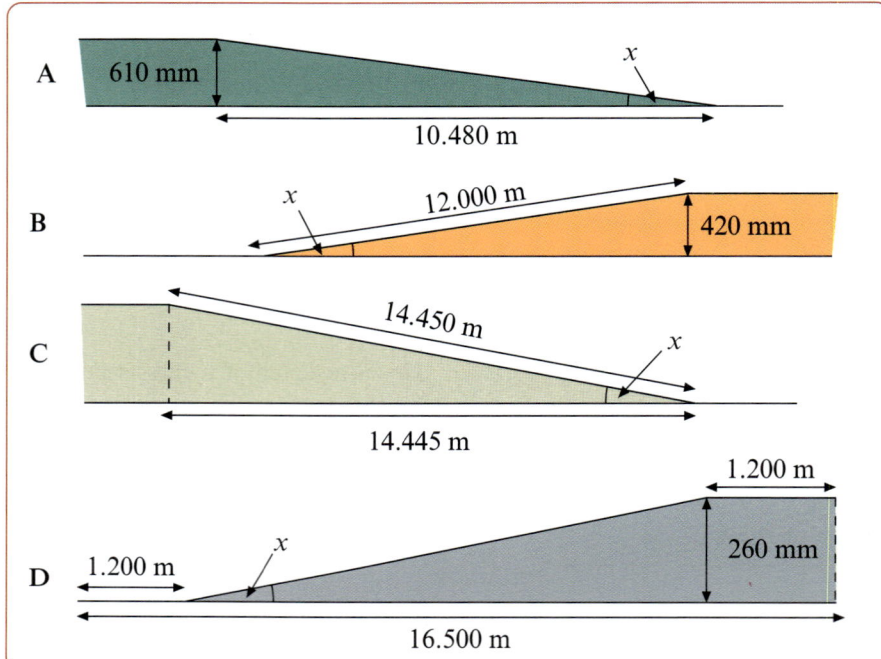

7 The maximum space for a ramp beside this building is 14.260 metres.

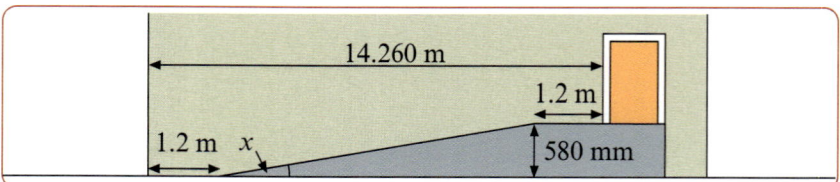

If they have to include a landing 1.2 metres long at each end of the ramp, what is the smallest pitch, x, they can use?

> Think carefully whether you need to round each answer up or down.

8 These are the regulations for the rise and going on any staircase.

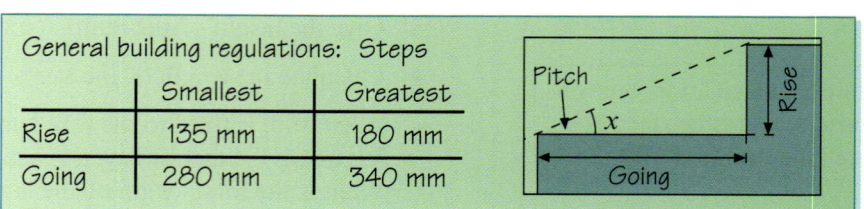

General building regulations: Steps	Smallest	Greatest
Rise	135 mm	180 mm
Going	280 mm	340 mm

Calculate:

a the maximum pitch
b the minimum pitch.

End points

You should be able to ...   ... so try these questions

A Recognise similar triangles

A1

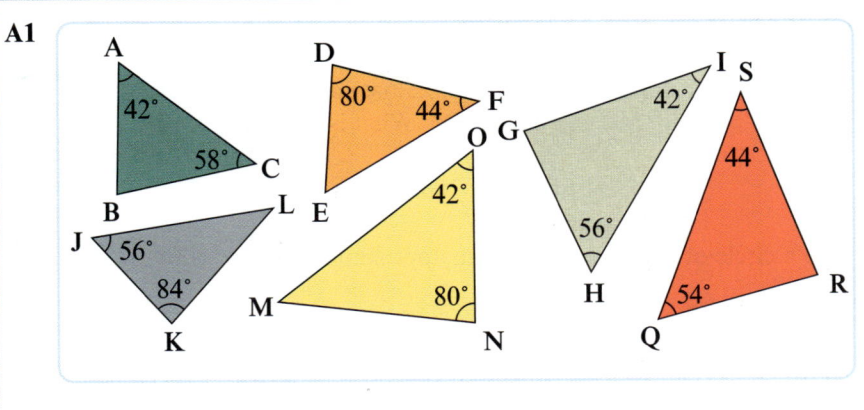

Which pair of triangles is similar?

B Calculate an angle in a
right-angled triangle

B1

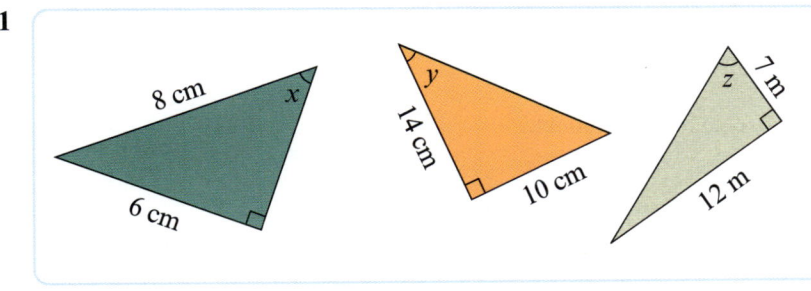

Calculate the angles x, y and z, each correct to 2 dp.

C Calculate the length of a side
in a right-angled triangle

C1

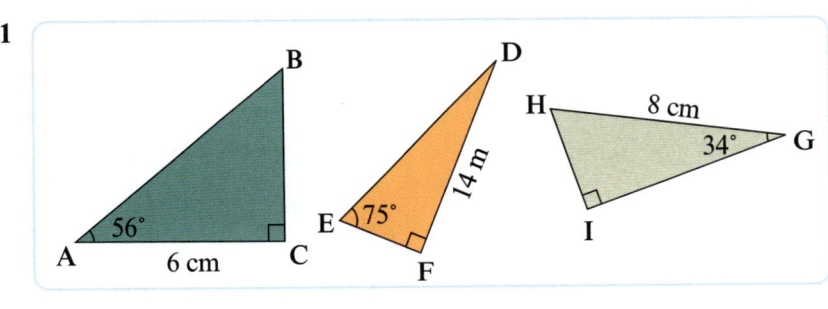

Calculate the lengths AB, EF and GI, each correct to 3 sf.

Some points to remember

- For a right-angled triangle, these trigonometric ratios apply:

$$\sin x = \frac{\text{opp}}{\text{hyp}} \qquad \cos x = \frac{\text{adj}}{\text{hyp}} \qquad \tan x = \frac{\text{opp}}{\text{adj}}$$

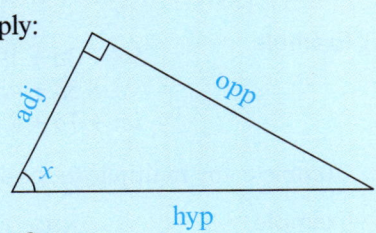

Some mnemonics to remember these by are:

Skive **O**ff **H**omework, **C**heat **A**t **H**omework, **T**elling **O**ff **A**fter.
Should **O**ld **H**arry **C**atch **A**ny **H**errings **T**rawling **O**ff **A**merica?

Starting points
You need to know about ...

... so try these questions

A Substituting into simple formulas

Example   A formula for the approximate area (A) of a circle with radius r is: $A = 3r^2$. What is A when $r = 10$?

$$A = 3 \times 10^2 = 3 \times 100 = 300$$

So a circle of radius 10 cm has an approximate area of 300 cm².

B Flowcharts

Flowcharts can be used to solve problems.

Example   I think of a number. I subtract 5 and then multiply by 6. My answer is 42. What number was I thinking of?

◆ A flowchart for the problem: Number → -5 → $\times 6$ → 42

◆ A reverse flowchart: 12 ← $+5$ ← $\div 6$ ← 42

So the number I was thinking of is 12.

C Balancing equations

Balancing can be used to solve problems.

Example   I think of a number. I subtract 5 and then multiply by 6. My answer is 42. What number was I thinking of?

◆ An equation for the problem: $6(n - 5) = 42$

◆ Solve the equation by balancing:

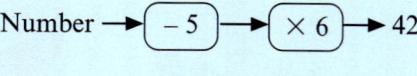
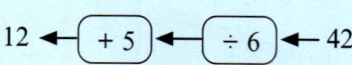

$6(n - 5) = 42$

$6n - 30 = 42$

$+30$ $\Big($ $6n = 72$ $\Big)$ $+30$

$\div 6$ $\Big($ $n = 12$ $\Big)$ $\div 6$

So the number I was thinking of is 12.

D Simplifying expressions

◆ Collect together any like terms.

Example
$$5a^2 + 3a + 10 + 4a - 2$$
$$= 5a^2 + 3a + 4a + 10 - 2$$
$$= 5a^2 + 7a + 8$$

◆ Complete any multiplications as far as possible.

Examples
◆ $p \times p \times 5 = 5p^2$
◆ $3a \times 5b = 15ab$
◆ $3x \times 5x = 15x^2$

A1   What is the approximate area of a circle with radius 4 cm?

A2   For each formula, find the value of A when $x = 5$.
  a $A = 4x - 1$ **c** $A = 7(x + 2)$
  b $A = \dfrac{x}{2}$ **d** $A = 30 - 2x$

B1   Use flowcharts to solve these number puzzles.

  a   I think of a number. I multiply by 4, then add 5. My answer is 53. What number was I thinking of?

  b   I think of a number. I subtract 8, then multiply by 3. My answer is 93. What number was I thinking of?

C1   Solve these equations:
  a $4(n - 1) = 76$
  b $2n + 5 = 5n - 1$

C2   I think of a number. I multiply by 7, then add 9. My answer is 65. What number was I thinking of?

  a   Write an equation for this puzzle.

  b   Solve it to find the number.

D1   Simplify these expressions:
  a $4x + 3 + x + 5$
  b $2a + 3b + 7a - b$
  c $2x^2 + 4x + 3x^2 - 2x$
  d $2a + 5a^2 + 4a - 7$
  e $t \times t \times 3$
  f $5c \times 2d$
  g $4k \times 3k$
  h $w \times 7w$

Thinking ahead to ...
rearranging formulas

A   A cook book gives this formula to find the time to cook a piece of lamb.

Allow 30 minutes per pound and an extra 20 minutes.

a   How long would it take to cook 6 pounds of lamb?

b   What weight of lamb would be cooked in 1 hour and 50 minutes?

Rearranging formulas

The formula that links cooking time (T) with weight (W) can be written as:

$$T = 30W + 20$$

♦ The formula shows how to find T when you know W, but it can be rearranged to show how to find W when you know T.

Addition 'undoes' subtraction and vice versa. For example:

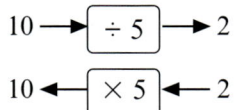

Multiplication 'undoes' division and vice versa. For example:

Flowchart method for rearranging formulas

♦ Draw a flowchart for the formula.

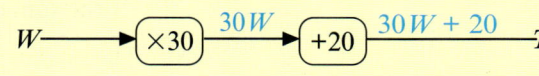

♦ Reverse the flow chart to rearrange the formula.

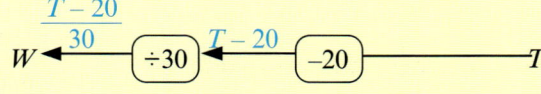

♦ The rearranged formula is: $W = \dfrac{T - 20}{30}$

Balancing method for rearranging formulas

♦ Add, subtract, multiply or divide **both** sides of the formula by equal amounts.

$-20 \Big($
$\div 30 \Big($

$T = 30W + 20$
$T - 20 = 30W$
$\dfrac{T - 20}{30} = W$

$\Big) -20$
$\Big) \div 30$

♦ The rearranged formula is: $W = \dfrac{T - 20}{30}$

Exercise 15.1
Rearranging linear formulas

1   Make p the subject of each formula:

a $t = p + 1$ **b** $s = 7p$ **c** $y = 5p - 2$ **d** $v = 10 - p$

e $h = 12 - 9p$ **f** $r = p + q$ **g** $t = pq$ **h** $x = 2p + f$

i $k = lp - m$ **j** $g = \dfrac{p}{9}$ **k** $m = \dfrac{p}{n}$ **l** $s = \dfrac{p}{2} + 1$

2   The formula that gives the cost in pence (c) of placing an advertisement in a local paper, where n is the number of words is:

$$c = 15n + 50$$

a   Find the cost of a 65-word advert.

b   Rearrange the formula so that it begins, $n = \dots$.

For the formula you need to write £19.85 in pence.

c   How many words were used in an advert that cost £19.85?

d   With £10.00, what is the maximum number of words you could use?

3 You can estimate the distance between you and a storm using the formula:

$$d = \frac{t}{5}$$

where d is the distance in miles and t is the number of seconds between the lightning and the thunder.

a Make t the subject of the formula.
b For a storm 1.5 miles away, how many seconds will be between the lightning and the thunder?

> When a formula begins,
> $t = \ldots$, then t is the subject
> of the formula.

4 The formula for the sum (S) of the interior angles of a polygon is:

$$S = 180(n - 2)$$

where n is the number of sides.

a What is the sum of the interior angles of a hexagon?
b Make n the subject of the formula.
c The sum of the interior angles of a polygon is 3240°.
   How many sides has the polygon?
d Explain why it is not possible to draw a polygon where
   the sum of the interior angles is 600°.

> The interior angles of this
> polygon are marked in red.
>
>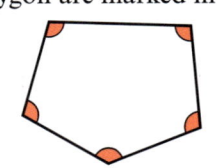

5 The formula for the number of vitamin pills (n) that Jim has left after d days is:

$$n = 100 - 3d$$

a How many vitamin pills would Jim have left after 14 days?
b How many vitamin pills does he take each day?
c Copy and complete this diagram to make d the subject of the formula.

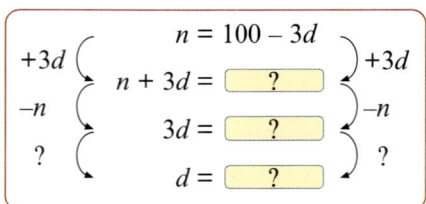

d After how many days does Jim have 10 pills left?

6 In the summer, an ice-cream seller uses this formula to estimate the number of ice-creams (n) she will sell in a day at x pence each.

$$n = 800 - 5x$$

a About how many ice-creams will she sell in a day at 50p each?
b Make x the subject of the formula.
c What price will she need to charge to sell 500 ice-creams?
d Explain why she is unlikely to charge £1.70 for an ice-cream.

7 The formula that links distance (d), speed (s) and time (t) is:

$$d = st$$

a How far will a car travel in 2 hours at a speed of 47 mph?

With s as the subject, the formula $d = st$ can be written:

$$s = \frac{d}{t}$$

b Calculate the speed of a plane that travels 1800 miles in 3 hours.
c Copy and complete this flow chart for $d = st$:

$$t \longrightarrow \boxed{} \longrightarrow d$$

d Reverse the flow chart and make t the subject of the formula.
e Find the time it takes to travel 125 km at a speed of 50 km/h.

> To make s the subject of
> the formula, $d = st$:
>
>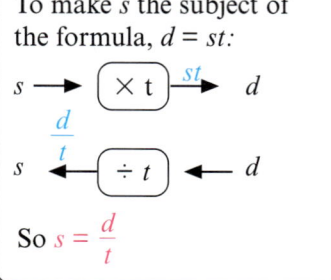
>
> So $s = \dfrac{d}{t}$

A formula for converting kilometres (k) to miles (m) is: $m = \dfrac{5}{8}k$

♦ Making k the subject gives
a formula for converting miles to kilometres.

$$\times 8 \left(\begin{array}{c} m = \dfrac{5}{8}k \\ 8m = 5k \end{array} \right) \times 8$$

$$\div 5 \left(\begin{array}{c} 8m = 5k \\ \dfrac{8m}{5} = k \end{array} \right) \div 5$$

♦ So $k = \dfrac{8m}{5}$

Exercise 15.2
Rearranging formulas
with fractions

1 Make w the subject of each formula.

a $k = \dfrac{3}{5}w$ **b** $\dfrac{3}{4}w = x$ **c** $\dfrac{3}{5}w + 1 = k$ **d** $3x = 4 + \dfrac{1}{2}w$

e $3y - 2 = \dfrac{1}{3}w + 1$ **f** $\dfrac{1}{3}w = 2k - 4$ **g** $\dfrac{3}{5}w + x = 4$ **h** $\dfrac{w+1}{3} = k$

i $\dfrac{1}{2}(w - 2) = 3k$ **j** $\dfrac{2}{3}w = \dfrac{1}{2}k + 1$ **k** $\dfrac{3}{4}w + 1 = 1 - k$ **l** $\dfrac{1}{2}w - 2 = \dfrac{1}{2}k$

2 A formula for converting kilograms (k) to pounds (p) is:

$$p = \dfrac{11}{5}k$$

a How many pounds are in 8 kilograms?
b Make k the subject of the formula.
c Convert 3 pounds to kilograms.

3 The formula for the area, (A), of a triangle with base length b and height h is:

$$A = \dfrac{1}{2}bh$$

a Make h the subject of the formula.
b Find the height of a triangle with area $100\,\text{cm}^2$ and base length $2.5\,\text{cm}$.

4 Temperature in °F can be converted to °C using this formula:

$$C = \dfrac{5(F - 32)}{9}$$

a Change 68 °F to °C.
b Copy and complete this flowchart:

$$F \longrightarrow \boxed{-32} \xrightarrow{\;F-32\;} \boxed{} \longrightarrow \boxed{} \longrightarrow C$$

c Reverse the flowchart to give a formula that converts °C into °F.
d Change 24 °C to °F.

5 Make y the subject of each formula:

a $z = \dfrac{3}{4}y$ **b** $m = \dfrac{1}{3}xy$ **c** $d = \dfrac{1}{2}y + 4$ **d** $k = \dfrac{4}{7}y - h$

$\boxed{\dfrac{18d}{5t} = 18d \div 5t}$

6 When d metres are travelled in t seconds, a formula for average speed (s) in km/h is:

$$s = \dfrac{18d}{5t}$$

a In 1992, Linford Christie ran 100 metres in 9.96 seconds.
What was his average speed in km/h, correct to 2 dp?
b Make d the subject of the formula.
c Make t the subject of the formula.

A formula for the total area (A) of glass in this window is:

$$A = 4x^2$$

♦ A flow chart for finding A is:

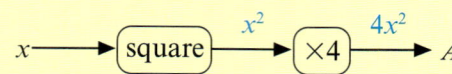

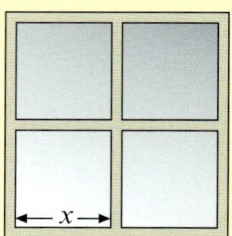

♦ Reversing the flow chart makes x the subject of the formula:

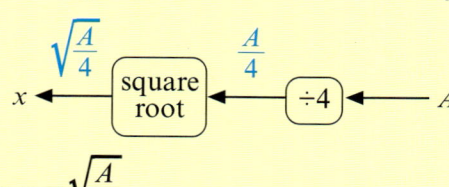

♦ So, $x = \sqrt{\dfrac{A}{4}}$

Exercise 15.3
Using roots

Accuracy
In this exercise,
♦ give answers correct to 2 dp
♦ use the π button on your calculator.

1 The formula for the surface area (S) of a cube with an edge length of x is:

$$S = 6x^2$$

a Find the surface area of a cube with an edge length of 8.2 cm.
b Make x the subject of the formula.
c What is the edge length of a cube with a surface area of 120 cm²?

2 The formula for the area (A) of a circle with a radius of r is:

$$A = \pi r^2$$

a Find the area of a circle with a radius of 1.5 cm.
b Which of these is the correct formula for the radius r?

   A $r = \sqrt{(A - \pi)}$ **B** $r = \sqrt{\dfrac{A}{\pi}}$ **C** $r = \dfrac{\sqrt{A}}{\pi}$

c What is the radius of a circle with an area of 200 cm²?

3 When an object is dropped, a formula for the approximate distance (d) travelled in metres in a time of t seconds is:

$$d = \frac{49}{10}t^2$$

a A stone is dropped. About how far will it fall in 6.5 seconds?
b Rearrange the formula to make t the subject.
c Sears Tower is 443 m high.
About how long will it take an apple to fall from the top to the ground?

4 Make k the subject of each formula:
  a $m = k^2 + 1$ **b** $m = 3k^2 + 1$ **c** $m = k^2 + n$

5 Make y the subject of each formula:
  a $k = 4y^2 + 2$ **b** $3x = 1 - 2y^2$ **c** $ax = 5y^2$

  d $ax + 4 = 3y^2$ **e** $x + 4 = \dfrac{1}{2}y^2$ **f** $ax^2 = by^2$

  g $3x + 5y^2 = a^2$ **h** $a^2 + x^2 + y^2 = 3$ **i** $4a^2 - 3x + 2y^2 = x^2$

  j $3y^2 + 4 = ay^2$ **k** $2w + xy^2 = 3y^2$ **l** $4 - ay^2 = 5y^2$

  m $ay^2 + bx^2 = cy^2$ **n** $3 + 5y^2 = x^2 + ay^2$ **o** $2w + 3x^2 = xy^2$

  p $\dfrac{1}{4}y^2 + ay^2 = x$ **q** $\dfrac{1}{3}y^2 + \dfrac{2}{3}x^2 = 1$ **r** $\dfrac{1}{4}x^2y^2 = a^2$

Substitution into non-linear expressions

♦ For the formula, $a = (b + 5)(b - 1)$, find the value of a when $b = 3$.

$$a = (3 + 5) \times (3 - 1)$$
$$= 8 \times 2$$
$$= 16$$

♦ For the formula, $y = x^2 + 4x + 1$, find the value of y when $x = 7$.

$$y = 7^2 + (4 \times 7) + 1$$
$$= 49 + 28 + 1$$
$$= 78$$

Exercise 15.4
Substitution

Accuracy
In this exercise, give each answer correct to 2 dp.

1 A $\boxed{y = x^2 + 5x - 8}$ B $\boxed{y = x^2 + 3x + 2}$

    C $\boxed{y = (x + 1)(x + 2)}$ D $\boxed{y = (x - 2)(x + 9)}$

For each formula find the value of y when:

a $x = 5$ **b** $x = 3$ **c** $x = 2$ **d** $x = 2.5$

2 A formula that gives the stopping distance of a car on a dry road is:

$$d = \frac{v^2}{200} + \frac{v}{5}$$

where d is the stopping distance in metres and v is the speed in km/h.

Calculate the stopping distance of a car travelling at 110 km/h on a dry road.

3 The net for an open square-based box is made by cutting four square corners from a 12 cm by 12 cm square piece of card as shown.

A formula for the volume, (V cm³), of the box is:

$$V = 4x(x^2 - 12x + 36),$$

where x is the length shown on the diagram.

a When $x = 1.5$, find the volume of the box.
b Calculate V when $x = 4$.
c What value for x gives the box with the greatest volume?

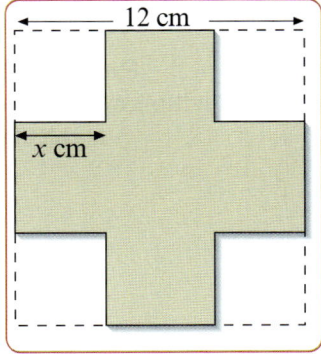

4 When an object is dropped, a formula for the approximate speed (v) in m/s after travelling s metres is:

$$v = \sqrt{20s}$$

The Empire State Building is 381 m high.
A coin is dropped from the top. Estimate its speed as it hits the ground.

5 The formula $v = \sqrt{u^2 + 2as}$ gives the speed of an object (v) in m/s, given the speed at the start (u) in m/s, the acceleration (a) in m/s² and the distance (s) travelled in metres.

Find v when:

a $u = 30$, $a = 5$ and $s = 6.8$ **b** $u = 20$, $a = 9.8$ and $s = 2.4$

Multiplication

The area of this shape is $28 \times 43 \, \text{cm}^2$.

One way to calculate the area is as follows.

- Split the shape into 4 rectangles.
- Find the area of each rectangle in cm^2.

$$20 \times 40 = 800$$
$$20 \times 3 = 60$$
$$8 \times 40 = 320$$
$$8 \times 3 = 24$$

- Add the areas:

$$800 + 60 + 320 + 24 = 1204$$

- The total area is $1204 \, \text{cm}^2$.

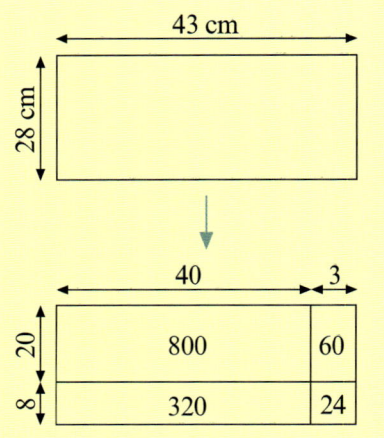

> This is one way to multiply without a calculator.

Any multiplication can be calculated in a similar way.

Example $42 \times 19 = (40 + 2) \times (10 + 9)$

- Work in a table.

$\times$	40	+ 2
10	400	20
+ 9	360	18

- The total is:
$$400 + 20 +$$
$$360 + 18 = 798$$
- So $42 \times 19 = 798$.

Exercise 15.5
Multiplication without a calculator

1 A shape has been split into four rectangles.
  a Calculate the area of each rectangle.
  b Find the total area of the shape.

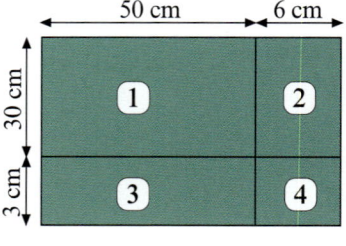

2 **a** Make a sketch of this diagram and fill in any missing lengths and areas.
  b Find the total area of the shape.

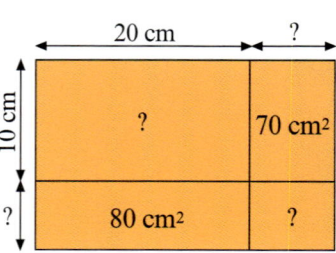

3 Copy and complete:

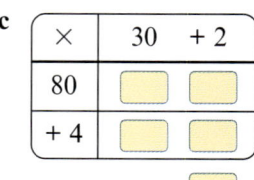

a

$\times$	40	+ 5
60	2400	
+ 1		

So $45 \times 61 = \square$

b

$\times$	20	+ 5
20		
+ 7		

So $25 \times 27 = \square$

c

$\times$	30	+ 2
80		
+ 4		

So $32 \times 84 = \square$

Thinking ahead to ...
multiplying out brackets

A This shape is split into four rectangles.

Find an expression for:

a the area of each rectangle
b the total area of the shape.

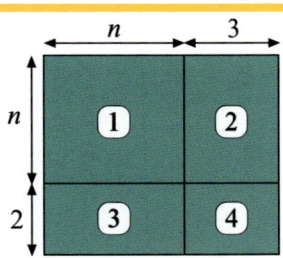

Multiplying out brackets

Example When $n = 4$,

$(n + 2)(n + 3)$
$= (4 + 2)(4 + 3)$
$= 6 \times 7$
$= 42$

$n^2 + 5n + 6$
$= 4^2 + (5 \times 4) + 6$
$= 16 + 20 + 6$
$= 42$

We can multiply out brackets from $(n + 2)(n + 3)$ like this:

Work in a table:

×	n	$+ 2$
n	n^2	$2n$
$+ 3$	$3n$	6

◆ The total is: $n^2 + 2n + 3n + 6$
 which simplifies to: $n^2 + 5n + 6$

◆ This shows that $(n + 2)(n + 3) = n^2 + 5n + 6$ for any value of n so $(n + 2)(n + 3)$ and $n^2 + 5n + 6$ are **equivalent expressions**.

Exercise 15.6
Multiplying out brackets

1 For each of these, multiply out the brackets and simplify.

a $(a + 1)(a + 3)$ **b** $(b + 2)(b + 4)$ **c** $(c + 8)(c + 5)$
d $(d + 5)(d + 9)$ **e** $(e + 1)(e + 7)$ **f** $(f + 2)(f + 11)$
g $(x + 3)(x + 4)$ **h** $(x + 8)(x + 3)$ **i** $(x + 9)(x + 9)$
j $(x + 2)(x + 7)$ **k** $(x + 9)(x + 15)$ **l** $(x + 10)(x + 16)$
m $(x + 3)(x + 1)$ **n** $(x + 8)(x + 7)$ **o** $(x + 6)(x + 21)$

$(x + 1)^2$
is shorthand for
$(x + 1)(x + 1)$

2 a For each of these, multiply out the brackets and simplify.
  i $(x + 1)^2$ **ii** $(x + 2)^2$ **iii** $(x + 3)^2$
 b Comment on any pattern you see in your results.

3 Copy and complete:

×	$2x$	$+ 4$
$3x$		
$+ 1$		

So $(2x + 4)(3x + 1) = $ ⬚

4 For each of these, multiply out the brackets and simplify.

a $(2a + 1)(a + 4)$ **b** $(b + 3)(3b + 5)$ **c** $(2c + 7)(3c + 2)$
d $(4d + 3)(3d + 4)$ **e** $(2e + 3)(2e + 3)$ **f** $(5f + 1)(5f + 1)$
g $(5x + 4)(2x + 1)$ **h** $(3x + 7)(4x + 5)$ **i** $(3x + 2)(3x + 2)$
j $(7x + 1)(8x + 5)$ **k** $(4x + 1)(5x + 1)$ **l** $(8x + 6)(2x + 1)$

5 Which of these expressions is equivalent to $(4x + 1)(5x + 2)$?

A $33x^2 + 2$ **B** $20x^2 + 13x + 2$ **C** $9x^2 + 13x + 3$ **D** $9x + 3$

6 Jo and Liz investigate the sequence: 4, 10, 18, 28, 40, ...
Each finds an expression for the *n*th term.

Jo: $n(n + 3)$ Liz: $(n + 1)(n + 2) - 2$

  a Show that each expression gives 40 for the 5th term in the sequence.
  b Show that these two expressions are equivalent.

7 Show that these two formulas are equivalent.

  A $y = 4x(x + 2)$ B $y = (2x + 1)(2x + 3) - 3$

Expressions that involve subtraction can also be multiplied using a table.

Example $(2y + 5)(3y - 1) = (2y + 5) \times (3y - 1)$

◆ Work in a table:

×	$2y$	$+ 5$
$3y$	$6y^2$	$15y$
$- 1$	$- 2y$	$- 5$

◆ The total is: $6y^2 + 15y - 2y - 5$
which simplifies to: $6y^2 + 13y - 5$

◆ So $(2y + 5)(3y - 1) = 6y^2 + 13y - 5$.

Exercise 15.7
Multiplying out brackets

1 For each of these, multiply out the brackets and simplify.

  a $(x + 3)(x - 4)$ **b** $(x + 5)(x - 4)$ **c** $(x + 2)(x - 7)$
  d $(x + 4)(x - 6)$ **e** $(x + 7)(x - 8)$ **f** $(x + 8)(x - 4)$
  g $(x + 6)(x - 7)$ **h** $(x + 9)(x - 2)$ **i** $(x + 7)(x - 3)$
  j $(x + 12)(x - 8)$ **k** $(x + 6)(x - 9)$ **l** $(x + 8)(x - 1)$

2 For each of these, multiply out the brackets and simplify.

  a $(x - 2)(x + 3)$ **b** $(x - 5)(x + 6)$ **c** $(x - 8)(x + 4)$
  d $(x - 7)(x + 5)$ **e** $(x - 8)(x + 9)$ **f** $(x - 7)(x + 2)$
  g $(x - 9)(x + 4)$ **h** $(x - 7)(x + 4)$ **i** $(x - 9)(x + 5)$
  j $(x - 10)(x + 8)$ **k** $(x - 12)(x + 5)$ **l** $(x - 8)(x + 7)$

3 For each of these, multiply out the brackets and simplify.

  a $(x - 3)(x - 8)$ **b** $(x - 4)(x - 6)$ **c** $(x - 5)(x - 8)$
  d $(x - 9)(x - 1)$ **e** $(x - 3)(x - 7)$ **f** $(x - 4)(x - 9)$
  g $(x - 12)(x - 7)$ **h** $(x - 9)(x - 10)$ **i** $(x - 5)(x - 6)$
  j $(x - 15)(x - 12)$ **k** $(x - 7)(x - 12)$ **l** $(x - 25)(x - 4)$

4 For each of these, multiply out the brackets and simplify.

  a $(3x + 4)(2x + 1)$ **b** $(3x + 2)(4x + 5)$ **c** $(6x + 3)(5x + 4)$
  d $(4x + 3)(2x - 1)$ **e** $(5x + 6)(3x - 2)$ **f** $(9x + 4)(6x - 3)$
  g $(6x - 2)(3x + 1)$ **h** $(4x - 3)(7x + 2)$ **i** $(8x - 6)(4x + 2)$
  j $(7x - 3)(6x - 4)$ **k** $(9x - 3)(2x - 6)$ **l** $(4x - 5)(5x - 4)$

Thinking ahead to ...
factorising

A Which pair of expressions multiply to give $x^2 + 8x + 12$?

| $x + 1$ | $x + 2$ | $x + 3$ | $x + 4$ | $x + 6$ | $x + 12$ |

B **a** Draw up this table so that it gives a total of $p^2 + 5p - 14$.

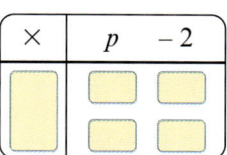

Total: $p^2 + 5p - 14$

b Copy and complete: $(p - 2)(\ldots) = p^2 + 5p - 14$.

Factorising

$48 = 12 \times 4$
so 12 and 4 are factors of 48.

$n^2 + 5n + 6 = (n + 2)(n + 3)$
so $(n + 2)$ and $(n + 3)$ are factors of $n^2 + 5n + 6$.

To **factorise** an expression is to write it as a multiplication of its factors.

Example Factorise $n^2 + 2n - 8$:

♦ Think about a table that gives a total of $n^2 + 2n - 8$ and fill in the parts you are sure about.

×	n
n	n^2
	-8

♦ Try values that give -8 in the correct position.

×	n	-4
n	n^2	$-4n$
$+2$	$2n$	-8

❖ The total for this table is $n^2 - 2n - 8$ which **is not** what you want.

×	n	$+4$
n	n^2	$4n$
-2	$-2n$	-8

❖ The total for this table is $n^2 + 2n - 8$ which **is** what you want.

♦ So $n^2 + 2n - 8$ factorises to give $(n + 4)(n - 2)$ or $(n - 2)(n + 4)$

Exercise 15.8
Factorising

1 Factorise these expressions.

a	$x^2 + 8x + 7$	**b**	$x^2 + 3x + 2$	**c**	$x^2 + 4x + 3$
d	$x^2 + 8x + 12$	**e**	$x^2 + 6x + 9$	**f**	$x^2 + 13x + 36$

2 Factorise these expressions.

a	$x^2 + 6x + 5$	**b**	$x^2 + 2x - 48$	**c**	$x^2 + 5x - 14$
d	$x^2 + 4x - 21$	**e**	$x^2 + 6x - 27$	**f**	$x^2 + 7x - 18$
g	$x^2 + 9x - 10$	**h**	$x^2 + 7x - 60$	**i**	$x^2 + 8x - 20$

3 Factorise these expressions.

a	$x^2 - 6x - 7$	**b**	$x^2 - 4x - 45$	**c**	$x^2 - 3x - 40$
d	$x^2 - 4x - 60$	**e**	$x^2 - 4x - 5$	**f**	$x^2 - 7x - 60$
g	$x^2 - 2x - 63$	**h**	$x^2 - 5x - 14$	**i**	$x^2 - 9x - 36$

4 Factorise these expressions.

a	$x^2 - 9x + 14$	**b**	$x^2 - 9x + 20$	**c**	$x^2 - 6x + 8$
d	$x^2 - 7x + 10$	**e**	$x^2 - 11x + 28$	**f**	$x^2 - 14x + 45$
g	$x^2 - 11x + 30$	**h**	$x^2 - 15x + 56$	**i**	$x^2 - 20x + 91$

End points
You should be able to ...   ... so try these questions

A   Substitute in formulas

A1 When a stone is thrown straight up with a speed of u m/s, the formula:

$$h = ut - 5t^2$$

gives its approximate height, h metres, after t seconds.

A stone is thrown upwards with a speed of 60 m/s.
What is its height after:
a 2 seconds   **b** 5 seconds   **c** 10 seconds?

A2 The formula for the volume (V) of a cone is:

$$V = \tfrac{1}{3}\pi r^2 h,$$

where r is the radius of the base and h is the height.

A cone has a base of radius 2.5 cm and a height of 8 cm.
Calculate its volume in cm³.

B   Rearrange formulas

B1 A car hire company uses this formula to calculate
the cost in pounds, C, of hiring a car for n days:

$$C = 30n + 20$$

a Make n the subject of the formula.
b With £470, for how many days could you hire a car?

B2 Asif uses this formula to calculate the cost of running his car for
a week:

$$C = 10 + \tfrac{1}{5}m,$$

where C is the cost in pounds and m is the number of miles he drives.

a Make m the subject of the formula.
b Asif wants the cost of running his car to be £30 or less per week.
     What is the maximum number of miles he can drive in a week?

B3 Make k the subject of each formula:

a $h = 5k^2$ **b** $m = \dfrac{n+k}{5}$ **c** $A = k^2 + 9$

C   Multiply out brackets

C1 Which of these expressions is equivalent to $(t + 5)(t - 3)$?

A $t^2 + 8t + 15$ **B** $t^2 + 8t - 15$ **C** $t^2 + 2t - 15$

C2 For each of these, multiply out the brackets and simplify.
a $(a + 1)(a + 8)$ **b** $(b + 4)^2$ **c** $(5c + 2)(c + 3)$
d $(d + 2)(d - 1)$ **e** $(2e + 5)(3e - 10)$ **f** $(f - 7)(3f - 2)$

D   Factorise expressions

D1 Which pair of expressions multiply to give $k^2 + 5k - 14$?

$k + 2$ $k - 7$ $k - 2$ $k + 14$ $k + 7$ $k - 1$

D2 Factorise these expressions.
a $x^2 + 12x + 11$ **b** $x^2 + 9x + 14$ **c** $x^2 + 4x - 5$
d $x^2 + x - 6$ **e** $x^2 - x - 20$ **f** $x^2 - 5x + 6$

Starting points
You need to know about ...

... so try these questions.

A Finding the median of a frequency distribution

- You can always find the median of ungrouped data by:
  - listing the data in order
  - finding the middle value (or the middle pair of values).

1996 Olympic Games – Pole Vault Final
Best height cleared by each finalist

Height (metres)	5.60	5.70	5.80	5.86	5.92	Total
Frequency	4	3	1	3	3	14

5.60 5.60 5.60 5.60 5.70 5.70 5.70 5.80 5.86 5.86 5.86 5.92 5.92 5.92

The median is halfway between the middle pair, so
median height = **5.75 m**

- For a distribution with a large total frequency, using cumulative frequencies is quicker than listing the data.

1996 Olympic Games – Pole Vault Final
Heights cleared by the 14 finalists

Height (metres)	5.40	5.60	5.70	5.80	5.86	5.92	Total
Frequency	6	10	5	4	6	3	34
Cumulative Frequency	6	16	21	25	31	34	

The middle pair (the 17th and 18th) are both 5.70, so median height = **5.70 m**

B Calculating the mean of a frequency distribution

- To calculate the mean of ungrouped data:
  - multiply each value by its frequency
  - calculate the total of all the values

1996 Olympic Games Pole Vault Final
Heights cleared by finalists

Height (metres)	Frequency		Total at each height
5.40	6	5.40 × 6	32.4
5.60	10	5.60 × 10	56
5.70	5	5.70 × 5	28.5
5.80	4	5.80 × 4	23.2
5.86	6	5.86 × 6	35.16
5.92	3	5.92 × 3	17.76
Totals	34		193.02

- divide the total of all the values by the total frequency.

Mean height = $\frac{193.02}{34}$ = **5.68 m** (to 3 sf).

**1996 Olympic Games
Men's High Jump Final**

Height (metres)	Frequency	
	Best height cleared	All heights cleared
2.15	–	3
2.20	–	9
2.25	4	12
2.29	3	8
2.32	4	6
2.35	1	3
2.37	1	1
2.39	1	1
Totals	14	43

A1 For the best-height distribution, find the median height cleared by listing the data.

A2 For the all-heights distribution, use cumulative frequencies to find the median height cleared.

**1996 Olympic Games
Women's High Jump Final**

Height (metres)	Frequency	
	Best height cleared	All heights cleared
1.80	–	8
1.85	–	14
1.90	–	13
1.93	5	14
1.96	4	9
1.99	2	5
2.01	1	3
2.03	1	2
2.05	1	1
Totals	14	69

B1 For the best-height distribution, calculate the mean height cleared.

B2 For the all-heights distribution, calculate the mean height cleared.

C Measures of spread

♦ The range is the simplest measure of how spread out a set of data is.

1996 Olympic Games – Pole Vault Final
Best height cleared by each finalist

Height (metres)	5.60	5.70	5.80	5.86	5.92	Total
Frequency	4	3	1	3	3	14

$$\text{Range} = \text{Highest value} - \text{Lowest value}$$
$$= 5.92 - 5.60$$
$$= \mathbf{0.32\,m}$$

♦ The interquartile range measures the spread of the middle 50% of the distribution by ignoring the lowest 25% and the highest 25%.

5.60 5.60 5.60 5|60 5.70 5.70 5.70|5.80 5.86 5.86 5|86 5.92 5.92 5.92

$$\text{Interquartile range} = \text{Upper quartile} - \text{Lower quartile}$$
$$= 5.86 - 5.60$$
$$= \mathbf{0.26\,m}$$

D Comparing sets of data

♦ You can compare sets of data using two types of value: an average, and a measure of spread.

1996 Olympic Games – Pole Vault Finals
Best heights cleared by finalists

	1992	1996
Median	5.625 m	5.75 m
Interquartile range	0.35 m	0.26 m

On average, finalists jumped 12.5 cm higher in 1996 than in 1992. The finalists were more closely matched in 1996 because the best heights cleared were less spread out than in 1992.

E Using grouped data

♦ Data which is collected in groups, or grouped to make it easier to present, is called a **grouped frequency distribution**.

1996 Olympic Games – Great Britain's Athletics Team

Age	18–22	23–27	28–32	33–37	38–42	Total
Frequency	9	37	27	5	4	82

♦ Each group of data is a **class**: the size of a class is the **class interval**.

The 18–22 class has a class interval of 5 years.

♦ You can show grouped data on a **grouped frequency diagram**.

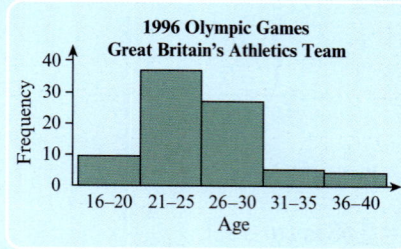

1996 Olympic Games
Great Britain's Athletics Team

C1 Use the data for the 1996 high jump finals on page 185 to calculate the range of the best heights cleared for:
a men **b** women.

C2 Calculate the interquartile range of the best heights cleared for:
a men **b** women.

1992 Olympic Games High Jump Finals
Best heights cleared by finalists

Median – men	2.295 m
Mean – women	1.91 m
Interquartile range	
– men	0.06 m
– women	0.085 m

D1 Use your answers to Questions **A1**, **B1**, and **C2** to compare the best heights cleared in 1992 and 1996 by:
a men **b** women.

1996 Olympic Games Great Britain's Athletics Team

Women		Men	
Age	Frequency	Age	Frequency
18–23	8	18–21	3
24–29	13	22–25	17
30–35	10	26–29	14
36–41	3	30–33	8
Total	34	34–37	5
		38–41	1
		Total	48

E1 Give the class interval for:
a women **b** men.

E2 Draw a grouped frequency diagram to show the ages of:
a women **b** men.

Histograms and frequency polygons

♦ A grouped frequency diagram is also called a **histogram**.
♦ The class with the highest frequency is called the **modal class**.
♦ A **frequency polygon** is a set of straight lines joining the middle points on the top of each bar of a histogram.

The frequency is the number of golfers.

153 golfers played in the 1st and 2nd rounds. These scores are for the top 77 golfers after the 2nd round. Only these 77 golfers played in the 3rd and 4th rounds.

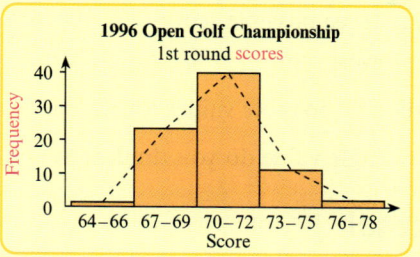

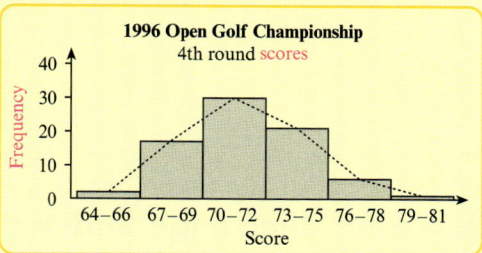

♦ You can compare frequency distributions on the same diagram by plotting the frequency polygon for each distribution.

The middle scores of each of the classes are 65, 68, 71, 74, 77, 80.

These can be called the **mid-class values**.

The frequencies are plotted against the middle scores.

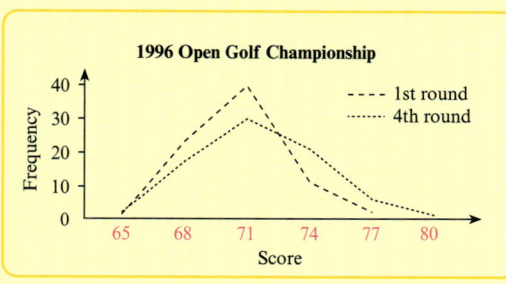

Exercise 16.1
Histograms and frequency polygons

1 Use the histograms to give the modal class for:
a the 1st round scores   b the 4th round scores

2 Golfers aim for as low a score as possible in a round.
a Do you think the golfers did better in the 1st round or the 4th round?
b Use the frequency polygons to explain your answer.

3 These are the ages of the 153 golfers who played in the 1996 Championship.

Age Frequency	17–27	28–38	39–49	50–60	Total
Qualifiers	18	34	22	3	77
Non-qualifiers	14	52	9	1	76

a List the eleven different ages in the 17–27 class.
b Which age is the mid-class value?
c List the mid-class values for each class in the age data.

Qualifiers are those golfers who played in all 4 rounds. Non-qualifiers only played in the first 2 rounds.

4 a Copy these axes on to squared paper.
  b Label your horizontal axis.
  c Draw frequency polygons to compare the ages of the qualifiers and non-qualifiers.

5 Do you think frequency polygons would give a good comparison if there were 97 qualifiers and 56 non-qualifiers? Give reasons for your answer.

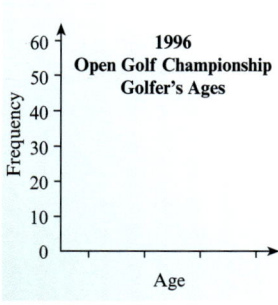

Thinking ahead to ...
estimating the range

1996 Open Golf Championship 4th round scores						
Score	64–66	67–69	70–72	73–75	76–78	79–81
Frequency	2	17	30	21	6	1

A List the three possible scores which could be:

    a the lowest score   **b** the highest score.

B List the five possible values for the range of the 4th round scores.

C Which single value do you think is the best one to use for the range? Why?

Estimating the range and the total

You cannot calculate the exact range of a set of grouped data because:
- the lowest value could be any value in the first class
- the highest value could be any value in the last class.

- The best estimate of the range is given by:
  - the difference between the first and last mid-class values

1996 Open Golf Championship 1st round scores – Qualifiers					
Score	64–66	67–69	70–72	73–75	76–78
Frequency	1	23	40	11	2

This value could be
64, 65, or 66.
The first mid-class value is:
65

This value could be
76, 77, or 78.
The last mid-class value is:
77

The best estimate of the range is **12** (77 – 65).

Exercise 16.2
Estimating the range and the total

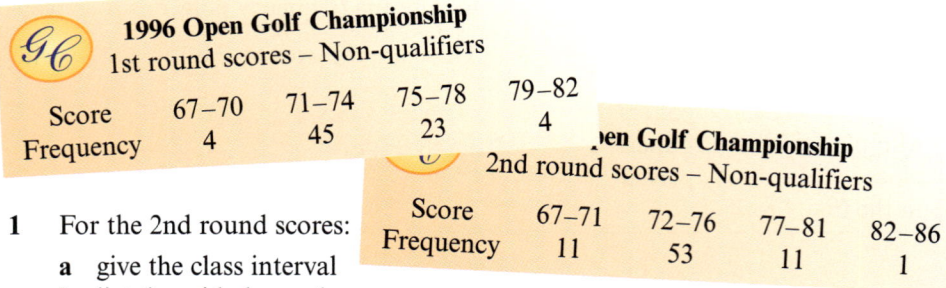

1996 Open Golf Championship 1st round scores – Non-qualifiers				
Score	67–70	71–74	75–78	79–82
Frequency	4	45	23	4

1996 Open Golf Championship 2nd round scores – Non-qualifiers				
Score	67–71	72–76	77–81	82–86
Frequency	11	53	11	1

1 For the 2nd round scores:

    a give the class interval

    b list the mid-class values

    c estimate the range of the scores.

The mid-class values here are not integers.

2 Repeat Question **1** for the 1st round scores.

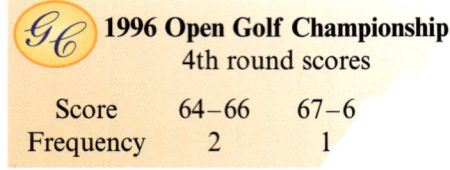

1996 Open Golf Championship 4th round scores		
Score	64–66	67–6
Frequency	2	1

3 **a** List the five possible totals of the scores in the 64–66 class.

    b Multiply the mid-class value by the frequency.

    c Why is this the best estimate of the total of the scores for the class?

The **decathlon** is held over two days and consists of 10 events.

Day 1	Day 2
100 m	110 m hurdles
Long jump	Discus
Shot	Pole vault
High jump	Javelin
400 m	1500 m

1996 Olympic Games – Decathlon (Day 2)
Points scored in each of the 5 events by the top 10 decathletes

Points	400–499	500–599	600–699	700–799	800–899	900–999	1000–1099	Total
Frequency	1	0	8	10	17	10	4	50

♦ To estimate the total of a grouped frequency distribution:
  ❖ write the data in vertical columns [A]
  ❖ find the mid-class value for each class [B]
  ❖ estimate the total for each class [C] (Frequency × Mid-class value)
  ❖ add up the estimated class totals [D]

	[A]		[B] Mid-class value	[C] Estimated class total
	Points	Frequency		
	400–499	1	450	450
	500–599	0	550	0
	600–699	8	650	5200
	700–799	10	750	7500
	800–899	17	850	14450
	900–999	10	950	9500
	1000–1099	4	1050	4200
	Totals	50		**41300** [D]

The **exact** mid-class value of the 400–499 class is 449.5
You can use the simpler value of 450 because you are only calculating an **estimate** of the total points score.

Exercise 16.3
Estimating the total

1

1996 Olympic Games – Decathlon (Day 1)
Points scored in each of the 5 events by the top 10 decathletes

Points	700–799	800–899	900–999	1000–1099	Total
Frequency	8	23	16	3	50

Calculate an estimate of the total points scored on day 1.

Alex Kruger (GB) retired on day 1 of the decathlon with a knee injury.

2

1996 Olympic Games – Decathlon
Total points scored by each decathlete

Points	6500–6999	7000–7499	7500–7999	8000–8499	8500–8999	Total
Frequency	1	1	7	16	6	31

Calculate an estimate of the total points scored by the decathletes.

3 How many points do you think each decathlete scored on average?

4

The throws in the class 68– are greater than, or equal to, 68 m but less than 70 m.

1996 Olympic Games – Discus Finals

Distance (metres)	56–	58–	60–	62–	64–	66–	68–	Total
Frequency (of throws) Women	2	5	10	13	13	4	2	49
Frequency (of throws) Men	2	3	8	16	13	1	2	45

Use mid-class values: 57, 59, 61, 63, 65, 67, 69.

Calculate an estimate of the total distance thrown by:
a women **b** men

Estimating the mean

When comparing sets of data, you can only use the total of each set when the sets have the same total frequency.

When the total frequencies are different, you should compare the means of the sets.

1996 Olympic Games – Women's Discus Final

Distance (metres)	56–	58–	60–	62–	64–	66–	68–	Total
Frequency	2	5	10	13	13	4	2	49

♦ To estimate the mean of a grouped frequency distribution:
  ❖ calculate an estimate of the total of the data

Distance (metres)	Frequency	Mid-class value	Estimated class total
56–	2	57	114
58–	5	59	295
60–	10	61	610
62–	13	63	819
64–	13	65	845
66–	4	67	268
68–	2	69	138
Totals	49		3089

  ❖ divide the estimate of the total by the total frequency.

Estimate of mean distance $= \dfrac{3089}{49} = $ **63.0 m** (to 3 sf)

**Exercise 16.4
Estimating the mean**

1

1996 Olympic Games – Men's Discus Final

Distance (metres)	56–	58–	60–	62–	64–	66–	68–	Total
Frequency	2	3	8	16	13	1	2	45

Calculate an estimate of the mean distance thrown in the men's discus final.

2 Do you think the women or the men threw better in their final?
Give reasons for your answer.

3

The 16.0– class has a mid-class value of 16.5

1996 Olympic Games – Women's Shot Final

Distance (metres)	16.0–	17.0–	18.0–	19.0–	20.0–	Total
Frequency	1	4	23	11	2	41

Calculate an estimate of the mean distance thrown in the women's shot final.

4

1996 Olympic Games – Men's Shot Final

Distance (metres)	19.0–	19.5–	20.0–	20.5–	21.0–	21.5–	Total
Frequency	5	11	17	8	0	1	42

Calculate an estimate of the mean distance thrown in the men's shot final.

5 Use your answers to Questions **3** and **4** to compare the women and men.

6 **a** Investigate the effect of different sizes of class interval on the estimate of the mean.
   b Does your investigation support the statement:
   "The smaller the class interval, the better the estimate of the mean"?

Make up your own data and group it in different ways.

Testing a hypothesis

The reaction time is the length of time between the starter's gun firing and the rear foot leaving the starting block.

s is the abbreviation for seconds.

◆ A statement which can be tested by analysing data is a **hypothesis**.

For example, the following data can be used to test the hypothesis: "Athletes' reaction times are faster in finals than in semifinals"

Men's 100 m Final

	Reaction time (s)
Bailey	0.174
Fredericks	0.143
Boldon	0.164
Mitchell	0.145
Marsh	0.147
Ezinwa	0.157
Green	0.169
Christie	DISQ

Men's 200 m Final

	Reaction time (s)
Johnson	0.161
Fredericks	0.200
Boldon	0.208
Thompson	0.202
Williams	0.182
Garcia	0.229
Stevens	0.151
Marsh	0.167

110 m Hurdles Final

	Reaction time (s)
Johnson	0.170
Crear	0.124
Schwarthoff	0.164
Jackson	0.133
Valle	0.179
Swift	0.151
Vander-Kuyp	0.167
Batte	0.160

Linford Christie (GB), the 100 m Gold Medal winner in 1992, was disqualified after a second false start when his recorded reaction time was 0.086 s.

Scientists believe a reaction time under one-tenth (0.100) of a second is impossible.

Women's 100 m Final

	Reaction time (s)
Devers	0.166
Ottey	0.166
Torrence	0.151
Sturrup	0.176
Trandenkova	0.151
Voronova	0.133
Onyali	0.174
Pintusevych	0.176

Women's 200 m Final

	Reaction time (s)
Perec	0.174
Ottey	0.194
Onyali	0.231
Miller	0.172
Malchugina	0.198
Sturrup	0.165
Cuthbert	0.175
Guidry	0.207

100 m Hurdles Final

	Reaction time (s)
Engquist	0.132
Bukovec	0.164
Girard-Leno	0.133
Devers	0.189
Rose	0.179
Freeman	0.181
Shekhodanova	0.175
Goode	0.160

1996 Olympic Games – Reaction Times
Semifinals – 100 m, 200 m, 100 m/110 m Hurdles

Reaction time (s)	0.120–	0.140–	0.160–	0.180–	0.200–	0.220–	0.240–	0.260–	Total
Frequency	7	24	30	22	4	3	3	1	94

Exercise 16.5
Testing a hypothesis

1 Group the data for the six finals in the same way as the semifinals data.

2 a Calculate an estimate of the range of the reaction times for:
    i the semifinals   **ii** the finals.
  b Calculate an estimate of the mean reaction time for:
    i the semifinals   **ii** the finals.

3 Do you think the hypothesis is true or false? Why?

There were exactly twice as many semifinalists as finalists.

4 a Copy your grouped frequency distribution from Question **1**, and double each of the frequencies.
  b Draw a frequency polygon to show this new distribution.
  c On the same diagram, draw a frequency polygon for the semifinalists.
  d Do you think your frequency polygons give a fair comparison of the reaction times in the finals and semifinals? Explain your answer.

Thinking ahead to ...
measuring and accuracy

A Which of these numbers could be rounded to 12.7?

| 12.73 | 12.76 | 12.68 | 12.749 | 12.75 |

B A number has been rounded to 1 dp to give 9.3.
Give three possible values for the number:

   a to 2 dp   **b** to 3 dp.

Measuring and accuracy

◆ Data which results from measuring, such as heights, distances, weights
and times, can be given to different degrees of accuracy.
For instance, times are often given to:
       the nearest one-tenth (0.1) of a second, or
       the nearest one-hundredth (0.01) of a second, or
       the nearest one-thousandth (0.001) of a second.

◆ The **limits of accuracy** of a measurement are the values between which
the exact measurement must lie.
For instance, the limits of accuracy of a time measured as 9.83 seconds are:
       9.825 seconds and **9.835** seconds.

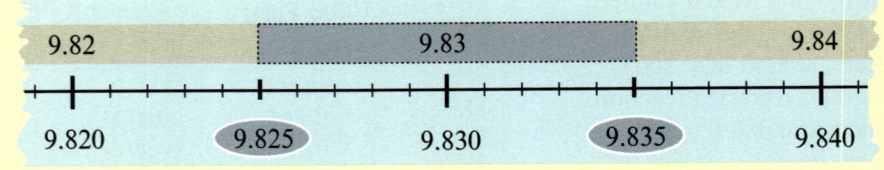

Exercise 16.6
Measuring and accuracy

1 Draw a diagram to show the limits of accuracy of:

   a 9.84 seconds   **b** 7.12 metres   **c** 0.82 metres   **d** 0.174 seconds

2 Give the limits of accuracy of these measurements.

   a 12.38 seconds   **b** 9.2 metres   **c** 7.3 seconds   **d** 7.30 seconds

◆ Times and distances in athletics events are not measured to the **nearest** unit.
All times are rounded up to the **next** one-hundredth of a second.
For instance, the limits of accuracy of a time given as 9.83 seconds are:
       9.820 seconds and **9.830** seconds.

Exercise 16.7
Measuring in athletics

1 Gail Devers and Merlene Ottey were given the
same time in the final of the women's 100 m.
Give the limits of accuracy of their time.

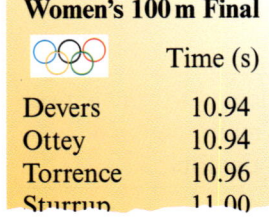

Women's 100 m Final

	Time (s)
Devers	10.94
Ottey	10.94
Torrence	10.96
Sturrup	11.00

All distances are rounded
down to the next centimetre.

2 Jackie Joyner-Kersee jumped 7.00 metres in
the women's long jump final.
Give the limits of accuracy of her distance.

3 Explain why it is appropriate in athletics events to round times up to the
next unit, and distances down to the next unit.

Drawing a cumulative frequency curve

1996 Olympic Games – Decathlon (Day 2)
Points scored in each of the 5 events by the top 10 decathletes

Points	400–499	500–599	600–699	700–799	800–899	900–999	1000–1099	Total
Frequency	1	0	8	10	17	10	4	50

♦ To draw a cumulative frequency curve:
  ❖ construct a cumulative frequency table

Points	<400	<500	<600	<700	<800	<900	<1000	<1100
Cumulative frequency	0	1	1	9	19	36	46	50

If you include a cumulative frequency of 0 in your table then you have a point to start the curve from: (400, 0)

  ❖ plot the cumulative frequencies on a graph
  ❖ join the points with a smooth curve.

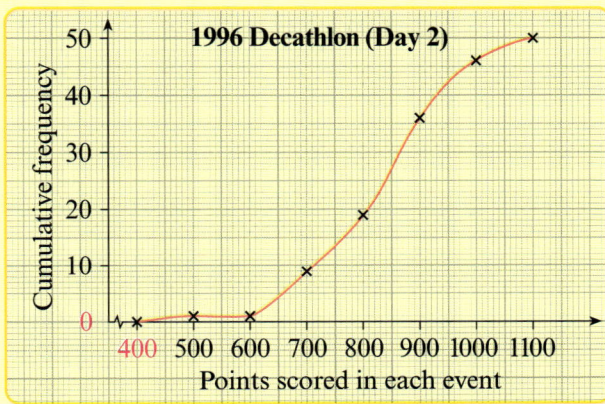

Exercise 16.8
Drawing cumulative frequency curves

1

1996 Olympic Games – Decathlon (Day 1)
Points scored in each of the 5 events by the top 10 decathletes

Points	700–799	800–899	900–999	1000–1099	Total
Frequency	8	23	16	3	50

a Copy and complete the cumulative frequency table below.

Points	<700	<800	<900	<1000	<1100
Cumulative frequency	0				

b Use your table to draw a cumulative frequency curve.

2

Steve Backley (GB) won the javelin silver medal with the very first throw of the final.

1996 Olympic Games – Men's Javelin Final

Distance (metres)	76–	78–	80–	82–	84–	86–	88–	Total
Frequency	2	4	14	13	8	6	1	48

a Copy and complete the cumulative frequency table below.

Distance (metres)	<76	<78	<80	<82	<84	<86	<88	<90
Cumulative frequency	0	2	6					

b Draw a cumulative frequency curve for the men's javelin final.

3 Using your answers to Question **2**, can you think of a way to find the median distance thrown in the final? Explain your method.

Estimating the median and the interquartile range

There were 48 throws in the final, so the **exact median** distance is halfway between the 24th and 25th longest.

It is impossible to find these distances from the table, so you can only **estimate** the median.

As you are only estimating distances within the classes, you can divide the 48 throws into four quarters, using the **12th**, **24th**, and **36th** longest (cumulative frequencies 12, 24, and 36).

◆ A cumulative frequency table shows which class the median is in.

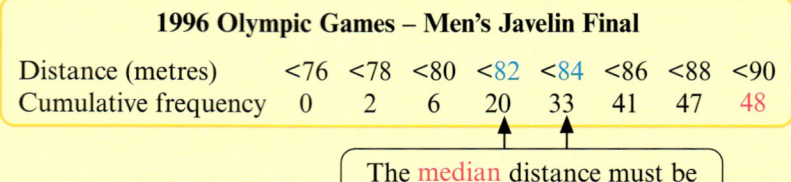

1996 Olympic Games – Men's Javelin Final								
Distance (metres)	<76	<78	<80	<82	<84	<86	<88	<90
Cumulative frequency	0	2	6	20	33	41	47	48

The median distance must be between 82 m and 84 m.

◆ To estimate median and quartiles from a cumulative frequency curve:
   ❖ divide the cumulative frequency into four quarters
   ❖ go across to the curve, then go down and read off each value.

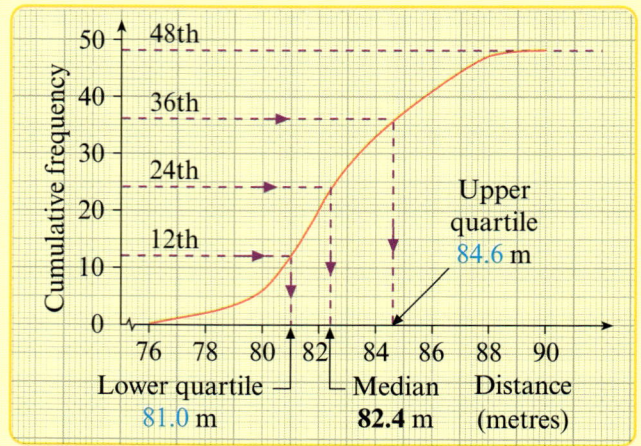

Estimate of median distance = **82.4 m**

Estimate of interquartile range = 84.6 – 81.0
= **3.6 m**

Exercise 16.9
Estimating the median and interquartile range

1

⬤⬤⬤	1996 Olympic Games – Men's Hammer Final					
Distance (metres)	72–	74–	76–	78–	80–	Total
Frequency	1	11	23	13	4	52

a Copy and complete the cumulative frequency table below.

Distance (metres)	<72	<74	<76	<78	<80	<82
Cumulative frequency	0	1	12			

b Draw a cumulative frequency curve for the men's hammer final.

2 Use your cumulative frequency curve to estimate:
   a the median distance thrown   b the interquartile range.

3 Compare the distances thrown in the men's hammer final and the men's javelin final.

4 a From page 193, copy the cumulative frequency table and the cumulative frequency curve for the decathlon (day 2).
   b Estimate the median points score.
   c Estimate the interquartile range of the scores.

Use cumulative frequencies 12.5, 25, and 37.5 to divide the data into four quarters.

Use cumulative frequencies 12.5, 25, and 37.5 to divide the data into four quarters.

5 Use your answers to Exercise 16.8, Question **1** to estimate:

    a the median points score on day 1 of the decathlon

    b the interquartile range of the scores.

6 Compare the points scored on days 1 and 2 of the decathlon.

7

1996 Olympic Games – Women's Javelin Final

Distance (metres)	56–	58–	60–	62–	64–	66–	Total
Frequency	7	11	10	10	7	1	46

    a Make a cumulative frequency table for the women's javelin final.

    b Draw a cumulative frequency curve.

8   **a** Decide how to divide the data into four quarters.

    b Estimate the median distance thrown and the interquartile range.

Estimating cumulative frequencies

♦ You can estimate cumulative frequencies from a cumulative frequency curve.

Example   Estimate how many throws were:

    a less than 81.5 m   **b** greater than 85 m.

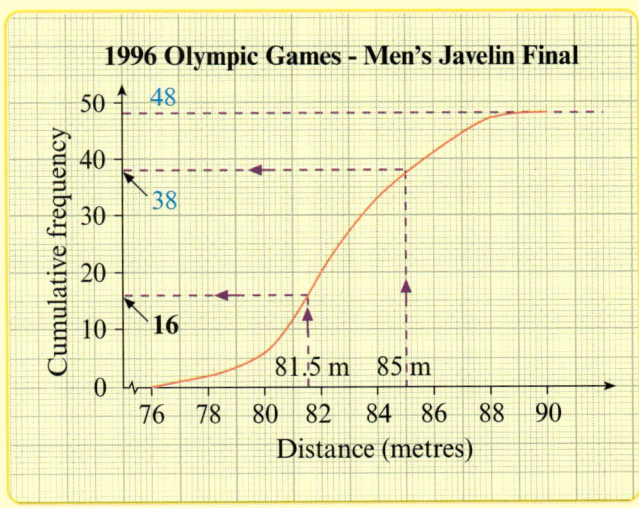

    a Estimated number of throws less than 81.5 m = **16**

    b Estimated number of throws greater than 85 m = 48 – 38

                                                    = **10**

Exercise 16.10
Estimating cumulative frequencies

1 Use your cumulative frequency curve from Exercise 16.9, Question **1** to estimate how many throws were less than 75 m.

2 The winning throw in the 1976 men's hammer final was 77.52 m. Estimate how many throws were greater than this in 1996.

3

Jonathan Edwards (GB) won the silver medal in the triple jump.

1996 Olympic Games – Men's Triple Jump Final

Distance (metres)	15.5–	16.0–	16.5–	17.0–	17.5–	18.0–	Total
Frequency	1	7	14	6	2	1	31

    a Draw a cumulative frequency curve for the men's triple jump final.

    b Estimate how many jumps were greater than 16.8 m.

Moving averages

A **moving average** is used to smooth out the changes in a set of data that varies over a period of time.

A moving average can often give a better idea of any trend shown in a set of data.

Example

This data gives the umbrella sales per week for a department store.

Week	1	2	3	4	5	6	7	8	9	10
Umbrellas sold	4	8	3	7	5	0	16	2	6	10

This set is the raw data gained from sales.

What can you say about the trend for umbrella sales?

Taking 3-week moving averages:

Average for weeks 1 to 3 : $(4 + 8 + 3) \div 3 = 5$
2 to 4 : $(8 + 3 + 7) \div 3 = 6$
3 to 5 : $(3 + 7 + 5) \div 3 = 5$
4 to 6 : $(7 + 5 + 0) \div 3 = 4$
5 to 7 : $(5 + 0 + 16) \div 3 = 7$
6 to 8 : $(0 + 16 + 2) \div 3 = 6$
7 to 9 : $(16 + 2 + 6) \div 3 = 8$
8 to 10 : $(2 + 6 + 10) \div 3 = 6$

This set of data is smoother than the raw data set.

These are called

3-point moving averages

because they each use these three data values.

So the 3-weekly moving averages of sales are:

5, 6, 5, 4, 7, 6, 8, 6

You can see the effect of moving averages on a graph.

Each moving average is plotted at the end of the group of 3 pieces of data.

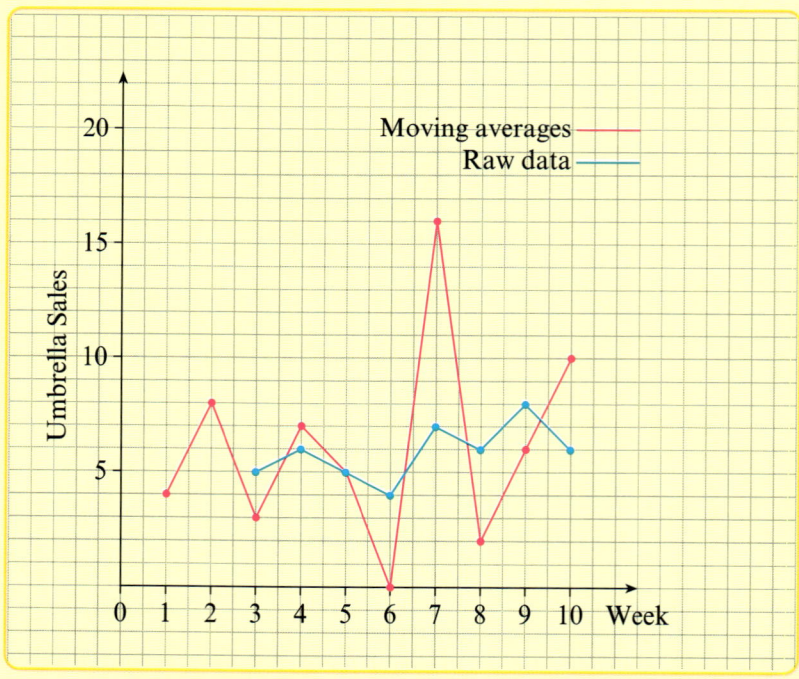

As for the trend, if anything sales of umbrellas seem to be rising.

Exercise 16.11
Moving averages

A moving average can become a decimal value.

1 Calculate the three-point moving averages for each data set.

a 24, 7, 5, 20, 2, 14, 8, 11, 8, 11, 23
b 8, 12, 16, 11, 3, 13, 2, 12, 1, 23, 3, 19
c 11, 14, 2, 20, 8, 5, 5, 35, 2, 15, 4, 11

2 Calculate the five-point moving averages for each data set.

a 18, 20, 27, 34, 16, 7, 52, 30, 24, 12, 8, 6
b 12, 9, 23, 14, 6, 8, 12, 15, 32, 40, 7
c 16, 22, 31, 18, 12, 4, 6, 19, 22, 25
d 14, 24, 0, 7, 17, 6, 23, 37, 2, 14, 18

3 Calculate the four-point moving averages for each data set.

a 12, 23, 14, 16, 7, 4, 18, 24, 35, 16, 7, 14
b 7, 24, 16, 35, 4, 18, 19, 18, 23, 6, 52
c 19, 6, 8, 7, 62, 34, 25, 32, 28, 46, 70
d 56, 14, 22, 34, 6, 7, 19, 0, 42, 57, 62, 31

4 This data gives walking boot sales per week.

Week	1	2	3	4	5	6	7	8	9	10	11	12
Sales (pairs)	5	12	14	1	3	6	10	5	6	15	6	5

a Calculate the four-point moving average for the data.
b On one pair of axes draw graphs to show:
　i the raw data
　ii the four-point moving averages.
c Comment on any trend you can identify in boot sales.

5 This data gives watch sales per week from a catalogue shop.

Week	1	2	3	4	5	6	7	8	9	10	11	12	13	14	15	16	17
Watches sold	23	28	14	16	9	23	8	14	21	19	3	18	19	1	9	13	8

a Calculate the four-point moving average for the data.
b On one pair of axes draw graphs to show:
　i the raw data
　ii the four-point moving averages
c Comment on any trend you can identify in watch sales.

End points

You should be able to ...   ... so try these questions

A Use histograms and
frequency polygons

1996 Olympic Games – Reaction Times
Semifinals & Finals – 100 m, 200 m, 100 m/110 m Hurdles

Reaction time (s)	0.12–	0.14–	0.16–	0.18–	0.20–	0.22–	0.24–	0.26–	Total
Frequency Men	7	22	23	10	3	2	2	1	70
Women	5	10	30	17	5	3	1	0	71

A1 Draw a histogram to show the reaction times for:
   a men   **b** women.

A2 Give the modal class for:
   a men   **b** women.

B Estimate the median and
the interquartile range

B1 **a** Make a cumulative frequency table for the men's reaction times.
   b Draw a cumulative frequency curve.
   c Use your cumulative frequency curve to estimate:
     i the median reaction time   **ii** the interquartile range.

C Calculate estimates of the
mean and the range

C1 Estimate the range of the reaction times in:
   a the women's hurdles   **b** the men's hurdles.

C2 Calculate an estimate of the mean reaction time in:
   a the women's hurdles   **b** the men's hurdles.

D Compare grouped
frequency distributions

D1 Compare the reaction times in the women's and men's hurdles.

E Estimate cumulative frequencies

E1

1996 Olympic Games – Women's Triple Jump Final

Distance (metres)	13.5–	14.0–	14.5–	15.0–	Total
Frequency	15	21	10	1	47

   a Draw a cumulative frequency curve for the women's triple jump final.
   b Use your curve to estimate how many jumps were:
     i less than 13.8 m   **ii** greater than 14.4 m

F Use moving averages

F1 This data shows numbers of students late for school.

Week	1	2	3	4	5	6	7	8	9	10	11	12	13	14
Lates	12	6	3	18	9	6	6	15	9	12	0	18	18	3

   a Calculate the three-point moving average for the data.
   b On a pair of axes draw graphs to show:
     i the raw data
     ii the moving average data.
   c Comment on any trend you can identify in the data.

Some points to remember

- ♦ You cannot find the modal value of a set of data when it is grouped,
but you can identify the modal class (the class with the highest frequency).

- ♦ You cannot find **exact** values of the mean, the median, the range and the interquartile range,
of grouped data, but you can calculate an **estimate** of each.

- ♦ When you estimate the mean of grouped data, usually the smaller the class interval the closer
the estimate is to the exact mean.

Starting points
You need to know about ...   ... so try these questions

A Writing fractions as decimals and percentages

◆ To write a fraction as a decimal:
  ❖ divide the numerator by the denominator.

$\frac{3}{8} = 3 \div 8 = 0.375 = 37.5\%$ $\frac{5}{6} = 5 \div 6 = 0.83... = 83.\dot{3}\%$

$\frac{7}{4} = 7 \div 4 = 1.75 = 175\%$ $\frac{5}{3} = 5 \div 3 = 1.6... = 166.\dot{6}\%$

B Calculating a percentage of a given amount

◆ To calculate 50% of an amount simply:
  calculate half the amount by dividing by 2 (50% = $\frac{1}{2}$)
  To calculate 25% of an amount simply:
  calculate a quarter of the amount by dividing by 4.
  Some other percentages can be calculated in a similar way.

  To calculate 65% of an amount is not as easy.
  This is one way: calculate 1%, and then use it to find 65%.

Example Find 65% of 420 kg:

    100%   420
    Divide both sides by 100:
    1%   4.20
    Multiply both sides by 65:
    65%   273

So 65% of 420 kg is 273 kg.

C Interpreting a calculator display in calculations with money

Interpreting a calculator display correctly is important, particularly when you are dealing with units of money.

For example, when you calculate 24% of £15 the calculator may display the answer as:

    `3.6`

The calculation is in pounds so the answer is: **£3.60**
        (£3.60 is 24% of £15)
When you calculate 24% of £1.50 the calculator displays the answer as:

    `0.36`

The calculation is in pounds so the answer is: **£0.36**
        (£0.36 is 24% of £1.50)
But an amount of money is not usually written as £0.36.
It is more likely that the amount is given as 36 pence.
So 24% of £1.50 is 36 pence.

A1 Write each of these as a decimal and as a percentage:

a $\frac{3}{4}$ b $\frac{5}{8}$ c $\frac{8}{5}$ d $\frac{5}{4}$

e $\frac{6}{5}$ f $\frac{13}{20}$ g $\frac{8}{32}$ h $\frac{9}{6}$

i $\frac{7}{16}$ j $\frac{16}{7}$ k $\frac{9}{8}$ l $\frac{18}{10}$

B1 Calculate:
a 50% of 650 miles
b 75% of £400
c 10% of 20 marks
d 72% of 3450
e 25% of 40 marks
f 36% of 475 kg
g 58% of £415
h 12% of 3 kg
i 7% of £2.50
j 4.5% of 360 metres
k 38% of £4.50
l 15% of 12 cm
m 25% of £25
n 80% of 3 miles
o 40% of 25 km
p 15% of £3200
q 18% of 615 miles
r 36% of 25 cm
s 60% of 14 metres
t 12% of 3.65 km.

C1 In each of these calculations interpret your calculator display to give the answer in the units asked for.
a Calculate 16% of £85 (answer in pounds).
b Calculate 14% of 4 cm (answer in millimetres).
c Find 44% of 3.5 kg (answer in grams).
d Find 9% of £2.27 (answer in pence).
e What is 78% of 65 mm:
  i in millimetres
  ii in centimetres?
f What is 12.5% of 72 pence:
  i in pence
  ii in pounds?
g Give 35% of 4 tonnes in:
  i tonnes   ii kilograms.

Writing one number as a percentage of another

When you compare two numbers, you can think of:
one number as a percentage of the other.

For instance, with the numbers 18 and 36 one comparison is:
18 is half of 36, or 18 is 50% of 36.

So 18 as a percentage of 36 is 50%.

When you compare numbers you might have to approximate one number as a percentage of the other.

Example

With the numbers 17 and 60, 17 is a little more than one quarter of 60.
So 17 as a percentage of 60 is a little more than 25%.

You can also compare the numbers the other way round, e.g.

36 as a percentage of 18

As 36 is twice 18:
36 as a percentage of 18 is 200%

Exercise 17.1
Comparing numbers

1 What is 35 as a percentage of 70?

2 Give 64 as a percentage of 16.

3 Is 22 as a percentage of 28 a little more, or a little less than 75%? Explain your answer.

4 A number p as a percentage of 80 is 10%.
 i What is the number p?
 ii Explain how you calculated your answer.

5 Roughly, what is 15 as a percentage of 32?

6 Two numbers k and j are chosen so that this rule is true:

k as a percentage of j is 75%

 a Jo chooses the value 12 for k. What is the value of j?
 b Rashid chooses the value 28 for j. What is the value of k?
 c List four values you choose for k, and for each give the value of j that makes the rule true.

7 Ian chooses a whole number n and describes it in this way:
 n as a percentage of 40 is a little more than 20% but not as much as 25%.
 a What is the number chosen by Ian?
 b Explain how you calculated Ian's number.

8 This table shows the amount Jim spent each month on travel and food.
 For each month give the amount spent on travel as a percentage of the total amount Jim spent.

	Amount spent on travel	Amount spent on food	Total amount spent
Oct	£6	£5	£24
Nov	£8	£20	£32
Dec	£12	£15	£60
Jan	£12	£16	£48
Feb	£12	£2	£16
Mar	£2	£5	£20

9 For the amount spent on food as a percentage of the total spent:
 a In which months was this exactly 25%?
 b In which month was it a little more than 30%?
 c What was it in November?

With the two numbers 18 and 36:
to find 18 as a percentage of 36, think of the comparison in this way:

- 18 is half of 36   or $18 \div 36 = 0.5$
- 18 is 50% of 36   or $0.5 \times 100 = 50\%$

This can be a single calculation:
$$(18 \div 36) \times 100 = 50\%$$
So 18 as a percentage of 36 is 50%

This method can be used to give any number as a percentage of another.
With any two numbers p and t:

You can **calculate p as a percentage of t in this way**: $(p \div t) \times 100$

Example

Jenny and Bruce walked from John O'Groats to Land's End for charity.
John O'Groats to Land's End is 868 miles.
By the end of day 4 they had walked 113 miles.
What percentage of the total distance is this?

To calculate 113 as a percentage of 868

$$(113 \div 868) \times 100 = 13.018 \ldots$$

By the end of day 4, they had travelled 13.0% (1 dp) of the total distance.

Exercise 17.2
Calculating one number as a percentage of another

1   Calculate, giving your answer correct to 2 dp:

a   16 as a percentage of 36   **b**   52 as a percentage of 80
c   14 as a percentage of 48   **d**   85 as a percentage of 184
e   12 as a percentage of 15   **f**   1.5 as a percentage of 12
g   0.75 as a percentage of 20   **h**   16 as a percentage of 12
i   35 as a percentage of 21   **j**   15 as a percentage of 8.

2   A driving school uses this as part of their advertising:

> **70% of our students pass their test first time !**

In July 1997, they had 55 people who took their test for the first time, and 38 of them passed the test.

a   What percentage of those people who took their test for the first time in July 1997, passed the test?
b   How accurate is the advertising for the driving school?
Explain your answer.

3   A sports club was sent a bill for repairs to its video camera.
The bill was for a total of £65.70, and only £9.20 of this was for parts.

What is the charge for parts as a percentage of the total bill?
Give your answer correct to the nearest whole number.

4   An old stadium had seating for 23 500 spectators.
The stadium was rebuilt, with seating in the new stadium for 40 000.

a   Give the seating of the new stadium as a percentage of the old seating.
Give your answer correct to the nearest whole number.
b   Is your answer about what you expected?
Give reasons for your answer.

Thinking ahead to ...
percentage changes

Often percentages are used to describe
a change in an amount.
For example, in supermarkets special offers
can be shown as:

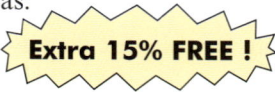

For the customer:

A How much extra is free?
Is it more or less than a quarter of the
amount for the normal price?

B What fraction would you use to
describe the extra free amount?

C How many ml of the product are in the special offer pack?

Percentage changes

When you increase an
amount by a percentage, you
will end up with **more than**
the **100%** you started with.

For example:
with an increase of 25%
there is the 100% you start
with, plus the 25% increase.

100% + 25% = 125%

As a percentage of the start
value, the end value is 125%.

When a value increases by a percentage, the final value can be calculated in
different ways. Two methods are shown here.

Example

Toothpaste is sold in tubes containing 150 ml.
In a special offer, 12% extra toothpaste is put in the tube at the same price.
How much toothpaste is in the special offer tube?

Method 1

❖ Calculate 12% of 150 ml.
❖ Add the extra amount to
the 150 ml.

So 150 ÷ 100   gives 1%
150 ÷ 100 × 12 gives 12%
So 12% of 150 is 18

The special offer tube contains
168 ml of toothpaste
(168 = 150 + 18)

Method 2

Think of the toothpaste in this way:

100% of the contents is 150 ml

The special offer tube has 12%
extra, so it must contain:

112% of 150 ml

112 % as a fraction is $\frac{112}{100}$

112 % as a decimal is 1.12

Calculate 112% of 150:
150 × 1.12 = 168

The special offer tube contains
168 ml of toothpaste.

Exercise 17.3
Increasing a value by
a certain percentage

1 In a special offer, the 440 grams of coffee in a jar is to be increased by 10%.
a What percentage of the 440 grams of coffee are in the special offer jar?
b Calculate the amount of coffee in the special offer jar.

2 **a** Increase 350 kg by 20%   **b** Increase 447 km by 15%
c Increase 4.50 metres by 22%   **d** Increase £35 000 by 8%

3 In 1992 in the USA there were a total of 143 081 443 registered cars.
By the year 2010 it is estimated that this total will increase by 44%.
Estimate the number of registered cars in the USA in 2010.

4 A crisp manufacturer sells 25 gram bags of crisps. They decide to increase the
weight of crisps in a bag by 4%. Give the new weight of crisps per bag.

When you decrease an amount by a percentage, you will end up with **less than** the **100%** you started with.

Example

With a decrease of 25% there is the 100% you start with, minus the 25% decrease.

$$100\% - 25\% = 75\%$$

As a percentage of the start value, the end value is 75%.

You can also decrease an amount by a percentage using methods 1 and 2.

For example:

A fast food store decided to decrease the weight of packaging for their regular meals, which weighed 40 grams, by 18%.
Calculate the weight of the new packaging.

Method 1
❖ Calculate 18% of 40 grams.
❖ **Take** this weight from the 40 grams.
So $40 \div 100$ gives 1%
      $40 \div 100 \times 18$ gives 18%
   So 18% of 40 is 7.2

The new regular meal packaging weighs **32.8 grams**
      $(32.8 = 40 - 7.2)$

Method 2
Think of the packaging in this way:

   100% of the contents weighs 40 g

The new packaging weighs 18% less, so it must weigh:

      82% of 40 grams

82 % as a fraction is $\frac{82}{100}$

82 % as a decimal is 0.82

Calculate 82% of 40:

      $40 \times 0.82 = 32.8$

The new regular meal packaging weighs **32.8 grams**.

Exercise 17.4
Decreasing a value by a certain percentage

1   a   Decrease 380 kg by 30%.   b   Decrease 416 km by 25%.
    c   Decrease 22 metres by 5%.   d   Decrease £338 000 by 15%.
    e   Decrease £25.50 by 28%.   f   Decrease 28 600 tonnes by 42%.
    g   Decrease 5 300 000 by 11%.   h   Decrease 1.6 cm by 25%.

2   Anya grows strawberries for supermarkets.
    Last year she used a total of 1560 kilograms of fertilizer.
    Next year, she wants to decrease the amount of fertilizer used by 6%.

    How much fertilizer would you expect to be used next year?
    Give your answer correct to the nearest kilogram.

3   The car ferry *Vista* was built to carry a maximum of 1210 cars.
    New safety rules mean that the number of cars must be reduced by 7%.

    a   What is the maximum number of cars for the *Vista* with the new rules?
    b   Explain the degree of accuracy you used, and why.

4   Before a bypass was built, an estimated 41 000 cars a day passed through the town of Ashington.
    The bypass was supposed to reduce the cars in Ashington by 35%.

    a   Estimate how many cars passed through Ashington, per day, after the bypass had been built.
    b   How many fewer cars per day is this?

5   Last year the fishing boat *Emma K* landed 9210 kg of shell fish.
    This year they expect shell fish landings to be reduced by 17%.

    How much shell fish does the *Emma K* expect to land this year?

6   In a sale items were reduced by 20%. Before the sale a kettle cost £24.50.

    a   What was the sale price of the kettle?
    b   How much was saved by someone who bought the kettle in the sale?

Looking ahead to ...
percentages and VAT

VAT is short for:
Value-added tax.

VAT is a tax added to the price of goods or services. It was introduced on: 1 April 1973 at a standard rate of 10%.

On 18 June 1979 the standard rate was increased to 15%.

On 1 April 1994 the standard rate was increased to 17.5%.

Shopkeepers, traders, and customers have had to calculate VAT since 1973.

When it was first introduced at 10%, one quick method was: "Divide by 10 and add it on."

When VAT was increased to 15%, a quick method was: "Divide by 10, half the answer, and add both amounts on."

A What quick method can you think of to calculate VAT at 17.5%?

B In 1995 Ria fitted a stair carpet and charged £200 + VAT at 17.5%. What was the total charge to fit the carpet?

C Peter bought a cycle tyre and was charged £12.50 + VAT at 17.5%.
  i What did he pay in total for the tyre?
  ii Explain any rounding you did when calculating the total.

Percentages and VAT

To calculate the total price of goods or services including VAT is the same as: increasing the cost price by the percentage VAT.

Example A garden shed is advertised for: **£114.99** + VAT at 17.5% Calculate the total charge for the shed.

117% as a decimal is 1.17.
118% as a decimal is 1.18.

117.5% is half-way between 117% and 118%.

As a decimal, 117.5% must be half-way between 1.170 and 1.180

which is 1.175.

Think of the total charge (cost price + VAT) for the shed in this way:

100% of the cost price + **17.5%** of the price (VAT)

The total charge for the shed is: **117.5%** of its cost price

117.5% as a decimal is 1.175

Calculate 117.5% of £114.99:
$$114.99 \times 1.175 = 135.113 \ldots$$

The total charge for the shed (including VAT) is:

£135.11 (to the nearest penny)

These examples are with VAT at 17.5%.

In all questions that involve VAT you must use the standard rate of VAT at the time.

If you are unsure, ask for the rate of VAT.

Or: you can calculate 17.5% of £114.99 and add this to £114.99

To calculate 17.5% of £114.99
$$114.99 \times 0.175 = 20.123 \ldots$$

The total charge (including VAT) is: £114.99 + £20.123 ...

£135.11 (to the nearest penny)

Exercise 17.5
Calculating prices that include VAT

1 The cost of these items is given without VAT (ex VAT). Calculate the charge, including VAT, for each item.

a camera £16	**b** trainers £44.25	**c** bike £185
d toaster £19.40	**e** pen 72 pence	**f** TV £368.42
g fridge £262	**h** mower £24.55	**i** CD £9.35

2 Jenny was told that repairs to her car would be £245. When she paid she found that the £245 did not include VAT.

  a How much did she pay, including VAT?
  b How much VAT was added to her bill?

Percentages and interest

> From the *Oxford Mathematics Study Dictionary*.

Interest The interest is the amount of extra money paid in return for having the use of someone else's money.

When you borrow, or save money with a Bank, Building Society, The Post Office, a Credit Union, or a finance company, interest is *charged* or *paid*.

Interest is: *charged* on money you borrow and *paid* on money you save.

Interest is: *charged* or *paid* pa (per annum) and *charged* or *paid* at a fixed rate e.g. 6% pa.

> per annum is a term that means each year.

> You borrow £350 for one year at 9%. How much do you pay back in total?
>
> At the end of a year you will pay back:
>
> 100% of the £350 + 9% interest
> You will pay back a total of 109% of £350
>
> Calculate 109% of £350:
> 350 × 1.09 = 381.5
>
> You will pay back a total of **£381.50**

> You save £120 for one year at 4%. How much in total will you have?
>
> At the end of a year you will have:
>
> 100% of your £120 + 4% interest
> You will have a total of 104% of your £120
>
> Calculate 104% of £120:
> 120 × 1.04 = 124.8
>
> You will have a total of **£124.80**

Exercise 17.6
Calculating interest paid for one year

1 To buy a TV, Ewan borrows £340 for one year at a rate of interest of 14% pa. How much, in total, does Ewan pay back for his loan?

2 To buy a new bike for £275, Jess decides to use £120 of her savings and to borrow the rest of the money for a year with an interest rate of 17% pa.
   a How much money will Jess have to borrow to buy the bike?
   b How much will she pay back in total for her loan?
   c In total, how much will she have paid for the bike?

3 Marie won £2500 in a competition.
   She decided to save the money, for a year, at an interest rate of 4.5% pa.
   a At the end of the year, with interest, how much will Marie have in total?
   b How much interest was she paid?

4 A hockey club was given a grant of £84 250. They did not spend the money straight away, but decided to save it for a year at 6.8% pa interest.
   By waiting a year, how much was the grant now worth in total?

Simple interest

> Simple interest is a type of interest not often used these days.

♦ Simple interest is fixed to the sum of money you actually borrow or save.

♦ When you borrow a sum of money:
   simple interest is *charged* for each year on the actual sum borrowed.

♦ When you save a sum of money:
   simple interest is *paid* for each year on the actual money saved.

This is known as the: simple interest formula.

The formula can be given in different forms.

If you use the formula in a different form, make sure you know how to use it to calculate the interest.

The amount of interest added to a loan, or added to a sum of money that is saved can be calculated using this formula:

$$I = P \times R \times T$$

In words, the formula is:

$$Interest = Principal \times Rate \times Time$$

Principal is the amount of money you borrow or save.
Rate is the interest rate pa **as a decimal**.
Time is for how long, usually in years.

Example

Calculate the interest charged on a loan of £750, for 4 years at 7% pa.
Use the formula

$$I = P \times R \times T \text{ with } P = 750, R = 0.07 \ (7\% = \tfrac{7}{100} = 0.07), T = 4$$

$$I = 750 \times 0.07 \times 4$$
$$I = 210$$
The interest charged on this loan is £210.

If the £750 had been saved for 4 years at 7% pa:
The interest paid on the savings would have been £210.

Exercise 17.7
Calculating simple interest

1   Calculate the simple interest charged, or paid for each of these:
   a   £450 borrowed for 6 years with interest at 12% pa
   b   £280 saved for 8 years with a rate of interest of 3% pa
   c   £1500 saved for 15 years at 6% pa. interest
   d   £6000 borrowed for 10 years at an interest rate of 17% pa
   e   £25 borrowed for 20 years with interest at 18% pa.

2   To buy new kit a band borrows £3500 over 10 years at 17% interest pa.
   a   Calculate the amount of interest paid on the loan.
   b   At the end of the loan, how much in total will have been paid for the kit?

3   Scot forgot about the £150 he had saved at an interest rate of 4.5% pa.
   He found it had been earning interest for twelve years.
   How much were these savings worth in total at the end of twelve years?

4   A new bridge will cost an estimated £44 million, and take six years to build.
   If, at the start, all £44 million is borrowed at a simple interest rate of 8.5%,
   estimate the total cost of the bridge.

Compound interest

Like simple interest, compound interest is either charged or paid, but not just on the original sum borrowed or saved. For example, £100 saved for 2 years at 4% can be thought of in this way:

◆ At the end of year 1, the total is £104 (£100 + £4 interest)

◆ At the end of year 2, the total is £108.16 (£104 + £4.16 interest)

In short, compound interest includes interest on interest already *paid* or *charged*.

Compound interest is used by banks, building societies, and shops.

> If money is invested it is saved.

Calculating compound interest this way is time consuming as the number of years increases. There is a formula you can use.

The compound interest formula is:

$$T = P\left(1 + \frac{r}{100}\right)^n$$ where T is the total of the amount plus interest

r is the rate of interest

n is the number of years

P is the amount an invested or borrowed

Example

Calculate the value of an investment of £2500 for 6 years at a compound rate of 4%.

Using the formula $T = P\left(1 + \frac{r}{100}\right)^n$ $r = 4$ and $n = 6$

$$T = 2500\left(1 + \frac{4}{100}\right)^6$$

$$T = 2500(1.04)^6$$

$$T = 2500 \times 1.2653.......$$

$$T = 3163.297.......$$

So after 6 years the investment is worth £3163.30 (nearest penny)

By working year-by-year, you can calculate compound interest and see that it builds to a greater total than simple interest over the same time span.

Example   Calculate the interest on £480 saved for 3 years at 7%

> Round the answers to each calculation to the nearest penny.

♦ Interest paid at the end of year 1 is: £33.60 (£480 × 0.07)
  Interest for year 2 will be calculated on £513.60 (£480 + £33.60)

♦ Interest paid at the end of year 2 is: £35.95 (£513.60 × 0.07)
  Interest for year 3 will be calculated on £549.55 (£513.60 + £35.95)

♦ Interest paid at the end of year 3 is: £38.47 (£549.55 × 0.07)

The total interest paid is: £33.60 + £35.90 + £38.47 = **£107.97**

Exercise 17.8
Calculating compound interest

1   Calculate the total interest paid on these savings at compound interest:

   a   £350 for 3 years at 9% pa   **b**   £1400 for 4 years at 3% pa
   c   £3600 for 5 years at 6% pa   **d**   £12 250 for 3 years at 4% pa
   e   £4050 for 2 years at 12% pa   **f**   £35 250 for 2 years at 12% pa

2   Shelly won £5000, and put it in a savings scheme for five years.
   The savings scheme pays interest at 8% pa compound.

   a   Calculate the total interest paid on this saving.
   b   At the end of five years what was the total in Shelly's saving scheme?
   c   **i**   What would the total be if the only rounding was done at the end?
       ii   What do you think the banks do with the rounding problem?

3   Calculate each of these for compound interest.

  a   £1500 is invested for 8 years at 3%. Find the value of the investment at the end of the term.

  b   Calculate the value of an investment of £4200 over 12 years at 5%.

  c   £6200 is invested for 8 years at 6%. What is the final value of the investment?

  d   Find the final value of an investment of £7800 for 5 years at 6%

4   To buy a car Marie takes out a loan of £4500 at 7% compound interest for 4 years.

  Find the total amount Marie will pay back on her loan.

5   A club borrows £21500 at 3% compund interest for a period of 12 years. Calculate the amount the club will actually pay back for this loan.

> For questions **3** to **5** use the formula.
>
> Give your answers correct to the nearest penny

Reverse percentages

> When you divide by 117.5, you are calculating 1%.
>
> You then multiply by 100% to calculate the 100%.
>
> This is for VAT at 17.5%
>
> Check on the current rate of VAT to find the number to divide by to calculate 1%.

Calculating the original value of something, before an increase or decrease took place, is called 'calculating a reverse percentage'.

Example   The total price of a bike (including VAT at 17.5%) is £146.85 Calculate the cost price of the bike without VAT.

$$\text{Total price of bike} = 100\% \text{ of cost price} + 17.5\% \text{ of cost price}$$
$$146.85 = 117.5\% \text{ of cost price}$$
$$146.85 \div 117.5 \times 100 = 100\% \text{ of cost price}$$
$$124.978... = \text{Cost price}$$

The price of the bike without VAT is £124.98 (to the nearest penny)

Exercise 17.9
Calculating reverse percentages

1   These prices include VAT, calculate each price without VAT. Give your answers correct to the nearest penny.

  a   CD player £135.50   **b**   camera £34.99   **c**   ring £74.99
  d   trainers £65.80   **e**   TV £186.75   **f**   phone £14.49
  g   calculator £49.99   **h**   PC £799.98   **i**   tent £98.99

2   Ella bought a pair of boots for £45 in a sale that made an offer '20% off!'.

  a   What was the non-sale price of the boots?
  b   How much did Ella save buying the boots in the sale?

3   When Mike sold his bike for £35, he said he made a profit of 35%. To the nearest pound, how much did Mike pay for the bike?

4   A 600 gram box of cereal is said to hold '35% more than the regular box'. How many grams of cereal are in the regular box?

End points

You should be able to ...

... so try these questions

A	Write one number as a percentage of another	

A1 What is 12 as a percentage of 48?

A2 Give 60 as a percentage of 20.

A3 Correct to 2 dp, what is 18 as a percentage of 64?

A4 In a 456-page book, 35 of the pages have pictures on them. Give the number of pages with pictures as a percentage of the total number of pages in the book, correct to 1 dp.

B Increase a value by a certain percentage

B1 Increase 480 km by 16%.

B2 A CJ regular size cola is 380 ml. CJ decide to increase the size of their regular cola by 12%. To the nearest ml, give the size of the new regular cola.

C Decrease a value by a certain percentage

C1 Decrease £55.80 by 6%. Give your answer to the nearest penny.

C2 In 1992, there were a total of 42 154 breakdowns on a motorway section. In 1993 the total number of breakdowns fell by an estimated 7%. Estimate the number of breakdowns in 1993 on this motorway section.

D Work with VAT

D1 What is the standard rate of VAT today?

D2 The cost of these items is given without VAT (ex. VAT)
  a a camping stove £34.75   **b** a body-board £158.40
  Calculate the cost, to the nearest penny, of each item including VAT.

E Calculate reverse percentages

E1 The price of a TV is £259.99 including VAT. Calculate the cost of the TV 'ex.VAT' (without VAT).

F Calculate simple interest

F1 Calculate the interest paid on £480 saved for six years, at a simple interest rate of 4%.

F2 £500 was borrowed at a simple interest rate of 9% over 4 years.
  a Calculate the total interest paid on this loan.
  b How much was paid back on the loan in total?

G Calculate compound interest

G1 Calculate the total of these savings at compound interest:
  a £680 for 3 years at 6%   **b** £2575 for 4 years at 3%
  c £3600 for 8 years at 4%   **d** £7150 for 12 years at 2%
  e £1500 for 20 years at 6%   **f** £318 for 4 years at 1%

Some points to remember

- When you work with a calculator it is easier if you think of, and use, percentages as decimals.

- Check that your answer to a calculation is of the right size, and to a sensible degree of accuracy.

- When you are calculating interest, paid or charged, make sure you know whether it is at a rate of simple interest, or compound interest.

Starting points
You need to know about ...

... so try these questions

A The equation of a straight line

♦ You can use the coordinates of points on a line to find the equation of the line.

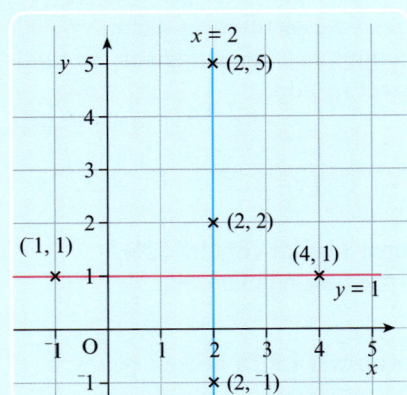

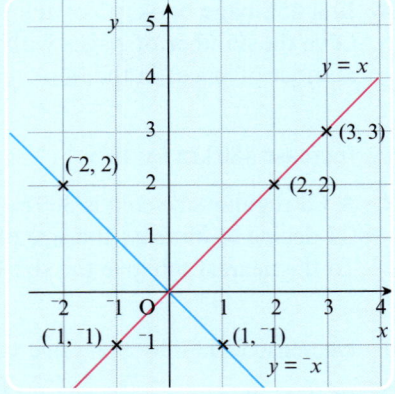

A1

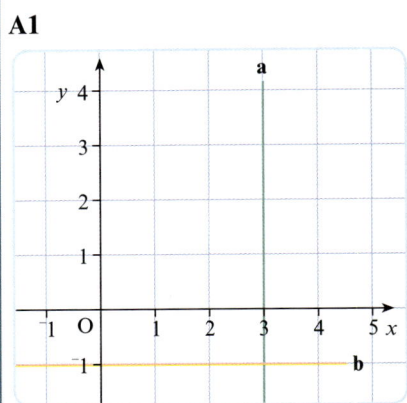

Write down the equation of the lines **a** and **b**.

B Reflections

♦ To describe a reflection you must give the mirror line.
On a grid you can give the equation of the mirror line.

Example

B is the image of object A after a reflection in $y = ^-x$.

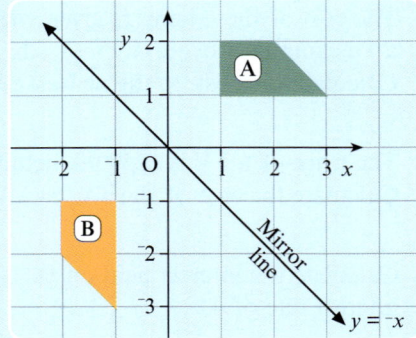

B1 **a** Copy this diagram.

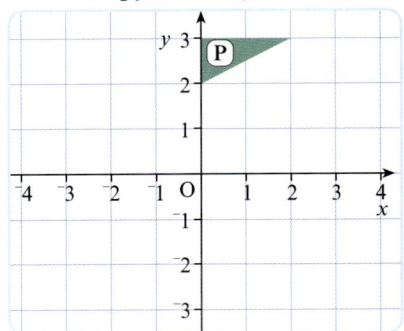

b Draw the image of object P after a reflection in:
  i $y = x$ **ii** $x = ^-1$

C Rotations

♦ To describe a rotation you must give:
  ❖ the angle and direction of rotation
  ❖ the centre of rotation.

♦ On a grid you can give the coordinates of the centre.

Example

  ❖ C is the image of object A after a rotation of $^+90°$ (90° anticlockwise) about (4, 1).
  ❖ D is the image of A after a rotation of $^-90°$ (90° clockwise) about ($^-1$, 2).

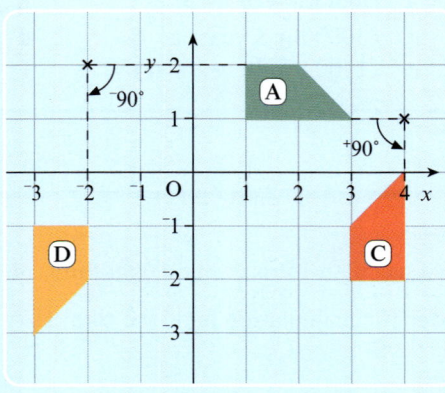

C1

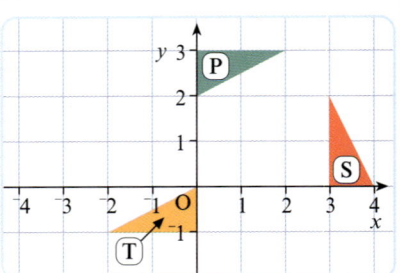

a Triangle S is the image of P after a rotation about (3, 3). What is the angle of rotation?

b Triangle T is the image of P after a rotation of 180°. Give the coordinates of the centre of rotation.

D Translations

♦ A translation only changes the position of a shape.

♦ On a grid you can use a vector to describe a translation.

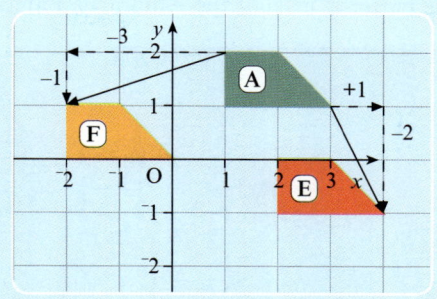

Example

❖ A is translated on to E by the vector $\begin{pmatrix} 1 \\ -2 \end{pmatrix}$.

❖ A is translated on to F by the vector $\begin{pmatrix} -3 \\ 1 \end{pmatrix}$.

D1 These translations map P on to U and V.

Object	Translation	Image
P	$\begin{pmatrix} 2 \\ 0 \end{pmatrix}$	U
P	$\begin{pmatrix} -1 \\ -3 \end{pmatrix}$	V

On your diagram, from Question **B1**, draw and label the image of object P after each of these translations.

E Enlargements

♦ If the scale factor (SF) of an enlargement is **greater than 1**:
   the image is **larger** than the object.

♦ If the scale factor of an enlargement is **less than 1**:
   the image is **smaller** than the object.

♦ If one shape is an enlargement of another:
   the two shapes are **similar**.

Example

G is an enlargement of A with centre (4, ⁻1) and SF 2.

H is an enlargement of A with centre (⁻1, 0) and SF $\frac{1}{2}$.

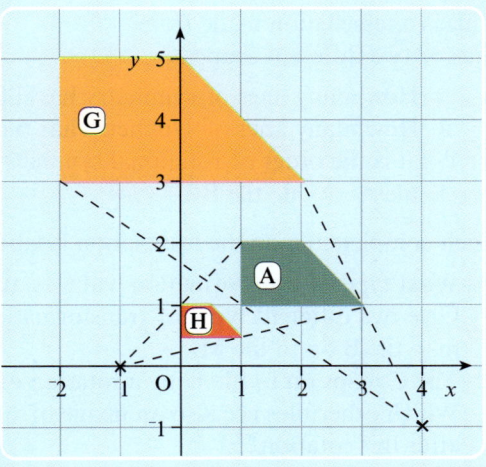

E1 Copy this diagram and enlarge triangle A with:
   a SF 2 and centre (⁻1, 4)
   b SF $\frac{1}{2}$ and centre (0, 5).

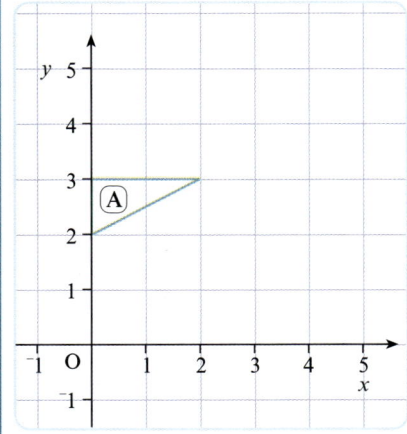

Identifying transformations

Exercise 18.1
Identifying single transformations

The Moors used tessellations of tiles to decorate walls and floors. These are patterns used at the Alhambra Palace in Spain.

> The Moors are Muslims of mixed Berber and Arab descent who live in North West Africa.

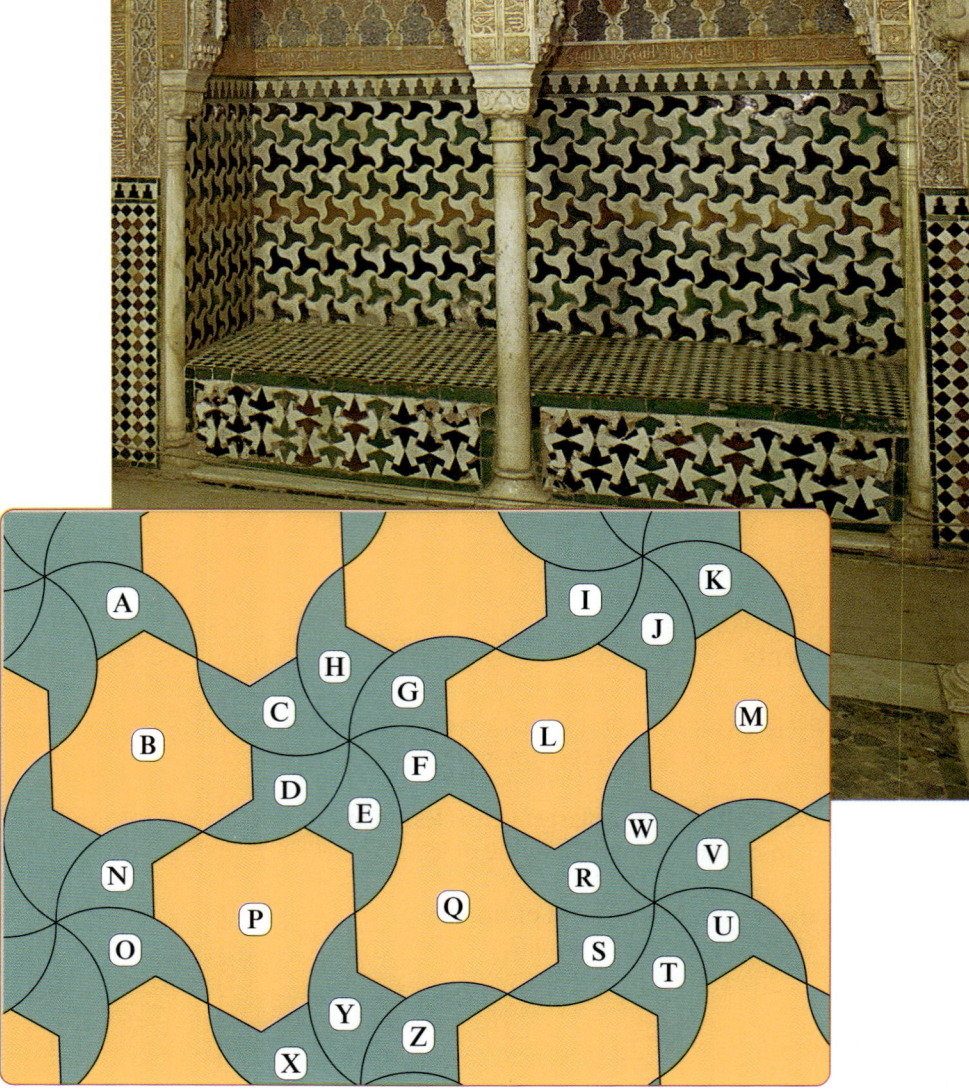

1   This is a tessellation made from tiles of two different shapes.

   a **i** How many lines of symmetry has tile A?
      ii How many lines of symmetry has tile B?
   b What is the order of rotational symmetry for:
      i tile A   **ii** tile B?

> When a shape is transformed the object is said to map on to the image.

2   Each complete tile on the tessellation is labelled.

   a What type of transformation will map tile C on to tile H?
   b Give two different types of transformation that will map tile B on to tile M.
   c Tile C maps on to tile U by a rotation of 180° clockwise. Which other tiles are also an image of tile C after this rotation?
   d Which tiles are an image of tile G after a translation?
   e List all the tiles that are an image of tile D after a rotation:
      i of 60° clockwise   **ii** of 60° anticlockwise.

Single transformations

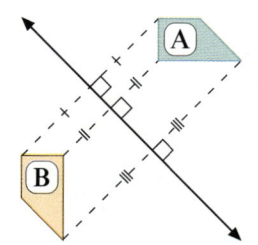

When an object is reflected in a mirror line:

* lines joining corresponding points on the object and image are perpendicular to the mirror line

* the object and image are the same distance from the mirror line.

♦ To describe a **reflection** fully you need to give the mirror line.

Example

A reflection can map triangle 1 on to triangles 3, 4 or 5.
* A reflection in the line GC maps triangle 1 on to triangle 3.
* A reflection in the line HD maps triangle 1 on to triangle 4.
* A reflection in the line HB maps triangle 1 on to triangle 5.

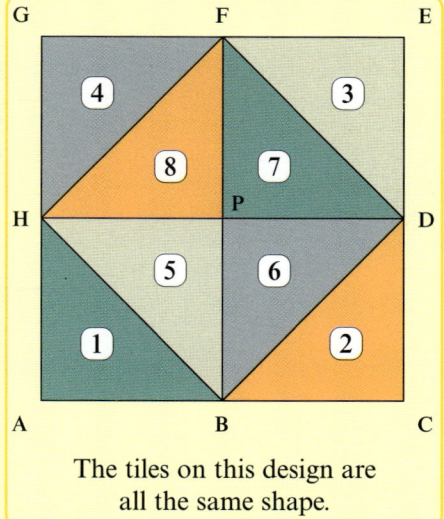

The tiles on this design are all the same shape.

♦ To describe a **rotation** fully you need to give:
* the angle and direction of rotation
* the centre of rotation.

Example

A rotation of 90° anticlockwise (⁺90°) can map triangle 1 on to triangle 2 or 8.
* A rotation of 90° anticlockwise about the point P maps triangle 1 on to triangle 2.
* A rotation of 90° anticlockwise about the point H maps triangle 1 on to triangle 8.

Exercise 18.2
Describing transformations

A rotation of 90° anticlockwise (⁺90°) and a rotation of 270° clockwise (⁻270°) about the same point both map an object on to the same image.

A rotation of 180° anticlockwise (⁺180°) and a rotation of 180° clockwise (⁻180°) about the same point both map an object on to the same image.

1   Use the triangles in square ACEG above to answer these questions.
  a Using a rotation of 90° clockwise (⁻90°) about the point P, what is the image of
    i triangle 5   **ii** triangle 2   **iii** triangle 4?
  b A rotation maps triangle 2 on to triangle 5.
    What is:
    i the centre of rotation
    ii the angle of rotation?
  c Describe fully a rotation that maps:
    i triangle 7 on to triangle 2
    ii triangle 3 on to triangle 1.
  d What is the mirror line for a reflection that maps:
    i triangle 2 on to triangle 6
    ii triangle 2 on to triangle 3?
  e Which triangle, after a rotation of ⁺90° about the point P, is the image of triangle 6?
  f Which triangle, after a reflection in the line FB, is the image of triangle 3?
  g What transformation maps triangle 7 on to triangle 4?
  h Describe two transformations that map triangle 4 on to triangle 2.
  i What transformation can map triangle 8 on to itself?

Exercise 18.3
Describing transformations

This pattern uses tiles of two different shapes.

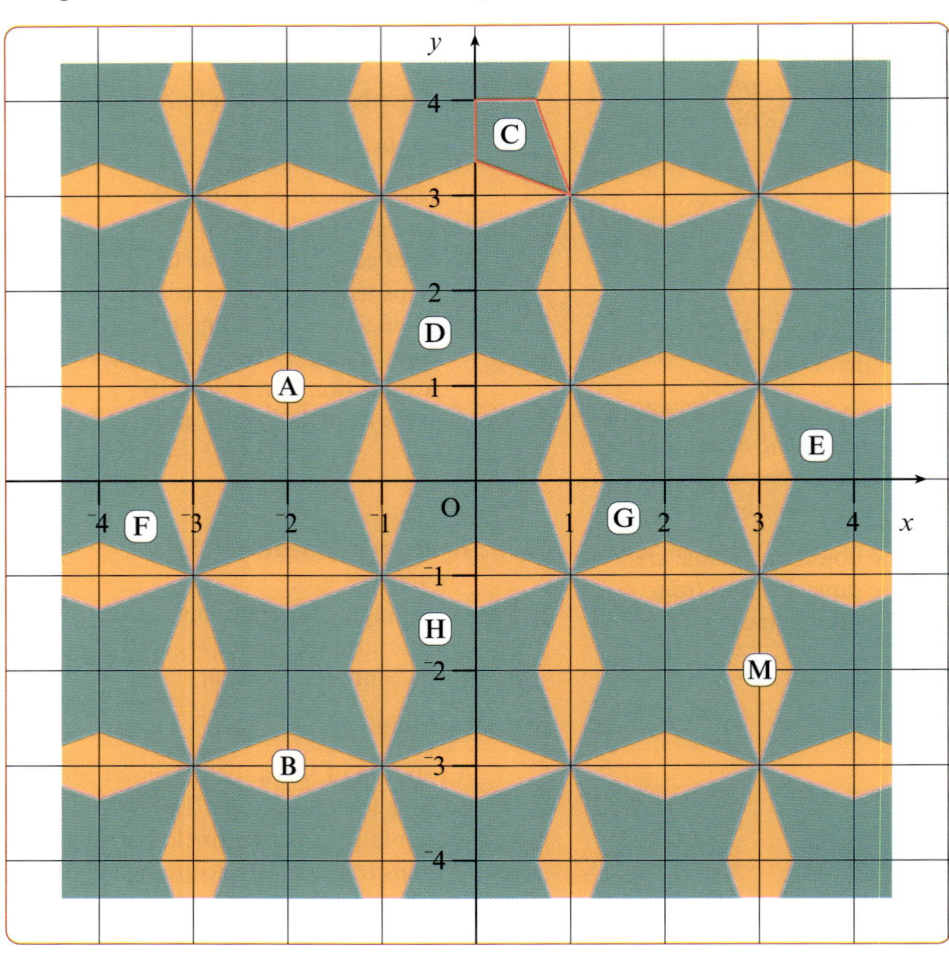

1   You can use vectors and coordinates to describe transformations on a grid.

   a   What vector translates tile A on to tile B?
   b   Tile A is mapped on to tile B by a rotation of 180°. What are the coordinates of the centre of rotation?
   c   What is the equation of the mirror line for the reflection which maps tile A on to tile B?

> You could trace the object and try different centres of rotation.

2   Six of the kites are labelled, C to H.

   a   **i**   Match kites C to H in three pairs so that:

   > in each pair one kite is a reflection of the other.

   ii   In a table like this show the equation of the mirror line for each pair.

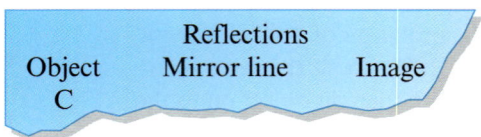

   b   **i**   Sort kites C to H into three pairs so that:

   > in each pair one kite is a translation of the other.

   ii   Make a table to show the vector for each translation.

3

Use this diagram for questions **3** to **8**.

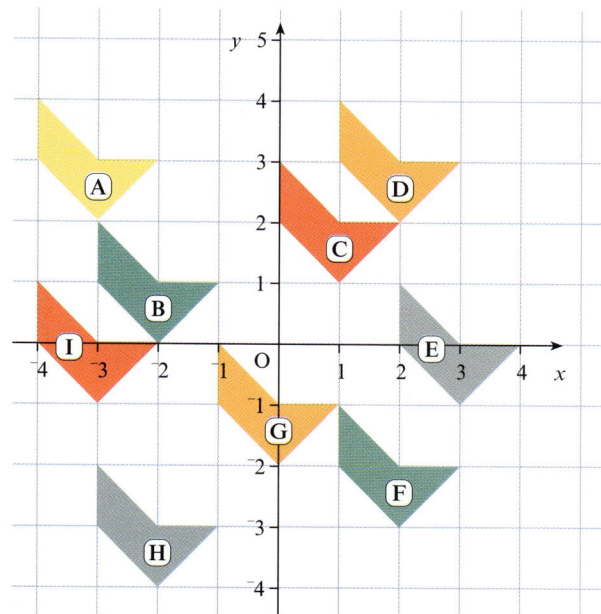

Describe each translation with a vector.
a	A to B	**b**	A to C	**c**	B to D
d	C to B	**e**	B to F	**f**	H to E
g	E to A	**h**	F to A	**i**	D to I

4 Shape B is translated by the vector $\begin{pmatrix} 0 \\ -4 \end{pmatrix}$.

Which shape does B map on to?

5 Shape C is mapped on to another shape by a translation described by $\begin{pmatrix} -3 \\ -5 \end{pmatrix}$.

Which shape is C mapped on to?

6 Shape F is mapped on to another shape by two translations. The translations
are $\begin{pmatrix} 1 \\ 2 \end{pmatrix}$ and $\begin{pmatrix} -5 \\ 1 \end{pmatrix}$.

a Which shape is F mapped on to?
b Give the vector that describes this mapping of F as a single translation.

7 **a** List the vectors that describe this set of translations:

   $D \rightarrow E \rightarrow F \rightarrow H \rightarrow B \rightarrow C$

b Describe the translation of D to C as a single vector.

8 Shape I is mapped on to shape E.
Give the vector that describes this translation.

Exercise 18.4
Transformations

1 a Draw triangle A on axes with
⁻5 ≤ x ≤ ⁺5 and ⁻5 ≤ y ≤ ⁺5.

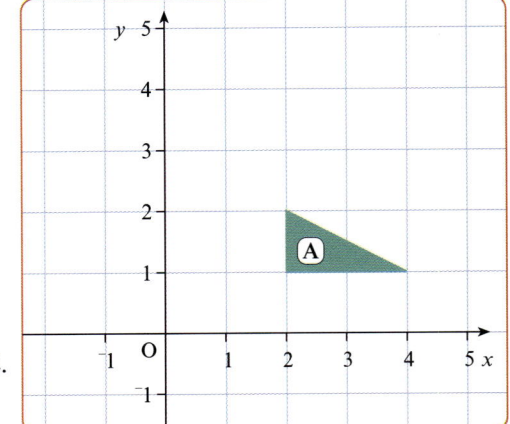

b These transformations map
triangle A on to B, C, D and E.

Object	Transformation	Image
A	Reflection in the line $y = x$	B
A	Reflection in the line $x = 1$	C
A	Rotation of 180° about the origin	D
A	Rotation of ⁻90° clockwise about the point (1, 0)	E

Draw and label the images B, C, D and E .
c Describe a transformation that maps B on to C.
d Describe a transformation that maps:
  i D on to E   **ii** E on to C.

2 Draw axes with ⁻5 ≤ x ≤ 5 and ⁻5 ≤ y ≤ 5.
a Plot the shape with vertices at (1, 2), (1, 4), (2, 3), (2, 2). Label the shape A.
b Reflect A in the line $x = 2$. Label the image B.
c Rotate A 90° anticlockwise about the point (0, 2). Label the image C.
d Translate A by $\begin{pmatrix} ^-2 \\ ^-3 \end{pmatrix}$. Llabel the image D.

3 Draw a pair of axes with
⁻5 ≤ x ≤ 5 and ⁻5 ≤ y ≤ 5.
Plot shape A with coordinates as
shown in the graph.
a Reflect A in the line $x = 0$.
Label the image B.
b Reflect A in the line $y = 1$.
Label the image C.
c Rotate A 90° clockwise about the
point (2, 2). Label the image D.
d Translate A by $\begin{pmatrix} ^-3 \\ ^-4 \end{pmatrix}$.

Label the image E.

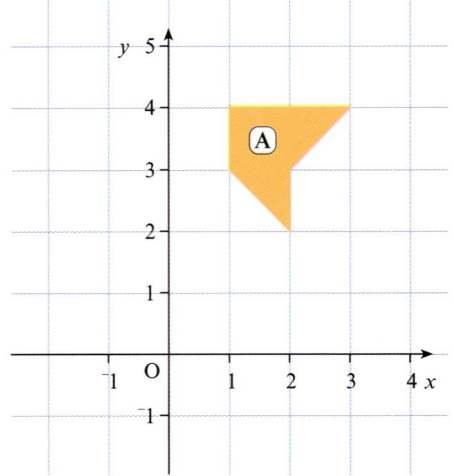

4 Draw a pair of axes with
$^-5 \leqslant x \leqslant 5$ and $^-5 \leqslant y \leqslant 5$.
Plot shape A with coordinates
as shown in the graph.
   a Rotate A 90° anticlockwise
about (0, 2).
Label the image B.
   b Compare A and B.
What can you say about the
area and perimeter of the
two shapes?
   c Reflect A in the line $y = 2$.
Label the image C.
   d Compare A and C.
What can you say about the
area and perimeter of the
two shapes?
   e Translate A by $\begin{pmatrix} ^-2 \\ ^-5 \end{pmatrix}$. Label the image D.

   f Compare A and D.
What can you say about the area and perimeter of the two shapes?

5 A rectangle 3.5 cm wide and 8 cm long is transformed by a rotation of
90° anticlockwise about the point (2, 3).
The image of the rectangle is labelled K.
Find the area and perimeter of rectangle K.

6 Triangle W is transformed by a reflection
in the line $y = ^-2$.
The image of W is labelled R.
Find the area and perimeter of R.

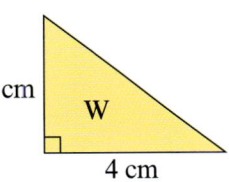

7 **a** Find the area and perimeter of
shape N.
   b Draw shape N on a pair of axes.
   c Rotate N 90° clockwise about (0, 0).
Label the image L.
   d Give the area and perimeter of L.
   e Reflect N in the line $y = ^-1$.
Label the image K.
   f Translate N by $\begin{pmatrix} ^-4 \\ ^-6 \end{pmatrix}$. Label the image J.

   g What can you say about the area and
perimeter of N, L K and J?

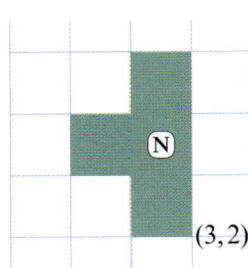

For the axes use:
$^-5 \leqslant x \leqslant 5$ and $^-8 \leqslant y \leqslant 8$.

Combined transformations

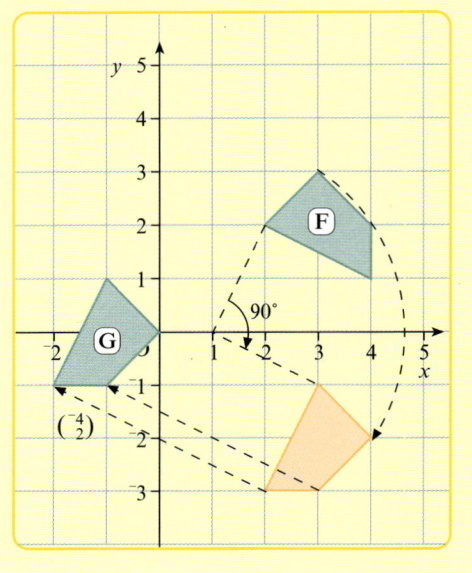

♦ To describe a mapping you can use a combination of transformations.

Example

To map F on to G you can use:

> a rotation of ⁻90° about (1, 0)

followed by

> a translation $\begin{pmatrix} ^-4 \\ 2 \end{pmatrix}$

Exercise 18.5
Combined transformations

1 On a pair of axes plot and draw shape F above.

a Transform shape F by:

• a translation of $\begin{pmatrix} ^-4 \\ 2 \end{pmatrix}$

followed by

• a rotation 90° clockwise about the point (1, 0).

Label the final image H.

b Compare shapes F and H.
  i List the things that are the same about the shapes.
  ii List the differences between the shapes.

2 **a** Transform shape A by:

• a reflection in the line $x = 0$

followed by

• a translation of $\begin{pmatrix} 0 \\ ^-4 \end{pmatrix}$.

Label the final image B.

b Transform shape A by:

• a rotation of 90° clockwise about (1, 1)

followed by

• a reflection in the line $y = ^-2$

Label the final image C.

c What can you say about the area and perimeter of shapes A, B and C?

For the axes use:
$^-5 \leqslant x \leqslant 5$ and $^-8 \leqslant y \leqslant 8$.

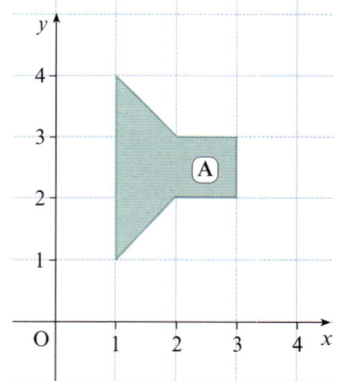

3   Each tile is a transformation of tile P.

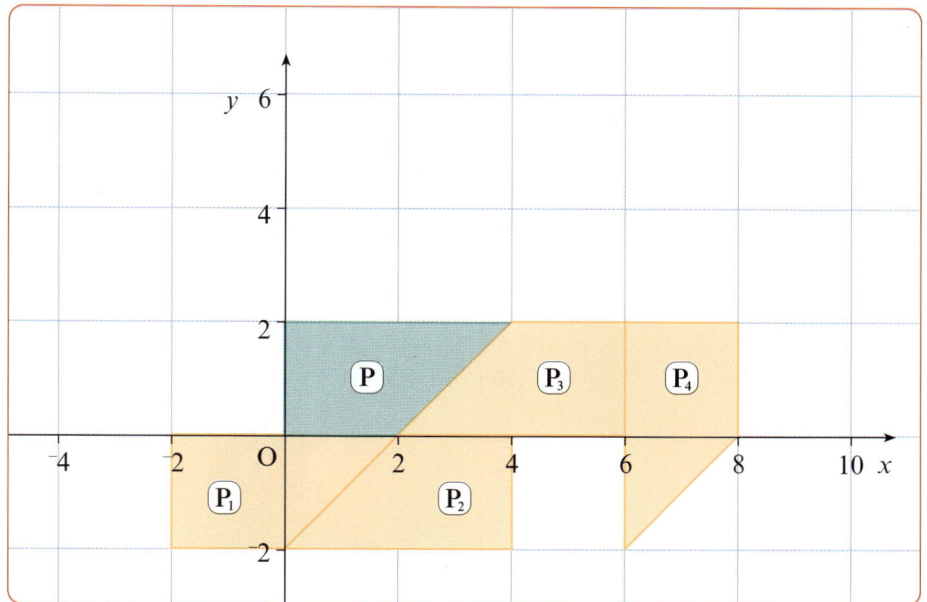

a   Draw these tiles on axes with $^-4 \leqslant x \leqslant 10$ and $^-4 \leqslant y \leqslant 8$.
b   Describe a single transformation that maps P on to:
     i P_1 ii P_2 iii P_3

4   P can be mapped on to P_4 by a rotation followed by a reflection.
a   Give the rotation.
     (You must give the angle, direction and centre of rotation.)
b   Give the reflection.

> You could translate and then rotate.

5   Give two transformations that combined will map P_1 onto P_4.

6   P_2 can be mapped on to P_4 with a translation followed by a rotation.
a   Give the translation.
b   Find the rotation.

7   P_3 can be mapped on to P_4 by reflection followed by a rotation.
a   Give the reflection.
b   Give the rotation.

8   P_1 can be mapped on to P_3 by a reflection followed by a different reflection.
     Give the two reflections.

Enlargements

Exercise 18.6
Enlargements

1 This is part of a pattern of tiles arranged in the shape of a Greek cross.

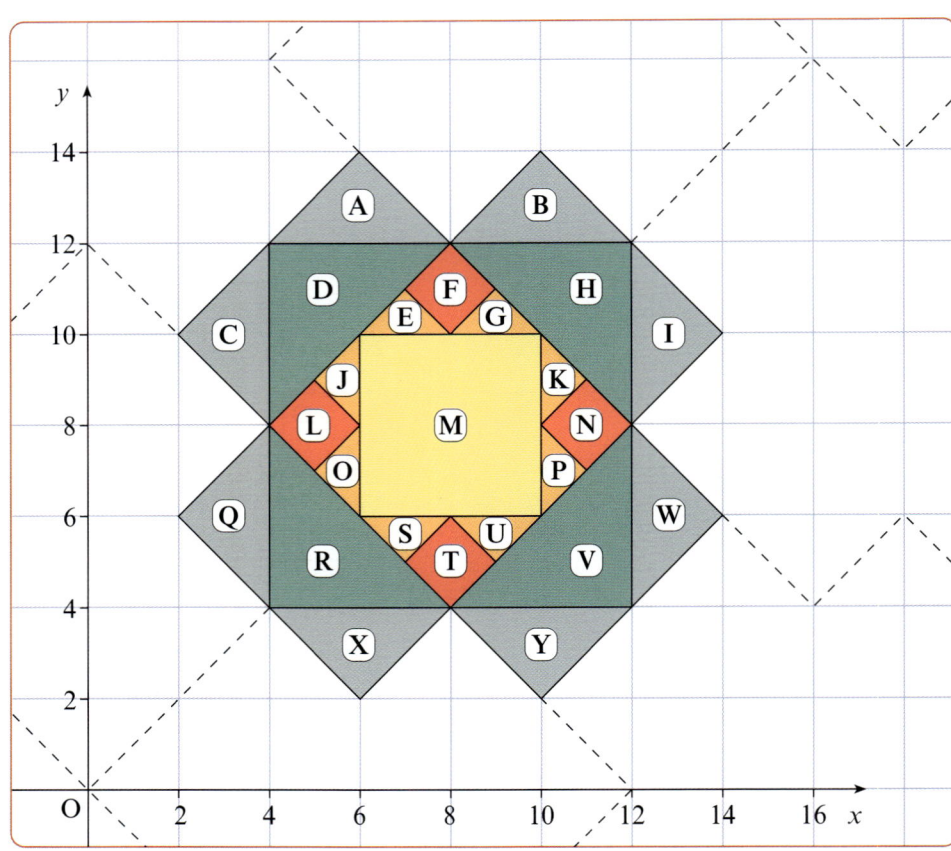

You could put tracing paper over your drawing for each question.

If you draw lines through corresponding points on the object and image, they will meet at the centre of enlargement.

a Copy the diagram and label the tiles A to Y.

b What is the image of E after an enlargement with centre (4, 8) and SF 2?

c Find the scale factor and centre for each of these enlargements.

Object	Image	SF	Centre
S	Y		
Y	S		
X	S		
X	U		

d Explain why tile T will not map on to M with just an enlargement.

e This table gives a pair of transformations for three mappings. Describe fully the second transformation for each of these mappings.

For an enlargement give the scale factor and the coordinates of the centre.

For a translation give the vector.

For a reflection give the equation of the mirror line.

For a rotation give the angle of rotation and the coordinates of the centre.

	Object	Image	Transformations	
			First	Second
i	G	W	Enlarge SF 2 with centre (8, 8)	
ii	Q	E	Rotate ⁻90° about (6, 6)	
iii	A	S	Enlarge SF ½ with centre (4, 8)	

Exercise 18.7
Drawing enlargements

1 Draw the triangle A on axes with:
$$^-6 \leqslant x \leqslant 10$$
$$^-6 \leqslant y \leqslant 10.$$

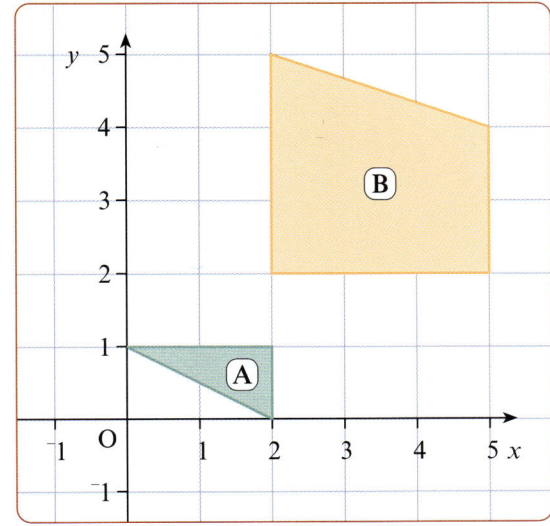

> The distance from the centre to a point on the object multiplied by the scale factor gives the distance from the centre to the corresponding point on the image.

These transformations map the triangle A on to A_1, A_2, A_3 and A_4.

Object	First transformation	Second transformation	Image
A	Enlarge SF 2 with centre (0, 0)	Translate $\begin{pmatrix} 1 \\ 2 \end{pmatrix}$	A_1
A	Rotate $^-90°$ about (2, 0)	Enlarge SF 2 with centre ($^-1$, 0)	A_2
A	Enlarge SF 2 with centre ($^-1$, 0)	Translate $\begin{pmatrix} 4 \\ 4 \end{pmatrix}$	A_3
A	Rotate $^+90°$ about (0, 1)	Enlarge SF 2 with centre ($^-3$, $^-2$)	A_4

On your diagram draw and label the images A_1, A_2, A_3 and A_4.

2 Draw shape B on axes with $^-1 \leqslant x \leqslant 5$ and $^-1 \leqslant y \leqslant 5$.

a Enlarge shape B with SF $\frac{1}{2}$ and centre ($^-1$, $^-1$).
Label the image B_1.

b Compare shapes B and B_1.
  i What can you say about lengths in B and B_1?
  ii Find the area of shape B.
  iii What is the area of shape B_1?
  iv Describe what happens to edge lengths and areas in an enlargement SF $\frac{1}{2}$.

3 **a** On you axes enlarge shape B with SF $\frac{1}{4}$ and centre (2, 2).

Label the image B_2.

b Compare shapes B and B_2.
  i Describe the changes in edge length.
  ii Describe any change in area.
  iii What can you say about the angles in B and B_2?

4 Each tile in this pattern is also a transformation of A.

a Copy this diagram.

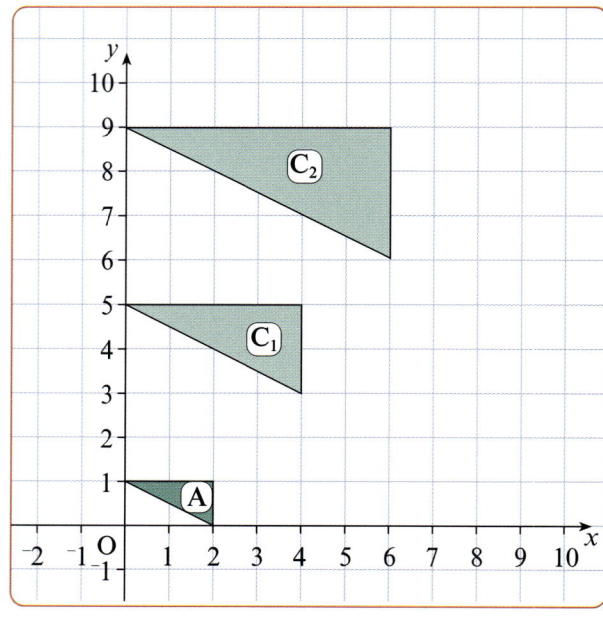

b Give the SF of enlargement for the mapping:
i C_2 to A   **ii** C_2 to C_1 **iii** C_1 to A.

c Find the area of C_2.

d How many times larger is the area of C_2 compared to the area of A?

e Explain how the perimeters of A and C_1 are linked.

f What can you say about the angle of A, C_1 and C_2?

> A six-sided polygon is a hexagon.

5 In this pattern each hexagon is an enlargement of E_1.
The consecutive enlargements touch but do not overlap.

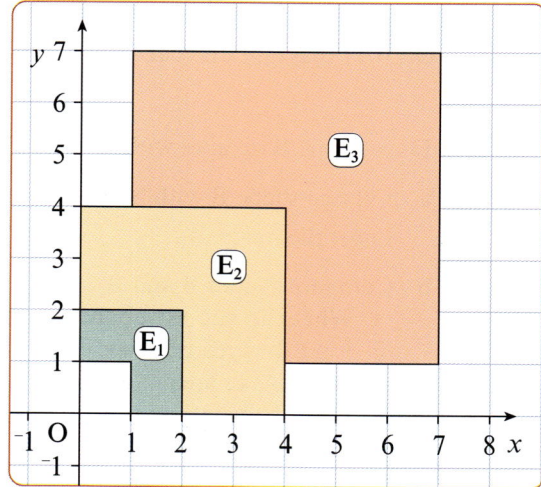

a Draw E_1, E_2 and E_3.

b Give the coordinates of the centre for the:
i enlargement SF 2   **ii** enlargement SF 3

c What scale factor of enlargement maps E_3 onto E_2?

d Find the area of E_3.

e E_3 is enlarged by a scale factor of $\frac{1}{2}$. The image is labelled E_4.
What is the area of E_4? Explain your answer.

f What will be the perimeter of E_4? Explain your answer.

End points

You should be able to ...

... try these questions

A Use single transformations

A1 A, B, C and D are congruent equilateral triangles.

Which triangles can A map on to after:
a a translation
b a rotation
c a reflection?

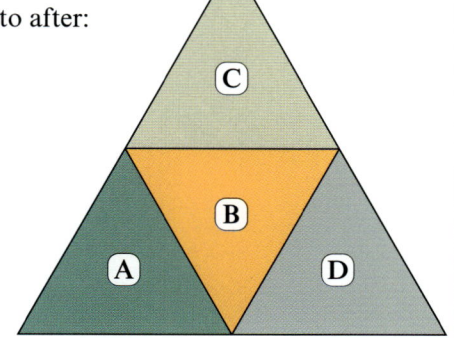

A2 a Draw the hexagon E on axes with:
$$^-4 \leqslant x \leqslant 8$$
$$^-6 \leqslant y \leqslant 6$$

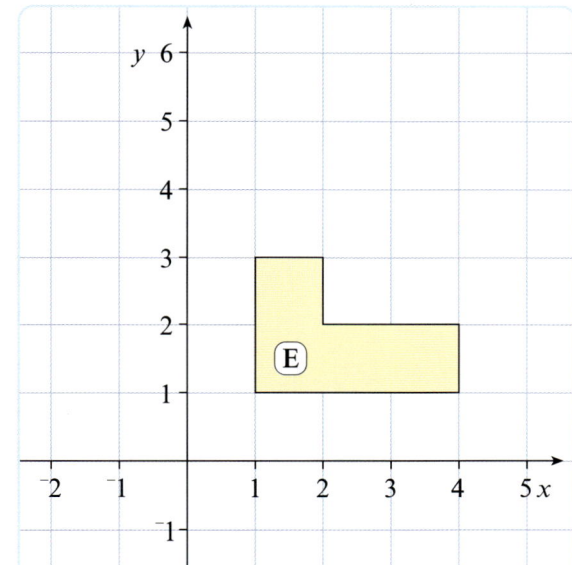

b These transformations map E on to F, G and H.

Object	Transformation	Image
E	Rotate 180° about the origin	F
E	Reflect in $x = 0$	G
E	Translate $\begin{pmatrix} 2 \\ ^-2 \end{pmatrix}$	H

Draw and label the images F, G and H.

c Describe a transformation that maps:
  i F on to E
  ii G on to E
  iii H on to E.
d Describe fully a transformation that maps:
  i F on to G
  ii H on to F.
e Rotate E 90° anticlockwise about (1, 3).
  Label the image I.
f Reflect I in the line $x = 0$.
  Label the image J.

223

A Use single transformations

A3 Draw the a pair of axes with:
$^-5 \leqslant x \leqslant 5$
$^-8 \leqslant y \leqslant 8$
Copy shape A.

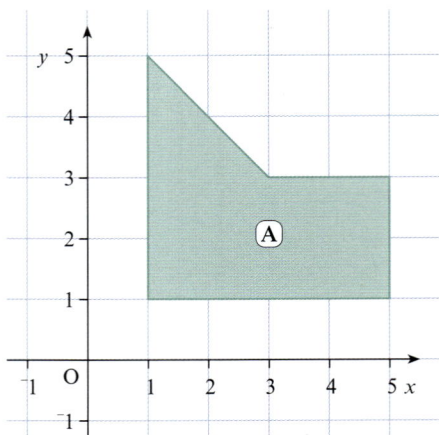

a Enlarge shape A with a SF $\frac{1}{2}$ centre (1, 1).

Label the image B.

b Find the area of shape A.

c What is the area of shape B?

d What fraction of the area of shape A is shape B?

e What can you say about the edge lengths in shape A and shape B?

f Reflect A in the line $y = ^-1$.

Label the image C.

g Compare shapes A and C.

What is the same about the shapes?

What is different about the shapes?

h Translate A by $\begin{pmatrix} ^-6 \\ ^-7 \end{pmatrix}$, and label the image D.

i Compare shapes A, B, C and D.

What is the same about each of the shapes?

j Shape A was enlarged by SF $\frac{1}{3}$ and the image was labelled E.

What fraction of the area of shape A is shape E?

k What can you say about the edge lengths in Shape A and shape E?

B Use combined transformations

B1 These transformations map the hexagon E, from Question **A2**, on to J, K and L.

Object	First transformation	Second transformation	Image
E	Reflect in $y = {}^-1$	Reflect in $x = 0$	K
E	Reflect in $y = {}^-x$	Reflect in $y = 0$	L
E	Rotate 180° about $(0, 0)$	Rotate 180° about $(0, {}^-2)$	M

a On a new diagram, on axes with ${}^-6 \leqslant x \leqslant 6$ and ${}^-6 \leqslant y \leqslant 6$, draw and label E and the images K, L and M.

b What single transformation maps E on to:
   i K   **ii** L   **iii** M?

c These transformations map K on to L, M and E. Describe the second transformation fully.

	Object	Image	First transformation	Second transformation
i	K	L	Rotate ${}^-90°$ about $({}^-1, {}^-3)$	
ii	K	M	Reflect in $x = 0$	
iii	K	E	Translate $\begin{pmatrix} 2 \\ 2 \end{pmatrix}$	

d What single transformation maps K on to:
   i L   **ii** M   **iii** E.

Some points to remember

♦ Transformations on a grid

Transformation	Describe by giving:	Object and image are:
Enlargement	the coordinates of the centre the scale factor	similar
Reflection	the equation of the mirror line	congruent
Rotation	the coordinates of the centre the angle of rotation the direction of rotation	congruent
Translation	the vector $\begin{pmatrix} \\ \end{pmatrix}$	congruent

Southampton Evening Chronicle   *Monday 15 April 1912*

TRAGEDY AT SEA

IT IS WITH great regret that we bring you the news that last night at 10:40 pm the 'unsinkable liner' the Titanic hit an iceberg on her way to New York. The Titanic later sank at 2:20 am with the loss of many lives. It was the liner's maiden voyage and on board were 331 first class passengers, 273 second class 712 third class and a full crew – only 32.2 % of those on board survived. Each first-class passenger had paid £870 for the privilege of making the voyage in this luxury floating palace. To reassure the passengers the orchestra was still playing as the liner was going down and many passengers were so sure the ship could not sink that they refused to board the lifeboats.

The captain had been given repeated warnings of icebergs ahead but chose to steam on at 22.5 knots. It was calm weather with good visibility but the lookouts had not been issued with binoculars. The iceberg is thought to have had a height of about 100 feet showing above the water and a weight of about 500 000 tons. The sea water temperature was only 28° Fahrenheit and this took its toll on those jumping overboard. It is thought that the capacity of the lifeboats was insufficient for the number of people on board.

Survivors were picked up by the liner Carpathia which had heard the SOS when it was 58 miles away. The Carpathia steamed at a staggering 17.5 knots to reach the sinking Titanic. The ship's engineer said this was 25% faster than her usual speed.

HOW FAIR WAS THE RESCUE?

Reports coming in give the final casualty figures from the Titanic. There was not enough lifeboat space for all on board because the Titanic was considered to be the first unsinkable ship. The owners White Star admit that lifeboats could only hold 33% of the full capacity of the liner and 53% of those on board on that fateful night. Breaking the survival figure down by class we find 203 first-class, 118 second-class and 178 third-class passengers were rescued. Nearly a quarter of all crew were saved. Analysis of these figures is taking place to see if all people on board had an equal chance of being rescued.

Strange BUT *true*

Fourteen years before the disaster, and before the Titanic had been built, a story was published which described the sinking of an enormous ship called the Titan after it had hit an iceberg on its maiden voyage.

The comparisons between ships is even more amazing.

	Titan (Fiction 1898)	Titanic (True 1912)
Flag	British	British
Month of sailing	April	April
Displacement (tons)	70 000	66 000
Propellers	3	3
Max. speed	24 knots	24 knots
Length	800 feet	882 feet
Watertight bulkheads	19	15
No. of lifeboats	24	20
No. on board (inc crew)	2000	2208
What happened?	Starboard hull split by iceberg	Starboard hull split by iceberg
Full capacity	3000	

Distances at sea are measured in nautical miles and ships' speeds are given in knots.
1 nautical mile is 1852 metres.
1 knot is a speed of 1 nautical mile/hour.

1 In 24-hour time give the time that:
  a the Titanic hit an iceberg
  b the Titanic sank.

2 How long did it take the Titanic to sink after hitting the iceberg?

3 **a** Calculate the number of crew on board the Titanic on her maiden voyage.
  b Calculate approximately how many people died.

4 If all those who survived were in lifeboats, what was the mean number of people per boat?

5 How much money in total was taken in fares for the first class passengers?

6 In 1912 the price of a small house was about £200. In 1996 the same house would cost about £68 000. If fares on a cruise liner increased in the same ratio, what would have been the first class ticket price in 1996?

7 About $\frac{7}{8}$ of an iceberg's height is below water level. Estimate the total height of the iceberg that the Titanic hit.

8 **a** What was the Carpathia's usual speed?
  b At 17.5 knots, how long would it take the Carpathia to steam the 58 nautical miles to the Titanic?
  c The Carpathia received the SOS message at 12:30 am. Approximately how long after the Titanic sank did she arrive at the scene?

9 You can convert a temperature from degrees Celsius (°C) to degrees Fahrenheit (°F) with this formula:
$$F = \tfrac{9}{5}C + 32$$
  a Make C the subject of the formula.
  b Calculate a water temperature of 28 °F in degrees Celsius.

10 **a** How many people could the lifeboats have held in total?
  b Use your answer to part **a** to calculate an approximate value for the full capacity of the Titanic.

11 Calculate the relative frequency of survival for:
  a first-class passengers
  b second-class passengers
  c third-class passengers.

12 Calculate the total percentage of the passengers aboard who were rescued.
(About 25% of the crew were rescued.)

13 What is the displacement of the Titanic to 1 sf?

14 The richest person on the boat was Colonel J.J. Astor who was thought to be worth £30 million. Write this number in standard form.

The drive shafts of the Titanic were 201 feet long and fell $3\frac{1}{2}$ feet over their length.

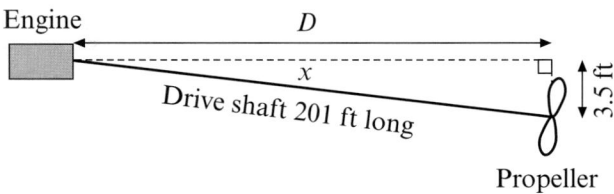

15 Calculate the angle x that a driveshaft made with the horizontal to the nearest degree.

16 Calculate the horizontal distance D to 2 dp.

The Titanic had cranes, known as derricks, for lifting the cargo on to the ship.
This diagram shows a derrick in one position.
The tower and part of the cable are vertical.

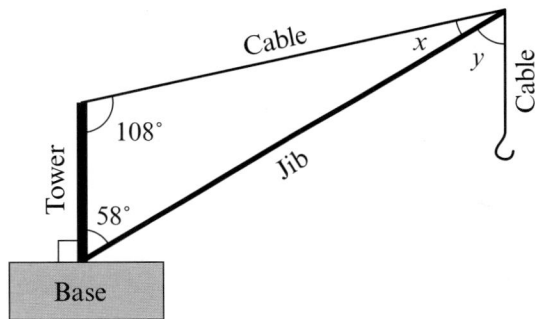

17 What is the angle y? Give your reasons.

18 Calculate the value of angle x. Give your reasons.

At the enquiry after the sinking, White Star, the owners, said that in the previous ten years they had carried 2179 594 passengers with the loss of only 2 lives.

19 Give the number of passengers carried to:
  a 1 sf   **b** 3 sf
  c 4 sf   **d** 5 sf

20 In the ten years between 1981 and 1990 about 1.2×10^7 people from the UK crossed the Atlantic by plane. With the same death rate as White Star gave, roughly how many people would have died in that time?

Bananananananananananananananananananananana
On 30 April 1988 in Selinsgrove Pennsylvania USA a banana split was made which was 7.32 km long.

On 31 May 1682 there was a cloudburst in Oxford which gave 24 inches of water in less than a quarter of an hour.
A slight shower!

Jo's always on hand
In 1900 Johann Hurlinger of Austria walked 871 miles from Vienna to Paris on his hands. His average speed was 1.58 mph and he walked for 10 hours each day.

OH! WHAT A LITTLE ONE-WHEEL
In March 1994 in Las Vegas Peter Rosendahl of Sweden rode a unicycle 20 centimetre high for a distance of 3.6 metres. The wheel diameter was only 2.5 cm.

Up the pole!
Mellissa Sanders lived in a hut at the top of a pole in Indianapolis USA for two years starting on 10 October 1986. Her hut measured 1.8 metres wide by 2.1 metres deep.

Stacks of cards!
In 1995 Brian Berg of Spirit Lake Iowa USA built a tower of playing cards with 83 stories. The tower was 4.88 metres high. In 1978 James Warnock of Canada held the previous record with 60 stories.

Number types
On 14 October 1993 Mikhail Shestov set a record when he had typed the numbers 1 to 795 on a PC by the time 5 minutes was up. He had made no errors.

Unique cycle on unicycle
Takayuki Koike of Japan road 100 miles on a unicycle in a record time of 6 hours, 44 minutes and 21 seconds on 9 August 1987.

Rail Trick
The Katoomba Scenic Railway in New South Wales in Australia is the steepest railway in the world. Its gradient is 1 in 0.8 but it is only 310 metres long. The ride takes about 1 minute 40 seconds and carries about 420 000 passengers a year.

The circumference of the Earth at the equator is 40 075 km and its mass is 5880 000 000 000 000 000 000 tons.

Weight watchers

Tall stories
The tallest man in the world was Robert Wadlow from Alton Illinois in the USA who was 8 feet 11.1 inches tall. The tallest man in Scotland was Angus Macaskill from the Western Isles who was 7 feet 9 inches tall. The highest mountain in the USA is Mount McKinley at 20 320 feet and the highest one in Scotland is Ben Nevis at 4408 feet.

Can beans be hasbeens?
Baked beans were first introduced into the UK in 1928. By 1992 they were selling at the rate of 55.8 million cans per year.

Piece on earth
A jigsaw with 1500 wooden pieces was made for the photograph on the cover of the BEEB magazine. The jigsaw was assembled in 1985 by students from schools in Canterbury. It measured 22.31 metres by 13.79 metres.

The swift hare and the XJ tortoise
In 1992 the Jaguar XJ220 set the land speed record for a road car of 217 miles per hour.
The spine-tailed swift has been recorded as flying at 220 miles per hour. Will this mean that Brands Hatch is converted for spine-tailed swift racing?

Can can or cannot
A square based pyramid tower of 4900 cans was built by 5 adults and 5 children at Dunhurst School, Petersfield on 30 May 1994 in a time of 25 minutes 54 seconds.

Barmy salami
A salami is usually about 9cm in diameter and about 35cm long, but at Flekkefjord in Norway in July 1992 a giant salami was made which had a circumference of 63.4 cm and was 20.95 metres long.

AMAZING FACTS
p9

1   Calculate the height of James Warnock's tower of cards in 1978.

2   Give the weight of the Earth in standard form.

3   Give the circumference of the Earth to:
a   4 significant figures
b   3 significant figures.

4   Calculate the diameter of the Earth at the equator.

5   For Peter Rosendahl's mini unicycle give:
a   the height of the bike in metres
b   the distance he travelled in millimetres
c   the circumference of the wheel in centimetres.

6   How many days was it between when Mellissa Sanders came down from her pole hut and Mikhail Shestov set his number typing record?

7   Convert the height of water that fell on Oxford in less than a quarter of an hour to metres.

8   What was the area of the Canterbury jigsaw?

9   If the Canterbury jigsaw had been out in the Oxford rain, what volume of water would have landed on it?

10 What was the average speed of Takayuki Koike's unicycle ride:
a   in miles per hour   b   in kilometres per hour?

11 How long would it take Takayuki Koike to unicycle along the length of the Selinsgrove banana split if he always unicycles at the same average speed?

12 How many of the giant banana splits would fit end to end round the equator? Give your answer in standard form to a suitable degree of accuracy.

13 A salami is shaped roughly like a cylinder. Use this to calculate the approximate volume of a normal salami.

14 a   Calculate the diameter of the Flekkefjord salami in centimetres.
b   Calculate the volume of the Flekkefjord salami in cm^3.
c   The density of salami is about 1.01 g/cm^3. Calculate the mass (weight) of the giant salami in kg.

15 If the capacity of Mellissa Sanders's hut on a pole was 6.62 metres^3, what was the height of her hut in metres? Give your answer to 3 sf.

16 Give the ratio of the length of the giant banana split to the length of the giant salami in the form 1 : n, to the nearest whole number.

17 a   What was Robert Wadlow's height …
i   inches (to the nearest inch)
ii   centimetres (to 1 dp)
iii   metres (to 2 dp)?
b   What is the ratio tallest man : highest mountain in the form 1 : n for:
i   the USA   ii   Scotland?

18 For the numbers 1 to 12, fifteen digits are used.
a   How many digits are in the numbers 1 to 100?
b   How many digits had Mikhail Shestov typed in by the time five minutes was up?
c   What was Mikhail Shestov's typing speed in digits per second (to nearest whole number)?

19 For how many days was Johann Hurlinger walking on his hands when he travelled between Paris and Vienna?

20 a   At what angle does the Katoomba Scenic Railway climb?
b   How many metres does the railway rise over its entire length?
c   What is the average speed of the train:
i   to the nearest metre per second
ii   in km per hour?

21 The spine-tailed swift and the Jaguar XJ220 race together at their maximum speeds over a course of 240 miles. How long will the swift have to wait for the Jaguar at the finish line?

22 A baked bean can has a diameter of 7.4 cm and a height of 10.5 cm.
a   What is the volume of one can of beans in cm^3?
b   What was the total volume of the cans in the Petersfield tower in m^3?

23 The pyramid of cans in Petersfield had one can on the top layer and each can in the stack rested on four cans below.
a   How many cans were in:
i   the second layer down from the top
ii   the third layer down?
b   There were 24 layers in the tower. How many cans were on the bottom layer?

24 Imagine that as baked bean cans are bought in the UK they are emptied then lined up end to end round the equator. After roughly how long would they encircle the Earth?

25 Land's End to John O'Groats is 886 miles.
a   How many giant salamis long is this?
b   If the giant salami was rolled along this distance how many rotations would it make?

26 How many times does a banana split go into a basin plant? Hint: think of a bananagram!

SKILLS BREAK 3B

about ...

...ncy

...ncy is a way of estimating a probability.

...used in equally likely situations but is the **only** way when

... not equally likely. It is found by experiment, by survey, or

by ... at data already collected.

Example   This data shows the colour of cars passing a factory gate
one morning.

Colour	Frequency
Red	68
Black	14
Yellow	2
Green	34
Blue	52
Grey	35
Other	23
TOTAL	228

The relative frequency of red cars is $\frac{68}{228} = 0.30$ (to 2 dp).

This probability is an estimate and can be used to predict the number of
red cars passing the gate on a different morning.

B Tree diagrams

A tree diagram has branches showing different events.
This is a tree diagram for spinning a coin then rolling a 1 to 6 dice.
The only square numbers on the dice are 1 and 4.

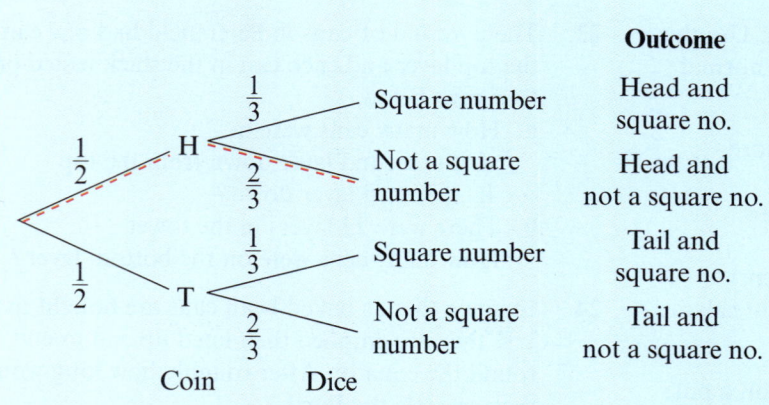

Outcome

Head and square no.

Head and not a square no.

Tail and square no.

Tail and not a square no.

Coin   Dice

To find the probability of a particular outcome you can multiply the
probabilities on the branches that lead to it.

Example   To find the probability of a head and a non-square number
you multiply the probabilities along the dotted branches.

Probability of a head and non-square no. is $\frac{1}{2} \times \frac{2}{3} = \frac{2}{6}$

A1 What is the relative frequency
of blue cars in the survey?
Give your answer to 2 dp.

A2 What is the relative frequency
of cars which are:
a not blue
b either yellow or green
c neither green nor red
d either black, grey or blue?

B1 What is the probability of a
head and a square number?

B2 What is the probability of a
tail and a non-square number?

B3 A red 1 to 6 dice and a
blue 1 to 6 dice are rolled.
a Draw a tree diagram to
show the probability of a
triangle number on the red
dice and an even or odd
number on the blue one.
b What is the probability
of getting a non-triangle
number on the red with
an even number on the
blue dice?

230

Thinking ahead to ...
experimental probability

Theoretical probability is based on the analysis of equally likely outcomes.

It is usually given as a fraction.

From the number of items you can calculate the theoretical probability of picking any one item at random when each item is equally likely.

Example   In this bag there are known to be 3 red cubes, 5 blue cubes and 1 yellow cube.

The probability of picking a particular cube is equally likely as picking any other cube.

So the probability of picking a red cube at random is $\frac{3}{9}$ (or $\frac{1}{3}$).

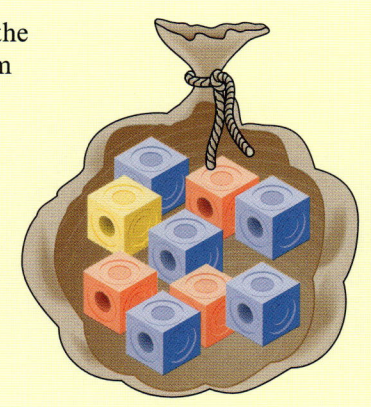

A   From this bag what is the probability of picking:

   a   a blue cube   **b**   a yellow cube

   c   a cube which is not blue   **d**   a cube which is not blue and not yellow?

Probability based on experiment

Relative frequency is usually given as a decimal or a percentage.

You might not know the number, or colour, of cubes in a bag.

The probability of a particular colour can be estimated by doing an experiment to find the relative frequency.

In this experiment a cube is picked at random, its colour recorded, and the cube replaced.

These results are from an experiment:

Colour	Frequency
Red	53
Blue	15
Yellow	67
Total	135

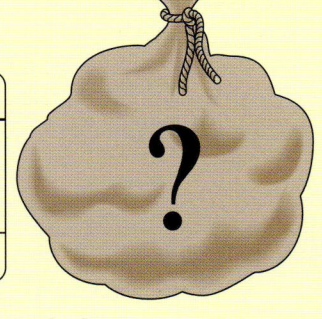

The relative frequency of picking red is $\frac{53}{135}$ = 0.39 (to 2 sf).

This means that it is likely that about 39% of the bag's contents are red.

Exercise 19.1
Relative frequency

1   From the bag above what is the relative frequency of picking:

   a   a blue cube   **b**   a yellow cube?

2   What is the relative frequency of not picking blue?

3   Suppose that you know that the total number of cubes in the bag is 20. Estimate the number of:

   a   red cubes   **b**   blue cubes   **c**   yellow cubes.

4   In another experiment, red, green and blue cubes were in a bag. Cubes were picked out at random, examined and replaced. The relative frequency of red was 0.62 and of blue it was 0.21

   a   What was the relative frequency of green?
   b   Why is it impossible to know how many cubes were in the bag?

Finding relative frequency

◆ Traffic police need to know the probability of an accident at different road junctions, so that they can suggest changes.
◆ A railway company needs to know how likely trains are to arrive on time, so it can change its timetable.
◆ A fairground attendant needs to know your chances of getting a ping-pong ball into a glass jar, so he can make a profit.

In these cases **relative frequency** and not **theoretical probability** is used. Relative frequency can be found either from data that is recorded or by doing an experiment (**experimental probability**).

You will need three Multilink cubes and special dice for Exercise 19.2.

Exercise 19.2
Relative frequency experiments

1 When a Multilink L-shape is dropped it can come to rest in three positions.

Side down   Flat down   On its edges

a Make a frequency table like this to record how the shape lands.

Position	Tally	Frequency
Side down		
Flat down		
On its edge		

b Drop an L-shape fifty times and record the outcomes in your table.
c Calculate the relative frequency of the shape landing:
   i side down   ii flat down   iii on its edge.
d What should the answers to parts **i**, **ii** and **iii** add up to?
e If your L-shape is dropped 231 times how many times would you expect it to land side down?

For Questions **2** to **5** you need three dice with faces marked like this:

Dice A has faces 1, 1, 5, 5, 5, 5
Dice B has faces 3, 3, 3, 4, 4, 4
Dice C has faces 2, 2, 2, 2, 6, 6

2 a Roll dice A and B together thirty times and record which dice wins each time.
  b What is the relative frequency of dice A winning?

3 a Roll dice B and C together thirty times and record which one wins.
  b What is the relative frequency of B winning?

4 a If dice A and C were rolled together, which dice do you think would win most often?
  b Do an experiment and record your results.
  c What is the relative frequency of A winning?
  d Was the result of your experiment what you expected?

Some experiments in probability do not work out quite as you would expect them to.

5 Roll all three dice together and find the relative frequency that each wins.

Using a relative frequency tree diagram

This diagram shows the relative frequencies of cars leaving a roundabout when they have approached it in the direction of a red arrow.

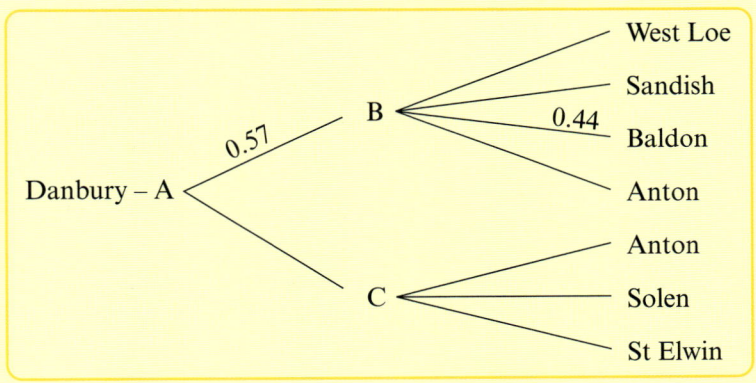

Sandish   Baldon   Solen

West Loe

0.2   B   0.44   0.36

0.15   0.21   Anton   0.24   C

St Elwin

0.57   A   0.43

Danbury

Assume motorists always take the shortest route and that they do not travel on the same road twice.

From this data you can estimate the probability that a car coming from Danbury will take the road to Baldon.

These relative frequencies were found from a survey.

A tree diagram can help to organise the data.

Example   Calculate an estimate of the probability that a car coming from Danbury will take the road to Baldon.

Danbury – A

0.57   B
West Loe
Sandish
0.44   Baldon
Anton

C
Anton
Solen
St Elwin

You can multiply the probabilities when you move along the branches of a tree diagram.

An estimate of the probability is 0.57 × 0.44 = 0.25 (to 2 dp)

So about 1 car in 4 coming from Danbury takes the Baldon road.

Exercise 19.3
Relative frequency and tree diagrams

1   Calculate an estimate of the probability that a car approaching roundabout C will take the road to St Elwin.
    Explain how you worked this out.

2   Copy the tree diagram and fill in all the probabilities.

3   Estimate the probability that a car from Danbury will take:
    a   the Sandish road   **b**   the West Loe road
    c   the Solen road   **d**   the St Elwin road.

4   Estimate the probability that a car will go from Danbury:
    a   to Anton via roundabout B
    b   to Anton via roundabout C
    c   to Anton by either route.

5   On Wednesday 342 cars approached roundabout A from Danbury.
    Estimate the number of these cars that took the road to:
    a   roundabout B   **b**   roundabout C   **c**   Baldon
    d   Solen   **e**   Sandish   **f**   Anton.

Deciding how to find probability

A If outcomes are equally likely, you can calculate the **theoretical probability**.

> The probability of throwing an even number with a fair dice is $\frac{3}{6}$ since there are six equally likely outcomes and three of them are even numbers. This is a **theoretical probability**.

B Sometimes you can only estimate a probability by doing a survey to see how often something happens, or by using data which already exists.

> To find the probability that a shopper will buy your brand of coffee you will need to count how many buy coffee (say 126) and of these how many buy your brand (say 45).
> So the probability that they buy your brand is $\frac{45}{126} = 0.36$
> This is the **relative frequency**.

C Sometimes you can do an experiment to find a probability.

> To find the probability that an elastic band will break when a weight is hung from it you could test 100 bands and find the number that break (say 17).
> So the experimental probability that a band breaks is $\frac{17}{100} = 0.17$
> Again this is the **relative frequency**.

D Sometimes there is not enough data to estimate a probability.

> There is not enough data to find the probability that a Chinese tennis player will win at Wimbledon in the next ten years.

Exercise 19.4
Probability decisions

1 For each of these situations say whether A, B, C or D best applies to it.

a The probability of winning the jackpot on the National Lottery with one ticket.

b The probability that this dice will land this way up.

c The probability that a train from London to Glasgow will arrive on time.

d The probability that a train from London to Glasgow will have a driver over 40.

e The probability that there is life on another planet.

f The probability that the next car which passes the school gates will be red.

g The probability that the next earthquake in San Francisco will happen on a Thursday.

h The probability that there will be an earthquake in San Francisco before the year 2010.

i The probability that this spinner will stop on 3.

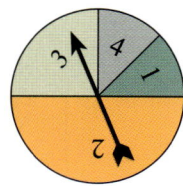

j The probability that you will live at the same address in 20 years time.

2 a What is wrong with this statement?

> You can calculate the probability that a person, chosen at random, is born in July.
> There are twelve months so the probability is $\frac{1}{12}$.

b Describe fully how you would find the probability that a person, chosen at random, was born in July.

3 Which of the following methods would a weather forecaster use to answer the question 'Will it snow on Christmas Day this year?'? Explain why.

a Carry out an experiment to find the relative frequency.
b Survey people's opinions and find the relative frequency.
c Look at past records to decide on the relative frequency.
d Use equally likely outcomes to calculate the theoretical probability.

4 Maria will open a new restaurant next week. She wants to find the probability that a person, chosen at random, is vegetarian.
Describe what is wrong with each of these surveys.

a She asks the first four people she meets on Friday if they are vegetarian.
b She interviews people who are leaving a butcher's shop.
c She asks every man she sees who has sandals and a beard.
d She asks all the members of her family.
e She asks one hundred babies who are under six months of age.

5 Andy had to answer this question 'What is the probability that a person chosen at random plays soccer?'. This is his answer:

> I asked twenty people and they all said they played soccer. So the probability is 1.

a Why must Andy's probability be incorrect?
b Give four reasons why Andy might have obtained the results he did.
c How would you do a reliable survey to decide on this probability?

6 The theoretical probability of getting a 5 with a fair 1 to 6 dice is $\frac{1}{6}$.

In an experiment you throw a dice to find the relative frequency of 5s.
How true is it that the more throws you have, the closer the relative frequency comes to the theoretical probability?

When can you add probabilities?

Exercise 19.5
Adding probabilities

This shows the set of males with criminal records in the town of Humbleton.

Give your probabilities as
fractions and do not cancel
them down.

1   What is the probability that a male criminal in Humbleton has:

   a   a beard   **b**   a necklace
   c   a beard or a necklace?

2   Compare your answers for the probabilities in Question **1**.
   What link can you find between them?

3   What is the probability that a male criminal in Humbleton has:

   a   an earring   **b**   a moustache?

4   **a**   Predict the probability of a criminal having
       either an earring or a moustache.
   b   Use the full set above to find the probability that a criminal has
       either an earring or a moustache.
   c   Why do you think your prediction may not be the
       same as the true answer?

5   What is the probability that a criminal, chosen at random, has:

   a   a hat   **b**   a beard   **c**   a beard or a hat?

Diagrams like these are known as **Venn diagrams** after the mathematician John Venn, who invented them in 1881 for his work on logic.

You can use diagrams to help you see why you can sometimes add probabilities, but at other times you must not.

Look at those criminals who have
♦ **dark hair**
♦ **earrings**.

All those with dark hair are in the red loop and all those with an earring are in the blue loop.

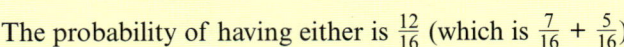

From this you can see that the probability of having dark hair is $\frac{7}{16}$

and of having an earring is $\frac{5}{16}$

The probability of having either is $\frac{12}{16}$ (which is $\frac{7}{16} + \frac{5}{16}$)

But now look at those with
♦ **a beard**
♦ **a moustache**.

This time the loops overlap because Alf, Liam and Jim have both beards and moustaches.

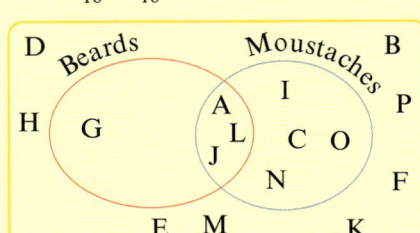

The probability of having a beard is $\frac{4}{16}$

and of having a moustache is $\frac{7}{16}$.

The probability of having either is $\frac{8}{16}$ (which is **not** $\frac{4}{16} + \frac{7}{16}$).

This shows that:
you can only add probabilities when the sets do not overlap.

Exercise 19.6
Probability and
Venn diagrams

1   **a**   Draw a Venn diagram to show the criminals with a necklace and those with fair hair.
   b   Can you add the separate probabilities to find the probability that a criminal has either a necklace or fair hair? Explain your answer.

2   Use your diagram to decide the probability that a criminal has:
   a   **either** a necklace **or** fair hair
   b   **both** a necklace **and** fair hair
   c   **neither** a necklace **nor** fair hair
   d   a necklace but not fair hair
   e   fair hair but not a necklace.

3   Draw Venn diagrams to help you decide on the probabilities of having:
   a   either a hat or an earing
   b   either a necklace or dark hair
   c   either glasses or fair hair
   d   neither glasses nor dark hair.

4   For each of these, say if you can add the separate probabilities to find the probabilities. Explain your reasons.
   a   a girl playing hockey in a school or a girl playing basketball
   b   a boy playing basketball in a school or a girl playing hockey
   c   a driver or a passenger
   d   a bus driver or a tall woman
   e   a dancer or a mechanic.

Outcomes from a biased dice

This table shows the probabilities of getting each number on a biased 0–9 dice.

Outcome	0	1	2	3	4	5	6	7	8	9
Probability	0.13	0.21	0.05	0.10	0.01	0.25	0.13	0.07	0	0.05

> A biased dice is one where each outcome is not equally likely.

Other probabilities can be found from these probabilities.

Example   What is the probability of getting a number less than 4 in one roll of the dice?

There are four numbers (0, 1, 2, and 3) less than 4.
There is no overlap between them (you can't get a 3 and a 1 in the same roll).
So you can add the probabilities for 0, 1, 2 and 3.

So the probability of a number less than 4 is 0.13 + 0.21 + 0.05 + 0.1 = 0.49

> Note that 0 is counted as the first even number.

A   What is the probability, with this dice, of getting:

a   a number greater than 6
b   an even number
c   a prime number
d   a number which is not prime
e   a number which is a multiple of 3 **and** an even number.

You must be careful with some probabilities where there is an overlap.

Example   What is the probability of getting either a multiple of 3 or an odd number?

> Here you can get a multiple of 3 **and** an odd number in the same roll of the dice.

The numbers 3 and 9 come in both sets but must only be counted once.

So the probabilities to add are those for
1,   5,   7,   3,   9 and 6.

$$0.21 + 0.25 + 0.07 + 0.1 + 0.05 + 0.13$$

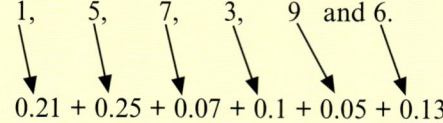

So the probability of a multiple of three or an odd number is 0.81

Exercise 19.7
Outcomes from a biased dice

1   Use the Venn diagram to calculate the probability of getting in one roll:
   a   a multiple of three **and** an odd number
   b   a number which is **neither** a multiple of 3 **nor** an odd number
   c   a number which **is not** a multiple of 3 but **is** an odd number.

2   Use a Venn diagram to find the probability of getting a number greater than 5 or an even number.

3   **a**   Why can you add all the probabilities when you find the probability that in one roll you get a number less than 5 or a number greater than 7?
   b   Calculate this probability.
   c   What is the probability of **not** getting a number less than 5 **nor** a number greater than 7?

End points

You should be able to ...

... so try these questions

A Calculate and use relative frequency

A1 Estimate the probability that a tree chosen at random, will be:
  a an ash   **b** a hornbeam.

A2 Estimate the probability that a tree is not an oak.

| Tree survey of Dudley Wood ||
Species of tree	Frequency
Oak	26
Ash	18
Lime	12
Hornbeam	7
Beech	25
Other	18

B Use relative frequency with tree diagrams

B1 The probability that Maria remembers her chain saw is 0.7, the probability that Pete remembers the petrol is 0.6
  a Finish the tree diagram to show all the probabilities and outcomes.
  b What is the probability that they remember both the saw and the petrol? Is this above or below an even chance?

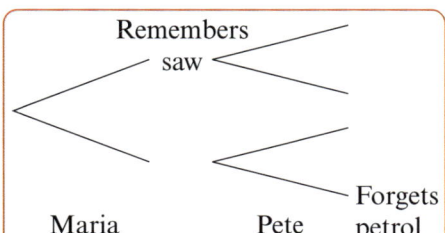

C Decide when you can use theoretical probability and when relative frequency must be used

C1 For each of the following say whether theoretical probability can be used or if relative frequency must be found. Give your reasons.
  a The probability that in any book a page chosen at random will be even.
  b The probability that it will rain on 1 May.
  c The probability that you open any book at random and see a photograph.

D Decide when you can find a probability by adding the separate probabilities

D1 **a** In the Dudley Wood survey estimate the probability that a tree is either an oak or a lime.
  b Why is it possible to add the probabilities in this case?

D2 In the apple tree survey why is it not possible to add the probabilities to find the probability of a tree having either canker or deer damage?

| Apple tree survey of 100 trees ||
Condition	Number of trees
Aphid attack	43
Damage by deer	24
Powdery mildew	58
Canker	15
Healthy	41

Some points to remember

   ♦ You can only calculate theoretical probability where outcomes are equally likely.

   ♦ Before you add separate probabilities check that there is no overlap between the sets. A Venn diagram can sometimes help you decide this.

Starting points
You need to know about ...

... so try these questions

A Using the inequality signs > and <

- These signs can be used with inequalities:

  > stands for **is greater than ...**
  < stands for **is less than ...**

  $x > 5$ means x can have any value greater than 5 but **not 5** itself.
  Here there are an infinite number of values for x.
  This includes non-integer values, such as 15.23 or $8\frac{3}{4}$

- Inequality signs can also be used to order numbers.

  For example: ¯23 < ¯4.23 < 1.5 < 37 < 100
  This can also be written as 100 > 37 > 1.5 > ¯4.23 > ¯23
  (Note. ¯23 is less than ¯4.23, but 23 is greater than 4.23)

- A range of values can be shown as an inequality.

  ¯4 < p < 2 means p has any value greater than ¯4 but less than 2.
  The numbers ¯4 and 2 are **not** included.
  ¯4 < p < 2 can also be written as **2 > p > ¯4**

- Sometimes you are only interested in the integer values.

  The integer values of x described by the inequality ¯3 < x < 5 are ¯2, ¯1, 0, 1, 2, 3 and 4.

B Sketching linear graphs

- Equations of straight line graphs can be expressed in the form

  $y = mx + c$ where m is the **gradient** of the graph and
  c is the **y-intercept**

  For example:
  $$y = \frac{1}{2}x + 1 \qquad\qquad y = 8 - 2x$$

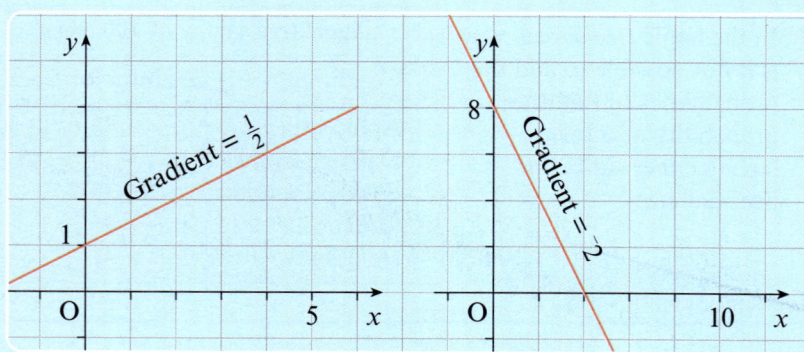

Gradient is $\frac{1}{2}$ Gradient is ¯2
y-intercept is 1 y-intercept is 8

A1 Replace the ☐ with either > or < to make each of these correct.
a 34 ☐ 12   b 2 ☐ 15
c ¯12 ☐ ¯56   d 5 ☐ ¯2
e ¯2 ☐ 0   f ¯17 ☐ ¯5

A2 Order this set of numbers:
9, 56, ¯24, 5, ¯4, 28, ¯30
a using the sign >
b using the sign <

A3 Which of these numbers is not a possible value for d, where $d < 5$?

12, ¯35, 3.217, ¯2, 5, 4.99, 6.2

A4 What are the integer values of t, where:
a ¯6 < t < 3
b ¯1 > t > ¯5?

A5 Explain why there are no integer values for x where:
¯2 < x < ¯1?

B1 Sketch the graphs of:
a $y = 2x - 2$
b $y = 12 - x$
c $y = 3x$
d $y = ¯4$

Inequalities

> means **is greater than**
>
> < means **is less than**
>
> ⩾ means **is greater than or equal to**
>
> ⩽ means **is less than or equal to**

♦ Often a quantity must be kept within a range of values.
For instance, the speed on a motorway must stay between 30 mph and 70 mph.

The legal speed can be shown in shorthand like this:

30 < S < 70   where S is the speed in mph.

This is called an **inequality**.

This inequality can also be written as   **70 > S > 30**

Since you can also drive at speeds equal to 30 mph and 70 mph this is more accurately shown as:

30 ⩽ S ⩽ 70   or   **70 ⩾ S ⩾ 30**

So all speeds **on or between** 30 mph and 70 mph are legal.

♦ Sometimes numbers at either end of a range are not included.

Example   A firm charges £3.00 each if you buy one switch,
£2.50 each if you buy between 1 and 10 switches,
and £1.75 each if you buy 10 or more.
For the £2.50 switches this can be written as:

1 < N < 10   where N is the number of switches (1 and 10 are not included).

Exercise 20.1
Writing ranges as inequalities

You sometimes need to choose letters to stand for the variables, such as page number or age.

1 For tomatoes a greenhouse temperature from 45 °F to 80 °F is recommended.
Let T stand for the temperature in °F.
Write the recommended temperature range as an inequality.

2 Write these page ranges as inequalities.

  a Pages 17 to 52 of a book are in colour.
  b Pages after 10 and before 64 have photographs.
  c The contents finish on page 4, followed by the features pages then the index which starts on page 164. Give the range for the features pages.

3 These are labels for two different drugs.

FERMATOL

For safety reasons this drug must not be used by people older than 65 or younger than 5.

Hypaticain

For safety reasons this drug must not used by the over 65s or by children 5 years and under.

  a Write the safe age range for Fermatol as an inequality.
  b Write the safe age range for Hypaticain as an inequality.
  c What is the difference between the safe age range for each drug?

Remember that ⁻18 is greater than ⁻25.

4 A freezer cabinet must be kept between ⁻25 °C and ⁻18 °C.
Write this temperature range as an inequality.

Inequalities with integer solutions

◆ For speeds on a motorway where $30 \leqslant S \leqslant 70$ there is an infinite number of values for S between 30 mph and 70 mph.

◆ The floors with executive suites in a hotel are given by $3 \leqslant F < 8$, where F is the floor number.
In this case there can only be integer values, since floor 3.6 cannot exist.

◆ The inequality $3 \leqslant F < 8$ can be shown on a number line like this:

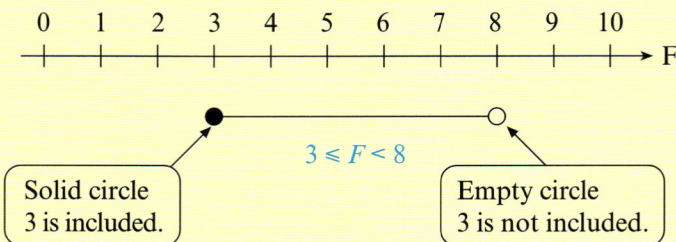

The integer values which **satisfy** this inequality are 3, 4, 5, 6, 7

An integer is a whole number. It can be positive, negative or zero.

**Solid circle
3 is included.**

$3 \leqslant F < 8$

**Empty circle
3 is not included.**

**Exercise 20.2
Integer solutions
to inequalities**

1 Which of the following inequalities can only have integer solutions?

A the level in a fuel tank given by $5 \leqslant h \leqslant 45$, where h is the height in cm
B the cost of sweets given by $32 \leqslant c \leqslant 42$, where c is the cost in pence
C the flats in a block given by $7 \leqslant h \leqslant 12$, where h is the flat number
D the weight of a lorry given by $12 \leqslant w \leqslant 42$ where w is the weight in tonnes
E the length of a car given by $300 \leqslant c \leqslant 450$, where c is the length in cm

2 **a** Show the inequality $1 < x < 5$ on a number line.
b What integer values for x satisfy the inequality?

3 **a** Show the inequality $^-1 \leqslant p < 4$ on a number line.
b What integer values for p satisfy the inequality?

4 For each of these, what integer values of n satisfy the inequality?
Show each answer on a number line.

a $4 < n \leqslant 9$ **b** $^-2 \leqslant n \leqslant 3$ **c** $^-4 < n \leqslant 5$
d $71 > n \geqslant 68$ **e** $^-4 \geqslant n > ^-12$ **f** $^-6 < n \leqslant 0$
g $^-3 \leqslant n < 7$ **h** $2163 < n < 2167$ **i** $^-2 \leqslant n \leqslant 3$

5 Which one of these inequalities describe a different set of integer values from the others?
Explain using a number line.

$^-4 \leqslant h < 6$, $^-5 < h < 6$, $^-4 \leqslant h \leqslant 5$, $^-4 < h < 6$, $^-5 < h \leqslant 5$

6 Write two different inequalities in g which are satified by these integers:

$^-3, ^-2, ^-1, 0, 1$

7 For each pair of inequalities, what integer values satisfy both?
Show each answer on a number line.

a $5 \leqslant k < 8$, $6 \leqslant k \leqslant 9$ **b** $^-7 < k < 2$, $^-5 < k \leqslant 1$
c $17 > k > 12$, $14 \geqslant k > 7$ **d** $^-1 \leqslant k \leqslant 3$, $^-3 \leqslant k < 1$
e $^-3 \leqslant k < 5$, $7 \geqslant k > ^-3$ **f** $^-4 < k \leqslant 1$, $^-2 \leqslant k < 0$

Two different inequalities can describe the same set of integers. For example:

$^-1 < x \leqslant 5$

$^-2 \; ^-1 \; 0 \; 1 \; 2 \; 3 \; 4 \; 5 \; 6$

$0 \leqslant x < 6$

Both of these ranges describe the integers 0, 1, 2, 3, 4, 5.

Inequalities and regions

Rules which have to be met are called conditions.

♦ Two inequalities can give a region on a graph.

For example, on airline flights there are limits on the luggage you can take.

For hold luggage, the condition on the weight W can be shown on a number line:

virgin atlantic

Hold luggage
Two pieces allowed, each to have:
maximum weight 32 kg
maximum dimension 158 cm

Hand luggage
One item only, to have:
maximum weight 6 kg
maximum dimension 114 cm

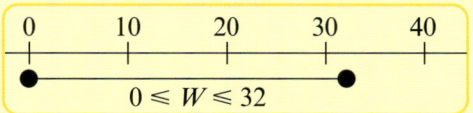

$$0 \leqslant W \leqslant 32$$

and the dimensions D as:

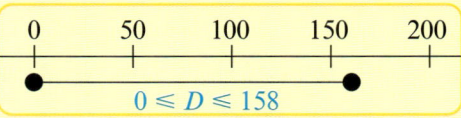

$$0 \leqslant D \leqslant 158$$

These can both be shown on one sketch graph.

The lines $W = 32$ and $D = 158$ split the graph into four regions.

The shaded region matches both conditions.

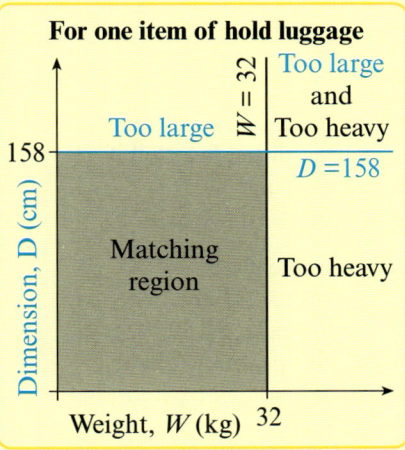

For one item of hold luggage

Too large and Too heavy

Too large

Matching region

Too heavy

$W = 32$

$D = 158$

Dimension, D (cm)

158

Weight, W (kg) 32

Exercise 20.3
Regions on graphs

1 These are some bags taken on the airline above.

Bag **A** $W = 24$ kg, $D = 175$ cm   Bag **B** $W = 5.3$ kg, $D = 109$ cm
Bag **C** $W = 28$ kg, $D = 142$ cm   Bag **D** $W = 5.3$ kg, $D = 123$ cm
Bag **E** $W = 35$ kg, $D = 136$ cm   Bag **F** $W = 7$ kg, $D = 99$ cm

a Which of these bags could not be hold luggage? Explain your answer.
b Draw a sketch graph to show the matching region for hand luggage. Label each region.
c Which bags could be taken as hand luggage?

2 This is an extract from a letter to recruit pilots for Jumbo jets.

> highly successful career with us flying 747 aircraft. Applicants
> should have at least 2000 hours flying experience and be aged
> between 28 and 50. Please send a letter of application giving
> your full CV to Mark Southgate at Flight Operations
> Recruitment, to arrive before 15 March

This sketch graph shows the limits of the regions described in the letter.

a Copy the sketch graph.
b Label each region.
c Shade in the region that satisfies all the conditions in the letter.

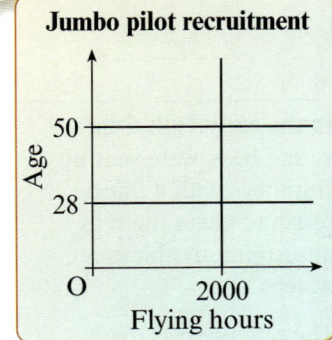

Jumbo pilot recruitment

Age

50

28

O   2000
Flying hours

3 These extracts are from notices or notes.

[A] *Rona Kennedy's exercise bike programme*

You should cycle a distance of between 10 and 20 miles
or for at least 15 minutes before breakfast each day.

[B] **PEUGEOT 405 SERVICE NOTE**

Your first service should take place within 6000 miles
or 6 months, whichever is the sooner.

[C] *Sungrow Vegetarian Restaurant*

Our party rates apply to groups of between 8 and 20 people.
Meal prices range from £6 to £12 a head.

[D] *JENNY'S 15TH BIRTHDAY*

Bike with 14 or more gears. Must cost less than £400.

[E] **Offkit flea spray**

Use only on cats older than 12 weeks
and heavier than 2 kilograms.

[F] **Kansas Car Hire**

Prices range from £36 to £62 per day.
Maximum five people per vehicle.

[G] **Mike's Supplies**

Dave, we need a scaffold tower next week. It needs to be 15
foot or more tall but must not cost more than £50 per day.
Thanks, Mike

[H] **Sandford Superstore**

We are looking for keen sales staff between the ages of 25 and 35
with at least four years experience of vegetarian beef sales.

It is often difficult to tell if the signs $>$ and $<$, or $\geq$ and $\leq$ fit a situation.

For example, does 'prices up to £40' include a price of exactly £40? It should not, but probably does.

Assume for these extracts that they all mean $\geq$ and $\leq$.

In the early 19th Century young boys were sent up chimneys with a hand brush to clean them by unscrupulous chimney sweeps.

The practice was made illegal by parliament in 1833.

---

a For extracts A, B and C
  i decide what axes you need and sketch a graph
  ii shade the region which meets all the conditions
  iii label the other regions on each graph.
b For extracts D to H sketch a graph and shade the matching region.
  Do not label the other regions.

4 Sketch a graph which **could** describe suitable applicants for
the job of a Victorian chimney sweep's assistant.
You will need to decide on what axes to use and the
inequalities which apply.

Shading regions on graphs

◆ An inequality such as **x > 4 or ⁻5 < y < 2** can be shown as a region on a graph.

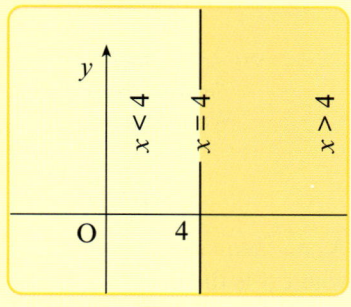

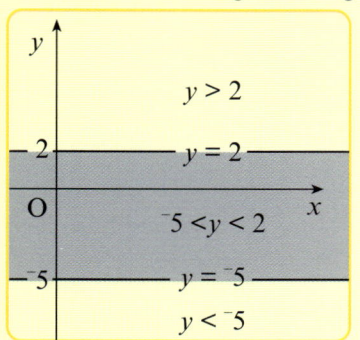

Region shaded ▢ is $x > 4$ Region shaded ▢ is $^-5 < y < 2$

◆ The conditions satisfied by both inequalities can also be shown on one graph.

Example

Show the region where $x > 4$ and $^-5 < y < 2$.

In this case the shaded region is where the region for $x > 4$ overlaps with the region for $^-5 < y < 2$.

This is the part shaded ▢

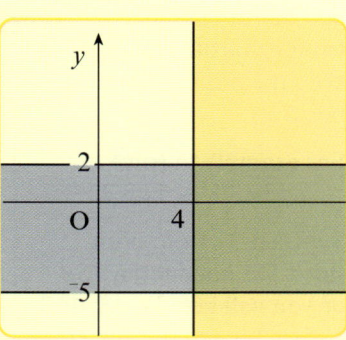

Exercise 20.4
Sketching regions

1 Sketch graphs to show these regions:

   a $x < 5$ **b** $y > ^-3$ **c** $3 < y < 7$
   d $0 < x < 4$ **e** $x < ^-2$ **f** $^-4 < y < ^-1$

2 Write an inequality for each shaded region on these graphs.

a
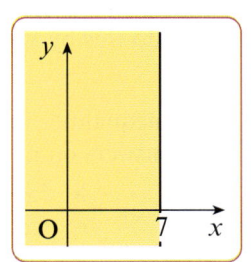

b
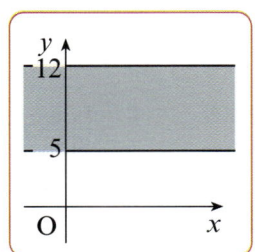

c
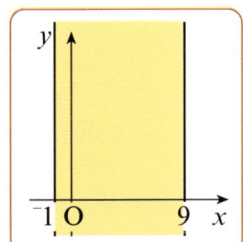

3 Sketch graphs to show the region where

   a $y > 5$ and $x > 3$ **b** $y < ^-1$ and $x > 4$
   c $^-1 < y < 5$ and $x > 2$ **d** $y > 3$ and $2 < x < 8$
   e $0 < x < 5$ and $1 < y < 6$ **f** $^-2 < y < 6$ and $^-5 < x < 0$

4 Give the inequalities which describe the shaded region on this sketch graph.

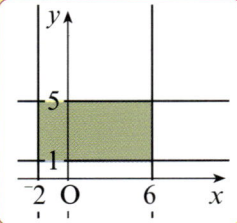

Solving inequalities

♦ An inequality can have an unknown on both sides – for instance:

> **Danfield Nurseries**
> We sold 5 trays of roses and 3 single rose plants on Monday.
> On Tuesday we sold 3 trays and 13 single rose plants.
> We sold more rose plants on Monday than on Tuesday.

> How many rose plants are on a tray?

As an inequality this can be written as $5x + 3 > 3x + 13$
where x is the number of rose plants on a tray.

♦ To solve an inequality you can treat it like an equation.

Example 1

Solve this inequality	$5x + 3 > 3x + 13$
Subtract $3x$ from both sides	$2x + 3 > 13$
Subtract 3 from both sides	$2x > 10$
Divide both sides by 2	$x > 5$

So the number of rose plants on a tray must be greater than 5, i.e. at least 6.

Example 2

Solve this inequality	$3x - 8 \leqslant 5x - 2$
Subtract $3x$ from both sides	$^-8 \leqslant 2x - 2$
Add 2 to both sides	$^-6 \leqslant 2x$
Divide both sides by 2	$^-3 \leqslant x$
So	$x \geqslant ^-3$

So x can have any values greater than or equal to $^-3$.

The only difference from solving an equation is that if you multiply or divide both sides of an inequality by a negative number the inequality signs will reverse. You can avoid having to do this by keeping the coefficient of x positive.

When you rearrange an inequality try to keep the number in front of the variable positive. (i.e. try to keep the coefficient of x positive).

For example to solve
$$3x - 8 \leqslant 5x - 2$$

Subtract $3x$ from both sides. Do not subtract $5x$ or you will get $^-2x - 8 \leqslant ^-2$ and will have the problem of dividing through by $^-2$ and changing the sign.

**Exercise 20.5
Solving inequalities**

1   Solve each of these inequalities to find the possible values of x.
    Show each of your answers on a number line.

a	$2x > 8$	**b**	$5x \leqslant 20$	**c**	$9x \leqslant 27$
d	$7x \geqslant ^-28$	**e**	$5x \geqslant 12$	**f**	$12x \geqslant 108$
g	$20x \leqslant 10$	**h**	$20 > 4x$	**i**	$25x \leqslant 400$
j	$150 \leqslant 10x$	**k**	$7x \geqslant 49$	**l**	$16x \leqslant 64$

2   Solve each of these inequalities. Show each answer on a number line.

a	$3y + 6 < 27$	**b**	$13 \geqslant s + 5$	**c**	$12 > k - 17$
d	$5p - 3 \geqslant 27$	**e**	$10 < 2b + 3$	**f**	$14 < 2x + 8$
g	$6a + 50 \leqslant 2$	**h**	$5x - 2 > 38$	**i**	$6x - 8 < 22$
j	$2t + 12 \leqslant 5t$	**k**	$4a - 6 \geqslant 22$	**l**	$5c - 15 \geqslant 25$
m	$13 - 4d > 33$	**n**	$3a + 7 \leqslant ^-32$	**o**	$6x + 4 > 40$

3   Solve each of these inequalities. Show each answer on a number line.

a	$2a - 5 \geqslant a + 6$	**b**	$7k + 4 < 2k - 6$
c	$3x + 7 > x - 11$	**d**	$4n - 9 \geqslant 2n - 2$

End points

You should be able to ...   ... so try these questions

A Find integer values which satisfy an inequality

A1 For each of these, what integer values of g satisfy the inequality? Show each answer on a number line.
  a $^-2 < g < 0$ **b** $^-5 \leqslant g < 4$
  c $7 \geqslant g \geqslant 4$ **d** $^-56 \geqslant g > ^-60$
  e $^-77 < g \leqslant ^-72$ **f** $^-3 < g < 4$

B Draw a region to illustrate an inequality

B1 Sketch a graph for Bargain Exhausts.
  a Shade in the region that matches these conditions.
  b Label the other regions.

B2 Sketch a graph to show the region where $^-1 < x < 3$.

B3 Sketch and shade the region on a graph to show where $^-1 < x < 2$ **and** $y > 3$.

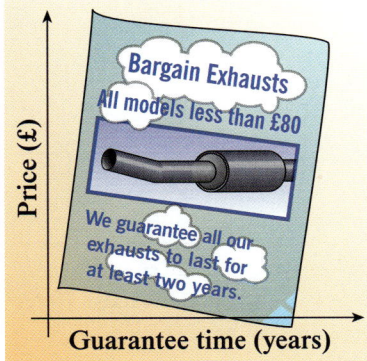

B4 Write inequalities for the shaded regions in graphs **a** and **b**.

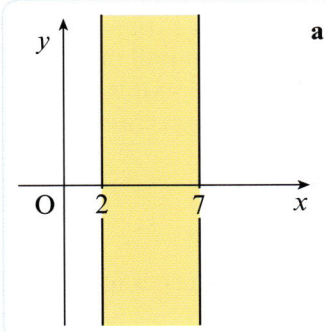

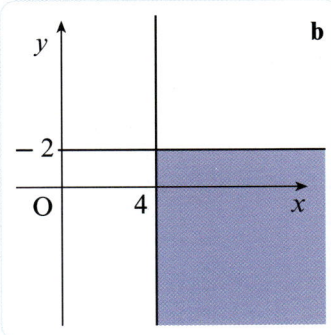

B5 Sketch the graph of $y = 2x - 3$
Shade in the region of your graph where $y < 2x - 3$.

C Solve an inequality to find the range of values for x

C1 Solve these inequalities. Show each answer on a number line.
  a $5x \leqslant 30$ **b** $12 < 4f$
  c $4a + 5 \geqslant 53$ **d** $5k + 28 > 3$
  e $7a - 6 \geqslant 36$ **f** $7d - 10 > 5 - 3d$

Some points to remember

- Always check carefully to see if $\geqslant$ rather than $>$, or $\leqslant$ rather than $<$, is used.
- When you solve an inequality you can treat it like an equation except that:
  - if you multiply or divide both sides by a negative value you must reverse the signs.

Starting points

You need to know about ...

A Names of common solids

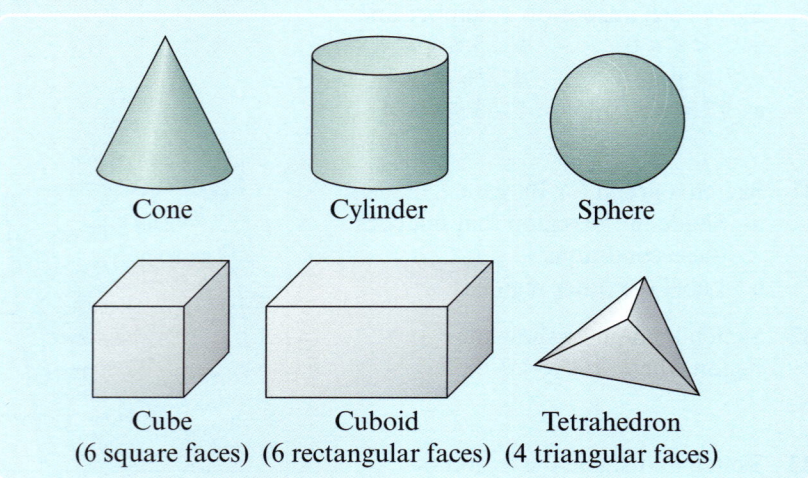

Cone

Cylinder

Sphere

Cube
(6 square faces)

Cuboid
(6 rectangular faces)

Tetrahedron
(4 triangular faces)

B Prisms and pyramids

♦ A **prism** is a solid with a uniform cross-section.

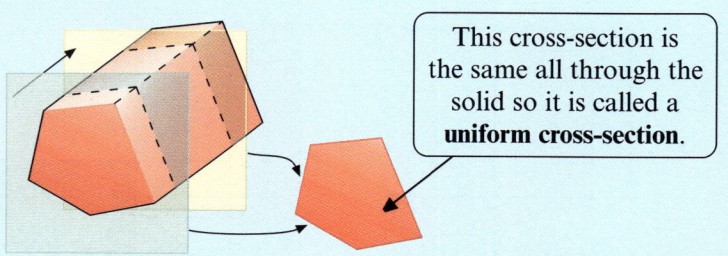

This cross-section is the same all through the solid so it is called a **uniform cross-section**.

♦ The mathematical name for a prism usually depends on the shape of its uniform cross-section.

Example

Triangular prism

Pentagonal prism

♦ For a **pyramid**, all faces except one meet at a common vertex.

♦ The mathematical name for a pyramid usually depends on the shape of its base.

Example

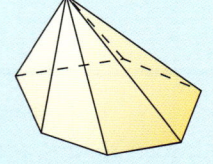 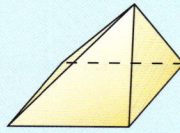

Hexagonal pyramid

Square pyramid

A1 For each object, name the solid that most closely matches its shape.
  a sugar lump
  b tennis ball
  c pound coin
  d matchbox
  e can of beans

A2 Name four objects that are in the shape of a cuboid.

A3 Why do you think that containers are usually in the shape of cuboids or cylinders?

B1

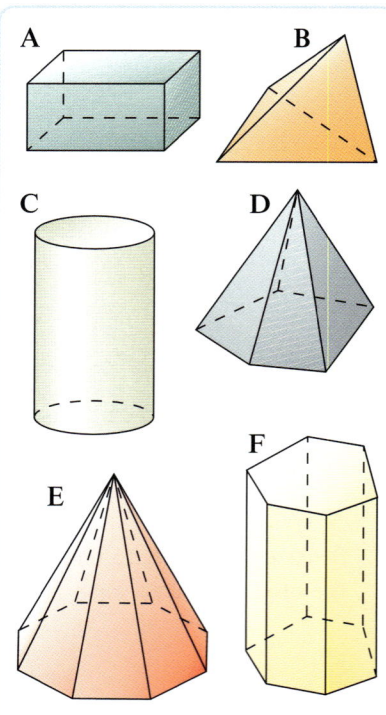

A B

C D

E F

a Each of the solids is either a prism or a pyramid.
  Make a list of:
    i the prisms
    ii the pyramids.

b Give a mathematical name for each solid.

C Faces, edges and vertices

- On a solid:
  - ❖ a flat surface is called a **face**
  - ❖ an **edge** is where two faces meet
  - ❖ a **vertex** is where two or more edges meet.

Example

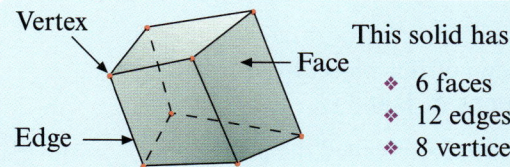

This solid has:
- ❖ 6 faces
- ❖ 12 edges
- ❖ 8 vertices

D Area and volume

- Area of a rectangle = Length × Width

- Area of a triangle = $\dfrac{\text{Base} \times \text{Height}}{2}$

- Area of a circle = πr^2

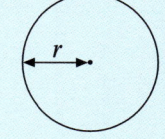

- Area of a trapezium = $\dfrac{h(a + b)}{2}$

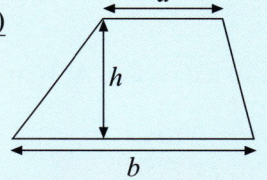

- Length, width and height are called the **dimensions** of a cuboid.

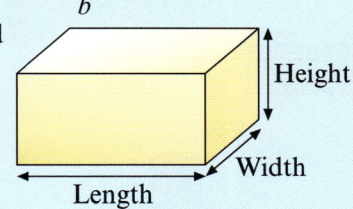

- Volume of a cuboid = Length × Width × Height

Example

Volume of this cuboid
= 4 × 3 × 2
= 24 cm³.

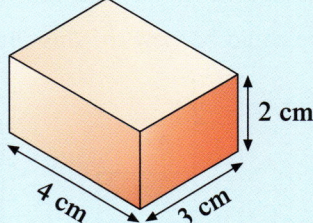

E Nets

- A **net** is a flat shape that can be folded up to make a solid.
  For example:
    A net of a cube

C1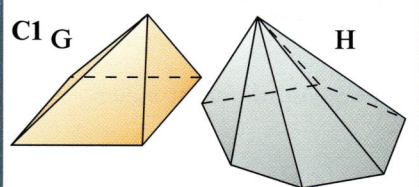

G   H

For each pyramid G and H, find the number of:
a faces   **b** edges
c vertices.

D1 Calculate the area of a circle with a radius of 4.5 cm, correct to the nearest cm².

D2 Find the area of this trapezium.

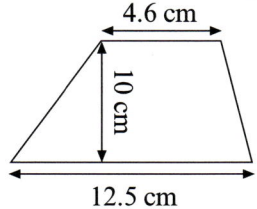

4.6 cm

10 cm

12.5 cm

D3 Estimate the volume of these objects in cm³.
a a sugar lump
b an apple

D4 Estimate the volume of air inside a car in m³.

D5 Calculate the volume of this cuboid in cm³.

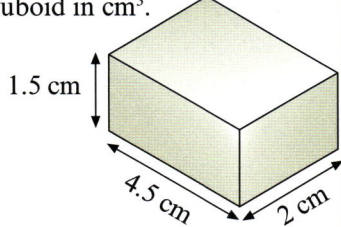

1.5 cm

4.5 cm

2 cm

D6 Sketch three different cuboids with a volume of 72 cm³. Show the dimensions clearly on each sketch.

E1 Draw three different nets for a cube where the length of each edge is 4 cm.

E2 Draw a net for a cuboid with:
- ❖ a length of 5 cm
- ❖ a width of 3 cm
- ❖ a height of 2.5 cm.

2D Views

Showing different views

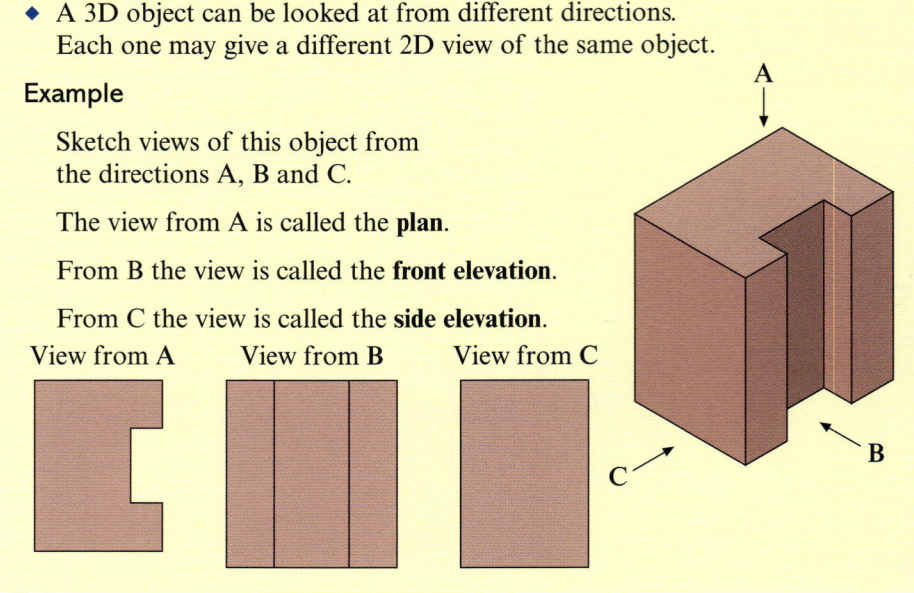

◆ A 3D object can be looked at from different directions. Each one may give a different 2D view of the same object.

Example

Sketch views of this object from the directions A, B and C.

The view from A is called the **plan**.

From B the view is called the **front elevation**.

From C the view is called the **side elevation**.

View from A   View from B   View from C

Exercise 21.1
Identifying views

1 Match each view to a direction A, B or C.

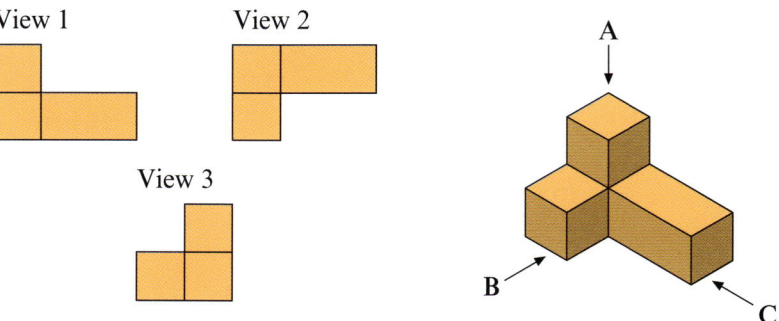

View 1   View 2

View 3

2 Which of these could be views of this church?

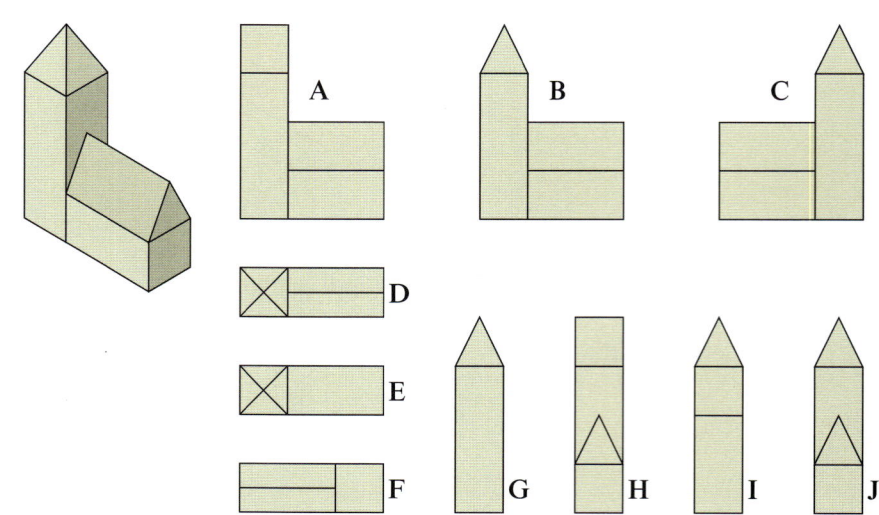

3   Object 1 is made from cubes.
On squared paper, sketch views of the object
from the direction of each arrow.

Object 1

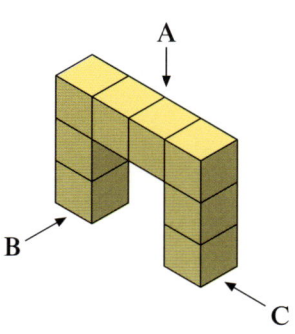

4   **a**   How many cubes make up object 2 ?
b   On squared paper, sketch the three views
of this object.

Object 2

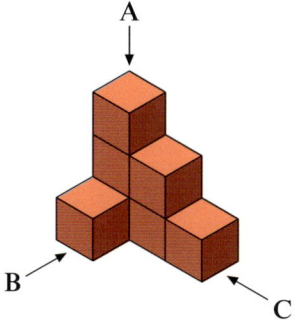

5   **a**   If object 3 were made of cubes of this size

how many would there be ?
b   On squared paper, sketch the
three views of this object.

Object 3

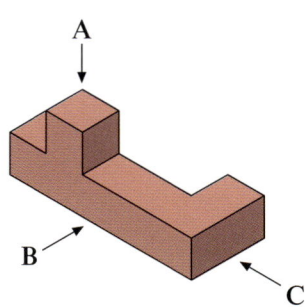

6   Sketch a view of object 4 from each direction
A, B and C.

Object 4

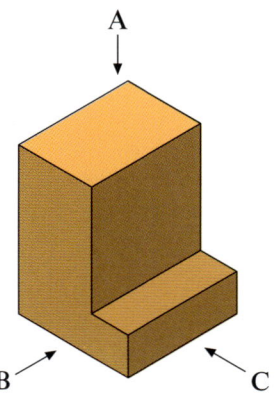

Planes of symmetry

Some solids have **plane symmetry**.
For example, this solid has plane symmetry.

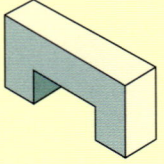

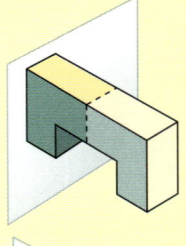

In the mirror, look at half of the solid like this and you see the other half.

So the mirror shows the position of a **plane of symmetry**.

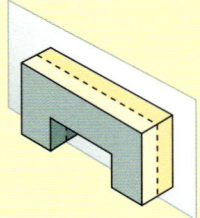

This mirror shows the position of another **plane of symmetry**.

Exercise 21.2
Properties of solids

1 Each of these solids is made from four cubes.

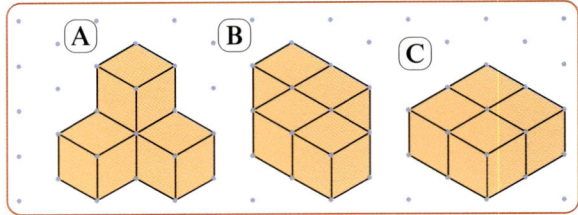

a Which of these solids are prisms?
b How many planes of symmetry has solid B?
c Which solid has exactly 3 planes of symmetry?

2 Draw a prism made from four cubes that has 2 planes of symmetry.

3 Draw all the solids made from four cubes that have 5 planes of symmetry.

4 Draw a solid made from four cubes that has no planes of symmetry.

5 Draw a solid made from five cubes that has:

a 2 planes of symmetry   b 0 planes of symmetry.

6 Each of these prisms has a regular polygon as a uniform cross-section.

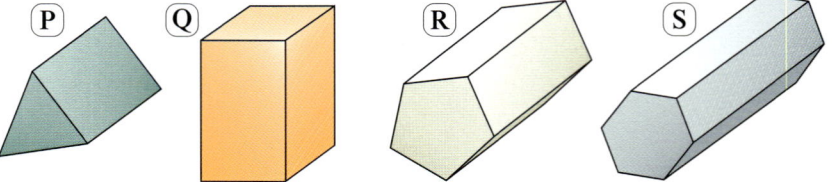

Make a table to show the number of faces, edges, vertices and planes of symmetry for each prism.

7 The uniform cross-section of a prism is a regular polygon with 100 sides. How many faces, edges, vertices and planes of symmetry does it have?

8 The uniform cross-section of a prism is a regular polygon with n sides. How many faces, edges, vertices and planes of symmetry does it have?

9 Each of these nets A to D is for a prism or a pyramid.

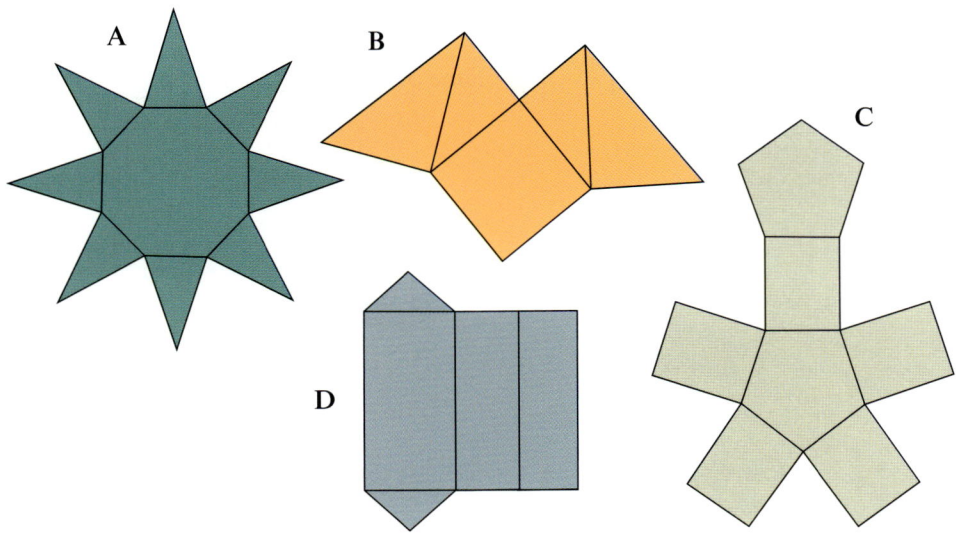

a Which of the nets is for:
   i a prism   **ii** a pyramid?
b Give the mathematical name for each prism or pyramid.

10 Each of the diagrams X, Y and Z is part of a net for a pyramid.

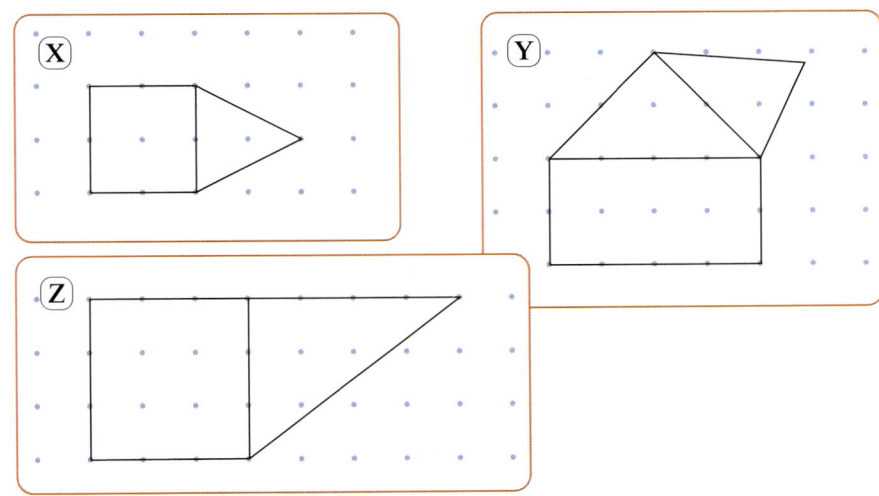

> You could cut out your nets and check that each makes a pyramid.

a Copy and complete each diagram to make the net of a pyramid.
b How many planes of symmetry would each pyramid have?

> For a regular tetrahedron, all faces are equilateral triangles.

11 This net gives a regular tetrahedron.
Each edge is 2 cm in length.

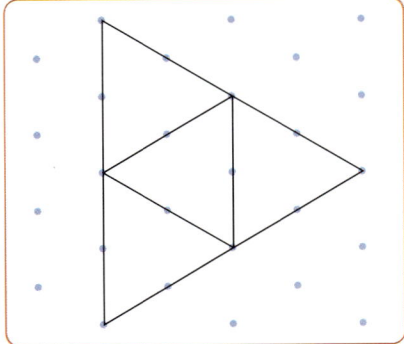

a Draw all the different nets you can find that would give this tetrahedron.
b How many planes of symmetry has a regular tetrahedron?

Volume of a prism

The depth of a prism is sometimes called the length or height.

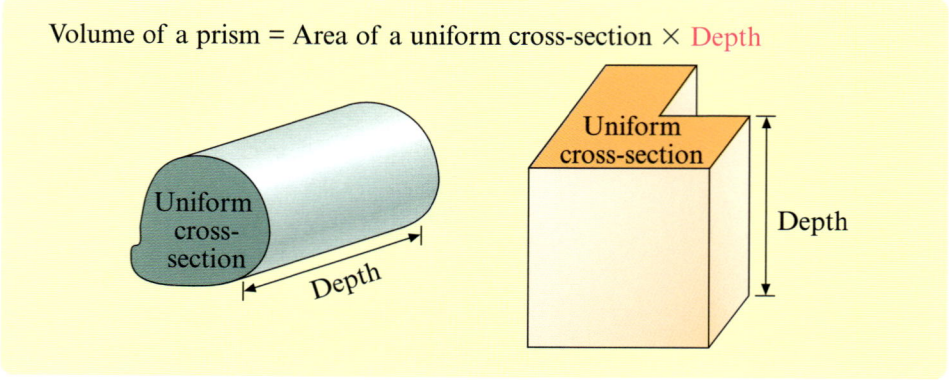

Volume of a prism = Area of a uniform cross-section × Depth

Exercise 21.3
Volume of a prism

1 Calculate the volume of each prism.

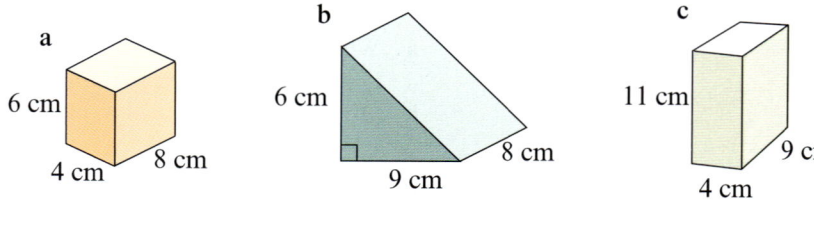

a 6 cm 4 cm 8 cm

b 6 cm 9 cm 8 cm

c 11 cm 9 cm 4 cm

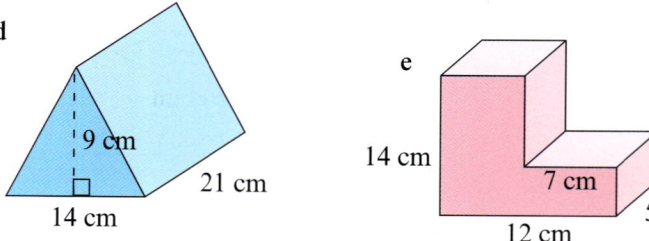

d 9 cm 14 cm 21 cm

e 14 cm 7 cm 4 cm 12 cm 5 cm

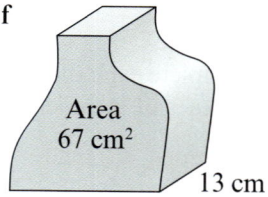

f Area 67 cm² 13 cm

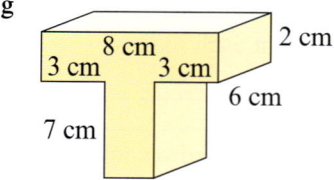

g 8 cm 3 cm 3 cm 2 cm 6 cm 7 cm

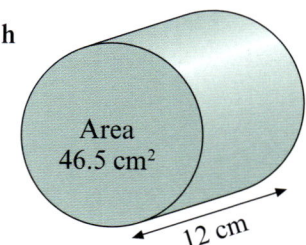

h Area 46.5 cm² 12 cm

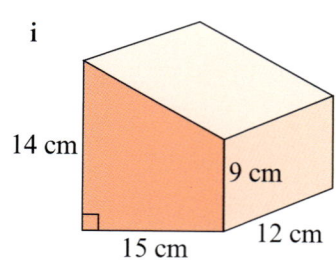

i 14 cm 9 cm 15 cm 12 cm

2 This toy is made from a clear plastic cuboid filled with blue liquid.
The depth of liquid is 3.5 cm.

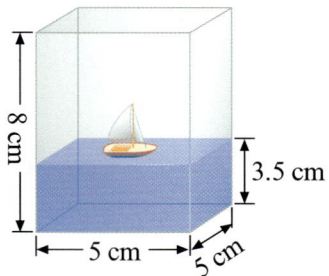

a Calculate the volume of the cuboid.
b Calculate the volume of blue liquid in the cuboid.

3 A box for drawing pins is to be designed in the shape of a cuboid.
It is to have a volume of 64 cm³. This sketch shows one possible box.

Give two other possible sets of
dimensions for the box.

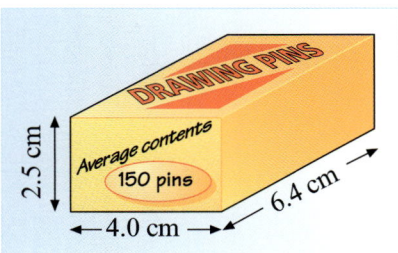

4 A box in the shape of a cuboid is to have a volume of 100 cm³.

a Give two different sets of dimensions the box might have.

b The manufacturer decides that the box should have a cross-section that is
a square.

What might be the dimensions of the box in this case?
Explain your answer using a diagram.

5 The diagram shows a triangular prism that
is 12 cm long.

The prism has a volume of 756 cm³.

a Calculate the area of the shaded cross-
section.

b Make a sketch of the prism giving
possible values for a and b.

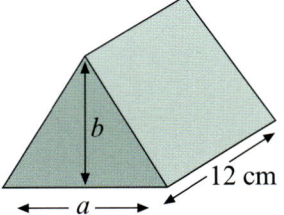

6 A cylinder has a volume of 345 cm³ and has a height of 15 cm.

a Calculate the area of the cross section of the cylinder.

b Write an expression in π for the radius of the cylinder.

Exercise 21.4
Maximising volume
of cylinders

1 This collecting box is in the shape of a cylinder.
The diameter of the base is 9.3 cm
and its capacity is 1120 cm³.

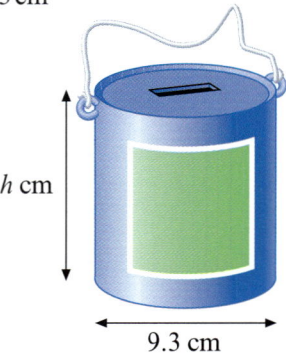

h cm

9.3 cm

Calculate *h* correct to 1 dp.

2 A designer in a plastics company is sent this letter.

> **World Nature Fund – fighting for the environment**
>
> 18 Westbury Road
> LONDON
> W2 4ZP
>
> Peterfield Plastics
> Unit 381
> Cupar Trading Estate
> SOUTH KILBRIDE
> Lanarkshire
>
> 22 August 1996
>
> Dear Ms Barnes,
>
> Our charity has been given a large number of plastic sheets.
> Each sheet measures 500 mm by 200 mm.
>
> We would like to use these sheets of plastic to make collecting
> boxes. Our collectors have found that a cylinder is the easiest
> shape to handle. Of course, we would like the boxes to have the
> maximum possible volume.
>
> Could you please provide us with your proposed design as soon
> as possible, including the dimensions of the collecting box.
>
> I look forward to your ideas.
>
> Yours sincerely,
>
> Vita A Green.

a Design a suitable collecting box for the charity.
b What is the diameter and height of your box?
c Write a short report to explain how you decided on these dimensions.
d For your design, what percentage of each sheet of plastic is not used?

> The sum of two numbers is
> found by adding:
> for example,
> the sum of 2 and 5 is 7.

3 a Sketch two cylinders where the sum of the radius and height is 12 cm.
b Find the volume of each of your cylinders.
c When the sum of the radius and height is 12 cm, what do you think
is the maximum possible volume?

4 Choose a different value for the sum of the radius and height of a cylinder
and investigate the maximum possible volume.

Exercise 21.5
Volume problems

A new park is planned.
It is to have a children's play area and an open-air swimming pool.

1 This is the design for a sandpit to be dug in the children's play area.

It is in the shape of a cylinder with radius 2.6 m and depth 0.8 m.

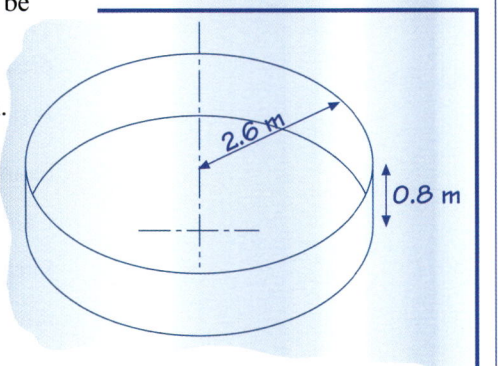

a Calculate the volume of the sandpit in m³.
b The sandpit is to be filled with sand to a depth of 0.6 m.
Calculate the volume of sand needed.
c The weight of 1 m³ of the sand is about 1.2 tonnes.
Find the weight of sand needed for the sandpit.

2 This is a plan of the space to be used for swings.

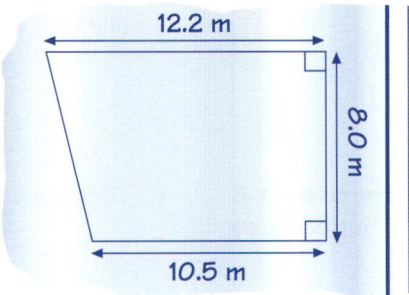

The space is to be dug to a depth of 0.3 m and filled with bark chippings.
a Calculate the volume of bark chippings needed.
b Forestry Products sell bark chippings in bags.
Each bag costs £5.12 and contains 0.07 m³ of chippings.
  i How many bags of chippings should be bought?
  ii What is the cost of these bags of chippings?

3 The diagram shows the uniform cross-section of the open-air swimming pool.

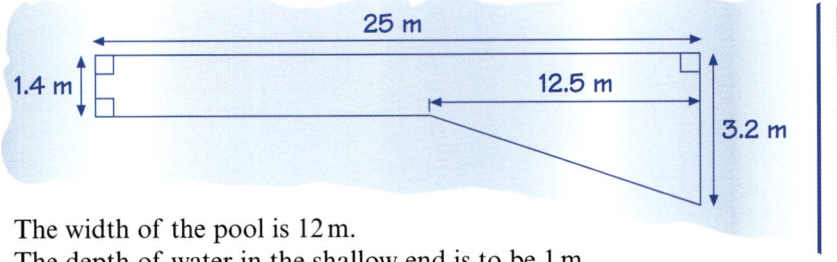

The width of the pool is 12 m.
The depth of water in the shallow end is to be 1 m.

a Calculate the area of the cross-section.
b What is the capacity of the swimming pool?
c Calculate the volume of water in the pool.
d The amount of chlorine added to the water in this pool is 1 cubic centimetre (cm³) per cubic metre (m³) of water.
How much chlorine will be added to the water in this pool?

Surface area of a prism

The surface area of a prism is the total area of all the faces (surfaces) of the prism:

It may help to draw a net to the prism.

Example

Calculate the surface area of this prism.

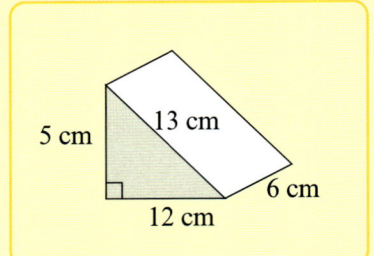

Surface area = total area of all the faces

Area of A:	5×6	$= 30 \text{ cm}^2$
Area of B:	12×6	$= 72 \text{ cm}^2$
Area of C:	13×6	$= 78 \text{ cm}^2$
Area of D:	$0.5 \times 12 \times 5$	$= 30 \text{ cm}^2$
Area of E:	$0.5 \times 12 \times 5$	$= 30 \text{ cm}^2$
	Total area	$= 240 \text{ cm}^2$

The surface area of the prism is 240 cm²

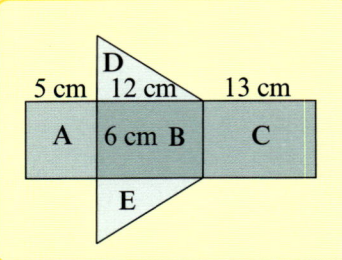

Exercise 21.6
Surface area

1   Calculate the surface area of each prism

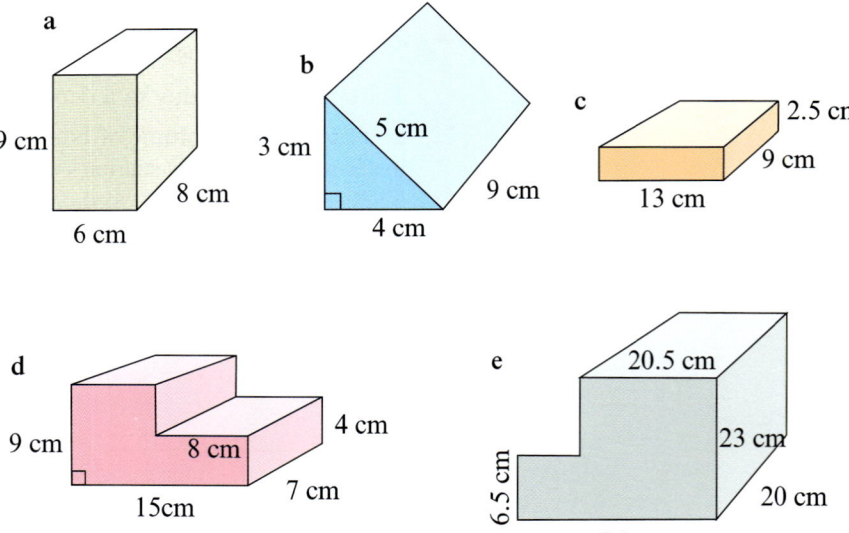

2   A cube has a surface area of 294 cm².
     What is the size of the cube? Explain your answer.

3   Cuboids A and B are similar with each length on B twice the corresponding length on A.
     What can you say about the surface area of A and B?
     Explain your answer.

♦ For liquids, volume or capacity is often measured in litres (l) or millilitres (ml).

   1 litre = 1000 millilitres
   1 millilitre is equivalent in volume to $1\,cm^3$.

♦ Mass (often called weight) is measured in kilograms (kg) or grams (g).

   1 kilogram = 1000 grams
   1 ml of water weighs 1 gram.

♦ Density can be measured in grams per cubic centimetre (g/cm^3).

$$\text{Density} = \frac{\text{Mass in grams}}{\text{Volume in } cm^3}$$

Exercise 21.7
Capacity, volume
and density

1   Write down one choice from each bracket to complete the sentence.
   a   The volume of an orange is about ($30\,cm^3$, $300\,cm^3$, $3000\,cm^3$).
   b   The capacity of a wine glass is about (75 litres, 7.5 litres, 75 millilitres).
   c   The weight of an apple is about (17 grams, 170 grams, 1.7 kilograms).

2   Estimate the capacity of a tea cup in millilitres.

3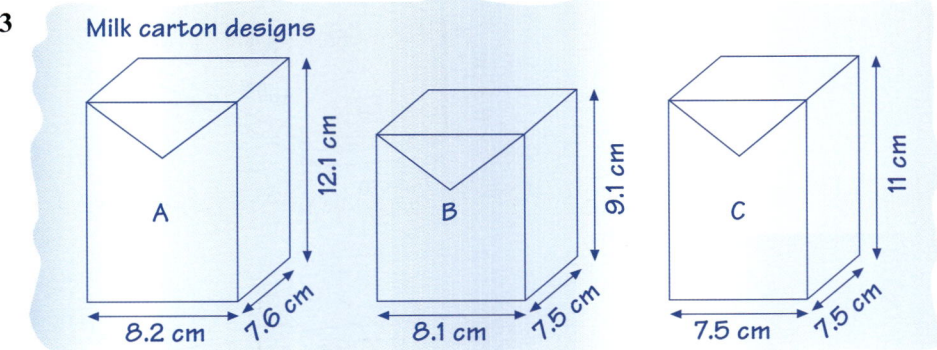

Milk carton designs

A dairy plans to sell cartons containing 550 ml of milk.
Which of these three cartons do you think the dairy should use?
Explain your decision.

4   Find the capacity of this carton in litres.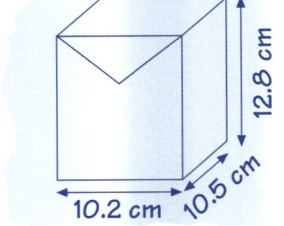

5   A storage tank on a milk lorry is a cylinder of radius 1.2 m and length 4.8 m.
Calculate the capacity of the tank in:
   a cm^3 **b**   litres.

6   Each of these pieces of cheese is in the shape of a prism.

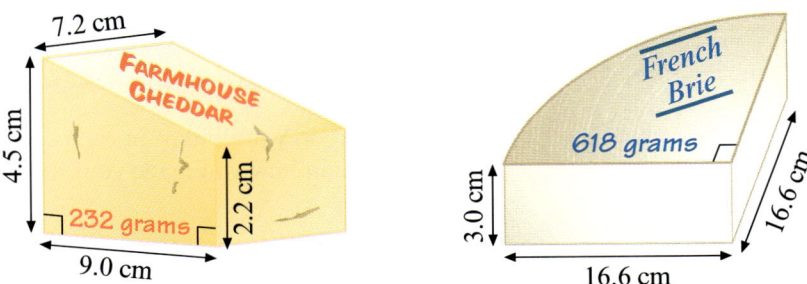

Find the volume and density of each piece of cheese.

7 Each of these containers is marked with the weight of its contents.

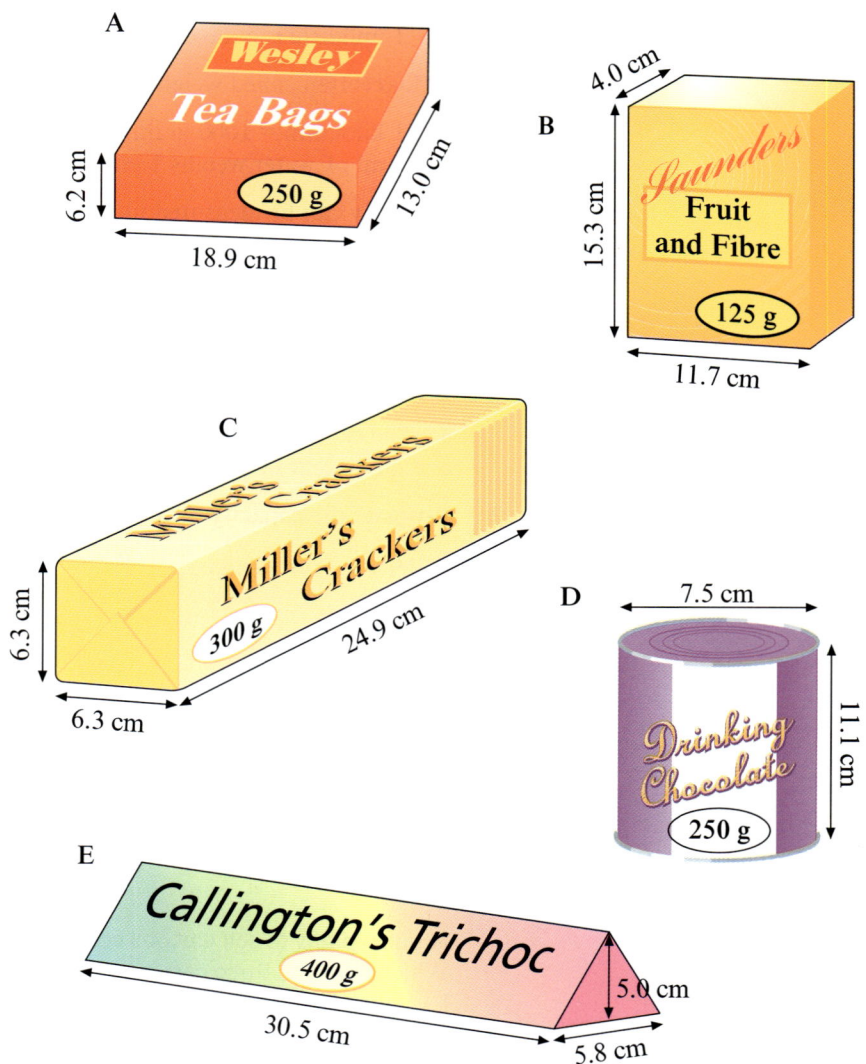

Calculate the volume of each container in cm³ correct to 2 sf.

8 This table shows the volume and weight for five more containers.
Each contains one of the five food products above but is a different size.

Container	Volume (cm³)	Weight (g)
P	640	200
Q	2700	500
R	770	125
S	230	200
T	930	500

What do you think is in each container?
Explain how you made your decision.

Formulas for length, area and volume

When you calculate with length, area and volume:

- ◆ a length added to or subtracted from a length gives a length
- ◆ an area added to or subtracted from an area gives an area
- ◆ a volume added to or subtracted from a volume gives a volume
- ◆ a length multiplied by a length gives an area
- ◆ a length multiplied by a length multiplied by a length gives a volume
- ◆ the square root of an area gives a length
- ◆ to add, subtract, multiply or divide by a number that is not a length does not change whether an expression gives a length, area or volume.

For example, if the letters a, b and c represent lengths:

$3a + b$ represents a **length**: the lengths $3a$ and b add to give a length

$\frac{1}{2}a(b - c)$ represents an **area**: the expression $(b - c)$ is a length; the lengths $(b - c)$ and a multiply to give an area

$5a(b^2 + 3c^2)$ represents a **volume**: the areas b^2 and $3c^2$ add to give an area; the length a and the area $(b^2 + 3c^2)$ multiply to give a volume.

Exercise 21.8
Formulas for length, area and volume

1 In the following expressions, l and w each represent a length.
For each expression, decide if it represents a length, area or volume.
   a $4(l + w)$ **b** $4lw^2$ **c** $lw + w^2$

2 One of these expressions gives the volume of the prism.

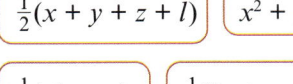 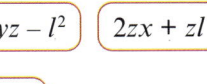

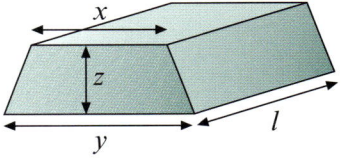

   $\frac{1}{2}(x + y + z + l)$ $x^2 + yz - l^2$ $2zx + zl$

   $\frac{1}{2}lz(x + y)$ $\frac{1}{4}l(x + y + z)$

   a Which is the correct expression for the volume of the prism?
   b Give reasons for your answer.

3 One of these expressions gives the perimeter
   of this shape and one gives the area.

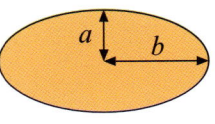

   πa^2b πab πb^2a^2 $\pi(a + b)$ $\pi b^2(a + 4)$

   Which is the correct expression for:

   a the perimeter   **b** the area?

π is a number, not a length.

4 In the following expressions, r and h each represent a length.
For each expression, decide if it represents a length, area or volume.
Give reasons for each answer.
   a $\frac{1}{4}rh$ **b** $\sqrt{r^2 + h^2}$ **c** $3(r + h) + \pi h$
   d $r^2(r + h)$ **e** $\frac{4}{3}\pi r^3$ **f** $\pi r(r + h)$

5 The letters x, y and z represent lengths.
Explain why $xy + xyz + y(x - z)$ cannot represent a length, area or volume.

End points

You should be able to ...   ... so try these questions

A   Describe properties of solids

A1   This solid is made from six cubes.

How many planes of symmetry has this solid?

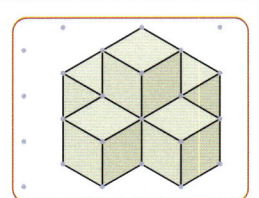

A2   This is a drawing of a hexagonal prism.

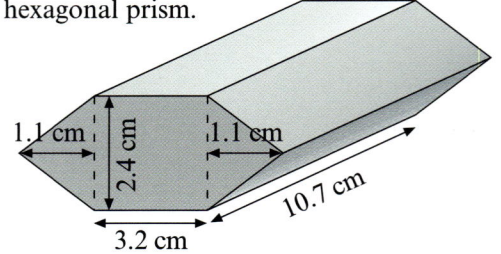

How many faces, edges and vertices does the prism have?

B   Find the volume of prisms

B1   Calculate the volume of the prism correct to the nearest cm³.

C   Use units for volume and density

C1   This tank is in the shape of a cylinder with radius 0.8 m and height 3.2 m.

Calculate the volume of the tank to the nearest:
a   cm³   **b**   litre.

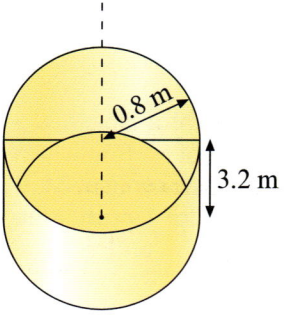

C2   A gold bar with a mass of 3180 grams is in the shape of a cuboid that measures 6.4 cm by 2.5 cm by 10.3 cm.

Calculate the density of the gold in g/cm³ correct to 1 dp.

D   Decide if a formula could be for length, area or volume

D1   One of these expressions gives the surface area of this cone and one gives its volume.

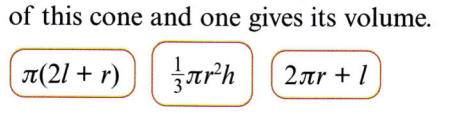

$\pi(2l + r)$ $\frac{1}{3}\pi r^2 h$ $2\pi r + l$

$\pi r(l + r)$ $\pi r^2 + 3l$

Which is the correct expression for:
a   the surface area   **b**   the volume?

Some points to remember

- ◆ Volume of a prism = Area of a uniform cross-section × Depth

- ◆ Density = $\dfrac{\text{Mass in grams}}{\text{Volume in cm}^3}$

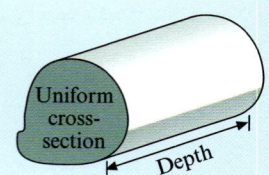

Starting points

You need to know about ...

... so try these questions

A Multiplying out brackets

- You can multiply out brackets using a table.

For example: Multiply out $(a - 5)(a + 3)$

- $(a - 5)(a + 3) = (a - 5) \times (a + 3)$

- Work in a table:

$\times$	a	-5
a	a^2	^-5a
$+3$	$3a$	$^-15$

- The total is: $a^2 - 5a + 3a - 15$
  which simplifies to: $a^2 - 2a - 15$

- So $(a - 5)(a + 3) = a^2 - 2a - 15$ for any value of a

B Factorising expressions

- When you factorise an expression you write it as a multiplication of its factors.

  For example: $a^2 - 2a - 15$ factorises to give $(a - 5)(a + 3)$

- To factorise $c^2 + 5c - 14$

  - think of a table that gives a total of $c^2 + 5c - 14$ and fill the parts you are sure of

$\times$	c	$\square$
c	c^2	$\square$
$\square$	$\square$	-14

  - try values that multiplied give $^-14$ in the correct positions in the table

  as $7 \times 2 = 14$ then $^-7 \times 2 = ^-14$ or $7 \times ^-2 = ^-14$
         $14 \times 1 = 14$ then $14 \times ^-1 = ^-14$ or $^-14 \times 1 = ^-14$

Try $^-7$ and $^+2$

$\times$	c	$^-7$
c	c^2	^-7c
$^+2$	$2c$	$^-14$

Remember: $^+2$ just means 2.

The total for this table is: $c^2 - 5c - 14$
which is **not** what you want.

Try $^+7$ and $^-2$

$\times$	c	$^+7$
c	c^2	^+7c
$^-2$	^-2c	$^-14$

The total for this table is: $c^2 + 5c - 14$
which **is** what you want.

There is no need to try other multiplications that give $^-14$.

$c^2 + 5c - 14$ **factorises to give** $(c + 7)(c - 2)$

A1 For each of these, multiply out the brackets, and simplify.
  a $(c - 4)(c + 7)$
  b $(a + 3)(a + 5)$
  c $(y + 5)(y - 8)$
  d $(h + 1)(h - 6)$
  e $(t - 5)(t + 5)$
  f $(y - 3)(y - 3)$
  g $(p + 5)^2$
  h $(k - 6)(k - 5)$

B1 Factorise these expressions.
  a $n^2 + 4n - 5$
  b $y^2 - 3y - 10$
  c $k^2 - 7k + 6$
  d $h^2 + 9h + 20$
  e $w^2 - 9w + 20$
  f $v^2 + 3v - 28$
  g $d^2 - 7d + 10$
  h $a^2 + 12a - 13$

B2 Factorise these expressions.
  a $p^2 + 6p + 9$
  b $m^2 - 10m + 25$
  c $s^2 - 4s + 4$
  d $g^2 + 8g + 16$

Solving quadratic equations by factorising

A quadratic equation is:
- an equation with one variable (only one letter, e.g. n)
- an equation where the highest power of the variable is 2 (e.g. n^2)

$$n^2 + 3n - 4 = 0 \qquad 3(a^2 + 1) = 0 \qquad 2v^2 + 3v - 40 = 0$$

are all examples of quadratic equations.

An equation is solved when you find a value, or values for the variable that satisfy the equation.

One way to think of a value that will satisfy an equation, is to think of a value for the variable that makes the equation true.

The value of p that satisfies the equation
$$3p = 12$$
is 4

The solution to the equation
is $p = 4$
and
$p = 4$ satisfies $3p = 12$

One way to solve a quadratic equation is to factorise it.

Example   To solve $n^2 + 3n - 4 = 0$

$n^2 + 3n - 4$ factorises to give: $(n + 4)(n - 1)$

So $\qquad\qquad\qquad$ $(n + 4)(n - 1) = 0$

That means: $\qquad$ either $(n + 4) = 0$ ... (as $0 \times (n - 1) = 0$)
$\qquad\qquad\qquad\qquad$ or $\quad (n - 1) = 0$ ... (as $(n - 4) \times 0 = 0$)

For $(n + 4) = 0$ $\quad$ n must have a value of $^-4$
For $(n - 1) = 0$ $\quad$ n must have a value of $^+1$

So $\quad n^2 + 3n - 4 = 0$ has two values for n that satisfy it.
$\qquad$ The values of n are: $n = {}^-4$ or $n = {}^+1$

The quadratic equation $n^2 + 3n - 4 = 0$ has two solutions $n = {}^-4$ or $n = {}^+1$.
All quadratic equations have no more than two solutions.

**Exercise 22.1
Solving quadratic
equations by factorising**

1 Factorise and solve these quadratic equations.

a $n^2 + 13n - 14 = 0$	**b** $n^2 - 5n - 14 = 0$	**c** $n^2 - 13n - 14 = 0$
d $b^2 + 2b - 15 = 0$	**e** $b^2 + 14b - 15 = 0$	**f** $b^2 - 2b - 15 = 0$
g $a^2 + 7a - 30 = 0$	**h** $a^2 + 13a + 30 = 0$	**i** $a^2 + a - 30 = 0$
j $x^2 + 4x - 5 = 0$	**k** $x^2 + 6x + 5 = 0$	**l** $x^2 - 6x + 5 = 0$
m $y^2 - 8y + 15 = 0$	**n** $y^2 - 16y + 15 = 0$	**o** $y^2 - 14y - 15 = 0$
p $k^2 + 2k - 63 = 0$	**q** $k^2 - 16k + 63 = 0$	**r** $k^2 + 62k - 63 = 0$

2 Match each quadratic equation with a pair of factors from List A, and solve the equation.

a $d^2 + 3d - 10 = 0$
b $d^2 - 5d - 6 = 0$
c $d^2 + 3d - 40 = 0$
d $d^2 - 8d - 33 = 0$
e $d^2 - 14d + 33 = 0$

List A

$(d + 8)(d - 5)$	$(d - 11)(d + 3)$
$(d - 2)(d + 5)$	$(d - 11)(d - 3)$
$(d + 4)(d - 10)$	$(d + 1)(d - 6)$

Factorising can be used to solve a quadratic equation when one side is equal to 0.

So, you might need to rearrange an equation before you factorise.

To solve $x^2 + 4x = 5$
rearrange to $x^2 + 4x - 5 = 0$
Now factorise and find the solution or solutions.

3 **a** Factorise $x^2 - 8x + 16$.
b Explain why $x = {}^+4$ is the only solution to $x^2 - 8x + 16 = 0$.

4 For each of these equations:
- rearrange   • factorise   • find two solutions

a $c^2 + 5c = 6$	**b** $y^2 = 7 - 6y$	**c** $p^2 + 9 = 6p$
d $x^2 = 3x + 10$	**e** $w^2 = 4w - 4$	**f** $b^2 + 14 = 15b$
g $k^2 - 3k = 70$	**h** $v^2 + 6v = {}^-9$	**i** $t^2 = 3t + 28$
j $y^2 - 7y = 18$	**k** $g^2 - 8g = {}^-16$	**l** $a^2 = 21 - 4a$
m $p^2 - 4 = 3p$	**n** $w^2 = 2w + 24$	**o** $u^2 + 10 = {}^-7u$

Solving quadratic equations by trial and improvement

The trial-and-improvement method can be used to find a solution to a quadratic equation.

♦ You can use this method if you find an equation difficult to factorise.

Example

Find a solution to this equation $x^2 - 3x = 8$

$$x^2 - 3x = 8$$

Try $x = 4$ $4 \times 4 - 3 \times 4 = 4$ $\neq 8$ ($\neq$ means *does not* equal)
Try $x = 5$ $25 - 15 = 10$ $\neq 8$
Try $x = 4.5$ $20.25 - 13.5 = 6.75 \neq 8$
Try $x = 4.6$ $21.16 - 13.8 = 7.36 \neq 8$
Try $x = 4.7$ $22.09 - 14.1 = 7.99$
Try $x = 4.71$ $22.18 - 14.13 = 8.05 \neq 8$ (too big)

As an exact value for x is often not possible by trial and improvement you have to accept an approximate answer.

So for $x^2 - 3x = 8$ $x = 4.7$ is a solution (correct to 1 dp)

You can, of course, try any value you like for x.
You must remember that trial and improvement is not like guessing one value, and then another, and so on.

Be systematic with the values you try. It might take a little longer than a lucky guess, but in the end you will find a solution.

Exercise 22.2
Solving quadratic equations by trial and improvement

1 Find a value for x from these quadratic equations, by trial and improvement.
   a $x^2 + 2x = 5$ **b** $x^2 - 2x = 6$ **c** $x^2 + 3x = 5$
   d $x^2 + 4x = 2$ **e** $x^2 - 3x = 1$ **f** $x^2 + 5x = 3$
   g $x^2 - x = 3$ **h** $x^2 + 6x = 2$ **i** $x^2 + 4x = 8$

2 Find by trial and improvement a value for x that satisfies the equation

   $x^2 + 4x - 7 = 0$

3 Find a value for x that satisfies the equation

   $x^2 - 5x = 2$

   Give your answer correct to 1 dp.

4 Find a solution to each equation by trial and improvemement.
   a $2x^2 - x = 5$ **b** $3x^2 + x = 7$ **c** $2x^2 - 3x = 4$

5 Explain why trial and improvement is not a good method to find a value for x that satisfies the equation:

   $x^2 + 4x - 5 = 0$

Quadratic graphs

The equation tells you that the graph is quadratic.

When you draw up a table of values you may be able to see symmetry in the y-coordinate values.

When you plot the points you should see that they lie in a curve, known as a **parabola**.

When you join the points you must draw a smooth freehand curve.

Do not use a ruler to join the points.

This is the part of the graph of $y = x^2$ for values of x between $^-4$ and $^+4$.

This graph of $y = x^2$:

♦ is a smooth curve

♦ is symmetrical about the y-axis

♦ passes through the origin (0, 0).

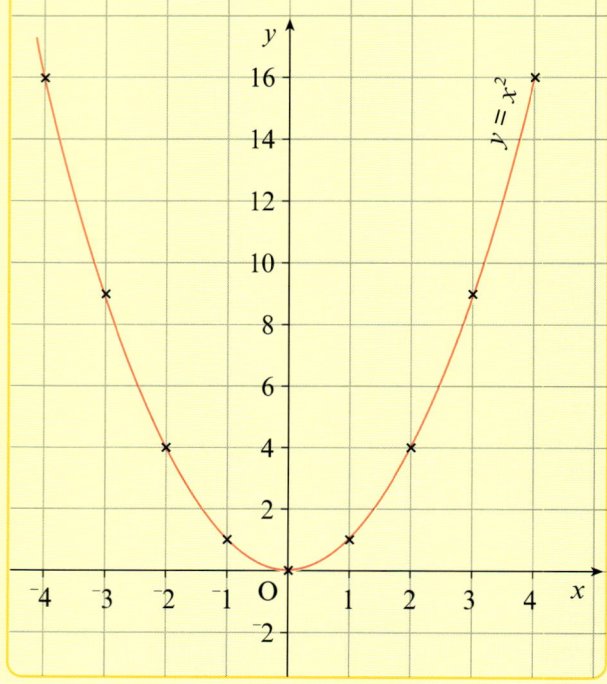

This symmetry can also be seen in the table of values for the graph.

x	$^-4$	$^-3$	$^-2$	$^-1$	0	1	2	3	4
x^2	16	9	4	1	0	1	4	9	16
y	16	9	4	1	0	1	4	9	16

Symmetry in the y-coordinate values

Exercise 22.3
Drawing quadratic graphs

1   To draw a graph of $y = x^2$ for values of x between $^-5$ and $^+5$:

a   Draw up a table of values with values of x between $^-5$ and $^+5$.
b   From your table:
   what are the largest, and smallest values you will need on the y-axis?
c   Draw a pair of axes for your graph.
d   From your table of values, plot points on the graph.
e   Join the points with a smooth curve.
f   Label the graph with its equation.

2   This is part of the table of values for the graph of $y = x^2 + 1$.

a   Copy and complete the table of values.
b   Draw the graph of $y = x^2 + 1$.

x	$^-4$	$^-3$	$^-2$	$^-1$	0	1	2	3	4
x^2	16	9							
$^+1$	$^+1$	$^+1$							
y	17								

c   Compare your graphs of $y = x^2$ and $y = x^2 + 1$.
   How are they the same? How are they different?

> When you sketch a graph you only show the shape of the graph and any other information you know – for example, information the equation gives you.
>
> You do not plot points on the graph, nor do any measurements on axes have to be accurate.
>
> Sketching a graph is not the same as drawing a graph.

3 This is a sketch of the graph of $y = x^2$.

   a Make a copy of this sketch
   b On your copy, sketch and label these graphs:
      i $y = x^2 + 2$
     ii $y = x^2 + 3$
    iii $y = x^2 + 6$
     iv $y = x^2 - 1$
      v $y = x^2 - 3$

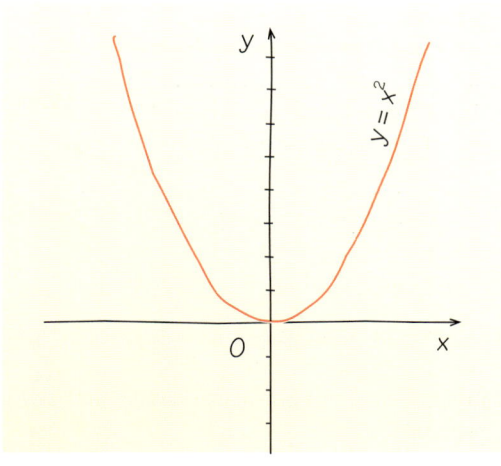

4 Misha sketched graph W, but did not label the graph with its equation.

   She had been asked to sketch these graphs:
       $y = x^2$
       $y = x^2 + 1.5$
       $y = x^2 - 2.5$

   a Which equation matches graph W?
   b Explain your answer.
   c Sketch and label all the graphs Misha was asked to sketch.

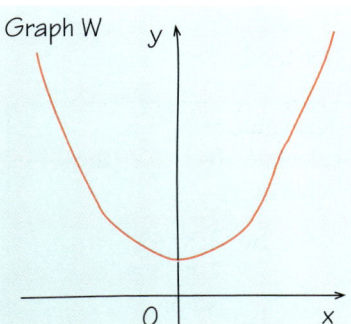

5 Draw a pair of axes with:
values of x from $^-3$ to $^+3$, and values of y from $^-4$ to $^+20$.

   a Copy and complete this table of values for the graph of $y = x^2 + x$.

x	$^-3$	$^-2$	$^-1$	0	1	2	3
x^2	9	4	1	0	1		
^+x	$^-3$	$^-2$	$^-1$	0			
y	6	2					

   b From your table of values:
     i Plot points on the graph.
    ii Draw and label the graph of $y = x^2 + x$.

> This $^-6$ is calculated from the ^+2x in the equation.
>
> That is:
> $^+2 \times$ (the x-coordinate)
>   i.e. $^+2 \times ^-3 = ^-6$
> $^+2 \times ^-2 = ^-4$
> and so on.

6 **a** Copy and complete this table of values, with x from $^-3$ to $^+3$.
   b On the same axes you used for Question **5** draw and label the graph of $y = x^2 + 2x$.
   c Complete a table of values for the graph of $y = x^2 + 3x$ (x from $^-3$ to $^+3$).
   d On the same axes, draw and label the graph of $y = x^2 + 3x$.
   e Compare your graphs of $y = x^2 + x$, $y = x^2 + 2x$ and $y = x^2 + 3x$ Explain how the graphs differ.

x	$^-3$	$^-2$
x^2	9	4
^+2x	$^-6$	$^-4$
y	3	0

> $x = 0$ is the equation of the y-axis.
>
> $y = 0$ is the equation of the x-axis.

7 The line $x = 0$ is the line of symmetry for the graph of $y = x^2$.
Give the equation of the line of symmetry for each of these graphs:

   a $y = x^2 + 1$ **b** $y = x^2 + 2x$ **c** $y = x^2 - 2$ **d** $y = x^2 + 3x$

8 **a** Copy and complete this table of values for the graph of $y = x^2 + 2x - 3$.

x	$^-4$	$^-3$	$^-2$	$^-1$	0	1	2	3
x^2	16	☐	4	1	☐	1	☐	9
$+2x$	-8	-6	☐	-2	0	2	4	☐
-3	-3	-3	-3	-3	-3	☐	-3	-3
y	5	0	$^-3$	$^-4$	$^-3$	☐	5	☐

b Explain any symmetry you can find in your table of values.
c Draw axes with:
values of x from $^-4$ to $^+3$, and values of y from $^-5$ to $^+25$.
d **i** Plot the points from your table of values.
  ii Draw and label the graph of $y = x^2 + 2x - 3$.

9 **a** Copy and complete this table of values for the graph of $y = x^2 - 2x - 3$
for values of x from $^-4$ to $^+3$.

x	$^-4$	$^-3$	$^-2$	$^-1$	0	1
x^2	16	9	4	1	0	1
$-2x$	$+8$	$+6$	$+4$	$+2$	0	-2
-3	-3	-3	-3			
y	21	12				

> The value of $-2x$ is calculated by : $^-2 \times$ (the value of x)
>
> So $^-2 \times {}^-4 = {}^+8$
> $^-2 \times {}^-3 = {}^+6$
> so on.

b On the same axes as Question **8**, draw the graph of $y = x^2 - 2x - 3$.
c Compare your graphs of $y = x^2 + 2x - 3$ and $y = x^2 - 2x - 3$
  i How are your graphs the same?
  ii How are your graphs different?

10 **a** Draw up a table of values for the graph of $y = x^2 + 3x - 4$
with values of x from $^-4$ to $^+4$.
b Draw a pair of axes with:
values of x from $^-4$ to $^+4$, and values of y from $^-10$ to $^+25$.
c On your axes, draw a graph of $y = x^2 - 3x - 4$.
d Give the coordinates of the points where your graph crosses the x-axis.

11 **a** Draw up a table of values for the graph of $y = x^2 - 3x - 4$
with values of x from $^-4$ to $^+4$.
b Draw a graph of $y = x^2 - 3x - 4$.
c Give the coordinates of the points where your graph crosses the x-axis.

12 **a** Draw up a table of values for the graph of $y = x^2 - 3x - 10$
with values of x from $^-5$ to $^+5$.
b From your table of values predict where the graph of $y = x^2 - 3x - 10$
will cross the x-axis.
Explain your prediction.
c Draw a pair of axes with:
values of x from $^-5$ to $^+5$, and values of y from $^-15$ to $^+30$.
d Draw the graph of $y = x^2 - 3x - 10$.

13 **a** Draw up a table of values for the graph of $y = x^2 + 2x - 8$,
with values of x from $^-5$ to $^+5$.
b Draw the graph of $y = x^2 + 2x - 8$.
c Give the coordinates where $y = x^2 + 2x - 8$ crosses the x-axis.

Solving quadratic equations graphically

This is the graph of $y = x^2 + 5x - 6$, for values of x from $^-8$ to $^+6$.

We can use the graph to solve the equation $x^2 + 5x - 6 = 0$.

The graph of $y = x^2 + 5x - 6$ crosses the graph of $y = 0$ where:

$x = {}^-6$ and where $x = 1$

At $x = {}^-6$, and $x = 1$ the two equations must be equal.
So, we can write:

$x^2 + 5x - 6 = 0$

From the graphs, the values of x that satisfy the equation are:

$x = {}^-6$ or $x = 1$

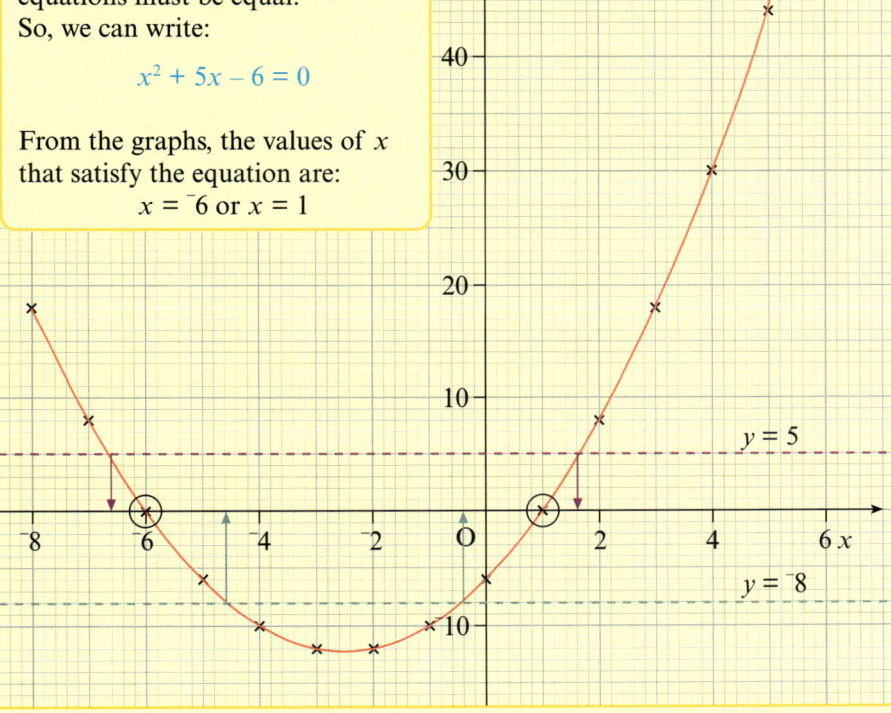

$y = x^2 + 5x - 6$

$y = 5$

$y = {}^-8$

$y = 0$ is the equation of the x-axis.

You can be sure of these solutions, as the values of x from the graph are both integer values.

These values of x, from the graph are not integer values. The values are not exact, they are a good approximation. They answer the question, from the graph.

For more accurate solutions, use the values from the graph as a starting point for the trial-and-improvement method.

The graph of $y = x^2 + 5x - 6$ can be used to solve many more equations.

Example From the graph solve $x^2 + 5x - 6 = 5$.
The graph of $y = x^2 + 5x - 6$ crosses the graph of $y = 5$
so we can say: $x^2 + 5x - 6 = 5$
The two equations are equal where $x \approx {}^-6.6$ and where $x \approx 1.6$

The solutions for $x^2 + 5x - 6 = 5$ are: $x \approx {}^-6.6$ and $x \approx 1.6$

Example From the graph solve $x^2 + 5x - 6 = {}^-8$.
The graph of $y = x^2 + 5x - 6$ crosses the graph of $y = {}^-8$
so we can say: $x^2 + 5x - 6 = {}^-8$

The solutions for $x^2 + 5x - 6 = {}^-8$ are: $x \approx {}^-4.6$ and $x \approx {}^-0.4$

Exercise 22.4
Solving quadratic
equations graphically

1 The graph of $y = x^2 + 5x - 6$ can also be used to solve the equation:
$$x^2 + 5x - 6 = 10$$

a What other graph would you draw on the axes to solve $x^2 + 5x - 6 = 10$?
b Explain how you would use the two graphs to solve the equation.
c From the graph on page 269, solve $x^2 + 5x - 6 = 10$.

2 The graph on page 269 can be used to solve these equations:
$$x^2 + 5x - 6 = 0$$
$$x^2 + 5x - 6 = 5$$
$$x^2 + 5x - 6 = {}^-8$$
$$x^2 + 5x - 6 = 10$$

a Give two other equations you think can be solved from the same graph.
b Explain how you would solve your equations.

3 Asif used the same graph to solve $x^2 + 5x - 6 = 22$,
but he could only find one solution.

a Explain why.
b Give the solution you think Asif did find from the graph.
c Is the other solution greater or less than $x = {}^-8$?
Explain your answer.

4 Draw the graph of $y = x^2 - 5x + 6$ for values of x from $^-2$ to $^+6$.

a Use your graph to solve $x^2 - 5x + 6 = 0$.
b On your axes draw and label the graph of $y = 4$.
c Use your graphs to solve $x^2 - 5x + 6 = 4$.
d From your graph solve $x^2 - 5x + 6 = 10$.

5 Draw the graph of $y = x^2 - 4x - 5$ for values of x from $^-3$ to $^+7$.

a Use your graph to solve:
   i $x^2 - 4x - 5 = 0$ **ii** $x^2 - 4x - 5 = 11$.
b From your graph solve the equation $x^2 - 4x - 5 = {}^-9$.
c Explain why $x^2 - 4x - 5 = {}^-9$ only has one solution.

> If an equation can be made
> by just rearranging another
> equation, then:
>
> the two equations are the
> same, and they will have the
> same solution or solutions.

6 **a** Show how each of these equations can be made by rearranging:
$$x^2 - 4x + 3 = 0$$

   i $x^2 + 3 = 4x$
   ii $x^2 = 4x - 3$
   iii $3 = 4x - x^2$

b For values of x from $^-3$ to $^+5$ draw the graph of $y = x^2 - 4x + 3$.
From your graph solve $x^2 - 4x + 3 = 0$.
c **i** For values of x from $^-3$ to $^+5$ draw the graph of $y = x^2 + 3$.
   ii On your axes draw and label the graph of $y = 4x$.
   iii Use your graph to show why the solutions of $x^2 + 3 = 4x$ are:
      $x = 1$ and $x = 3$.

7 To solve the equation $x^2 + 4 = 5x$:

a Rearrange the equation so that one side is equal to 0.
b For values of x from $^-2$ to $^+5$ draw a graph that will help solve
the equation.
c From your graph, solve the equation $x^2 + 4 = 5x$.

8 With values of x from $^-4$ to $^+4$:

a On the same axes draw the graphs of $y = x^2 - 2x - 3$ and $y = x + 1$.
b Use your graphs to solve the equation $x^2 - 2x - 3 = x + 1$.

Graphs of other equations

A quadratic equation, or expression, is said to be 'of order 2'.
This is because the highest power of the variable is 2, as in:
$$3a^2 + 4a \qquad p^2 + 3p - 4 = 0$$

So an equation or expression of higher order than a quadratic has a
highest power of the variable greater than 2, as in:
$$2a^3 + a \qquad a^3 + a^2 - 1 \qquad y^5 + y = 12 \qquad k^4 = 20$$

Example Draw the graph of $y = x^3 + 2$ for values of x from $^-3$ to $^+3$.

For the y-coordinate, substitute the values of x in the equation.

When $x = ^-3$	$y = (^-3)^3 + 2 = ^-27 + 2 = ^-25 \dots (^-3, ^-25)$
When $x = ^-2$	$y = (^-2)^3 + 2 = ^-8 + 2 = ^-6 \dots (^-2, ^-6)$
When $x = ^-1$	$y = (^-1)^3 + 2 = ^-1 + 2 = 1 \dots (^-1, 1)$
When $x = 0$	$y = (0)^3 + 2 = 0 + 2 = 2 \dots (0, 2)$
When $x = 1$	$y = (1)^3 + 2 = 1 + 2 = 3 \dots (1, 3)$
When $x = 2$	$y = (2)^3 + 2 = 8 + 2 = 10 \dots (2, 10)$
When $x = 3$	$y = (3)^3 + 2 = 27 + 2 = 29 \dots (3, 29)$

> $(^-3)^3$ is:
>
> $^-3 \times ^-3 \times ^-3 = ^-27$

The graph of $y = x^3 + 2$ for values of x from $^-3$ to $^+3$ is:

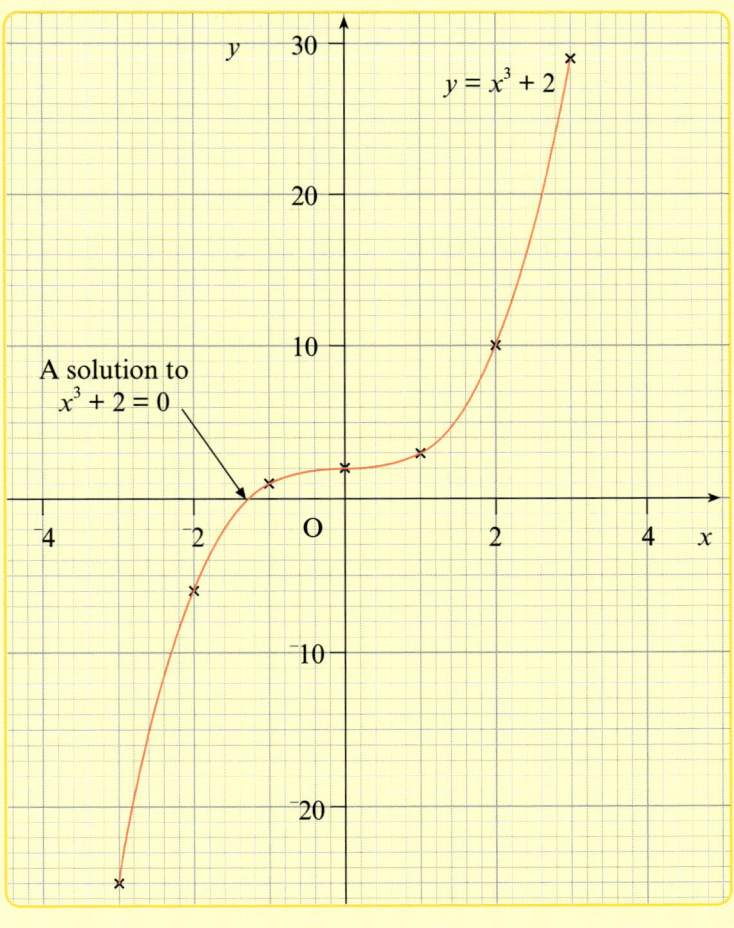

- This graph is a smooth curve.
- This graph will help you find a solution to the equation $x^3 + 2 = 0$:
  from the graph $x \approx ^-1.3$ (remember this is not exact).

The equation $x^3 + 2 = 0$ can be described as a cubic equation.

> In a cubic equation the
> highest power of the
> variable is 3.

Exercise 22.5
Drawing graphs of
other equations

1 **a** Draw the graph of $y = x^3 + 4$ for values of x from $^-3$ to $^+3$.
b From your graph, give a solution to $x^3 + 4 = 0$.

2 **a** Draw the graph of $y = x^3 - 2$ for values of x from $^-3$ to $^+3$.
b From your graph, give a solution to $x^3 - 2 = 0$.

3 Iain sketched four cubic graphs.
This is a copy of his sketch of these graphs:

$$y = x^3 + 6$$
$$y = x^3$$
$$y = x^3 - 5$$
$$y = x^3 + 10$$

a Make a copy of Iain's sketch.
b Label each graph with the equation that
best matches the sketch.

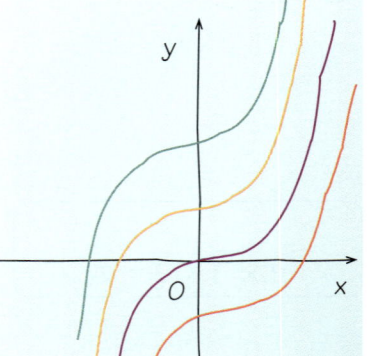

4 **a** Draw the graph of $y = x^3 + x$
for values of x from $^-3$ to $^+3$.
b On the same axes draw the graph of $y = 1$.
c Use your graphs to find a solution to the equation $x^3 + x = 1$.
d Give a solution to the equation $x^3 + x - 1 = 0$.
Explain how you were able to give a solution.

Solving other equations by trial and improvement

Example Find a value of x that satisfies the equation $x^4 = 20$
correct to 3 sf.

For $\qquad\qquad x^4 \qquad\qquad\qquad = 20$
Try $x = 2$ $\qquad (2)^4 \qquad = 16 \qquad \neq 20$
Try $x = 3$ $\qquad (3)^4 \qquad = 81 \qquad \neq 20$
Try $x = 2.5$ $\qquad (2.5)^4 \quad = 39.062.. \neq 20$
Try $x = 2.4$ $\qquad (2.4)^4 \quad = 33.177.. \neq 20$
Try $x = 2.3$ $\qquad (2.3)^4 \quad = 27.984.. \neq 20$
Try $x = 2.2$ $\qquad (2.2)^4 \quad = 23.425.. \neq 20$
Try $x = 2.1$ $\qquad (2.1)^4 \quad = 19.448.. \neq 20$
Try $x = 2.11$ $\qquad (2.11)^4 \quad = 19.821.. \neq 20$
Try $x = 2.12$ $\qquad (2.12)^4 \quad = 20.199.. \neq 20$
Try $x = 2.115$ $\qquad (2.115)^4 = 20.009.. \neq 20$

Here $x = 2.11$ is too small
2.12 is too large, so we try
2.115 which is half-way
between 2.11 and 2.12

The value of x must be between 2.11 and 2.115
So $x = 2.11$ (to 3 sf).

Exercise 22.6
Solving higher-order
equations by trial and
improvement

1 Find a value of x that satisfies each equation correct to 3 sf.

a $x^3 = 18$ $\qquad\qquad$ **b** $x^3 = 40$ $\qquad\qquad$ **c** $x^3 + 2 = 21$
d $x^4 - 5 = 20$ $\qquad$ **e** $x^3 - x = 21$ $\qquad$ **f** $x^3 - 2x = 30$
g $x^4 + x = 3$ $\qquad$ **h** $x^3 + 3x = 8$ $\qquad$ **i** $x^4 + x = 600$
j $x^3 - 4x = 10$ $\qquad$ **k** $x^4 - x^3 = 12$ $\qquad$ **l** $x^3 - 5x = 800$

End points

You should be able to ...   ... so try these questions

A   Describe a quadratic equation

A1   In words describe what makes an equation quadratic.

A2   Give three different examples of quadratic equations.

B   Factorise and solve
quadratic equations

B1   **a**   Factorise and solve $x^2 + 3x - 10 = 0$
  b   Factorise and solve $x^2 + 9x + 20 = 0$
  c   Factorise and solve $x^2 - 10x + 16 = 0$
  d   Factorise and solve $x^2 - 6x + 9 = 0$

C   Rearrange quadratic equations

C1   Rearrange these quadratic equations so that one side is equal to 0.
  a $x^2 + 5x = 24$
  b $p^2 - 2 = 3p$
  c $2 = 5c - c^2$
  d $y^2 = 2 + 6y$

D   Solve quadratic equations
by trial and improvement

D1   Use trial-and-improvement methods to solve these equations.
  a $v^2 - 3v = 3$
  b $b^2 + 2 = 4b$
  c $p^2 + 5p + 5 = 0$
  d $w^2 - 59 = 0$

E   Draw up a table of values
for a quadratic graph

E1   For the graph of $y = x^2 + 2$, draw up a table of values.
Use values of x from $^-4$ to $^+4$.

E2   Explain any symmetry you find in your table of values.

F   Draw a quadratic graph
from a table of values

F1   Draw the graph of $y = x^2 + 2x - 3$.
Use values of x from $^-4$ to $^+3$ in your table of values.

G   Use and draw sketch graphs
of quadratics

G1   This is the sketch of a quadratic graph.
Explain why this cannot be a
sketch of $y = x^2 + 1$.

Give a possible equation for
the graph.

G2   Sketch a pair of axes.
On these axes sketch, and label
these graphs:
  $y = x^2 + 5$
  $y = x^2 - 4$
  $y = x^2$

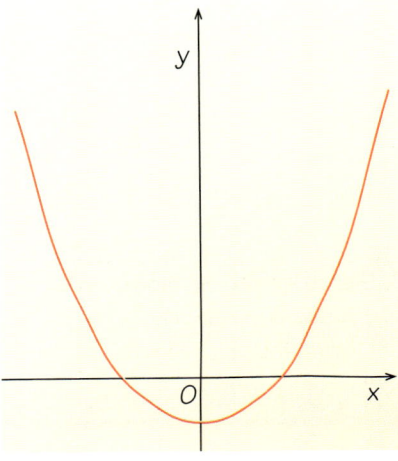

H Solve quadratic equations graphically

x	‾8	‾7	‾6	‾5	‾4	‾3	‾2	‾1	0	1	2	3	4	5	6
y	27	16	7	0	‾5	‾8	‾9	‾8	‾5	0	7	16	27	40	55

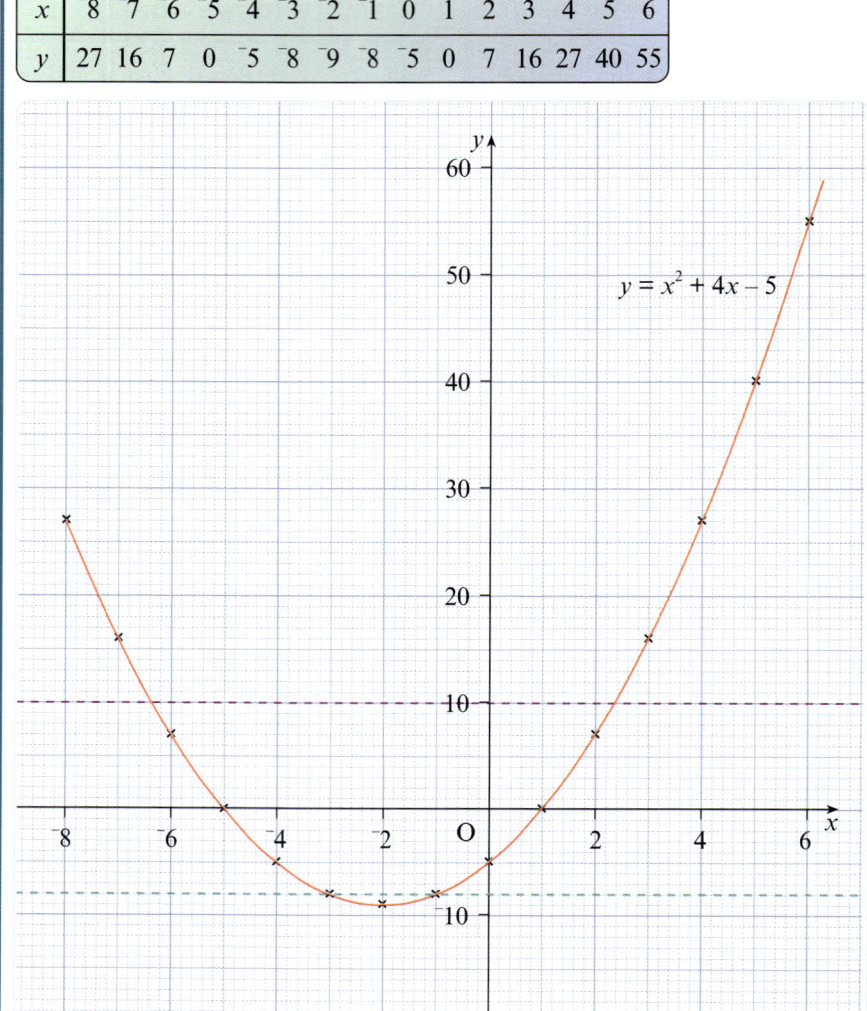

$y = x^2 + 4x - 5$

This is the graph of $y = x^2 + 4x - 5$ for values of x from ‾8 to ⁺6.

H1 Use the graph above to solve the equation $x^2 + 4x - 5 = 0$.

H2 Use the graph above to solve $x^2 + 4x - 5 = 10$.
Explain how you used the graph.

H3 From the graph above, $x = $ ‾3 or $x = $ ‾1 is the solution to an equation. What equation is this?

H4 Use the graph to explain why $x^2 + 4x - 5 = $ ‾9 only has one solution.

I Sketch graphs of higher-order equations

I1 Sketch the graph of $y = x^3$.

Some points to remember

- ◆ Quadratic graphs are smooth curves, and points should not be joined using a ruler.
- ◆ When you solve a quadratic equation, you should expect to find no more than two solutions.

Starting points
You need to know about ...

... so try these questions

A Investigations using data

◆ The reason for carrying out an investigation using data is to either test a **hypothesis** or answer a **question**.

The start of an investigation into sleep could be:

Hypothesis – most people sleep at least 7 hours a night
Question – how long do people sleep at night?

◆ You can carry out an investigation in four stages:

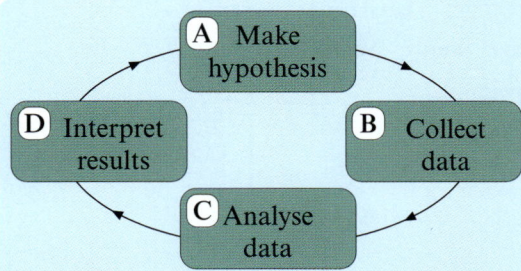

A, make a hypothesis (or ask a question)
B, collect the data you need
C, analyse the data you have collected
D, use the results of your analysis to decide whether the hypothesis is true or false.

◆ Stage D may give you an idea for a follow-up investigation.
For example, the start of a follow-up into sleep could be:

Hypothesis – people need more sleep in the winter
Question – is there a link between sleep and time of year?

A1 a Make your own hypothesis about sleep.
b Decide what data you need to collect to test your hypothesis.

B Collecting data

◆ A **data collection sheet** is any form or table used to collect data.

To the BDA,
10 Queen Anne Street, London W1M 0BD
Tel: 0171-323 1531

A charity helping people with diabetes and supporting diabetes research.

I enclose a cheque/postal order*
payable to the BDA   £ _____
Debit my Access/Visa* card
by the amount of   £ _____
Card number ☐☐☐☐☐☐☐☐☐☐☐☐☐☐☐☐
Expiry data   ☐☐☐☐

Please send me more information and membership details

Name _____
Address _____

Signature _____
*Delete which is inapplicable   Reg. Charity no. 215199

Body Matters

Name	Length of thumb (cm)	Length of foot (cm)	Height (cm)
Sam			
Wasim			
Liz			
Shane			
Des			
Linda			
Dean			
Nisha			

◆ A **questionnaire** collects data by asking questions.

Sleep Questionnaire

1 What is your name?   _____

2 How old are you?   _____ years

3 What time do you usually go to bed? _____

◆ A data collection sheet or questionnaire is also called a **survey**.

B1 Design a data collection sheet to use for a traffic survey.

C Types of question

- ◆ You can use different types of questions on a questionnaire:

  - ❖ multi-choice questions

    > **7** What do you sleep on?
    >
    > *Please tick one box only*   ☐ Back   ☐ Front   ☐ Side

  - ❖ multi-choice questions with a scale

    > **8** How heavy a sleeper are you?   HEAVY ——————— LIGHT
    >
    > *Please tick one box only*   Very   Fairly   Average   Fairly   Very
    >
    >   ☐   ☐   ☐   ☐   ☐

  - ❖ branching questions

    > **9** Do you suffer from regular sleepless nights?
    >
    > *Please tick one box only*   ☐ Yes   ☐ No
    >
    > *If your answer is NO then go to Question 12*

  - ❖ questions with more than one answer.

    > **12** What helps you get a good night's sleep?   ☐ A hot drink just before bed
    > *Put a 1 in the box for the most helpful,*   ☐ Eating just before bed
    > *a 2 for the next most helpful,*   ☐ Exercise during the evening
    > *and so on*   ☐ Relaxation breathing
    >   ☐ Other *(please state)*

D Using a scatter diagram

- ◆ You can use a scatter diagram to investigate if there is a link between two sets of data.

  Question – does the time taken to get to sleep depend on the amount of light in the room?

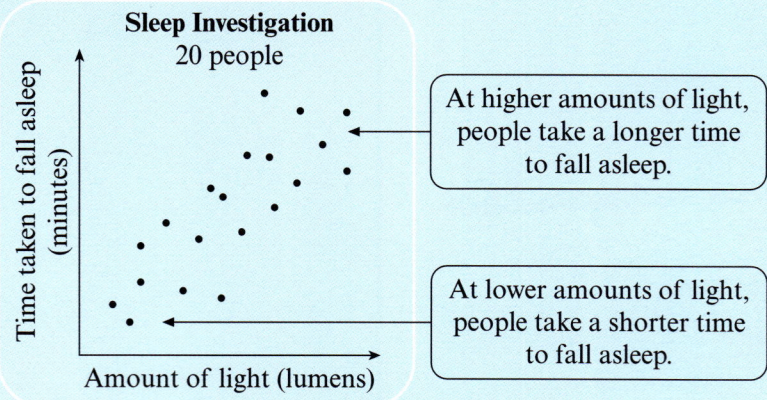

At higher amounts of light, people take a longer time to fall asleep.

At lower amounts of light, people take a shorter time to fall asleep.

The time taken to fall asleep does depend on the amount of light: as the amount of light increases, the time taken to fall asleep also increases.

C1 Design a questionnaire about sleep which includes different types of question.

D1

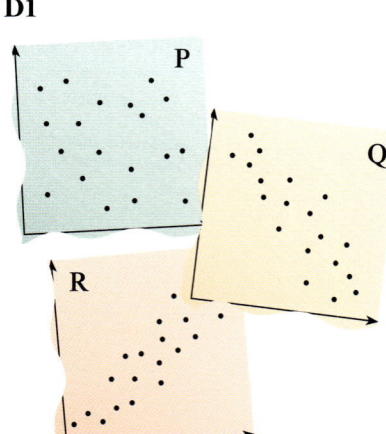

Match each scatter diagram to each pair of sets of data.

a	Time taken to get to sleep	against	Time since drank coffee
b	Amount of sleep	against	Room temperature
c	Amount of sleep	against	Height of bed off floor

Designing and criticising questions

◆ When you create a questionnaire, your questions must be carefully designed to:
  ❖ make them easy to answer
  ❖ make sure the answers give you the data you need.

Poor question

How much sleep did you get last night?

☐ Less than average
☐ About average
☐ More than average

This question is **not clear**: the words used need to be more exact.
(Different people are likely to have different ideas of what is meant by 'average'.)

Improved question

How much sleep did you get last night?

☐ Less than 8 hours
☐ About 8 hours
☐ More than 8 hours

Do you agree that we need at least 8 hours sleep each night?

This is a **leading** question: it leads people into giving a certain answer.
(The question seems to expect the answer 'Yes'.)

Do you think we need at least 8 hours sleep each night?

☐ Yes   ☐ No
☐ Not sure

What do you sleep on?

This question is **ambiguous**: it could have more than one meaning.
(The question is meant to be about sleeping position, but could be answered 'a bed'!)

What do you sleep on?

☐ Back
☐ Front
☐ Side

Exercise 23.1
Designing and criticising questions

1 **a** Explain why this question is not clear.
   b Write an improved question.

When do you usually go to bed?
☐ Early   ☐ Late

2 **a** Explain why this is a leading question.
   b Write an improved question.

You get a worse night's sleep on a soft bed, don't you?

3 **a** Explain why this question is ambiguous.
   b Write an improved question.

Where do you sleep best?

4

Leisure Centre Survey

1 Do you agree that the town needs a new leisure centre? ☐ Yes ☐ No

2 Would you be a frequent user of the centre? ☐ Yes ☐ No

3 Would you use the courts? ☐ Yes ☐ No

4 How much would you be prepared to pay to use the pool? ☐ Less than £1.50
☐ More than £2.50

This questionnaire has been written to survey local people about a new leisure centre.

a Criticise each of the questions.
b Write an improved question for each one.

◆ A survey asks people to give an opinion about something, or asks about facts which are easy to remember.

Example

> ### TV Survey
>
> 1 What is your favourite TV channel? ☐ BBC1 ☐ BBC2 ☐ ITV
>
>   ☐ Channel 4 ☐ Channel 5
>
> 2 Did you watch TV last night? ☐ Yes ☐ No

◆ The data needed for some investigations can only be collected:

  ❖ over a period of time

> How much time do you spend in a week watching each TV channel?

  ❖ by designing an experiment.

> People take longer to get to sleep the more light there is in the room.

The data collection sheet used for these types of investigation can be called an **observation sheet**.

Exercise 23.2
Experiments

> To design your experiment:
> ❖ decide what data you need
> ❖ decide how to collect it
> ❖ design an observation sheet.

1 Design an observation sheet to collect data on how much time people spend in a week watching each TV channel.

2 Design an experiment to test this hypothesis.

> **Body Matters**
> Your waist is roughly two times the distance around your neck.

3 **a** Carry out your experiment.
  b Analyse the data you collect.

4 Do you think the hypothesis is true or false? Explain why.

5 Design an experiment to answer this question.

> **Body Matters**
> How many times do people blink in a day?

6 **a** Carry out your experiment.
  b Analyse the data you collect.
  c Interpret your results to answer the question.

7 Design an experiment to test this hypothesis.

> **Body Matters**
> Taller people do not have as good a sense of balance as shorter people.

8 **a** Carry out your experiment.
  b Analyse the data you collect.

9 Do you think the hypothesis is true or false? Explain why.

Correlation

◆ You can describe the link between two sets of data using the term **correlation**.

Sleep Experiment 1
Does the length of time you take
to fall asleep depend on how light
the room is?

These results show **positive** correlation:
an increase in one set of data tends
to be matched by an increase in the
other set.

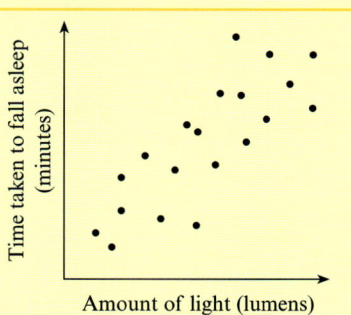

The result of experiment 2
may not be what you expect.
It happens because sleep is
part of your daily rhythm
of sleeping and waking.

Going to bed late means that
you will soon reach your time
for waking, and vice versa.

A daily rhythm, like this
sleep/wake example, is
called a **circadian rhythm**.

Sleep Experiment 2
Does the length of time you
sleep depend on the length of time
since you last slept?

These results show **negative** correlation:
an increase in one set of data tends to be
matched by an decrease in the other set.

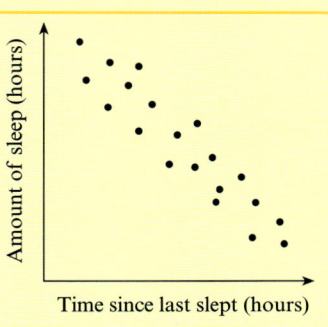

**Exercise 23.3
Correlation**

1

Mean semi-detached house prices in towns near London – 2nd Quarter 1996												
Distance from London (miles)	44	66	41	75	86	47	68	36	62	77	57	53
Mean house price (£000's)	93	78	98	72	63	97	71	104	86	64	78	88

a Draw these axes: horizontal 0 to 90, vertical 50 to 130.
b Plot the house price data on your diagram.
c Is the correlation positive or negative?

2

Eye Tests for 10 people										
Pressure in eye (mmHg)	12.1	11.7	15.2	19.1	11.2	18.9	15.9	17.3	13.0	16.6
Refractive power of lens (dioptres)	3.6	¯3.9	5.1	10.4	¯6.4	3.0	¯6.9	6.5	¯8.4	0.8

A negative number of
dioptres shows short-
sightedness; a positive
number of dioptres shows
long-sightedness.

a Draw these axes: horizontal 10 to 20, vertical ¯12 to 12
b Plot the eye test data on your diagram.
c Is the correlation positive or negative?

This is called drawing a line
by eye or **by inspection**.

3 For each of your scatter diagrams, draw a line
through the middle of the plots, like this:

4 Which scatter diagram did you find it
easier to draw the line on? Explain why.

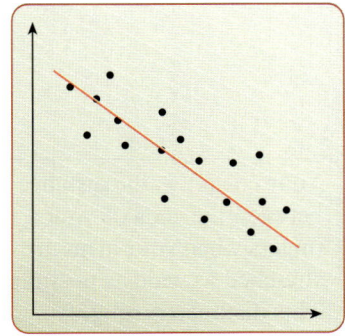

Using a line of best fit

◆ A line drawn through the middle of the plots on a scatter diagram is called a **line of best fit**. The stronger the correlation, the easier it is to draw this line.

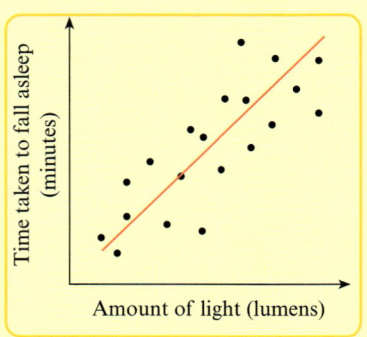

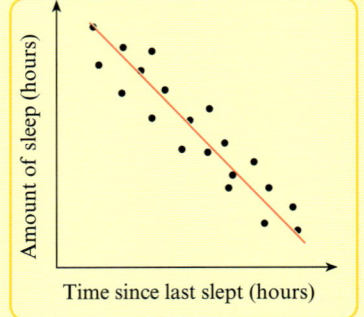

This is **moderate** positive correlation because the plots are well scattered around the line of best fit.

This is **strong** negative correlation because the plots are quite close to the line of best fit.

◆ When it is not possible to draw a line of best fit, there is no link between the two sets of data: there is **no correlation**.

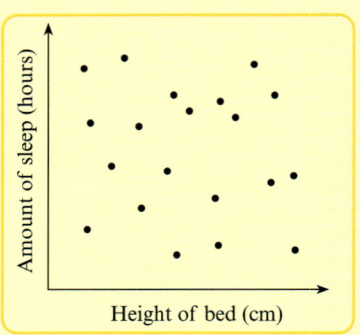

Exercise 23.4
Describing correlation

1   Use this scatter diagram to describe the correlation between income and percentage of income given to charity.

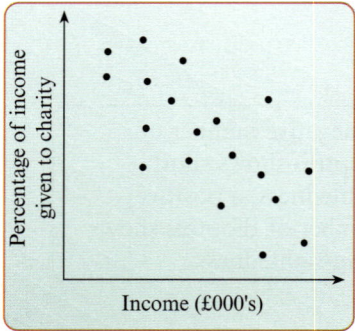

2   Use your scatter diagram from Exercise 23.3 Question **1** to describe the correlation between distance of town from London and mean house price.

3   Use your scatter diagram from Exercise 23.3 Question **2** to describe the correlation between pressure in eye and refractive power of lens.

4   Design an experiment to answer this question:
'Is there any correlation between your fathom and your height?'

5   **a** Carry out your experiment.
    b Plot the data you collect on a scatter diagram.
    c Draw a line of best fit.
    d Use your scatter diagram to describe any correlation.

Your fathom is the distance between the ends of your fingers when your arms are stretched as wide as possible. This distance is roughly six feet for an adult.

Estimating values from a line of best fit

♦ It is possible to estimate values from a line of best fit.

Example

a   Estimate the height of a person with head circumference 56 cm.
b   Estimate the head circumference of a person 195 cm tall.

In part **a** you are estimating within the range of data. This is called **interpolation**. In part **b** you are estimating outside the range of data. This is called **extrapolation**, and is less reliable.

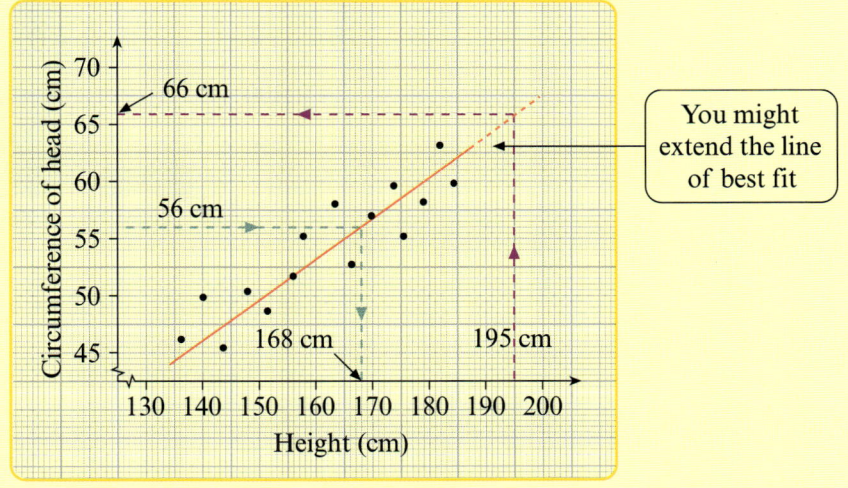

You might extend the line of best fit

a   Estimated height of person with head circumference 56 cm = **168 cm**
b   Estimated head circumference of person 195 cm tall = **66 cm**

Exercise 23.5
Estimating values

Use your scatter diagram from Exercise 23.3 Question **1** for Questions **1** to **3**.

1   Estimate the distance from London of a town with a mean house price of:
   a   £90 000   b   £65 000

2   Estimate the mean house price for a town:
   a   70 miles from London   b   55 miles from London

3   Extend your line of best fit to estimate:
   a   the distance from London of a town with a mean house price of £120 000
   b   the mean house price for a town 25 miles from London.

4

Natural Births – Length of Pregnancy & Weight of Baby									
Length of pregnancy (days) 271	287	283	274	271	279	263	276	283	270
Weight of baby (kg) 2.5	4.2	3.8	3.3	4.5	3.4	2.9	4.1	4.3	3.5

This data has been collected to test the hypothesis:
'A longer pregnancy leads to a heavier baby.'
Plot this data on a scatter diagram.

5   a   Draw a line of best fit on your scatter diagram.
   b   Describe the correlation between length of pregnancy and weight of baby.
   c   Use your line of best fit to estimate:
      i   the length of pregnancy for a baby that weighs 3.5 kg
      ii the weight of a baby with a length of pregnancy of 280 days.

6   Do you think it would make sense to extend this line of best fit? Why?

Misleading diagrams

◆ Diagrams which present data can be misleading in several ways, including:
  ❖ when the vertical axis does not start at 0

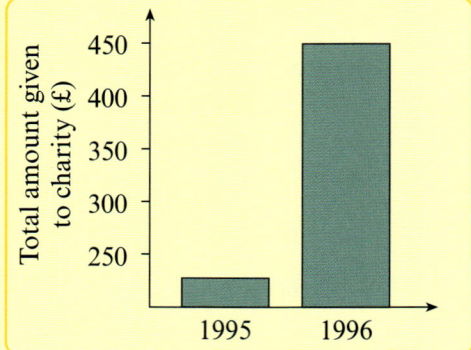

The vertical axis starts at £200.

This gives the impression that the amount given in 1996 was 10 times the amount given in 1995, not 2 times.

  ❖ when enlargements of a shape are used.

Total amount given to charity

The 1996 note is 2 times the height *and* 2 times the width of the 1995 note.

This gives the impression that the amount given in 1996 was 4 times the amount given in 1995.

Exercise 23.6
Misleading diagrams

1 These diagrams were used by Wyvern Water.

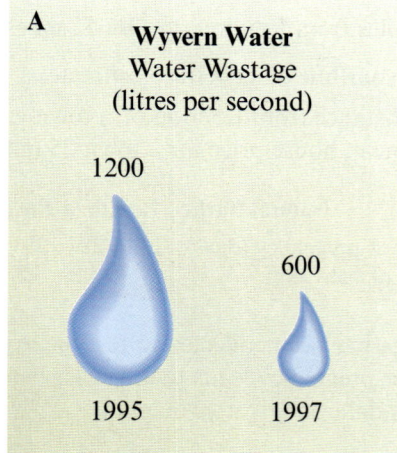

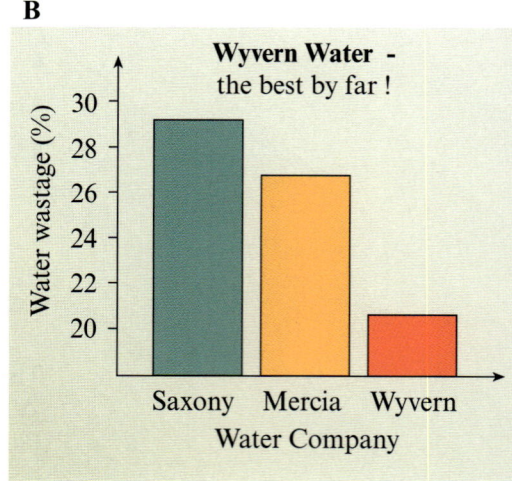

a Explain why diagram A is misleading.

b Explain why diagram B is misleading.

c Redraw diagram B as you think it should be.
Explain your answer giving reasons for any changes you have decided to make.

2   This diagram was used in a newspaper.

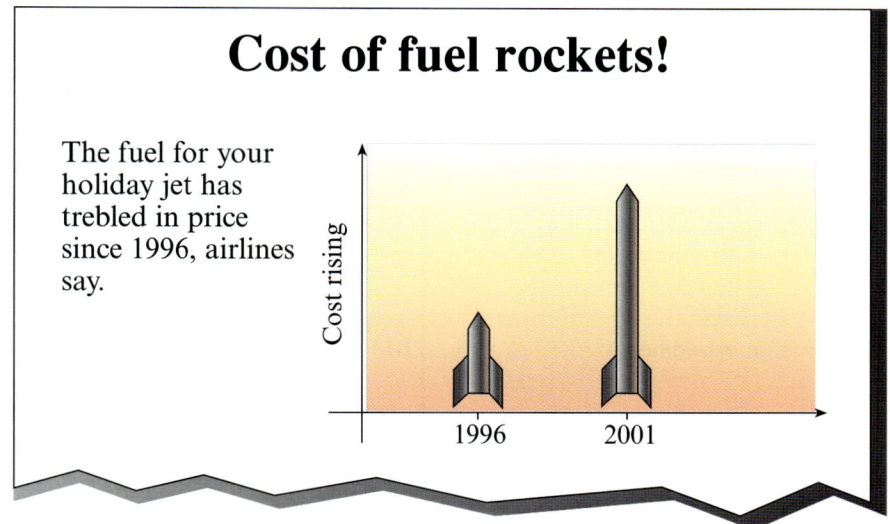

Cost of fuel rockets!

The fuel for your holiday jet has trebled in price since 1996, airlines say.

Cost rising

1996   2001

Explain why the diagram used for this story misleads the person who just looks at the diagram.

3   **Profits go through the roof this year (1998)**

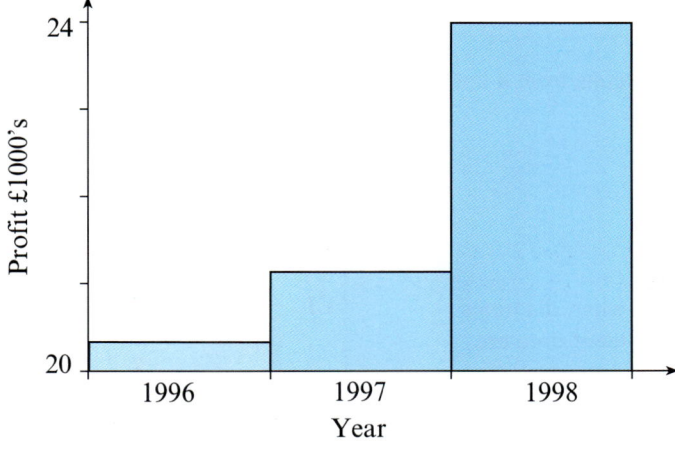

a   Explain why this diagram is misleading?
b   Why do you think the company chose to use a misleading diagram like this?
c   Draw the diagram as you think it should be drawn.
    Give the reasons for your answer.

4   Two images of a £1 coin were used in a diagram.
    Coin A had a radius of 1 cm, and coin B had a radius of 2 cm.
    The headline was: 'Savings doubled last year!'

a   Do you think the diagram of the coins is misleading?
    Explain your answer.
b   If the coin diagrams were accurate what do you think the headline should read?

End points

You should be able to ...

... so try these questions

A Design and criticise questions for a questionnaire

B Design experiments

C Use a scatter diagram and line of best fit to describe correlation

D Estimate values from a line of best fit

E Recognise when diagrams used to present data are misleading

A1

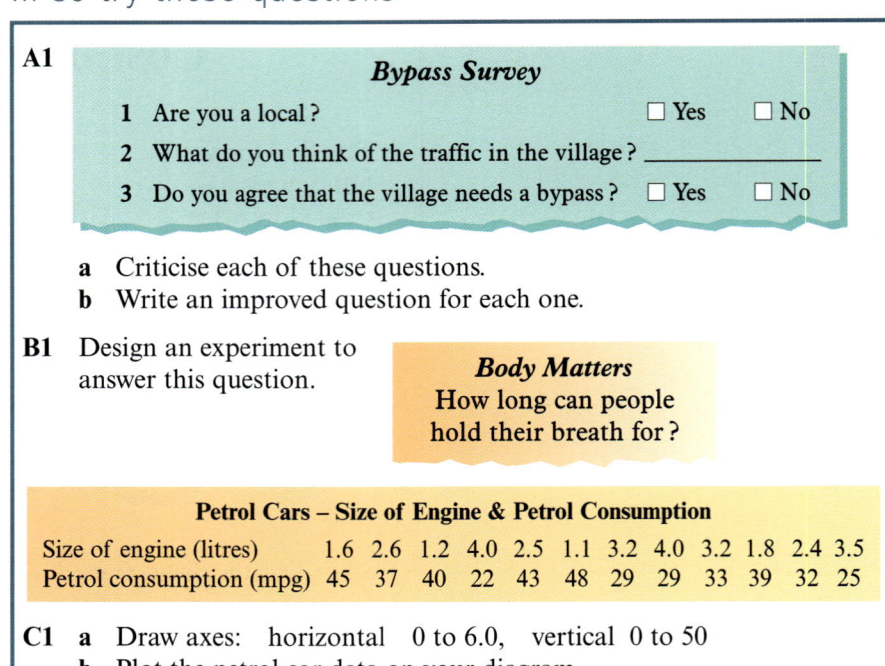

Bypass Survey

1 Are you a local? ☐ Yes ☐ No
2 What do you think of the traffic in the village? _____
3 Do you agree that the village needs a bypass? ☐ Yes ☐ No

a Criticise each of these questions.
b Write an improved question for each one.

B1 Design an experiment to answer this question.

Body Matters
How long can people hold their breath for?

Petrol Cars – Size of Engine & Petrol Consumption											
Size of engine (litres)	1.6	2.6	1.2	4.0	2.5	1.1	3.2	4.0	3.2	1.8	2.4	3.5
Petrol consumption (mpg)	45	37	40	22	43	48	29	29	33	39	32	25

C1 **a** Draw axes: horizontal 0 to 6.0, vertical 0 to 50
b Plot the petrol car data on your diagram.
c Draw a line of best fit.
d Describe any correlation between size of engine and petrol consumption.

D1 Use your scatter diagram to estimate:
a the petrol consumption of a car with a 3.0 litre engine.
b the engine size of a car with petrol consumption of 40 mpg.

D2 Extend your line of best fit to estimate:
a the petrol consumption of a car with a 5.0 litre engine.
b the engine size of a car with petrol consumption of 20 mpg.

E1

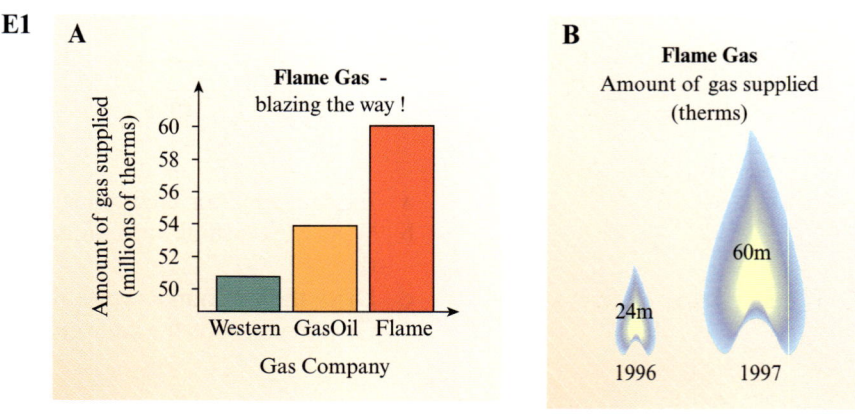

Explain why each of these diagrams is misleading.

Some points to remember

♦ When you draw a line of best fit on a scatter diagram, make sure the line goes through the middle of the points.

♦ In some cases, it does not make sense to extend the line of best fit on a scatter diagram.

Starting points
You need to know about ...

... so try these questions

A Bearings

- All bearings are:
  - measured clockwise from North
  - written using three figures.

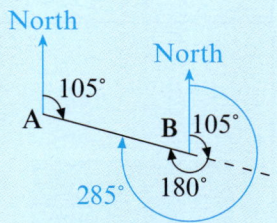

Example

The bearing of B from A is 105°.
The bearing of A from B is 285°.

- You can fix a position by:
  - giving a bearing and distance from one point
  - giving a bearing from two different points.

Example

The point B is on a bearing of 105° from A and is 10 km from A.

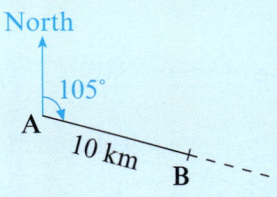

Example

The point C is on a bearing of 078° from A and on a bearing of 320° from B.

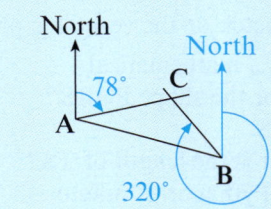

B Pythagoras' rule

- In any right-angled triangle, the area of the square on the hypotenuse is equal to the sum of the area of the squares on the other two sides.

  AB is the hypotenuse, so $AB^2 = AC^2 + BC^2$

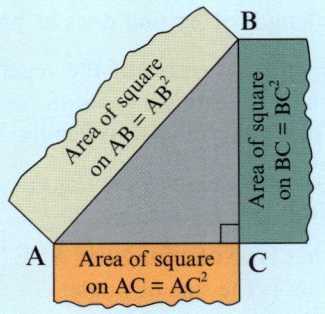

C Trigonometric ratios

- The trigonometric ratios can be used to calculate sides or angles in right-angled triangles.

- Each trig ratio can be written in different ways.

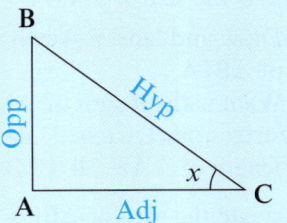

$$\sin x = \frac{\text{Opp}}{\text{Hyp}}$$

$$\text{Opp} = \text{Hyp} \times \sin x$$

$$\cos x = \frac{\text{Adj}}{\text{Hyp}}$$

$$\text{Adj} = \text{Hyp} \times \cos x$$

$$\tan x = \frac{\text{Opp}}{\text{Adj}}$$

$$\text{Opp} = \text{Adj} \times \tan x$$

A1 This diagram shows some towns on a radar screen positioned at H.

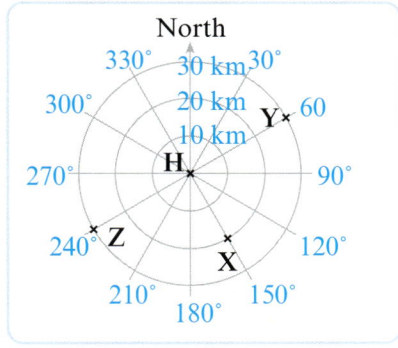

  a What is the bearing of:
  - **i** Y from H
  - **ii** H from Y
  - **iii** Z from Y?

  b Estimate the bearing of X from Y.

  c Point P is due North of X and on a bearing of 030° from H.
  How far from H is P?

B1 a **i** From the radar screen above sketch the triangle XHY.
  - **ii** What is the angle $\widehat{XHY}$?
  - **iii** What is the distance HX?
  - **iv** What is the distance HY?

  b Calculate the distance XY to the nearest 0.1 km.

C1 Using triangle XHY above calculate the angle $\widehat{HXY}$ to the nearest degree.

C2

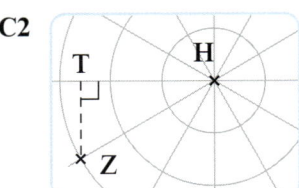

Calculate, to the nearest km:
  a TH   **b** TZ

285

Using trigonometry and Pythagoras' rule

Exercise 24.1
Using trigonometry and
Pythagoras' rule

Accuracy
For this exercise round
your answers to 3 sf.

1   This is Mike's plan for his garden.

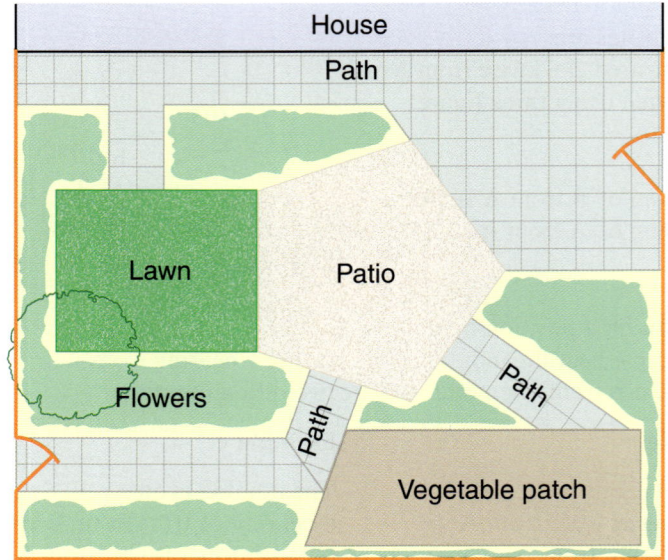

This is a sketch of the vegetable patch.

a   What is a mathematical
name for the shape PQRS?
b   In ΔPTS:
  i   what is the length of TS?
  ii use Pythagoras' rule to
       calculate the length of PS.

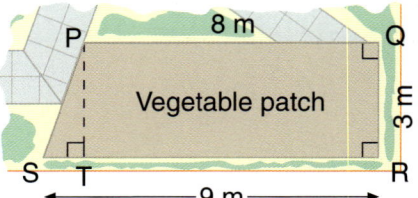

Use the full calculator value
and round your answer at
the end of each question.

2   Mike puts netting round the
edge of the vegetable patch.
What length of netting does he need?

3   **a**   Calculate the area of the vegetable patch.
b   A box of fertiliser costs 48 pence and treats 10 square metres.
What would it cost to fertilise the vegetable patch?

4   The patio is a regular pentagon.
Each side is 3 metres long.

a   What type of triangle is AFE?
b   What is the size of angle:
  i x **ii** y?

5   **a**   What is the size of FÂM?
b   **i**   Draw and label a sketch
        of ΔBFA.
     ii What is the length of AM?
c   Use trigonometry to calculate
the length of: **i** AF   **ii** FM

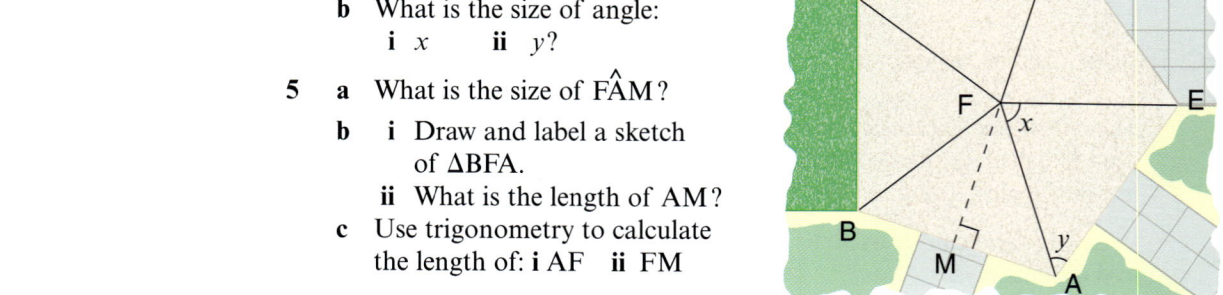

For area work in m².

6   **a**   Calculate the area of ΔBFA.
b   Find the total area of the patio.

7   Chippings 7 cm deep cover the patio. The chippings cost £14 per cubic metre.
What is the total cost of chippings for his patio, to the nearest 10 pence?

8 On the plan the lawn is a rectangle 3 metres by 5 metres.
The diagonals HB and GC are the same length.

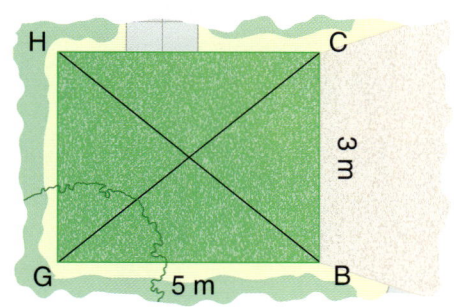

a Use Pythagoras' rule to calculate
the length of the diagonal GC.

b When Mike laid the lawn he found that GC was longer than HB.
Which of these statements is true?

A ⎡ CB̂G > 90° ⎤   B ⎡ CB̂G = 90° ⎤   C ⎡ CB̂G < 90° ⎤

Exercise 24.2
Choosing trigonometry or
Pythagoras' rule

Accuracy
For this exercise round
your answers to 2 dp.

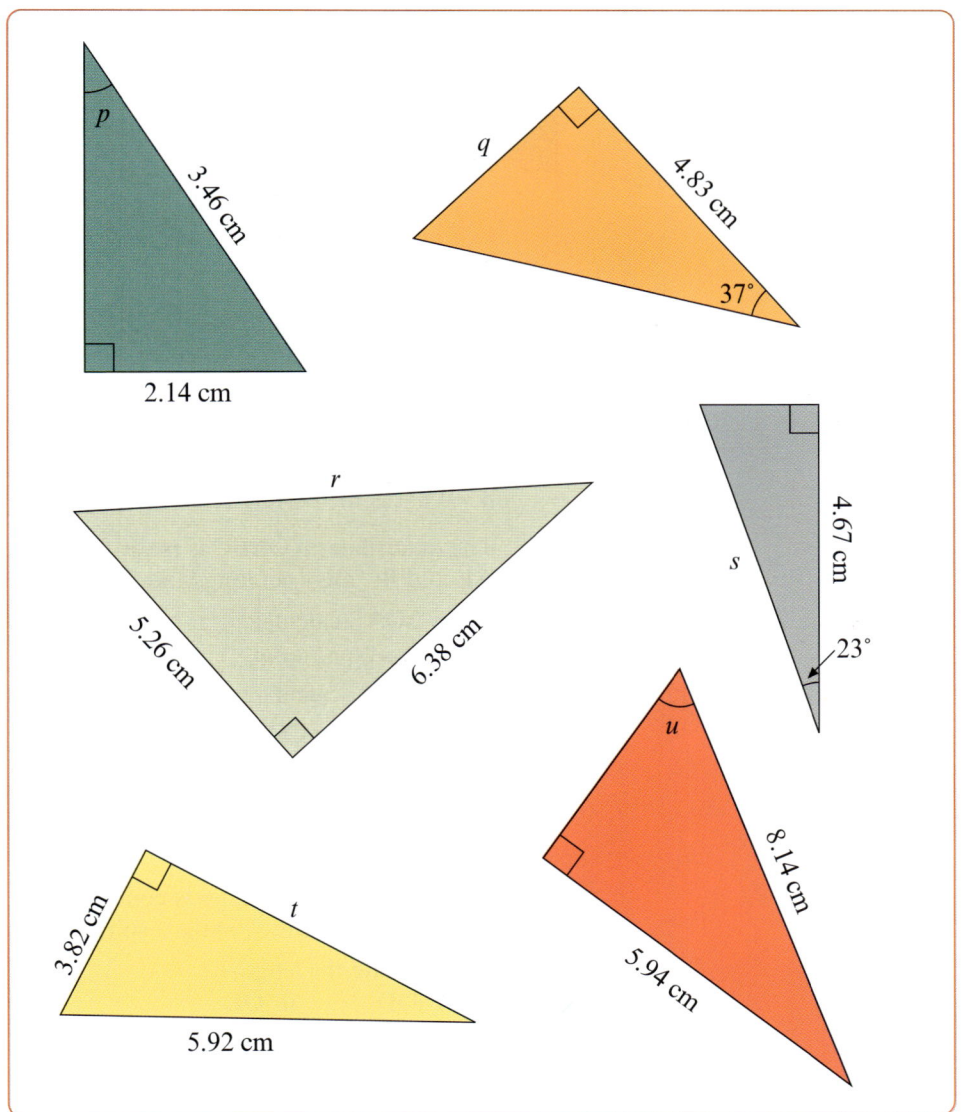

1 For each triangle:

a calculate the angle or side marked with a letter.

b say whether you used trigonometry or Pythagoras' rule and explain why.

- In any **right-angled triangle** you can use Pythagoras' rule and trigonometry.

  ❖ To **find** **an angle** when
  you **know** **two sides** → use **trigonometry**

  ❖ To **find** **a side** when
  you **know** an **angle** and **a side** → use **trigonometry**

  ❖ To **find** **a side** when
  you **know** **two sides** → use **Pythagoras' rule**

Exercise 24.3
Areas and perimeters

Accuracy
For this exercise round
your answers to 3 sf.

1 This is part of a calculation to find the area of △ABC.

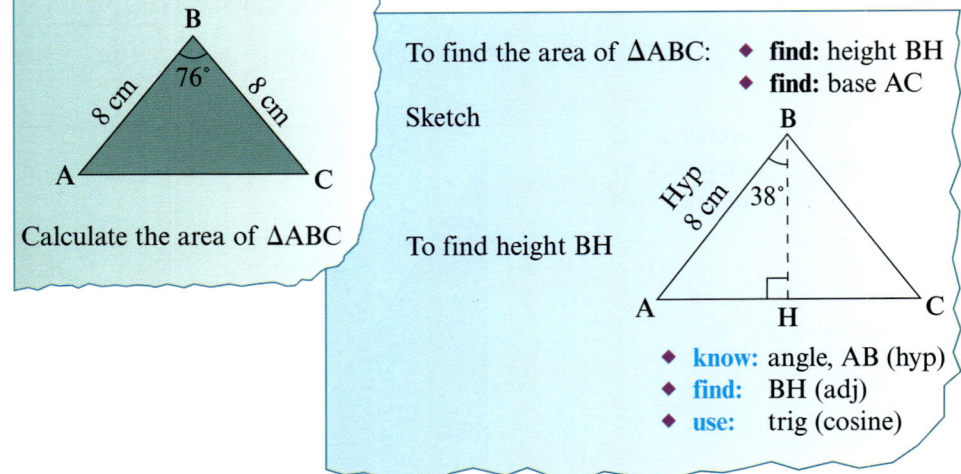

Calculate the area of △ABC

To find the area of △ABC: ◆ **find:** height BH
  ◆ **find:** base AC

Sketch

To find height BH

◆ **know:** angle, AB (hyp)
◆ **find:** BH (adj)
◆ **use:** trig (cosine)

   a Calculate the height BH.
   b Calculate the length of AC.
   c Calculate the area of △ABC.
   d What is the perimeter of △ABC?

2 The quadrilateral OABC is made from two right-angled triangles.
AB and BC are 4 cm long, and OB is 8 cm.

   a In △OAB calculate:

      i the angle AÔB   **ii** OA

   b In △OBC calculate:

      i the angle CÔB   **ii** OC

   c What is the perimeter of OABC?
   d Calculate the area of OABC.

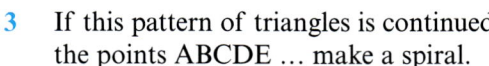

3 If this pattern of triangles is continued
the points ABCDE … make a spiral.

   a Calculate the lengths:
      i OD   **ii** OE

   b Calculate the angles:
      i DÔC   **ii** EÔD

   c As the pattern continues:
      i what happens to the length of
      lines from O?
      ii what happens at O to the angles
      in the triangles?

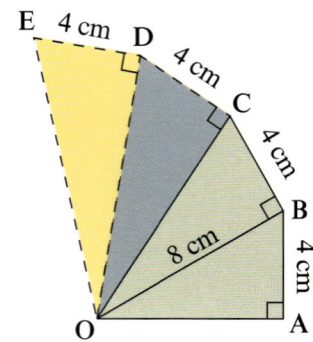

Angles of elevation and depression

◆ From a point A the **angle of elevation** of a point B is the angle between the horizontal and the line of sight from A to B.

Line of sight of B from A

B

A x Horizontal

x is the angle of elevation of B from A

◆ From a point A the **angle of depression** of a point C is the angle between the horizontal and the line of sight from A to C.

A Horizontal y

Line of sight of C from A

C

y is the angle of depression of C from A

Exercise 24.4
Angles of elevation and depression

Accuracy
For this exercise round your answers to the nearest whole number.

1 Jan measures the distance and the angle of elevation or depression to points A and B from T.

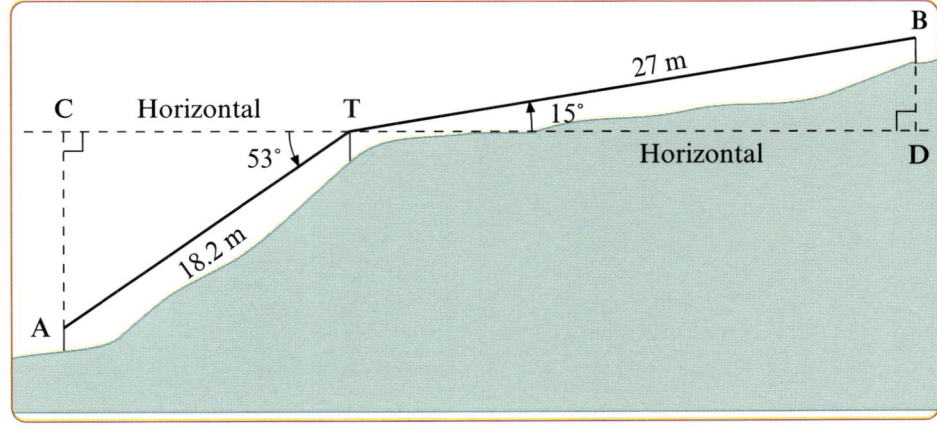

a The angle of depression from T to A is 53° and the distance from T to A is 18.2 metres. The point C is vertically above A.
  i Calculate the vertical height AC.
  ii Calculate the horizontal distance CT.

b The angle of elevation from T to B is 15° and the distance from T to B is 27 metres. The point D is vertically below B.
  i Calculate the vertical height BD.
  ii Calculate the horizontal distance DT.

c On this sketch AE shows the vertical height between A and B, and EB the horizontal distance between A and B. Calculate:
  i the vertical height AE
  ii the horizontal distance EB
  iii the distance AB.

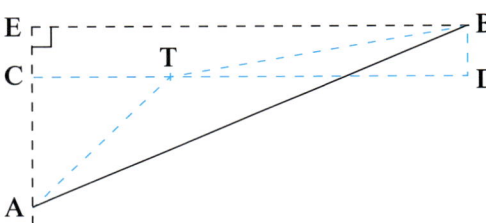

Bearings

Exercise 24.5
Using scale drawings

1 This sketch shows the position of three landmarks A, B and C.

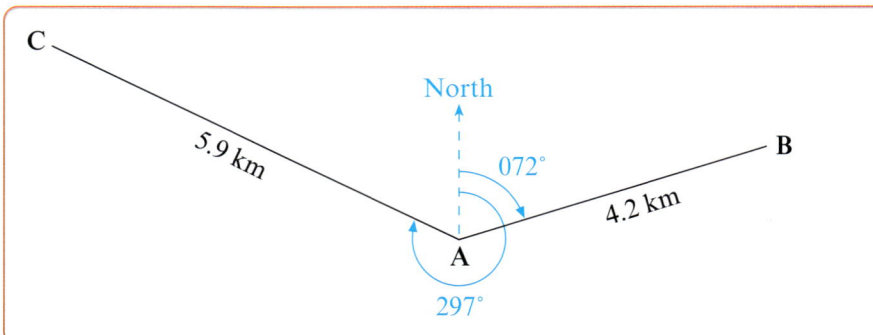

The bearing of B from A is 072° and the distance AB is 4.2 km.
The bearing of C from A is 297° and the distance AC is 5.9 km.

a Make a scale drawing to show the relative position of these landmarks.
b Calculate the bearing of A from B.
c **i** Measure the bearing of C from B.
   ii What is the bearing of B from C?
d Use your drawing to find the distance between B and C.

> Check that any north lines you draw on a diagram are parallel.

2 This sketch shows the position of three boats K, L and M.

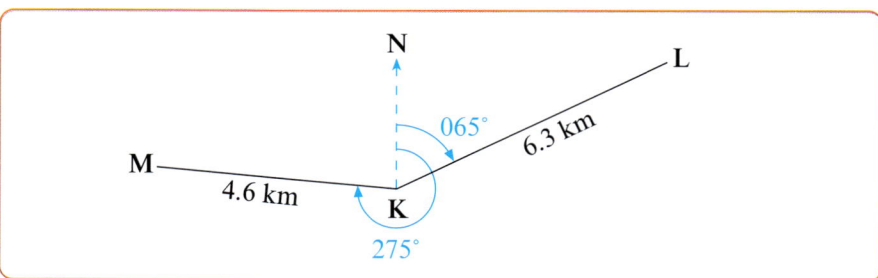

> Measure accurately – try to be no more than 1 mm out either way.

a Make an accurate scale drawing to show K, L and M using a scale of 1 cm to 1 km.
b What is the bearing of M from L?
c How far east of K is L?
d How far west of L is M?
e What is the distance LM?

3 The sketch shows the position of four buildings A, B, C and D using a scale of 1 cm to 1 km.

a Make a scale diagram showing A, B, C and D.
b How far is the distance CD?
c What is the bearing of B from D?
d How far west of D is C?
e How far south of B is D?

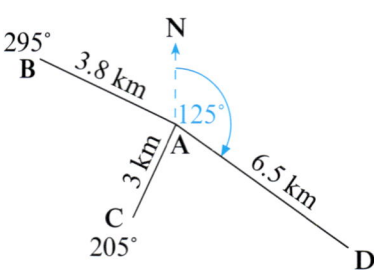

4 This shows the position of four buoys R, S, T and V.

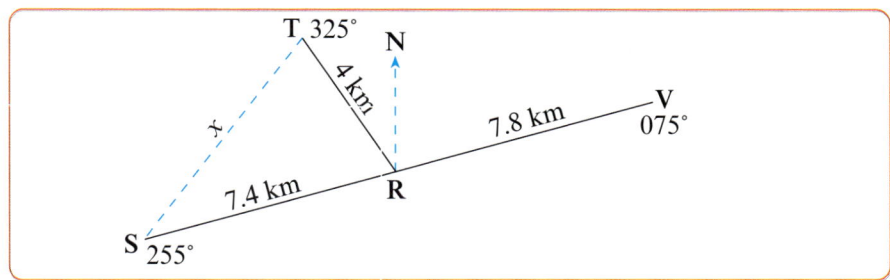

  a Make an accurate scale diagram to show R, S, T and V using a scale of 1 cm to 1 km.

  b Measure the bearing of S from V.

  c How far west of T is S?

  d How far east of R is V?

  e Measure the distance VT.

  f What is the distance marked x on the diagram?

5 This sketch shows the position P, Q and R of three buoys.

  a Make a scale drawing to show the relative positions of the buoys.

  b What is the bearing of Q from R?

  c **i** Use your drawing to find the distance between P and R.

    ii How accurate do you think your answer is?

  d In $\triangle$PQR:

    i What is the angle $P\hat{Q}R$?

    ii Calculate the distance between P and R.

  e **i** Calculate the angle $Q\hat{P}R$.

    ii What is the bearing of R from P?

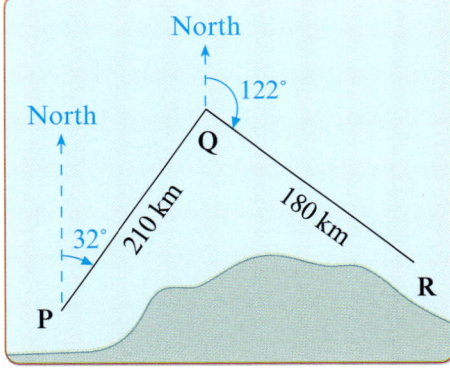

> You can measure each distance and angle from your scale drawing but it is more accurate to calculate them.

> **Accuracy**
> For Questions **5d** to **6d**
> ❖ round each angle to the nearest degree
> ❖ round each distance to the nearest km.

6 On this sketch TP shows how far P is south of Q, and TQ shows how far P is west of Q.

  a Calculate how far P is:

    i south of Q

    ii west of Q.

  b In $\triangle$QUR what is the angle $R\hat{Q}U$?

  c Calculate how far R is:

    i south of Q

    ii east of Q.

  d How far is P south of R?

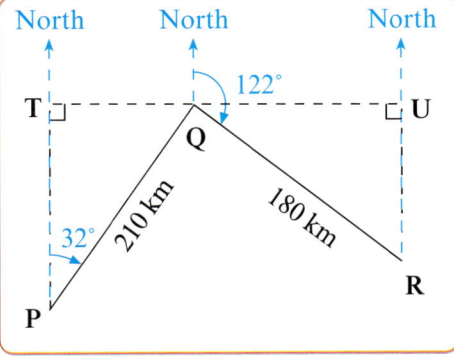

Exercise 24.6
Calculating bearings
and distances

Accuracy
For this exercise
❖ round each distance
   to the nearest 0.01 km
❖ round each angle
   to the nearest degree.

1 This diagram shows the relative positions of three schools:
Shaw, Castle and Greys.

   a **i** How far north of Castle is Shaw?
      ii How far east of Castle is Shaw?
   b Calculate:
      i the distance of Castle from Shaw
      ii the bearing of Shaw from Castle.
   c **i** How far south of Castle is Greys?
      ii How far east of Castle is Greys?
   d Calculate:
      i the distance from Castle to Greys
      ii the bearing of Greys from Castle.

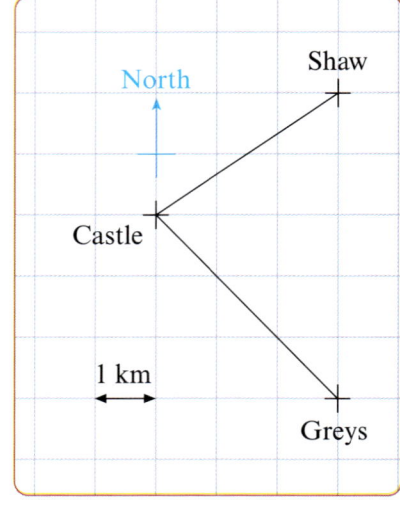

2 There is a lighthouse at Alington, Bandon and Colton.
Bandon is 78 km from Alington on a bearing of 320°.
Colton is due north of Alington and due east of Bandon.

   a Draw a sketch to show the relative positions of these lighthouses.
   b What is the bearing:
      i of Alington from Bandon
      ii of Colton from Bandon
      iii of Alington from Colton?
   c Calculate the distance from:
      i Alington to Colton
      ii Bandon to Colton.

3 This diagram shows the route taken by a boat which starts at P.
The boat sails on a bearing of 143°
for 24 km to a buoy at Q, then sails
due south for 18 km to a buoy at R.
The point S is due north of Q and
due east of P.

   a Calculate the distance:
      i PS   **ii** SQ
   b How far is R south of P?
   c Use your answers to
      Questions **3a** and **3b**:
      i to calculate the distance
         between P and R
      ii to calculate the bearing of
         R from P.
   d What is the bearing of P from R?

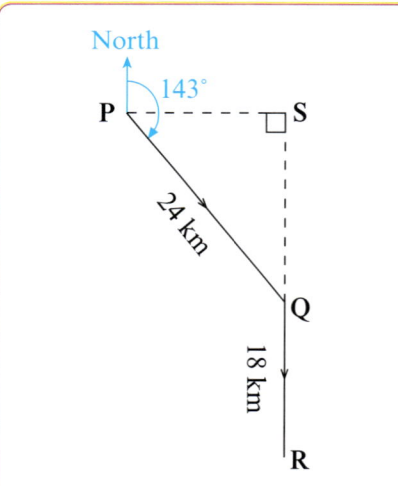

4 Another boat starts at P and sails 8 km
due east, then 17 km due north to a buoy at T.

   a Sketch a diagram to show the route taken by this boat.
   b Calculate:
      i the bearing of T from P
      ii the distance from P to T.

Solving problems

Exercise 24.7
Solving problems

Accuracy
For this exercise round your answers to 3 sf.

1 This is a sketch of the cross-section of a roof.

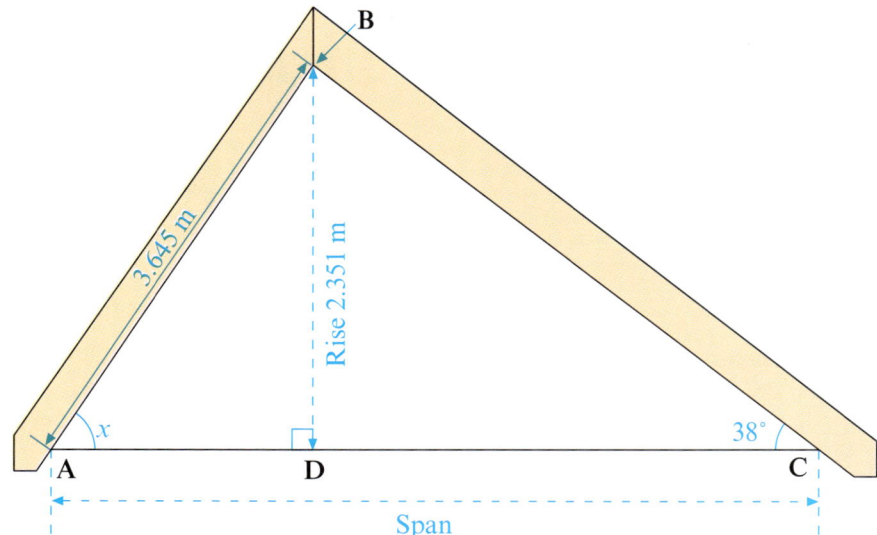

The pitch of the rafter BC is 38°. The rise of the roof is 2.351 metres.
The length AB is 3.645 metres.

a For the rafter AB calculate the pitch, x.
b Calculate the distance AD in metres.
c Calculate the length of BC.
d Calculate the distance DC.
e What is the span of the roof?

2 This diagram shows the cross-section of another roof with a dormer window.
The pitch of the roof at E is 47° and the distance EH is 2.449 m.

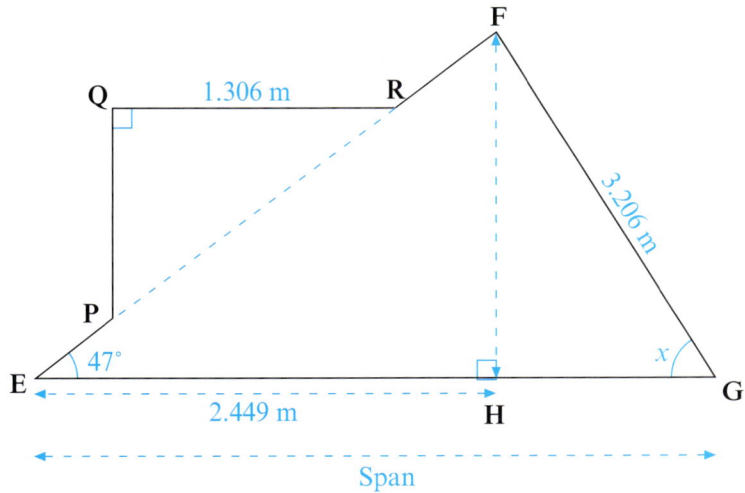

a Calculate the rise of the roof, FH.
b Calculate the distance GH.
c What is the span of the roof?
d For the rafter FG, calculate the pitch, x.
e What is the angle $Q\hat{R}P$?
f Calculate the height of the window, PQ.

End points

You should be able to ...   ... so try these questions

A Use Pythagoras' rule
      and trigonometry

For each of these questions give your answers correct to 1 dp.

A1

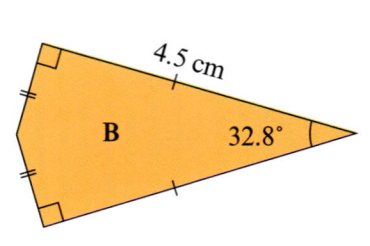

6.3 cm

4.8 cm

A

1.9 cm

4.5 cm

B

32.8°

Calculate the area and perimeter of shapes A and B.

B Use angles of elevation
      and depression

B1 This sketch shows a point A at the top of a vertical cliff
        200 metres above sea level.
        The boats at C and D are due west of A.
   a The angle of depression
          from A to D is 38°.
      i What is the angle ADB?
      ii Calculate the distance DB
            in metres.
   b The distance CB is 163 metres.
      i What is the distance between
            the two boats?
      ii Calculate the angle of x.

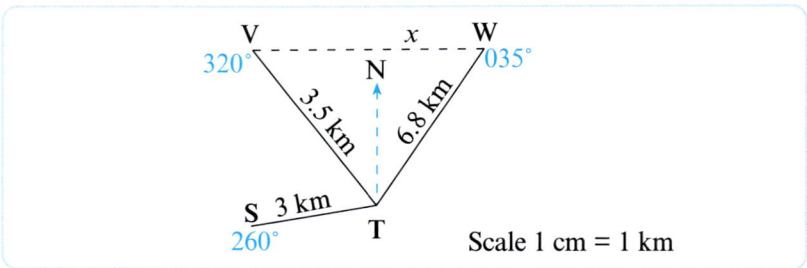

C Use bearings and distances

C1 The diagram shows 4 buildings T, S, V, W.

V
320°
3.5 km
S 3 km
260°
N
x
W
035°
6.8 km
T

Scale 1 cm = 1 km

Draw an accurate scale diagram and answer the questions:
a How far east of S is W?
b How far south of S is W?
c Give the distance x in km.
d What is the bearing of W from V?
e What is the distance VS in kilometres?
f Calculate how far T is south of V.

C2 A boat sailed 15 km due west from P, and then sailed 17 km
        on a bearing of 207° to R.
   a What is the bearing of Q from R?
   b Calculate these distances
          in kilometres:
      i TR   **ii** QT   **iii** TP
   c Calculate the distance RP.
   d Calculate the bearing:
      i of P from R
      ii of R from P.

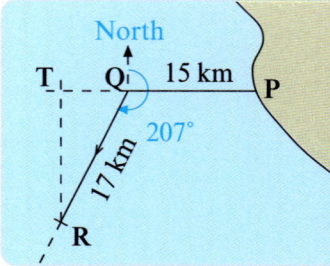

D   Solve problems

D1

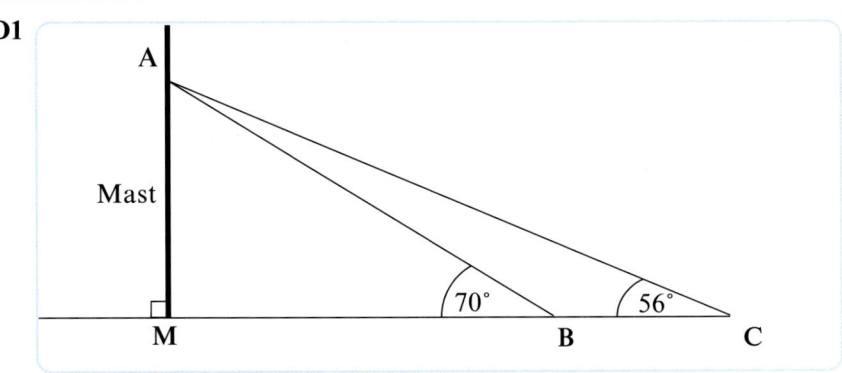

The diagram shows a TV mast.
Two guy ropes are fixed to the mast at A and to the ground at B and C.
The point A is 15.5 metres above the ground.

a Calculate the distance MB.
Give your answer correct to 1 dp.
b Calculate the length of AB.
c Calculate the distance BC.

D2 A laser beam is fixed 50 feet above the ground at the top of a verticle pole.
The beam can reach a distance of 120 feet from the foot of the tower.
The diagram shows the laser.

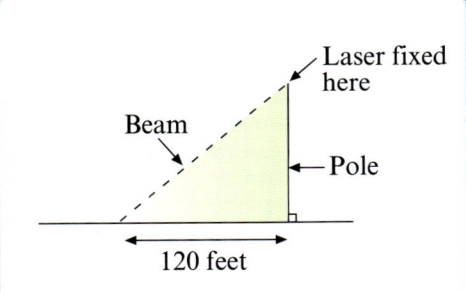

a Calculate the length of the laser beam.
b At what angle does the beam strike the ground?

Some points to remember

◆ Pythagoras' rule and trigonometric ratios can be used in a **right-angled triangle**.
It may help to draw and label a sketch of each triangle you use.

◆ Measured answers from scale drawings will not be as accurate as ones you calculate.

◆ **Angles of elevation** and **angles of depression** are measured from the horizontal.

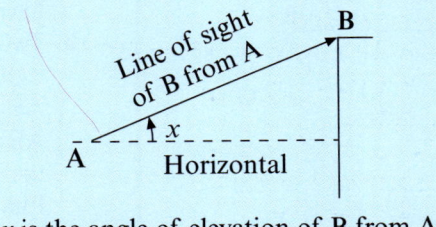

x is the angle of elevation of B from A.

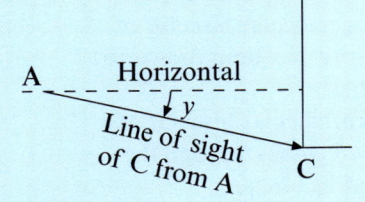

y is the angle of depression of C from A.

Starting points

You need to know about ...

... so try these questions

A The terminology for parts of a circle

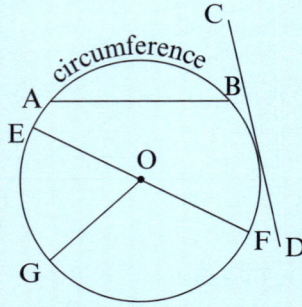

In this circle with centre O:

AB is a **chord**	A chord is a straight line that joins two points on the circumference of a circle.
CD is a **tangent**	A tangent to a circle is a straight line which touches the circle at one point only.
EF is a **diameter**	A diameter is a chord that passes through the centre.
Circumference	The circumference is the distance measured around the curved edge of a circle.
OG is a **radius**	A radius is any straight line from the centre to the circumference.

The curved part of the circle between A and B is known as an **arc**.
There are two **arcs** AB:
the larger is the **major arc**
the smaller is the **minor arc**.

The chord AB divides the circle into two **segments**.
There are two segments created by the chord AB: the **major segment** and the **minor segment**.

AB, PR, and GH are all chords.
The **perpendicular bisector** of each chord has been drawn to show that:
The perpendicular bisector of any chord passes through the centre of its circle.

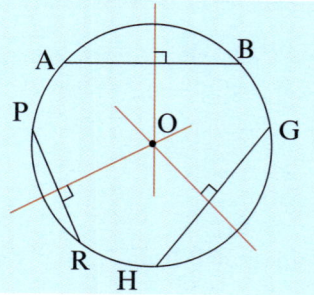

A1 ◆ Draw a circle, centre O with a radius of 4 cm.
◆ Mark any point P on your page 8 cm from O.
◆ From P draw two tangents to the circle.
◆ Label the point where one tangent touches the circle A, and where the other tangent touches B.

  a Measure the distance PA and PB.
  b What do you notice?
  c Is this the case for other points P?
◆ Join PO.
  d Measure angles APO and BPO.
  e What do you notice?
  f Is this always the case for other points P?
◆ Draw in the lines OA and OB.
  g Measure angles OAP and OBP.
  h What do you notice?
  i Is this always the case?
◆ Draw in the chord AB.
  j Is OP the perpendicular bisector of AB?
    Give reasons for your answer.

A2 Draw any circle centre O. The chord AB is such that the ratio
minor arc AB : major arc AB
        is 1 : 2.
  a Mark a position for A and B on your circle.
  b Explain how you were able to fix points for A and B.

A3 a Draw around a circular object. By drawing and bisecting chords find the centre of the circle you have drawn.
  b Two straight lines meet at right angles at B so that AB = 4 cm and BC = 6 cm. AB and BC are chords of the same circle, centre O. Find the distances OA, OB and OC.

Angles in the same segment

> For a chord AB and any point C on the circumference of a circle, joining A to C, and B to C gives angle ACB.
>
> Angle ACB is said to be subtended by the chord AB.

In this circle, AB is a chord. The points C, D, and E are other points on the circle.

The angles ACB, ADB, and AEB are all:

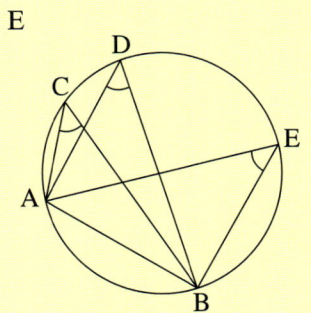

♦ subtended by the chord AB
♦ at the circumference of the circle
♦ in the same segment
  (in this case the major segment).

A geometrical property of the circle is that:
the angles ACB, ADB, and AEB are all equal in size.
The general statement is:
Angles subtended at the circumference by the same chord in the same segment are equal.

The angle subtended at the centre

In this circle, centre O, AB is a chord and C is a point on the circumference.
Angle ACB is:

♦ subtended by the chord AB
♦ at the circumference of the circle.

Angle AOB is:

♦ subtended by the chord AB
♦ at the centre of the circle.

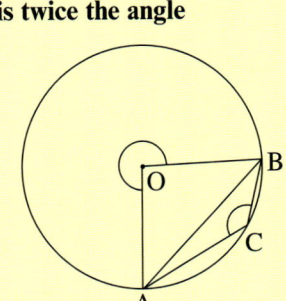

A geometrical property of the circle is that:
the angle AOB is twice the angle ACB ($A\hat{O}B = 2 \times A\hat{C}B$).

The general statement is:
For the same chord, the angle subtended at the centre is twice the angle subtended at the circumference in the major segment.

When the angle subtended at the centre is in the minor segment:

The general statement is:
For the same chord, the angle subtended at the circumference is half the reflex angle subtended at the centre.

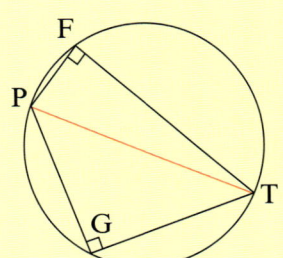

Angles in a semicircle

When a chord becomes a diameter, then each segment is a semicircle. The angle subtended at the circumference in a semicircle is a right angle.

In this case $P\hat{G}T = P\hat{F}T = 90°$

The general statement is:
Any angle subtended at the circumference in a semicircle is a right angle.

Tangents to a circle

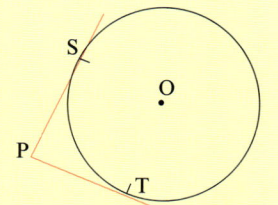

From any fixed point (P) outside a circle, two
tangents can always be drawn to that
circle, in this case PS and PT.
From point P to the point where either tangent
touches the circle, is the same distance, i.e. PS = PT.

The general statement is:
**From any point P, outside the circle, two tangents to the circle can
be drawn, the distance from P to the point of contact being the same
for each tangent.**

A radius drawn to the point of contact of a tangent
is at right angles to the tangent.

In this case: $P\hat{S}O = P\hat{T}O = 90°$

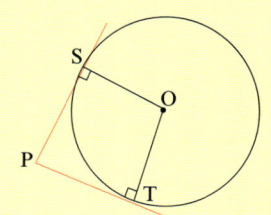

The general statement is:
**The angle between a radius, drawn to the point of
contact of a tangent, and the tangent is a right angle.**

Example

AB is a diameter of the circle with centre O.

a What is the size of $A\hat{B}C$?
b What is the size of $B\hat{O}C$?

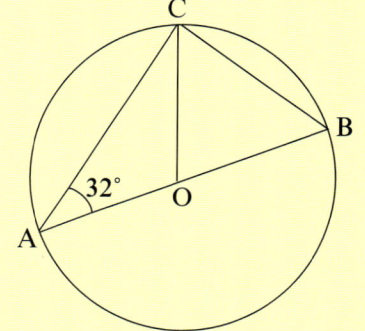

a The angle subtended at the circumference
in a semicircle is a right angle.
So $A\hat{C}B = 90°$.
Angles of a triangle add up to 180°.
So $A\hat{B}C = 180° - 32° - 90°$
$= 58°$.
b OC and OB are both radii, so they are equal lengths.
So △OCB is isosceles, and $O\hat{C}B = O\hat{B}C = 58°$.
So $B\hat{O}C = 180° - (2 \times 58°)$
$= 64°$.

Exercise 25.1
Using circle theorems

1 **a** Name two angles subtended at the circumference by the chord AC.
   b Name two angles subtended at the circumference by the chord CD.
   c What is the size of $C\hat{A}D$?
   d What is the size of $C\hat{E}A$?
   e BC is a tangent to the circle.
   What is the size of $B\hat{C}A$?
   Explain the reason for your answer.

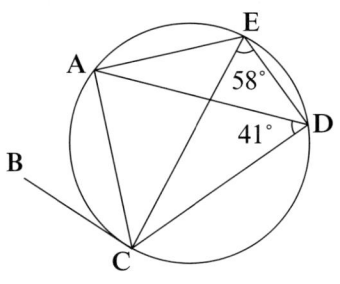

2 In this diagram, the circle has centre O.
   BE is a chord, AB a tangent, and $B\hat{D}E = 37°$.

   a What is the size of $E\hat{O}B$?
      Explain your answer.
   b What is the size of $B\hat{C}E$?
   c What is the size of $A\hat{B}O$?
      Explain the reason for your answer.
   d What is the size of $O\hat{B}E$?
      Explain the reason for your answer.
   e What is the size of $E\hat{B}A$?

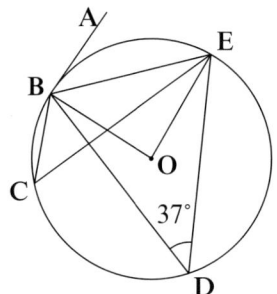

3 In the diagram, the circle is centre O.
   ED is a chord.

   a Name an angle at the circumference in the major segment subtended by ED.
   b Calculate the size of $E\hat{O}D$ (reflex).
   c Calculate the size of $D\hat{F}E$.
   d Calculate the size of $O\hat{E}D$.
      Give the reason for your answer.

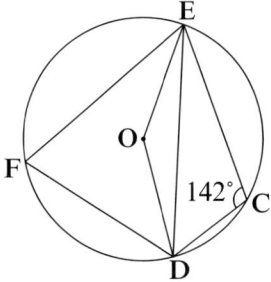

4 In the diagram, the circle has centre O.
   DE is a diameter, FE is a chord, and FC a tangent.

   a Calculate the size of $F\hat{G}E$.
      Explain the reasons for your answer.
   b Calculate the size of $O\hat{E}F$.
      Explain the reasons for your answer.
   c Explain why $E\hat{F}C$ is 41°.
   d DEH is 29°. Calculate the size of $H\hat{D}E$.
      Explain your answer.

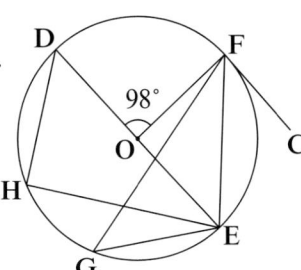

The cyclic quadrilateral

RSTU is a cyclic quadrilateral.
A quadrilateral is said to be cyclic when:

♦ the four vertices of the quadrilateral lie on the circumference of the same circle.

The other property all cyclic quadrilaterals have is that:

♦ the angles at opposite vertices are supplementary.

As RSTU is cyclic then:

$$R\hat{S}T + T\hat{U}R = 180°$$
$$\text{and} \quad S\hat{R}U + U\hat{T}S = 180°$$

Supplementary angles are a pair of angles that, when added together, give a total of 180°.

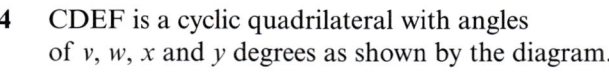

Exercise 25.2
Cyclic quadrilaterals

1 Explain why all rectangles are cyclic quadrilaterals.

2 Is it possible to draw a parallelogram that is cyclic? Explain your answer.

3 Apart from the rectangle, what other quadrilateral is always cyclic? Explain your answer with diagrams.

4 CDEF is a cyclic quadrilateral with angles of v, w, x and y degrees as shown by the diagram.

   a If $p = 15°$, explain why:
      $v + x = 12p$
   b If v, w, x and y are such that:
      $w : y = 2 : 7$ and $v : x = 13 : 17$
      find values for v, w, x, and y.

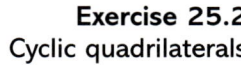

5 In the diagram the circle has centre O. The size of $N\hat{K}L$ is given as a.

   Use circle properties to explain why $N\hat{K}L$ and $L\hat{M}N$ are supplementary.

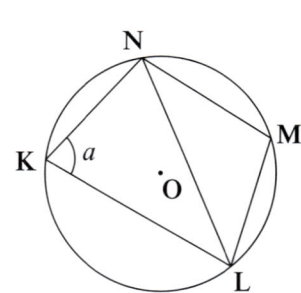

6 In the diagram the circle has centre O.
   EOG is the diameter
   Calculate $F\hat{G}H$.

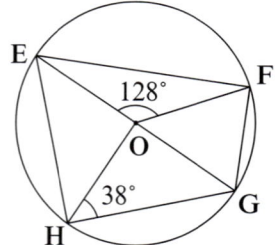

End points

You should be able to ...

... so try these questions

A Identify angles in the same segment

A1 **a** What is the size of $D\hat{A}B$?
 b What is the size of $E\hat{A}C$?

A2 Give an angle that is the same size as:
 a $E\hat{A}D$ **b** $C\hat{E}B$

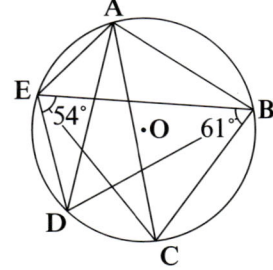

B Link angles at the centre and at the circumference on the same chord

B1 Calculate the size of:
 a $A\hat{B}C$ **b** $A\hat{D}C$

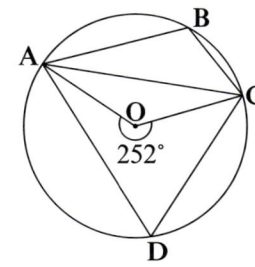

C Work with angles in a semicircle

C1 In the circle the centre is O and $C\hat{A}B = 28°$
 Calculate the size of $A\hat{C}B$.
 Explain your calculation.

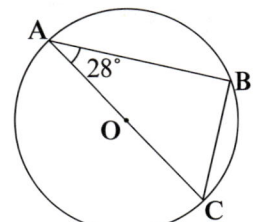

D Solve problems when tangents are involved

D1 Calculate the size of $D\hat{A}E$.
 Explain your working.

D2 Calculate the size of $D\hat{E}A$.
 Give reasons for your answer.

D3 Calculate the size of $D\hat{A}O$.
 Explain your answer.

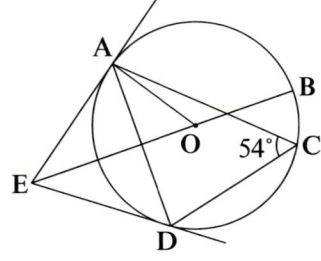

E Use the properties of cyclic quadrilaterals

E1 List the properties of all cyclic quadrilaterals.

E2 The angles of a cyclic quadrilateral are given as $3k$, $4k$, $5k$, and $6k$.
 a On a diagram show possible positions of these angles.
 b Write and solve an equation in k to find the size of each angle.

Eurotunnel

The first tunnel-boring machine (TBM) started work for Eurotunnel on the UK side in December 1987 and on the French side in January 1988. Each TBM bored about 4.4 metres per hour.

The two railway tunnels and the smaller service tunnel are each 50 km long with 38 km under the Channel, 3km on land in France and 9km on land in the UK.

The lining rings for the railway tunnels are 40 cm thick and 1.6 metres long. For the service tunnel the rings are 32 cm thick and 1.4 metres long. The ventilation system replaces 145 m³ of air per second.

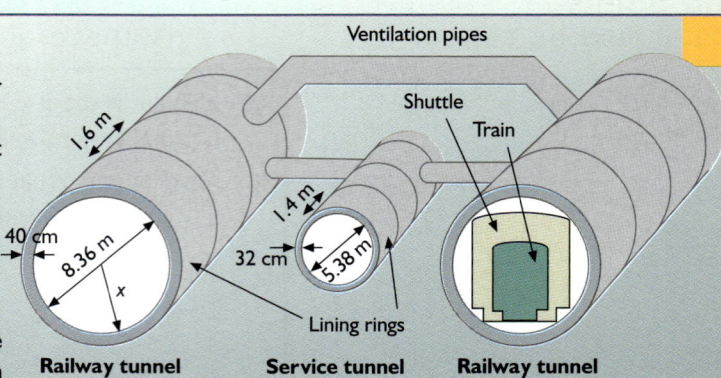

Ventilation pipes

Shuttle

Train

1.6 m

1.4 m

40 cm   8.36 m   x

32 cm   5.38 m

Lining rings

Railway tunnel   **Service tunnel**   **Railway tunnel**

In Eurotunnel
- Shuttles carry vehicles with their passengers between Folkestone and Calais
- Eurostar trains carry passengers between Waterloo and Paris and Brussels

Eurostar trains fact file

Each train on Eurostar consist of 2 engines, 2 buffet cars and 16 carriages:

- 6 first-class (2 smoking, 4 no-smoking)
- 10 second-class (4 smoking, 6 no-smoking)

These give a total of 794 seats (210 first-class, 584 second-class) and an additional 52 folding seats.

Each train is 394 metres long, 2.8 metres wide and weighs 752 tonnes.

Its maximum speed is 300 km/h on high-speed lines, 220 km/h on standard track and 160 km/h in the tunnel.

Check-in closes 20 minutes before departure.

Le Shuttle

The maximum speed for any shuttle is 140 km/h.

Passenger-vehicle shuttles

These carry passengers who remain with their car, coach, caravans, minibus, motorcycle or bicycle. Each shuttle can transport 120 cars and 12 coaches. The average journey time is about 35 minutes, of which 26 minutes are spent in the tunnel. At peak times there are up to 4 departures, and one departure per hour at off-peak night times.

Freight shuttles

These transport Heavy Goods Vehicles (HGV's). Their drivers are carried in a club car at the front of each shuttle, which carry up to 28 HGV's. At peak times there are up to 3 departures, and one departure per hour at off-peak night times. The average journey time from motorway to motorway is 80 minutes.

TIMETABLE

Eurostar trains	SUNDAYS					July to September		
London to Paris								
London Waterloo	08:10	10:10	11:57	12:53	13:57	15:10	16:23	16:53
Ashford	09:23	-	-	13:53	-	-	17:24	17:54
Calais-Frethun	-	14:29	-	-	17:56	-	-	-
Lille Europe	-	-	-	-	-	-	19:26	-
Paris Nord	12:23	14:17	15:56	16:53	17:56	19:23	20:29	20:56
Paris to London								
Paris Nord	08:07	10:19	11:43	13:04	15:19	16:07	17:10	18:18
Lille Europe	-	-	12:44	-	-	-	-	-
Calais-Frethun	09:34	-	-	14:31	-	17:34	-	-
Ashford	09:10	-	-	14:07	-	17:11	18:11	-
London Waterloo	10:30	12:30	13:47	15:26	17:13	18:13	19:13	20:13
London to Brussels								
London Waterloo	09:14	12:14	14:10	17:27	18:27	19:27		
Ashford	10:27	13:27	-	-	19:28	20:27		
Lille Europe	12:30	15:29	17:21	-	21:31	-		
Brussels Midi	13:44	16:44	18:34	21:38	22:45	23:46		
Brussels to London								
Brussels Midi	08:27	10:31	12:31	15:28	17:22	19:27		
Lille Europe	09:40	11:45	13:45	16:42	18:36	20:39		
Ashford	09:41	-	-	16:41	18:37	20:41		
London Waterloo	10:47	12:47	14:47	17:43	19:39	21:43		

Traffic in Eurotunnel 1995

	Passenger-vehicle shuttles	Freight shuttles	Eurostar trains	Freight trains
Jan/Feb	101 324 cars	40 328 lorries	728	733
Mar	72 618 cars	22 580 lorries	506	432
Apr	96 735 cars	22 648 lorries	542	404
May	80 995 cars	28 267 lorries	595	465
Jun	100 534 cars	32 657 lorries	730	596
Jul	112 060 cars 2402 coaches	7126 lorries	851	644
Aug	145 861 cars 2728 coaches	36 517 lorries	844	425
Sep	105 914 cars 3033 coaches	38 136 lorries	862	515
Oct	120 368 cars 3794 coaches	42 630 lorries	983	523
Nov	129 286 cars 5120 coaches	48 263 lorries	897	464
Dec	156 999 cars 6306 coaches	41 770 lorries	659	88

1 On which side of the Channel did they first start digging the tunnel?

2 What is the radius of the hole they bored for:
a the service tunnel
b a railway tunnel?

3 In which month in 1995 did Le Shuttle carry the greatest number of lorries?

4 How many minutes does the 08:10 take from London to Ashford on a Sunday?

5 Is the journey time from London to Paris the same for every train? Explain your answer.

6 How many minutes of each shuttle journey are not spent in the tunnel?

7 What is the difference between the maximum speed of a Eurostar train and a Shuttle in the tunnel?

8 If Amin reaches Waterloo at 3:50 pm on a Sunday, what is the first Eurostar train he can catch to Brussels?

9 **a** What is the last train Pia could catch from Waterloo to arrive in Paris by 6:00 pm?
b Which stations does this train stop at?

10 Which is the first train from Paris that stops at Lille?

11 What is the time of the first train from Lille to London on a Sunday?

12 **a** What is the latest time that Ethel can arrive at Waterloo to catch the 14:10 to Brussels?
b At what time should she arrive in Brussels?

13 What is the destination of the first Eurostar train that stops at Calais on a Sunday?

14 **a** Calculate the volume of material that was removed (spoil) when they bored the service tunnel. Give your answer in m³ correct to 3 sf.
b Rewrite your answer in standard index form.

15 **a** What total volume of spoil was removed to make the two railway tunnels?
b The total spoil from all the excavations was 8 million m³. What percentage of the spoil was from the two railway tunnels?

16 **a** What is the thickness of the lining ring for a railway tunnel in metres?
b What is the value of x, the internal radius of a railway tunnel?
c Calculate the volume inside a railway tunnel.

17 How many lining rings were used in:
a one railway tunnel?
b the service tunnel?

18 How many minutes would it take the ventilation system to replace all the air in the two railway tunnels?

19 In what month were coaches carried on Le Shuttle for the first time?

20 How many coaches were carried in total in 1995?

21 **a** Draw a graph to show the number of coaches carried during 1995.
b Sketch a graph to show the number of coaches you think were carried each month in 1996.

22 **a** What was the highest number of cars carried in one month in 1995? Write your answer in standard form correct to 2 sf.
b Calculate the mean number of cars carried per week.

23 Calculate the mean number of lorries carried per month.

24 If the average number of passengers on a Eurostar train was 635, estimate the total number of passengers carried on Eurostar trains in 1995. Give your answer in standard form to 2 sf.

25 What is the average number of seats in a first-class carriage?

26 Do you think that every second-class carriage has the same number of seats? Explain your answer.

27 What is the average number of seats per carriage on a Eurostar train?

28 **a** What is the maximum speed of a shuttle?
b What is the average speed of a shuttle in the tunnel?

29 What is the ratio of first-class to second-class seats in its simplest form?

30 Write the length of a Eurostar train as a percentage of the length of the tunnel.

31 If the average speed of a Eurostar train in the tunnel is 158 km/h, calculate how long it takes to go through the tunnel.

32 What percentage of Eurotunnel is on:
a French land? **b** UK land?

33 Estimate the number of passenger-vehicle shuttles used during September. Explain your answer.

Decorum Design

DDC is the favourite shop for trade and private buyers who want a new look for bathrooms, kitchens and bedrooms. Here are some of our items but come to the shop to see our full range.

■ WALLPAPER – rolls

width 53 cm, length 10 metres.

FLORAL DESIGN £5.49 per roll.

£4.97 each for 12 rolls or more

ANTIQUE EMBOSSED £7.99 per roll

WALLPAPER PASTE £4.99 – covers 10 sq metres

> TRY OUR ANTIQUE EMBOSSED PAPERS TO COVER THAT TATTY WALL

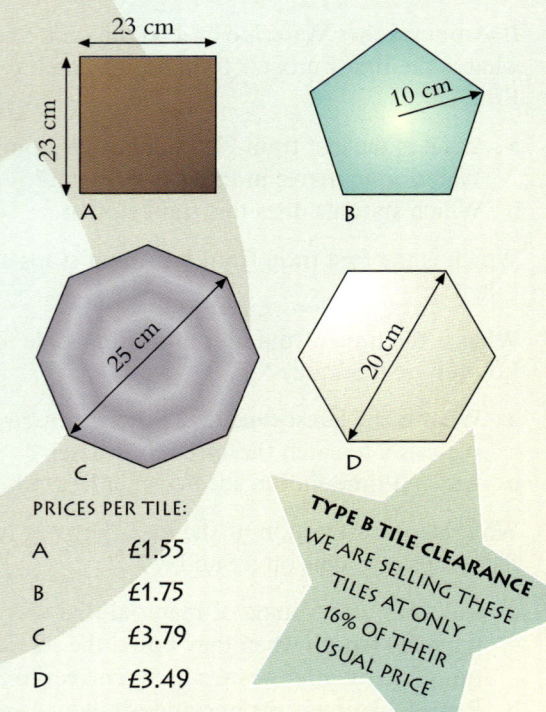

> ALL PAINT PRICES REDUCED BY 20% FOR NEXT THREE WEEKS

■ EMULSION PAINTS – Top quality own brand

BRILLIANT WHITE	1 litre	£3.42
	2 (1/2) litre	£8.45
	5 litre	£16.99
PASTEL SHADES	2 (1/2) litre	£11.99
	5 litre	£19.49

A litre tin will cover about 8 m² with a single coat. Two coats needed over very dark surfaces.

When calculating how much paint to order, do not subtract the area of doors and windows.

■ WALL TILES – imported Italian and French

JARDIN RANGE – Box of 10 tiles £1.79

ASSISI RANGE – Box of 10 tiles £1.99

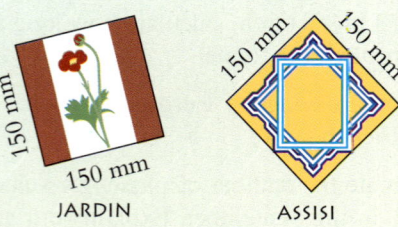

JARDIN   ASSISI

For those who have not caught up with metric units yet.
1 foot = 0.3048 metres   1 inch = 25.4 millimetres
All prices include VAT at 17.5%

TILE-FIX CEMENT – £7.99 a tub, covers 4 sq. metres

FORRET – Box of 10 wall tiles £2.14

BENETIA – Box of 10 tiles £2.14

200 mm
100 mm
FORRE

100 mm
200 mm
BENETIA

■ FLOOR TILES

QUARRY TILES – terracotta, 120mm x 120 mm, 31p each

CERAMIC REGULAR SHAPED TILES – choice of patterns

23 cm
23 cm
A

10 cm
B

25 cm
C

20 cm
D

PRICES PER TILE:

A	£1.55
B	£1.75
C	£3.79
D	£3.49

> TYPE B TILE CLEARANCE WE ARE SELLING THESE TILES AT ONLY 16% OF THEIR USUAL PRICE

We also sell small square tiles to fit with our type C tile. Pack of ten £5.75

TILE CEMENT – 12 kg bag – £17.89 – enough for 10 square metres of floor.

COLOURED GROUT £7.69 per tub – enough for 15 square metres of floor.

When calculating what to order, allow one complete tile for every part of a tile you need.

■ DESIGN SERVICE

We offer a free design service for bathrooms and kitchens. Just give us your plans to a scale of 1:50 and we will calculate how much paint, wallpaper or tiling you need.

1   What is the cost of a $2\frac{1}{2}$ litre can of pastel shade paint if you buy it within the next 3 weeks?

2   How much does it cost for 9 rolls of floral design wallpaper?

3   Spencer has decided to tile the wall above a bath with Jardin style tiles. He wants to cover an area 2 m by 60 cm.
   a   How many tiles will he need to buy?
   b   What will this cost him (including tile-fix)?

4   Sally decides to paper a wall in her bedroom with antique embossed paper. The wall is 3.1 metres long by 2.3 metres high.
   a   How many widths of antique paper are needed for the length of wall?
   b   How many rolls should she buy?
   c   What is the total cost of paper and paste?

5   What is the reduced price of one type B tile?

6   Mike uses Decorum's design service to calculate the number of quarry tiles he should buy for a floor which is 3.82 metres by 4.15 metres.
   a   Make a scale drawing of the floor to the correct scale.
   b   Calculate the number of tiles he needs.
   c   Decorum add 5% for tiles that might break. How many should Mike buy?

7   Kate's bedroom is cuboidal in shape.
   It is 2.84 metres long by 2.32 metres wide and 2.28 metres high. The door and windows occupy 2m².
   She wishes to paint all walls and the ceiling in brilliant white.
   a   Calculate the total area she wants to paint.
   b   Her room is dark blue now. What size tins of paint should she buy? Give your reasons.

8   For type D tiles:
   a   What is the mathematical name of the shape?
   b   What is the size of an internal angle?
   c   What is the length of a side?

9   Steve's floor is 3.95 metres by 2.37 metres and he wants to tile it with type A tiles. He works out that the area of the floor is 9.36 m² and that the area of one tile is 0.0529 m².
   He says the number of tiles he needs is 9.36 ÷ 0.0529 = 177 tiles.
   a   What is wrong with Steve's method?
   b   How many tiles does he really need?
   c   What is the cost including grout and cement?

10   This is a scale drawing (scale 1:50) of a wall.
   a   What is its true length in metres?
   b   What is its true height?

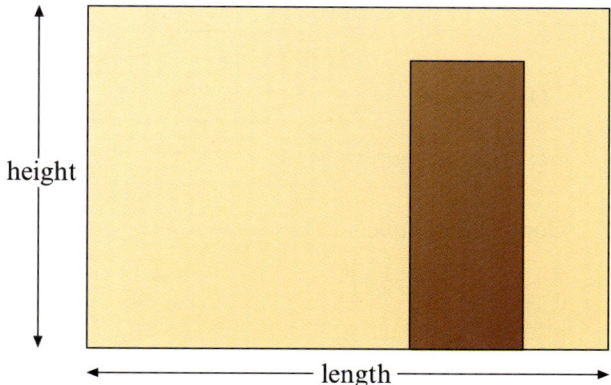

height

— length —

11   Which size tin of brilliant white paint is the best value for money? Explain your answer.

12   Shirley and Arthur measure their kitchen floor as 13 ft 6 inches by 11 ft 10 inches.
   a   Give these measurements in metres to the nearest centimetre?
   b   How much will they have to pay for tile cement and grout if they tile the floor?

13   Decorum are thinking of selling smaller bags of tile cement. They have chosen a 5 kg size as most useful. If they price it at the same unit cost as the 12 kg bag at what price will they sell it?
   Give your answer to the nearest penny.

14   An area 1.28 metres high by 2.32 metres wide is to be tiled with either Forret or Benetia tiles.
   Which tiles would it be cheaper to use and why?

15   Pragna runs her own business and can claim back the VAT she pays. She buys a 5 litre tin of pastel paint for £19.49 which includes VAT.
   a   What is the cost of this tin without VAT?
   b   How much VAT can she claim back?

16   Calculate the median price of a floor tile.

17   A new size of the tile-fix cement covers 15 m².
   This size is charged at the same rate as the tub that covers 4 m².
   Calculate the price of this new size tile-fix cement.
   Give your answer correct to the nearest penny.

18   On the first of next month all wallpaper prices must increase by 15%.
   Re-write the wallpaper data giving the new price.
   Give your answers correct to the nearest penny.

Number bites

1 Which of these calculations give a negative answer?

A $^-4 \times ^-3$ B $1.5 \times ^-8$ C $8 \div ^-2.4$

D $^-9 \div 3.7$ E $5.1 \times ^-1.1$ F $^-8 \div ^-7$

G $^-2.4 \times ^-3.8$ H $4.4 \div ^-1.9$

2 Copy and complete these calculations.

a $^-6 \times 3 = \square$ b $^-4 \times ^-2 = \square$ c $^-14 \div 7 = \square$
d $\square \div 6 = ^-3$ e $\square \times ^-4 = ^-20$ f $^-5 \times \square = 35$
g $10 \div ^-2 = \square$ h $8 \times ^-3 = \square$ i $^-15 \div ^-3 = \square$
j $\square \times 7 = ^-28$ k $\square \div ^-2 = 8$ l $^-36 \div \square = ^-4$

3 Find the values of each of these square roots to 2 dp.

a $\sqrt{73}$ b $\sqrt{94.36}$ c $\sqrt{6464}$ d $\sqrt{13.1044}$

4 Find the value of:

a 8^3 b 2.7^3 c 0.5^3 d $^-4^3$ e $^-1.8^3$ f $^-0.3^3$

g $\sqrt[3]{27000}$ h $\sqrt[3]{729}$ i $\sqrt[3]{1}$ j $\sqrt[3]{1.728}$

5 Find the reciprocal of:

a 2   b 10   c 0.4   d 0.6   e $0.\dot{3}$
f 0.16   g 0.01   h 5   i 12   j $0.\dot{2}$
k 11   l $0.\dot{2}\dot{7}$

6 Write as a fraction the reciprocal of:

a $\frac{2}{3}$ b $\frac{4}{5}$ c $\frac{11}{6}$ d $\frac{13}{8}$

e $1\frac{1}{3}$ f $2\frac{2}{5}$ g 1.4   h 1.6

7

5^{-3}	5^{-1}	5^0		5^2	
$\frac{1}{25}$			5		125

Copy and complete this powers of 5 table.

8 Make a powers of 6 table from 6^{-3} to 6^3.

9

$7^0 = 0$ ✗

Explain why this is wrong.

10 Write these as fractions.

a 4^{-1} b 2^{-3} c 3^{-2} d 6^{-2} e 10^{-3} f 10^{-4}
g 6^0 h 7^{-2} i 8^{-2} j 4^{-3} k 3^{-4} l 10^0

11 Calculate:

a 297×3^{-2} b 400×2^{-5} c 63×4^{-1}
d 2.7×10^{-2} e $13 \div 2^{-3}$ f $7.2 \div 5^{-2}$
g $0.9 \div 3^{-4}$ h $62 \div 10^{-2}$

12 Write these as decimals.

a 5^{-1} b 2^{-2} c 8^{-1} d 10^{-2} e 4^{-3}
f 3^{-2} g 8^{-2} h 5^{-3} i 10^{-3} j 100^{-1}
k 6^{-2} l 10^{-5}

13 Write these numbers in standard form.

a 6170 000   b 92 000 000 000   c 307 000
d 0.000 025   e 0.000 002 603   f 0.000 0001

14 Write these as ordinary numbers.

a 4.5×10^8 b 3.606×10^{12} c 1.24×10^{-9}
d 7×10^7 e 5.1×10^{-11} f 6.1047×10^{-6}

15 Give the answer to each of these in standard form.

a $75 \times (1.74 \times 10^9)$ b $144 \times (8.6 \times 10^{-6})$
c $(6.09 \times 10^{-7}) \div 7$ d $(4.446 \times 10^{12}) \div 52$
e $(3.2 \times 10^8) \times (5.95 \times 10^{-2})$
f $(2.92 \times 10^{-6}) \div (4 \times 10^7)$
g $(4.7 \times 10^{14}) + (3 \times 10^{15})$ h $(4.5 \times 10^9) - (4 \times 10^8)$

16 Give the answer to these using index notation.

a $4^{-7} \times 4^3$ b $3^8 \times 3^{-2}$ c $5^6 \div 5^2$ d $2^5 \div 2^{-3}$
e $5^0 \times 5^7$ f $7^{-4} \times 7^{-2}$ g $4^{-6} \div 4^3$ h $6^{-3} \div 6^{-8}$

17 Copy and complete these calculations.

a $3^8 \times 3^\square = 3^6$ b $6^\square \div 6^3 = 6^2$
c $2^{-4} \div 2^\square = 2^{-7}$ d $8^\square \times 8^{-3} = 8^2$
e $3^\square \div 3^{-7} = 3^5$ f $7^{-5} \times 7^\square = 7^{-8}$

18 Without using a calculator, give the answer to these in standard form.

a $(6 \times 10^7) \div (3 \times 10^4)$ b $(5 \times 10^{-2}) \times (8 \times 10^6)$
c $(2 \times 10^{-3}) \times (3.5 \times 10^{-5})$ d $(4 \times 10^{-2}) \div (8 \times 10^4)$

19

$p = \frac{1}{3}$ $q = \frac{3}{4}$ $r = \frac{2}{5}$ $s = \frac{5}{6}$

$w = 1\frac{2}{3}$ $x = 2\frac{1}{2}$ $y = 1\frac{4}{5}$ $z = 3\frac{1}{6}$

Evaluate:

a $p+r$ b $r+s$ c $q+s$ d $r-p$ e $q-r$
f $s-p$ g $w+z$ h $s+w$ i $x+y$ j $y-w$

20 Evaluate:

a pr b pq c rs d $\frac{r}{q}$ e $\frac{p}{s}$ f $\frac{s}{q}$

g ry h sw i xy j $\frac{q}{x}$ k $\frac{z}{w}$ l $\frac{z}{s}$

21 Evaluate:

a $3q$ b $8s$ c $12p$ d $\frac{20}{s}$ e $\frac{12}{q}$ f $\frac{3}{r}$

g $2w$ h $12z$ i $4y$ j $\frac{5}{w}$ k $\frac{36}{y}$ l $\frac{16}{x}$

22 Find the LCM of:

a 4 and 9   b 6 and 8   c 16 and 120
d 126 and 300   e 14, 20 and 175

23 Find the HCF of:

a 45 and 81   b 42 and 154   c 90 and 175
d 72 and 132   e 84, 105 and 175

Sequences

1 Each diagram shows the first three patterns in a sequence.

Sequence W

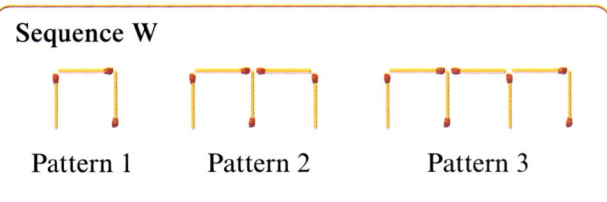

Pattern 1   Pattern 2   Pattern 3

Sequence X

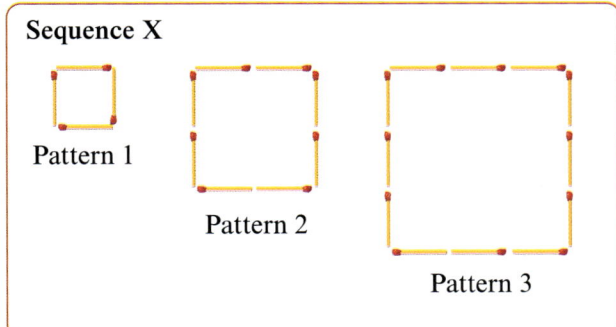

Pattern 1

Pattern 2

Pattern 3

Sequence Y

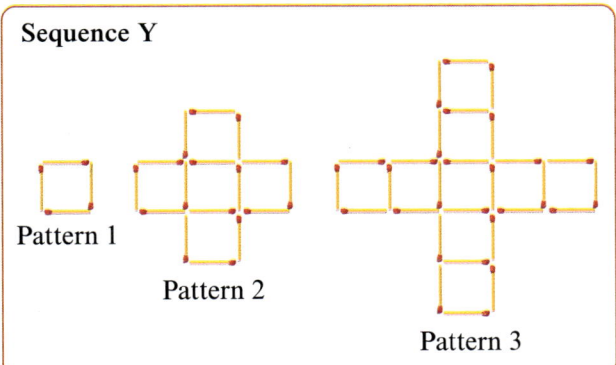

Pattern 1

Pattern 2

Pattern 3

For each sequence:

a Draw the 4th pattern.

b Find a rule for the number of matches (m) in the nth pattern.
Write it in the form $m = \ldots$.

c Show how you found your rule.

d Use your rule to calculate the number of matches in the 10th pattern.

2 Copy and complete each mapping diagram.

a
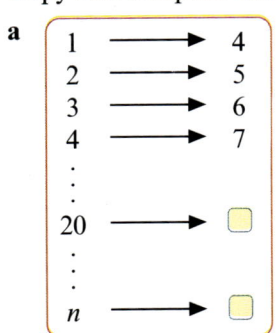

b
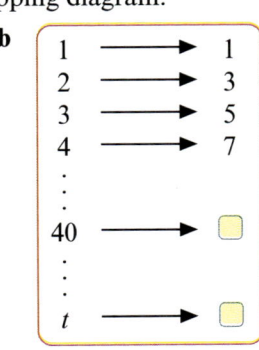

c
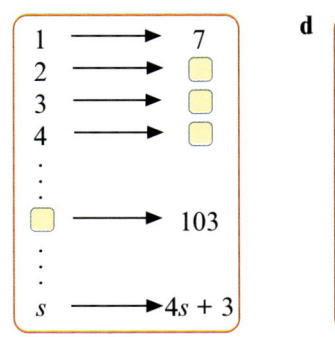

d
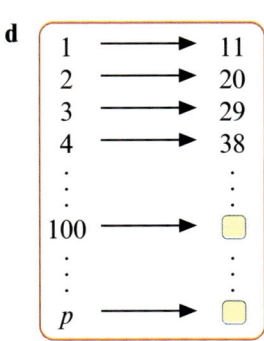

3 These are the first three patterns in a sequence.

Sequence Z

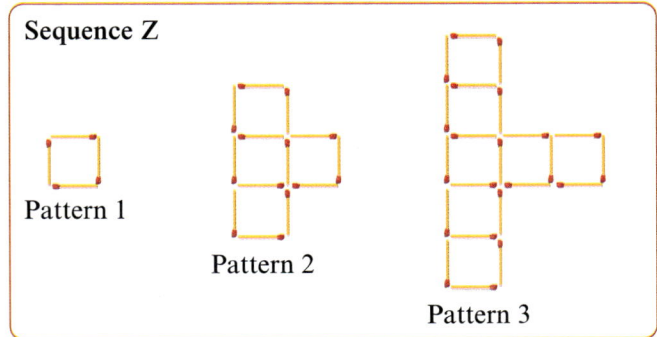

Pattern 1

Pattern 2

Pattern 3

a Find a rule for the number of matches in the nth pattern in the form $n \longrightarrow \ldots$.

b Show how you found your rule.

4 Copy and complete this mapping diagram.

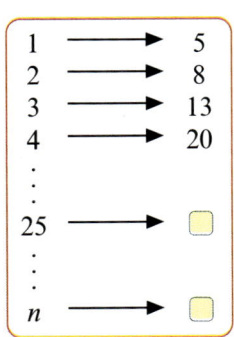

5 A 1, 4, 7, 10, 13, …   B 5, 8, 11, 13, …
C 7, 13, 19, 25, 31, …   D 4, 9, 14, 19, …
E 7, 12, 17, 22, 27, …   F 2, 6, 10, 14, 18, …
G 3, 9, 15, 21, …   H 9, 17, 25, 33, 41, …

For each of the sequences A to H:

a Find an expression for the nth term.

b Use your expression to find the 30th term.

6 Multiply out the brackets from:

a $5(p + 2)$ **b** $8(n - 3)$ **c** $4(c + 5)$
d $2(3r + 1)$ **e** $5(7 + x)$ **f** $7(2s - 3)$
g $3(1 - 5t)$ **h** $2(3m - 6)$ **i** $20(5k - 1)$
j $8(4 - z)$ **k** $10(5 - 2t)$ **l** $4(a + b)$
m $3(2g + h)$ **n** $5(x - y)$ **o** $7(3c - 2d)$

Properties of shapes

1

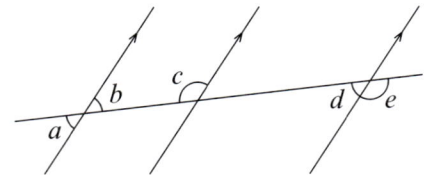

a If $a = 47°$, calculate angles b to e.
b If $a = 51.5°$, calculate the angles b to e.
c If $e = 169°$, calculate angles a to d.

2 These polygons are drawn on a grid
of parallel lines. The diagonals of PQRS
are marked in red.

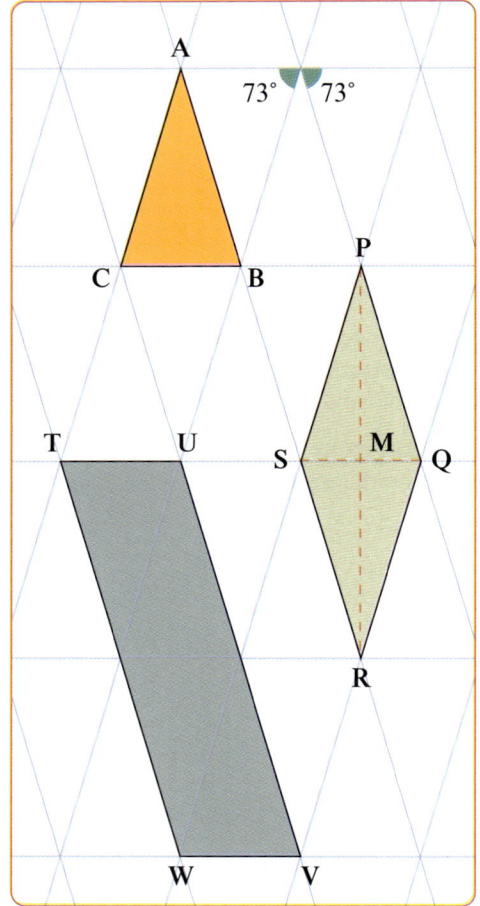

a Work out each interior angle of ΔABC.
b What type of triangle is ABC?
c Work out each interior angle of:
   i PQRS   **ii** TUVW.
d Explain why PQRS is a rhombus.
e Explain why TUVW cannot be a rhombus.
f The diagonals of PQRS intersect at M.
   Work out each of these angles:

   i PM̂Q   **ii** PQ̂M   **iii** MQ̂R
   iv QR̂M   **v** SR̂M

3 **a** Explain what the angle sum of the exterior
angles of a polygon is.

b Explain how you would prove your answer to
part **a**.

4 Explain why the exterior angle of a regular
octagon is 45°. Use a diagram in your answer.

5 Calculate the angles a to m in the diagram below.

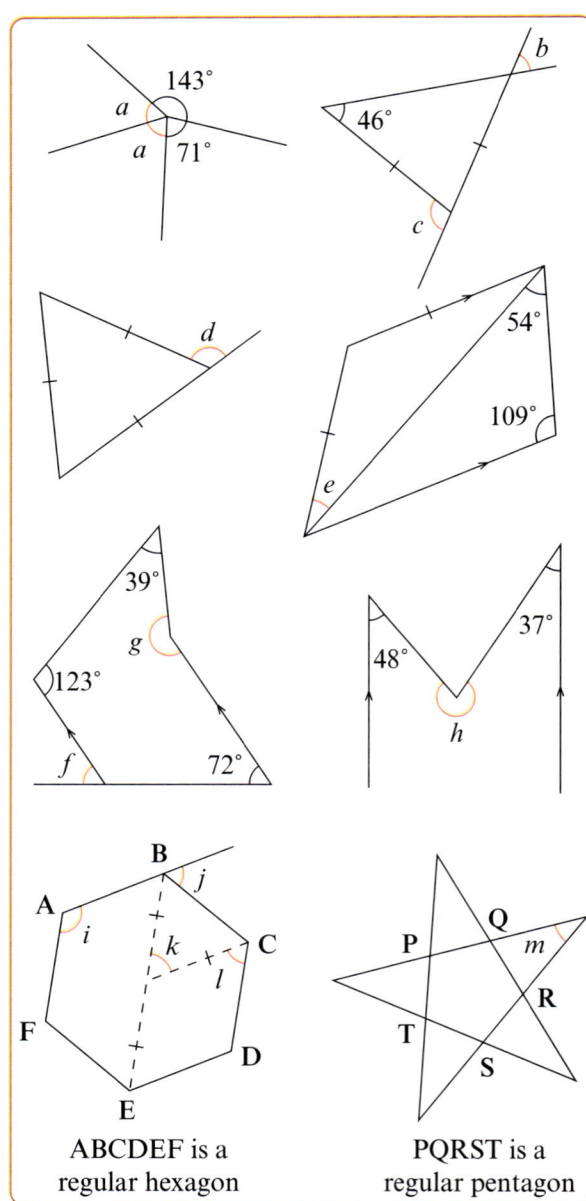

ABCDEF is a
regular hexagon

PQRST is a
regular pentagon

Linear graphs

1 Give the gradient of each of these lines.

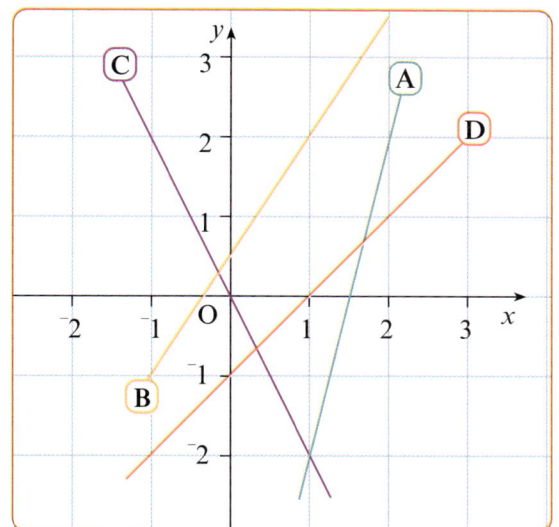

2 On a grid, draw lines with each of these gradients.

a 3 **b** $\frac{1}{2}$ **c** ‾2 **d** $\frac{3}{4}$

e 1 **f** $\frac{-1}{2}$ **g** $\frac{2}{3}$ **h** 4

i $\frac{1}{4}$ **j** $\frac{-2}{3}$ **k** $\frac{1}{5}$ **l** 5

3 Each of these is the equation of a linear graph:

$y = 2x + 1$ $y = 3x - 2$ $y = x - 2$
$y = 2x - 3$ $y = x + 2$ $y = x$

For each equation:

a Draw up a table of values.
Use values of x from ‾1 to 4.

b List the coordinates of points on the line.

c Draw the graph of the equation.

4 For each of these lines give the gradient and the y-intercept.

a $y = 4x - 3$ **b** $y = 1 + 2x$ **c** $y = 2 - 3x$
d $y = x$ **e** $y = ‾x$ **f** $y = 2 + 5x$
g $y = ‾1 - 2x$ **h** $y = x + 1$ **i** $y = 1.5x + 2$
j $y = 0.5 + x$ **k** $y = 3 - 0.5x$ **l** $y = 1$
m $y = \frac{1}{2}x$ **n** $y = \frac{3}{4}x - 1$ **o** $y = x - \frac{3}{5}$
p $y = 1.6 - x$ **q** $y = x + 1.2$ **r** $y = ‾3x$

5 Here is a description of five different lines.

Line A: has a gradient of 2 and a y-intercept of ‾8
Line B: gradient 3, y-intercept 4
Line C: gradient $\frac{3}{4}$, y-intercept ‾1
Line D: y-intercept 3, gradient ‾5
Line E: gradient ‾7

Give an equation for each of these lines.

6 For each of these lines give the gradient of a line at right angles to it.

a $y = 2x + 3$ **b** $y = 3x - 4$
c $y = 4 - 3x$ **d** $y = x + 1$
e $y = 5 - \frac{1}{2}x$ **f** $y = 5x + 1$

7 Rewrite each of these equations to read $y = \ldots$

a $2y = 4x - 8$ **b** $3y = 6x + 9$
c $5y = 10x - 15$ **d** $4y = 2 + 8x$
e $3y = 2x + 6$ **f** $2y = 5x + 7$
g $4y = 3x + 1$ **h** $3y = 6x - 3$

8 Write three different equations that can be rewritten as $y = 3 - 5x$.

9 Write three different equations that can be rewritten as $y = x + 0.5$.

10 For each of these equations draw a graph.

a $y = 2x$ **b** $y = 3x - 5$ **c** $y = \frac{1}{2}x$
d $2y = 3x - 4$ **e** $4y = 5x$ **f** $3y = 3x$
g $2y = 4 - 3x$ **h** $2y = 6$ **i** $5y = x$
j $5y = ‾x$ **k** $3y = 4x + 6$ **l** $y = 0.25x$

11

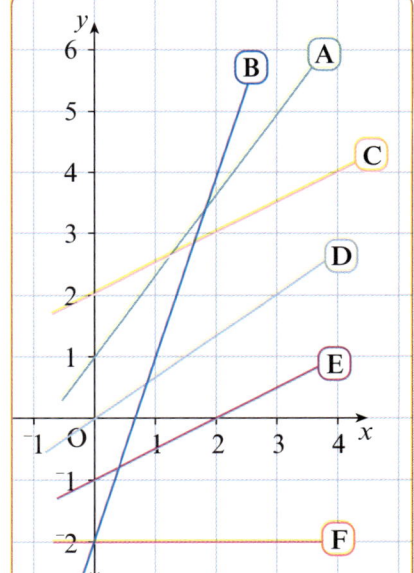

Give an equation for each of the lines A to F.

12 Without drawing a graph, which of these lines is drawn through the point (3, 2)?

a $y = x - 1$ **b** $3y = 7x - 15$
c $y = 2$ **d** $3y = x + 3$
e $3y = 2x$ **f** $y = 2x - 4$
g $3y = 12 - 2x$ **h** $y = 5 - 2x$

13 Without drawing a graph, give the equation of three different lines that are drawn through the point (1, 3).

Comparing data

1

CHILDREN IN CARE

Expenditure (1994–95)

	£m
Child Care	18.5
Fundraising	7.0
Administration	3.0
Other costs	1.5

CHILD SAFETY

Expenditure (1994–95)

	%
Child Care	65
Fundraising	20
Administration	10
Other costs	5

Draw a pie chart for each charity to show how their money is spent.

2

Chris

Ring score	6	7	8	9	10
Frequency	1	4	2	1	4

Dani

Ring score	3	4	5	6	7	8	9
Frequency	2	1	0	3	4	2	3

Sam

Ring score	1	2	3	4	5	6
Frequency	4	3	1	0	3	5

Find the median score for each archer by listing the data.

3

Bella

Ring score	2	3	4	5	6	7	8	Total
Frequency	3	4	7	6	11	6	6	43

Emily

Ring score	5	6	7	8	9	Total
Frequency	12	19	17	9	7	64

Gavin

Ring score	3	4	5	6	7	8	9	10
Frequency	4	2	5	7	8	11	9	6

For each distribution:

  a construct a cumulative frequency table
  b find the median score.

4 Calculate the mean score for each archer. Give your answers to 1 dp.

5

Asif

Ring score	3	4	5	6	7	8	9
Frequency	1	3	2	2	3	4	1

Suki

Ring score	4	5	6	7	8
Frequency	3	6	2	1	3

Compare these two distributions using the mode and the range.

6

Kurt

Ring score	3	4	5	6	7	8	9
Frequency	1	5	3	1	1	2	4

Jean

Ring score	4	5	6	7	8	9
Frequency	5	8	15	13	10	9

Compare these distributions using the median and the range.

7

Ring score	3	4	5	6	7	8	9	10
Frequency	1	7	10	10	5	4	1	2

Sally

Gavin

Ring score	2	3	4	5	6	7	8	9
Frequency	1	2	3	7	14	13	6	7

Compare these distributions using the mean and the range.

8

Peggy

Ring score	2	3	4	5	6	7	8
Frequency	1	0	2	3	5	4	1

Toby

Ring score	3	4	5	6	7	8	9
Frequency	2	4	5	4	6	0	1

Sue

Ring score	2	3	4	5	6	7	8	9
Frequency	2	1	3	2	3	6	4	3

  a For each distribution of scores, find the mean, median and range
  b Compare the distributions.

9

	Number of chips								
Restaurant	34	35	36	37	38	39	40	41	Total
P	8	10	14	9	6	5	4	4	60
Q	8	7	5	9	12	11	8	0	60
R	0	3	5	10	12	14	10	6	60

For each restaurant:

  a find the median number of chips
  b find the lower and upper quartiles
  c draw a box-and-whisker plot.

Working in 2D

1   Calculate the area of each shape below.

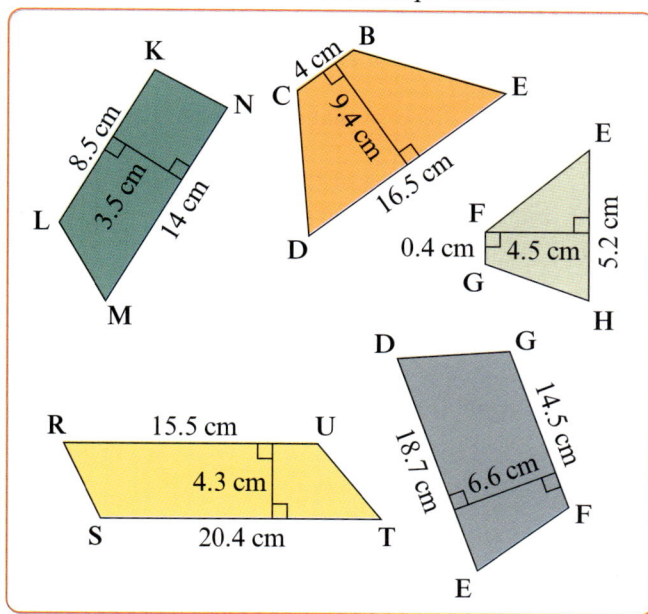

2   This is a sketch and a net of a magazine file.

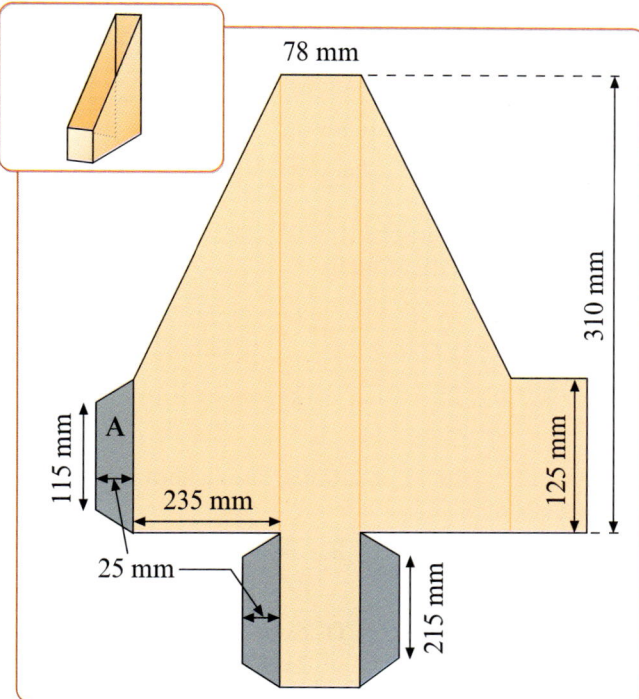

a   Calculate the area of glue tab A.
b   Calculate the total area of the net.

The net is cut from a rectangle of card which measures 550 mm by 575 mm.

c   Calculate the area of card wasted when the net is cut out.

3   Calculate the diameter of a circle that has the same perimeter as a 9 cm by 7 cm rectangle.

4   Calculate the perimeter and area of:
a   a circle of diameter 6.5 cm
b   a circle of radius 4.8 cm
c   a semicircle of diameter 12.4 cm.

5   Calculate the area of the shaded part.
Give your answer in terms of π.

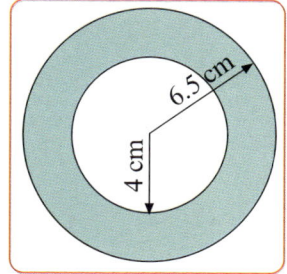

6   The design of this logo uses three shapes.

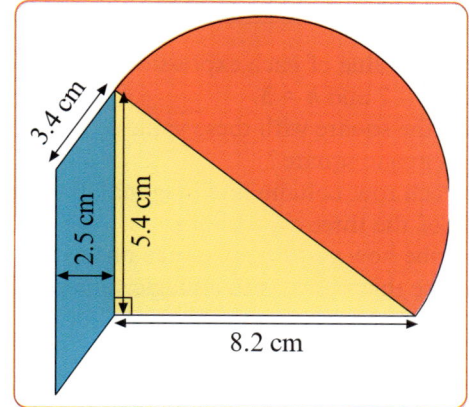

a   Calculate the total area of the logo.
b   Calculate the perimeter of the logo.

7   This diagram shows a plastic sail for a toy boat.

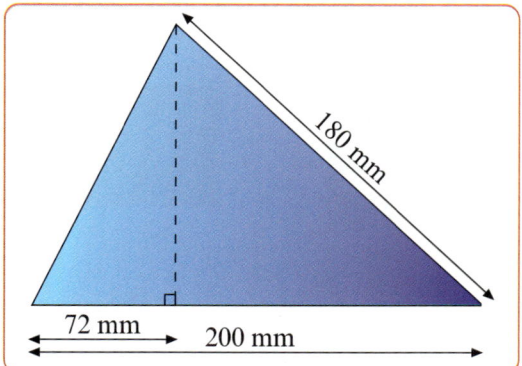

a   Calculate the height of the plastic sail.
b   Calculate the area of a sail.

Sails are cut from a roll of plastic 205 mm wide and 50 metres long.

c   How many sails can be cut from one roll?
d   Calculate the area of waste from one roll.

Solving equations

1 Simplify each of these expressions.

    a $7k + 8k - 9k$ **b** $2l + 3m + 4l - 2m$

    c $4n - 3p + n - 4p$ **d** $3q + 5q - 4q + 6r$

    e $5s - 7t - 4s + 7t$ **f** $6 + 5u + 3 + 7v$

    g $8w - 7x - 2w + 9x$ **h** $10 - 5y + 1 + 2y$

    i $3z + 5 + 6 - 8z - 9 + 10z$

2 Some expressions are arranged in a square.

$4g + h$	$3g - h$	$5g$
$5g - h$	$4g$	$3g + h$
$3g$	$5g + 2h$	$4g - h$

    a Find the value of each expression in the square when $g = 7$ and $h = 3$.

    b Draw the square with these values.

    c Is it a magic square?

    d For each row, column and diagonal, find the total of the three expressions.

    e Describe how you could change one expression to make this square into a magic square.

3 Solve these equations.

    a $7x + 1 = 3x + 21$ **b** $3x - 2 = x + 7$

    c $5(x + 1) = 24$ **d** $4(x + 3) + 9 = 13$

    e $3x - 2 = 2(x + 1)$ **f** $4x + 5 = 2x - 1$

    g $6x - 8 = 10 - 3x$ **h** $13 - 4x = 2x + 4$

    i $11 - x = 3(x - 1)$ **j** $2x + 11 = 4 - 5x$

    k $21 - 3x = 15 - 2x$ **l** $3x - 10 = 10 - x$

    m $4x + 23 = 2(4 - x)$ **n** $4(x - 1) = 2(x + 7)$

    o $3(5 - 2x) = 7(x + 4)$ **p** $5(10 - x) = 4(3x + 4)$

4

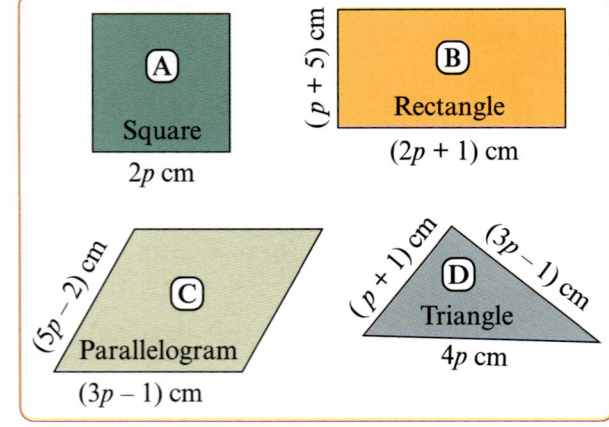

    a Find the perimeter of each shape when $p = 7$.

    b For each shape, write an expression for the perimeter in terms of p.

    c Find a value of p that gives shape D a perimeter of 24 cm.

    d Which value of p gives shape C a perimeter of 34 cm?

    e Find a value of p that gives shape B a perimeter of 21 cm.

    f Find a value of p so that the perimeters of shapes A and B are equal.

    g Find a value of p so that the perimeters of shapes A and C are equal.

    h Find a value of p so that the perimeters of shapes B and C are equal.

    i Explain why the perimeters of shapes A and D are equal for any value of p.

5 For each puzzle A and B, write an equation and solve it to find the number.

A I think of a number, double it and subtract 3. I get the same answer if I subtract my number from 12. What is my number?

B I think of a number, add 1 and multiply by 2. I get the same answer if I subtract 5 and multiply by 3. What is my number?

6 For each pair of equations:

    a On one set of axes, draw a graph for each equation.

    b Use your graphs to find values for x and y that fit both equations.

    A $y = x - 6$ **B** $y = 2x$
        $x + y = 8$ $y = 4x - 3$

    C $y = 2x + 9$ **D** $y = 6x - 10$
        $y = 3 - x$ $y - x = 4$

7 For each pair of equations, use algebra to find the values of x and y.

    A $4x + y = 9$ **B** $3x + y = 37$
       $2x + y = 6$ $2x + 5y = 29$

    C $3x + 5y = 20$ **D** $5x + 4y = 5$
       $2x + 3y = 14$ $2x + 6y = 13$

    E $3x + y = 18$ **F** $2x - 3y = 3$
       $11x - y = 10$ $6x + y = 29$

    G $3x - y = 16$ **H** $3x - 2y = 1$
       $x + 2y = 3$ $2x + 5y = 7$

    I $5x - 2y = 10$ **J** $2x - y = 7$
       $4x - 2y = 7$ $9x - 3y = 24$

Estimation and approximation

1 Round each number to the degree of accuracy given:

a	5.674	(2 dp)
b	12.652	(1 dp)
c	2143	(nearest ten)
d	534.687	(2 dp)
e	34.648	(nearest whole number)
f	2639.2	(nearest thousand)
g	13 468.284 52	(2 dp)
h	0.0666	(2 dp)
i	63.68	(nearest ten)
j	59.999	(1 dp)
k	8502	(nearest thousand)
l	68.499 99	(nearest whole number)
m	56.289 64	(3 dp)

2 Round each of these to the number of significant figures given.

a	56.83	(3 sf)
b	16 389	(3 sf)
c	2.456	(2 sf)
d	45.923	(1 sf)
e	15.777 77	(4 sf)
f	725 184	(2 sf)
g	94.56	(1 sf)
h	93 747 656	(5 sf)
i	564.23	(4 sf)
j	196.5	(1 sf)
k	6.7849	(2 sf)
l	15.682	(4 sf)
m	673 492.35	(4 sf)
n	0.035 62	(2 sf)
o	0.027 95	(3 sf)
p	3.0004	(3 sf)

3 By approximating each number to 1 sf work out approximate answers to each of these.

a 84.3×452.53
b 4.876×37.71
c 5683.2×9.372
d 458.12×518
e $734.6 \div 2.316$
f $56.8243 \div 7.8931$
g $41 952 + 77 442$
h $34.6296 + 87.3$
i $6834 - 1939.453$
j $45.95 \times 2.943 56$
k 74.68×4.87
l 7468×0.487
m 7.468×48.7

4 Work out approximate answers then calculate each of these exactly.

a 56.6×21.5
b 4924×3.7
c 246.3×9.67
d 54.27×21.6
e 17.63×1839
f 3.45×0.054
g 0.42×264
h 38.6×25.002

5 For each of these rectangles, estimate its area. Explain how you calculated your estimate.

a
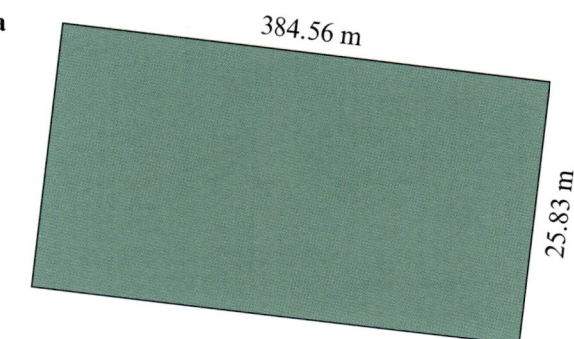
384.56 m
25.83 m

b
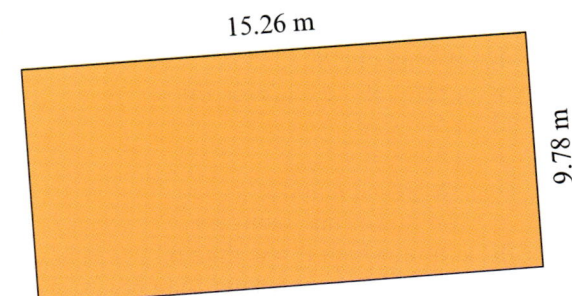
15.26 m
9.78 m

c
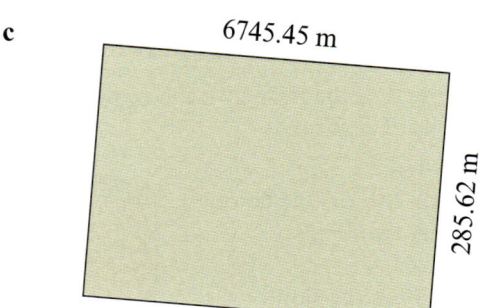
6745.45 m
285.62 m

Probability A

1 With a 1 to 6 dice what is the probability that you score:

    a 5   **b** an even number
    c a multiple of 3   **d** a prime number?

2 For this wheel, what is the probability that the pointer stops on:

    a B   **b** A
    c either A or D
    d C   **e** not C
    f N   **g** A or B

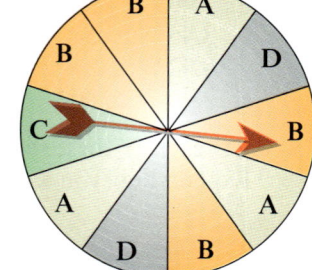

3 For the wheel give an event which has a probability of $\frac{3}{5}$.

4

Draw a sample space diagram to show the outcomes when rolling a 1 to 4 dice and a 1 to 8 dice together.

5 From your sample space diagram, give the probability of getting:

    a at least one 3
    b at least one 6
    c a total of 7 by adding the scores
    d a 5 and a 2
    e a 3 and a 4
    f two prime numbers
    g two non-prime numbers
    h a total which is less than 6
    i a total which is greater than 6
    j a total which is a multiple of 3
    k two numbers the same
    l two numbers which are different.

6 A cube dice has three red faces and three blue faces. Draw a tree diagram to show three rolls of the dice.

7 From your tree diagram give the probability in three rolls of getting:

    a three reds
    b exactly two of one colour
    c at least two blues
    d no blues
    e all colours the same
    f no colours the same.

8 This tree diagram is for two spinners. The outcomes are not all equally likely.

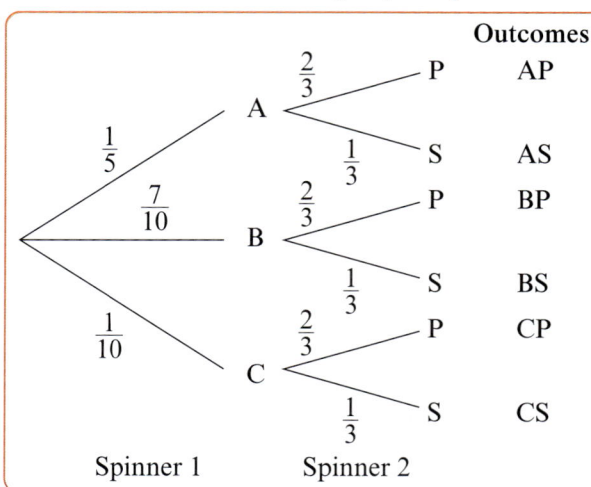

By multiplying probabilities give the probability of:

    a B and P   **b** C and S
    c A and P   **d** B and S
    e A and S   **f** C and P

9 Four CDs P to S are stacked in random order.

    a List all the different arrangements that are possible.
    b Give the probability that the CD at the bottom is red.
    c What is the probability that the top and bottom CDs are blue?
    d What is the probability that there is a blue CD at the top?
    e Give the probability that the two blue CDs are next to each other.

10 Five mugs are hanging from hooks. A pair of mugs are chosen at random.

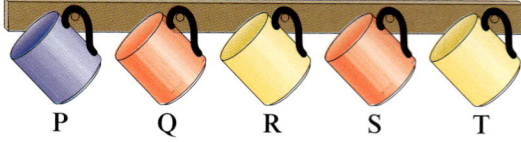

    a List every pair it is possible to pick.
    b What is the probability of picking:
        i a pair of the same colour
        ii a pair where only one mug is red
        iii a pair of yellow mugs?

Using algebra A

1 This shape is cut from a rectangle.

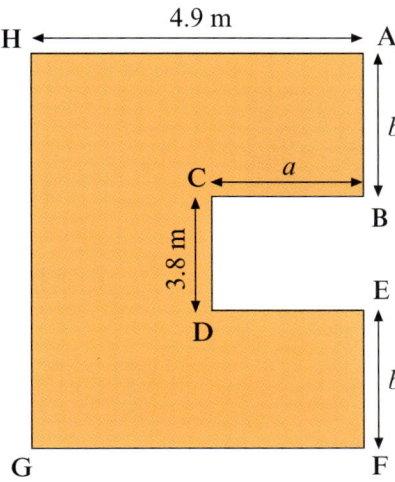

The length of AB is b metres and BC is a metres.
a Write an expression for the length of GH.
b Write an expression for the perimeter of this shape in terms of a and b.
c What is the perimeter if $a = 1.8$ and $b = 4.1$?

2 The dimensions of this trapezium are in centimetres.

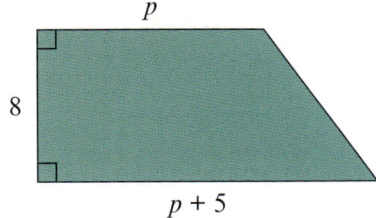

a Write an expression for the area of the trapezium.
b What is the area when $p = 6.4$?
c What value of p gives an area of $54\,cm^2$?

3 Find three pairs of equivalent expressions.

A $2(2a + 4)$ B $4a + 4$ C $2a + 6$

D $4(a + 2)$ E $2a + 8$ F $2(a + 2)$

G $2a + 5$ H $2(a + 3)$ I $4(a + 1)$

4 Multiply out these brackets.
a $3(a + 6)$ **b** $5(6 + z)$ **c** $3(n - 4)$
d $3(2n + 4)$ **e** $x(x + 2)$ **f** $4p(p - 3)$
g $3b(3 + b)$ **h** $a(2a - 8)$ **i** $4p(3p - 4)$

5 The length of a rectangle is 8 cm greater than its width.

a If the rectangle is w centimetres wide, write an expression for the length of the rectangle.
b Write an expression for the perimeter of the rectangle in terms of w.
c What value of w gives a perimeter of 72 cm?

6 The length of a rectangle is three times its width. The perimeter is 80 cm. What is the area?

7 Multiply these terms.
a $8p \times 6q$ **b** $9y \times y$ **c** $mn \times mn$
d $pq \times 4p$ **e** $8m \times mn$ **f** $9b \times 2a^2$
g $a^3 \times g^3$ **h** $4p \times 5p^2$ **i** $7a^3 \times 4b$

8 Multiply out these.
a $6(4a - 3b)$ **b** $p(n - p)$ **c** $s(s + t)$
d $6m(n - m)$ **e** $5x(x + y)$ **f** $u(4u + 3)$
g $p(5q + 3r)$ **h** $3n(4m + 7n)$ **i** $9a(3b + a)$

9 Which of these expressions is equivalent to $4a(2ab + 3a) + a(5b - a) + 4b(2a^2 + 5a)$?

A $35a^2b + 11a^2$

B $12a^2b + 14ab + 7a^2$

C $41ab + 10a$

D $16a^2b + 11a^2 + 25ab$

10 Simplify these.
a $8n + 4m - 6n + 3m$ **b** $12a^2 - a + 9a^2$
c $b^3 - b^3 + 2b$ **d** $8x - 4xy - 2y + 6xy$
e $p + 8q - 5q + 3p$ **f** $8p^2 + 3pq + 5q^2 - pq$
g $4m^2 + mn - 6n + nm$ **h** $x + 2xy + x - 4xy$

11 Multiply out these brackets and simplify.
a $6(7a - 2b) + 7(2a + b)$
b $9(3x + y) + 4(y - 2x)$
c $x(6x - 2) + 2x(5x + 3)$
d $8x(2y + 3x) + x(6x - 3y)$

12 Write an expression for the width of rectangles A and B.

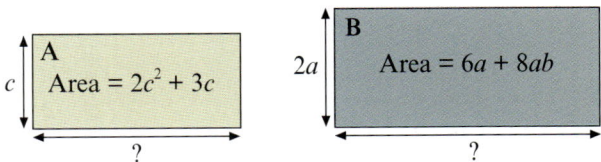

13 Factorise these fully.
a $7p + 3pq$ **b** $m + 3mn$ **c** $8pq + 4q$
d $3y - 12xy$ **e** $2b^2 + 10b$ **f** $18ab + 24bc$
g $x^2y - 2xy$ **h** $9a^2b + 6ab^2$ **i** $12m^2n - 9mn$

Constructions and Loci

1 Construct these triangles accurately.
Show all your construction lines.

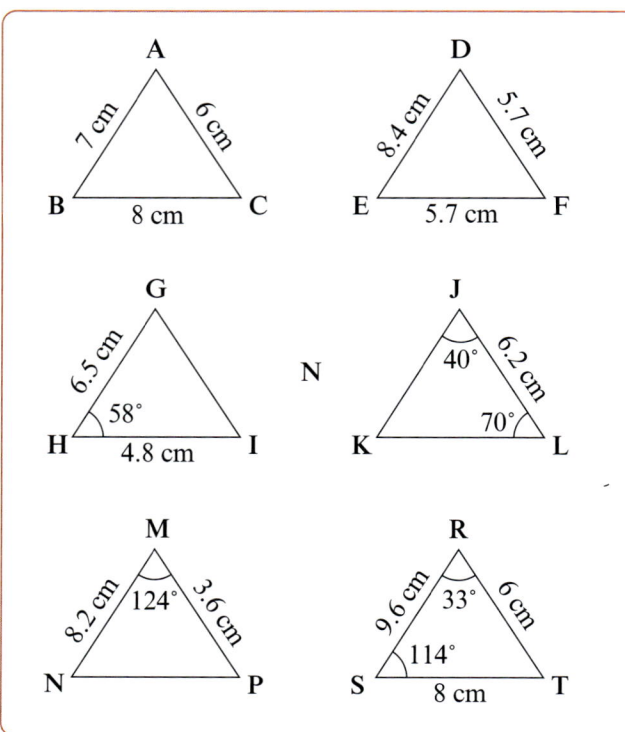

2 Copy each line full size and contruct a
perpendicular bisector of it.

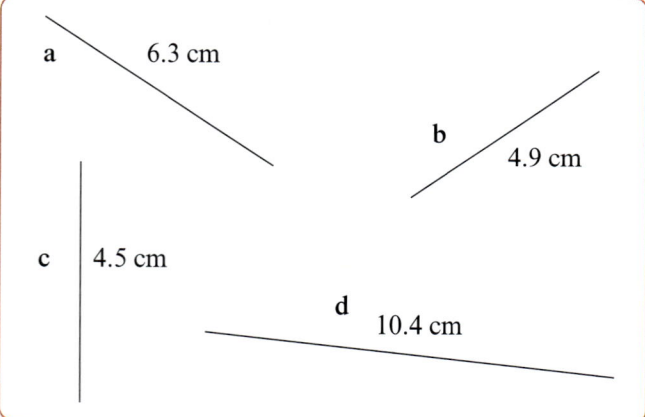

3 Draw each angle accurately and construct its
bisector.

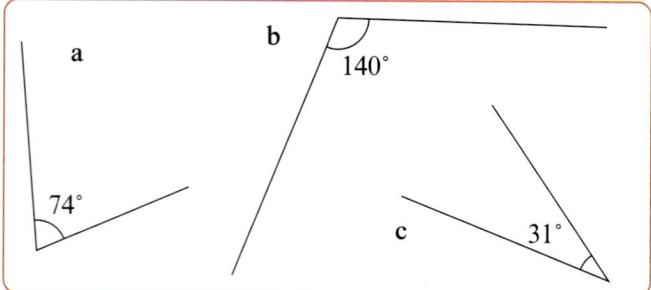

4 Draw each of the following shapes full size and
show the locus of points 2 cm from each one.

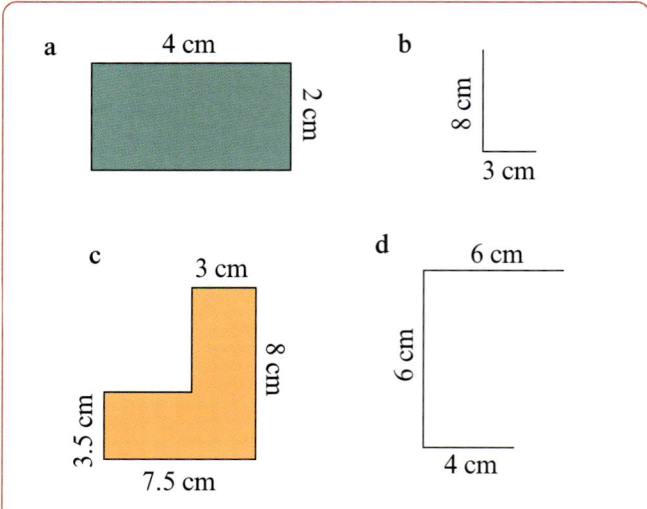

5 Explain how you would construct a perpendicular
from a point to a line, with compasses and a ruler.
Use a diagram in your answer.

6 A field for the village fete is shaped as a triangle
with these dimensions.

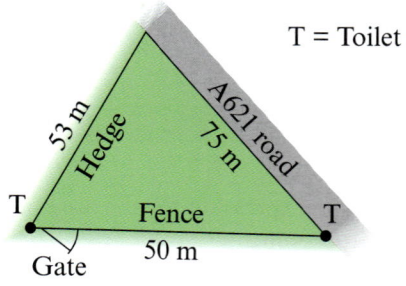

Colonel Briggs Shilton has set out these conditions
for the position of the drinks tent.

> **Condition 1** – The tent must be equidistant
> from the hedge and the fence.
> **Condition 2** – The distance from each toilet to
> the tent must be equal.

a Construct a scale drawing of the field to a
scale of 1 to 1000.
b Find by construction the position of the tent.
c Why is this position not a sensible one?
d Give some different conditions which you think
puts the tent in a better position.

Ratio

1 Green dyes are a mix of blue and yellow dyes.

Copy the table below and calculate the missing amounts.

Dye	Blue	:	Yellow
Grass	1	:	3
Lime	3	:	7
Pea	5	:	9

Dye	Blue	Yellow	Total
Grass	120 ml		
Lime	450 ml		
Pea		360 ml	
Lime		175 ml	
Pea	600 ml		
Grass		510 ml	
Lime	645 ml		
Pea		495 ml	
Grass		720 ml	
Lime			1200 ml
Grass			860 ml
Pea			700 ml
Grass			1280 ml
Lime			890 ml
Pea			1050 ml

2 Share each of these amounts in the given ratio.

a £84 2:5 **b** £248 5:3 **c** 195 g 9:4
d £120 2:1:3 **e** 200 g 3:4:1 **f** 360 ml 2:3:4
g 420 g 2:3:2 **h** £495 1:1:7 **i** 99 cm 4:5:2
j 2 m 5:2:1 **k** 1.4 kg 1:2:4 **l** 2.7 cm 1:6:2

3 This recipe makes 25 biscuits.

> **Chocolate Biscuits**
> 50 g icing sugar • 225 g margarine
> 100 g plain chocolate • 225 g flour

Calculate how much of each ingredient is needed to make:
a 45 biscuits   **b** 10 biscuits

4

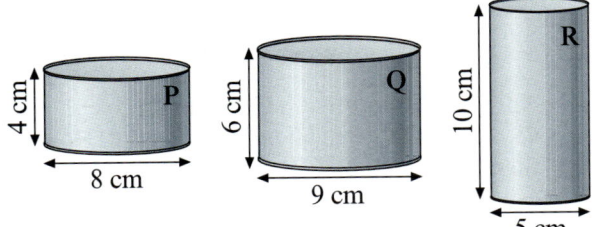

For each tin, give the ratio of height to diameter in its simplest terms.

5 A 2:5   B 3:1   C 4:3   D 1:3   E 3:5

These are the ratios of men to women in five self-defence classes.
What fraction of each class are men?

6

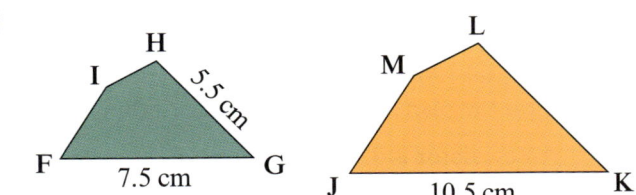

Shape JKLM is an enlargement of shape FGHI.
a Calculate the scale factor of the enlargement.
b Find the length LK.

7

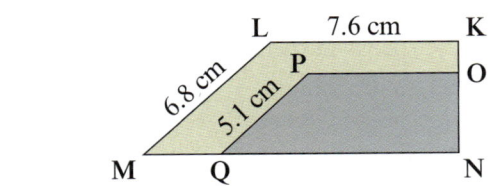

NOPQ is an enlargement of NKLM.
a Calculate the scale factor of the enlargement.
b Find the length PO.

8 A lottery win of £395 775 is shared by 4 winners in the ratio 2:3:4:6.
a Find the amount each winner received.
b Explain how you can check your answer to part **a**.

9 The length and width of a rectangle are in the ratio 3:5.
The perimeter of the rectangle is 56 cm.
a Find the length of the rectangle.
b Find the width of the rectangle.
c Calculate the area of the rectangle.

10 A paint colour called Zulu is mixed in this ratio:
125 ml of red:275 ml of blue:100 ml of yellow.
a Give this ratio in its simplest form.
b A mix uses 225 ml of yellow to make the colour called Zulu.
How much of each of the other colours is used?
c 1750 ml of Zulu paint is needed.
Give the amounts of each colour used for the mix.

Distance, speed and time

1 A chairlift travels at a constant speed of 2.6 metres per second.
What distance does one of the chairs travel in:

a 2 seconds **b** 5 seconds
c 9.3 seconds **d** 1 minute
e 3 minutes and 38 seconds?

2 At a constant speed, a plane takes 2 hours to travel 1208 miles.
Find the plane's speed in miles per hour.

3 An escalator is 32.8 metres long and travels at a speed of 0.75 m/s.
How long does it take to travel to the top of this escalator?

4 It takes 20 seconds to go up an escalator that is 14 m long. What is its speed in m/s?

5 Write these speeds in metres per second, to 1 dp.

a 129 metres per minute
b 2 kilometres per second
c 432 metres per hour
d 20 kilometres per hour

6 Write these speeds in kilometres per hour.

a 2 kilometres per minute
b 40 000 metres per hour
c 65 100 metres per hour
d 6 metres per second

7 Write these speeds in miles per minute, to 2dp where appropriate

a 60 mph **b** 30 mph
c 90 mph **d** 45 mph
e 25 mph **f** 42 mph

8 Write these speeds in miles per hour.

a 0.6 miles per minute
b 1.2 miles per minute
c 0.45 miles per minute
d 0.1 miles per second

9 How far does a plane travel in 3 h 15 min at a constant speed of 960 kilometres per hour?

10 How far can a car travel at a constant speed of 51 mph in:

a 1 h 30 min **b** 45 min
c 15 min **d** 1h 20 min
e 2 h 40 min **f** 3 h 45 min?

11 Calculate the time taken in hours and minutes, to the nearest minute, to travel 60 km at a speed of:

a 120 km/h **b** 30 km/h
c 75 km/h **d** 9 km/h
e 55 km/h **f** 49 km/h
g 50 km/h **h** 150 km/h.

12 This sketch graph shows two different vehicles starting from Bristol. One is a motorcycle the other a truck.

Journeys from Bristol to Glasgow

Distance from Bristol

O

Time

Which line do you think shows the truck?
Explain your answer.

13 Calculate the average speed
(in mph or km/h to 2 dp) of a car which travels:

a 90 km in 2 h
b 115 km in 3 h
c 40 miles in 50 min
d 50 km in 35 min
e 70 miles in 1 h 20 min
f 100 km in 1 h 12 min
g 400 miles in 7 h 30 min
h 500 metres in 1 min.

14 Mandy left Glasgow at 10:00 am and cycled 28 miles at a speed of 10 mph.
Amin left Glasgow at 10:20 am and travelled along the same route at a speed of 12 mph.

a Show both their journeys on one graph.
b About what time did Amin pass Mandy?

15 Andy left Taunton at 3:00 pm and took 45 minutes to drive 50 miles to Bristol. He stayed there for 2 h 30 min. He then returned along the same route and arrived home at 7:05 pm.

a Draw a graph to show Andy's complete journey.
b At what time did he begin his return journey to Taunton?
c Calculate his average speed for the journey to Bristol giving your answer to 1dp.
d What was his average speed on the return journey?

Trigonometry

1 Find angle x to the nearest degree when:

   a $\sin x = \frac{5}{6}$ **b** $\cos x = \frac{3}{8}$

   c $\tan x = \frac{9}{5}$ **d** $\sin x = \frac{11}{15}$

   e $\tan x = \frac{3}{13}$ **f** $\cos x = \frac{5}{8}$

2 In each of these triangles calculate the length marked with a letter, to the nearest millimetre.

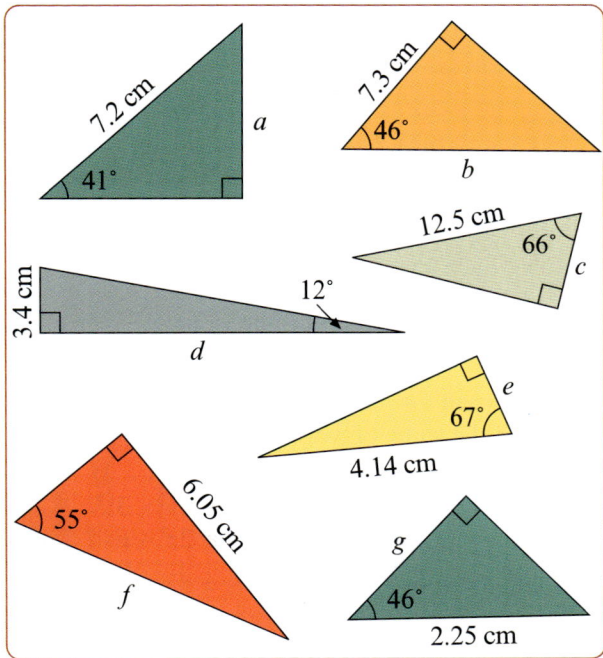

3 In each of these triangles calculate the angle x to the nearest degree.

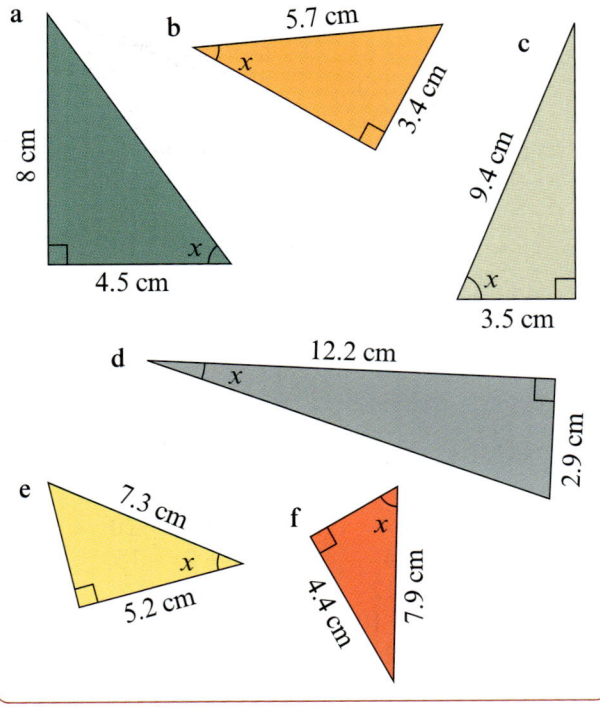

4 In this question give each answer correct to 3 sf.

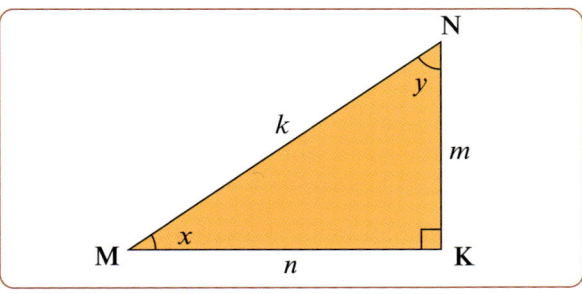

In $\triangle$NKM:

a calculate x when:
  i $n = 4.60$ cm   and $k = 10.60$ cm
  ii $m = 8.40$ cm   and $n = 4.62$ cm
  iii $k = 5.65$ cm   and $m = 3.00$ cm

b calculate y when:
  i $n = 6.62$ cm   and $k = 10.60$ cm
  ii $m = 15.00$ cm   and $n = 10.50$ cm

c calculate m when:
  i $x = 55°$ and $k = 3.58$ cm
  ii $y = 16°$ and $n = 41.50$ cm
  ii $x = 38°$ and $k = 4.05$ cm

d calculate k when:
  i $y = 25°$ and $m = 19.00$ cm
  ii $x = 61$ and $n = 2.40$ cm

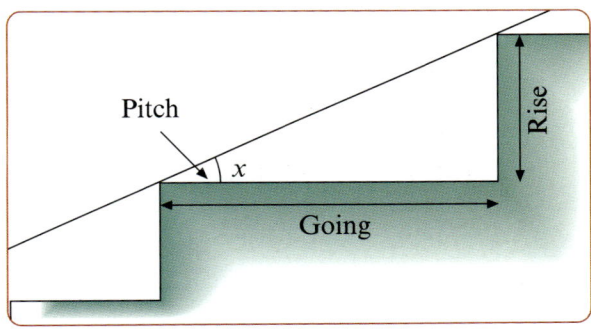

5 Calculate the pitch x for each of these stairs, correct to 2 dp.

Stair	Rise (mm)	Going (mm)
A	160	310
B	173	294
C	153	319

6 For safety the pitch x of these stairs must be between 32° and 22°.

  a For a going of 303 mm what is the greatest rise, to the nearest mm?

  b For a rise of 175 mm what is the greatest going, to the nearest mm?

Using algebra B

1 A formula for making tea in a pot for p people is:

$$t = p + 1$$

where t is the number of spoons of tea.

  a Find the number of spoons of tea needed for 5 people.

  b Rearrange the formula to make p the subject.

  c Calculate p when $t = 3$.

2 A formula for the perimeter of a square (P) of edge length x is:

$$P = 4x$$

  a Make x the subject of the formula.

  b Find the edge length of a square that has a perimeter of 41 cm.

3 A cook book gives this formula for the time in minutes (T) to cook a turkey that weighs W pounds:

$$T = 20W + 20$$

  a How long would it take to cook a turkey that weighs 12.5 pounds?

  b Make W the subject of the formula.

  c What is the weight of the largest turkey that is cooked in 2 hours?

4 A formula for converting litres (l) to pints (p) is:

$$p = \tfrac{7}{4}l$$

  a Use the formula to convert 5 litres to pints.

  b Make l the subject of the formula.

  c Convert 8.5 pints to litres, correct to 2 dp.

5 The formula for the area of a rectangle (A) with length l and width w is:

$$A = lw$$

  a Make w the subject of the formula.

  b Find the width of a rectangle 10 cm in length with an area of 62 cm².

6 The formula for the perimeter of a rectangle (P) with length l and width w is:

$$P = 2l + 2w$$

  a Make w the subject of the formula.

  b Find the width of a rectangle 10 cm in length with a perimeter of 28.4 cm.

7 Make k the subject of these formulas:

  a $V = 4k - 25$ **b** $y = 3(k - 1)$

  c $j = \dfrac{k}{7}$ **d** $w = \tfrac{3}{4}k$

  e $2v + k = 5$ **f** $k - b = 7$

  g $d = \tfrac{2}{5}k + 1$ **h** $h = \dfrac{k + 3}{8}$

  i $p = w + k$ **j** $q = 20 - 5k$

  k $v = 13 - \tfrac{1}{3}k$ **l** $v = s + tk$

  m $h = \dfrac{k + a}{b}$ **n** $h = 2gk$

  o $2y + 5k = 3$ **p** $3k - 2h = 9$

8 Make p the subject of these formulas:

  a $A = p^2$ **b** $v = 5p^2$

  c $g = p^2 + 1$ **d** $q = 3p^2 - 5$

  e $j = \tfrac{1}{4}p^2$ **f** $F = \tfrac{2}{3}p^2$

  g $p^2 + z = 2$ **h** $K = 2\pi p^2$

  i $A = 2p^2 + 1$ **j** $H = 3p^2 - 8$

9   **A** $\boxed{y = (x + 2)(x + 4)}$ **B** $\boxed{y = (x + 3)(x - 1)}$

  C $\boxed{y = 2x^2 + 3x - 5}$ **D** $\boxed{y = x^2 - x - 2}$

For each formula, find the value of y when:

  a $x = 6$ **b** $x = 3$ **c** $x = 1.5$

10 For each formula in x, y and z, find the value of x when $y = 5$ and $z = 2$.

  a $x = 3yz + y$ **b** $x = 4z(y - 3)$

  c $x = \sqrt{y^2 + 12z}$ **d** $x = 2y^2 - 3z^2$

  e $x = (5z + y)^2$ **f** $x = 2z^3 - y$

  g $x = \dfrac{y + 3z}{2y}$ **h** $x = \dfrac{1}{y} + \dfrac{1}{z}$

11 For each of these multiply out the brackets and simplify:

  a $(n + 2)(n + 5)$ **b** $(f + 1)(f + 10)$

  c $(w + 2)(w + 6)$ **d** $(2n + 1)(n + 3)$

  e $(4d + 7)(d + 3)$ **f** $(2y + 9)(2y + 5)$

  g $(x - 3)(x + 6)$ **h** $(v + 3)(v - 2)$

  i $(b + 3)(b - 5)$ **j** $(m - 7)(m - 1)$

  k $(h - 1)(h - 5)$ **l** $(t - 3)(t - 7)$

  m $(2p + 1)(3p - 1)$ **n** $(5x + 6)(3x - 2)$

  o $(6y + 3)(2y - 5)$ **p** $(3g - 7)(4g - 5)$

12 $\boxed{x + 1}$ $\boxed{x + 2}$ $\boxed{x + 3}$ $\boxed{x + 6}$ $\boxed{x + 4}$ $\boxed{x + 12}$

Which pair of expressions multiply to give:

  a $x^2 + 7x + 12$ **b** $x^2 + 8x + 12$

  c $x^2 + 7x + 6$ **d** $x^2 + 5x + 6$

  e $x^2 + 13x + 12$ **f** $x^2 + 6x + 8$?

13 Factorise:

  a $x^2 + 12x + 11$ **b** $x^2 + 8x + 15$

  c $x^2 + 4x + 4$ **d** $x^2 + 9x + 20$

  e $x^2 + 3x - 4$ **f** $x^2 + 2x - 15$

  g $x^2 + 4x - 12$ **h** $x^2 - x - 6$

  i $x^2 - 11x + 10$ **j** $x^2 - 2x - 8$

  k $x^2 - 3x + 2$ **l** $x^2 - 4x + 4$

Grouped data

Table A

1996 Olympic Games
Reaction Times
Sprint Hurdles
(Semi-Finals & Finals)

Reaction time (s)	Frequency Men	Frequency Women
0.120–	1	1
0.130–	4	2
0.140–	1	2
0.150–	4	2
0.160–	5	5
0.170–	4	7
0.180–	2	5
0.190–	3	0
Totals	24	24

Use Table A for Questions 1 to 12

1 Draw a histogram to show the reaction times for:

   a men   **b** women.

2 Give the modal class for:

   a men   **b** women.

3 Draw frequency polygons to compare the reaction times of men and women.

4 Calculate an estimate of the range of the times for:

   a men   **b** women.

5 Calculate an estimate of the mean reaction time for:

   a men   **b** women.

6 For these two distributions, the totals of the reaction times are just as useful for comparison as the means. Explain why.

7 Draw a cumulative frequency curve for:

   a men   **b** women.

8 Estimate how many reactions times were less than 0.135 s for:

   a men   **b** women.

9 Estimate how many reactions times were greater than 0.175 s for:

   a men   **b** women.

10 Estimate the median reaction time for:

   a men   **b** women.

11 Calculate an estimate of the interquartile range for:

   a men   **b** women.

12 Use your answers to Questions **4, 5, 10,** and **11** to compare the two distributions.

Table B

1996 Olympic Games
GB Athletics Team

Age	Frequency Track	Frequency Field
18–22	7	2
23–27	27	10
28–32	20	7
33–37	3	2
38–42	2	2
Totals	59	23

Use Table B for Questions 13 to 15

13 Draw a histogram to show the distribution of:

   a track athletes   **b** field athletes.

14 Explain why drawing frequency polygons on the same diagram would not give a good comparison of the ages of track athletes and field athletes.

15 Calculate an estimate of the mean age of:

   a track athletes   **b** field athletes.

C   **1996 Olympic Games – Men's 20 km walk**

Time (min)	80–	84–	88–	92–	96–	100–	Total
Frequency	18	21	9	2	1	1	52

D   **1996 Olympic Games – Men's 50 km walk**

Time (min)	220–	230–	240–	250–	260–	Total
Frequency	8	13	9	4	2	36

E   **1996 Olympic Games – Women's 10 km walk**

Time (min)	41–	42–	43–	44–	45–	46–	47–	48–	Total
Frequency	1	4	7	3	12	5	4	2	38

16 For each distribution in table C, D and E:

   a draw a histogram

   b calculate an estimate of the range of the times

   c calculate an estimate of the mean time

   d draw a cumulative frequency curve

   e estimate the median time

   f calculate an estimate of the interquartile range.

17 This data shows sales of burgers at a motorway service area.

Week	1	2	3	4	5	6	7	8	9	10	11	12
Burger sales	44	32	28	32	40	16	24	20	28	24	36	16

   a Calculate the 4-point moving averages for the data.

   b On a pair of axes draw graphs of the raw data and the moving average data.

   c Comment on any trend you can identify in burger sales.

Working with percentages

1 Calculate, giving answers correct to 2 dp where appropriate:

a 38 as a percentage of 60
b £14 as a percentage of £55
c 68 as a percentage of 24
d 1550 as a percentage of 2500
e 3500 km as a percentage of 75 000 km
f 12.5 kg as a percentage of 40 kg
g £125.50 as a percentage of £150
h 132 miles as a percentage of 868 miles
i £9.38 as a percentage of £56.45
j 15 650 km as a percentage of 1.5 million km.

2 Give answers to each of these correct to 2 dp where appropriate.

a Increase 25 kg by 18%
b Increase 1400 km by 65%
c Increase 3560 miles by 4%
d Increase 1350 yards by 35%
e Increase £25 645 by 6%
f Increase 137 500 tonnes by 5.5%
g Increase 0.7 cm by 50%
h Increase 42 mm by 12.5%
i Increase £35.99 by 7%
j Increase 365 ml by 15%.

3 Give your answers to these correct to 2 dp where appropriate.

a Decrease 485 ml by 12%
b Decrease £45.75 by 20%
c Decrease 65 mm by 65%
d Decrease 0.8 cm by 8%
e Decrease 15 875 tonnes by 75%
f Decrease £15 944 by 34%
g Decrease 1760 yards by 28%
h Decrease 5682 miles by 56%
i Decrease 3600 km by 35.5%
j Decrease 48 kg by 48%.

4 The price of each item is given ex. VAT. Calculate the price including VAT at today's standard rate.

a crash helmet £185.85
b cycle tyre £11.69
c fishing rod £44.86
d steam iron £21.75
e microwave oven £268.55
f VCR £159.99
g personal CD player £135.38
h CD £9.24
i phone £49.99
j multi-media PC £1499.

5 A printer is advertised for £132 + VAT. Calculate the total price of the printer.

6 Callum sees the same model TV advertised by two shops in this way:

TV World	£199.99 inc. VAT
Price busters	£169.99 ex. VAT

a From which shop would you advise Callum to buy the TV?
b Give a reason for your answer to part **a**.

7 Calculate the simple interest charged or paid on each of these:

a £675 borrowed for 5 years at 12% pa
b £12 400 borrowed for 2 years at 17% pa
c £170 saved for 4 years at 3% pa
d £65 saved for 9 years at 4% pa
e £1500 borrowed for 3 years at 18% pa
f £4600 borrowed for 4 years at 12.5% pa
g £25 saved for 3 years at 5% pa
h £150 saved for 6 years at 4.5% pa
i £200 borrowed for 2 years at 22.9% pa
j £175 borrowed for 3 years at 18% pa.

8 Calculate the compound interest charged or paid on each of these (use the formula):

a £500 saved for 3 years at 6% pa
b £1400 borrowed for 2 years at 19% pa
c £250 saved for 2 years at 7.5% pa
d £105 saved for 4 years at 3%
e £12 500 borrowed for 3 years at 16%
f £32 saved for 3 years at 4%
g £750 borrowed for 3 years at 21%
h £3675 borrowed for 2 years at 17%
i £125 saved for 3 years at 8%
j £500 saved for 3 years at 4.5%.

9 These prices include VAT. Calculate each price ex. VAT.

a freezer £299.99
b CD £12.99
c camera £44.99
d phone £9.99
e climbing boots £75
f tent £89.95
g kettle £26.99
h PC £129.99
i calculator £18.99
j TV £139.99

10 Jo bought a ski jacket in a 15% off sale for £85.45.

a What was the pre-sale price of the jacket?
b How much did she save in the sale?

11 In 1995 a ferry company made a profit of £3 600 000.
This was 14% more than the profit for 1994.

Calculate the profit made in 1994.

Transformations

1 Draw the pentagon A on axes with: $^-4 \leqslant x \leqslant 10$ and $^-6 \leqslant y \leqslant 8$.

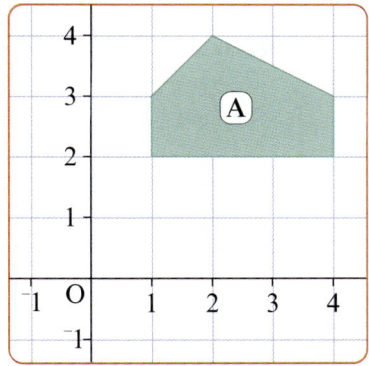

2 Draw the image of A after:

a an enlargement SF 2 with centre (6, 5)

b an enlargement SF $\frac{1}{2}$ with centre (7, $^-4$).

3 These transformations map A on to B, C, D, E and F.

Object	Transformation	Image
A	Rotate 90° anticlockwise about (0, 0)	B
A	Rotate 90° clockwise about (1, 1)	C
A	Reflect in $x = 5$	D
A	Rotate 90° anticlockwise about (6, 2)	E
A	Reflect in $y = ^-x$	F

a On a new diagram, on axes with: $^-6 \leqslant x \leqslant 10$ and $^-6 \leqslant y \leqslant 6$ draw and label the images B, C, D, E and F.

b Describe fully the transformation that maps:

  i C on to A   **ii** D on to A.

c For each of these mappings which pentagon is the image of B?

	Object	Transformation	Image
i	B	Rotate 180° about (0, 1)	
ii	B	Reflect in $y = 0$	
iii	B	Translate $\begin{pmatrix} 8 \\ ^-4 \end{pmatrix}$	

d Describe fully the transformation that maps C on to E.

4

Transformations			
Object	First	Second	Image
A	Rotate 90° anticlockwise about (0, 1)	Reflect in $y = 0$	G
A	Reflect in $y = 0$	Rotate 90° anticlockwise about (0, 1)	H

a On a new diagram draw the images G and H.

b Compare G and H and comment on any differences.

c In this table each pair of transformations maps G on to H. Describe the second transformation fully.

	Object	Image	First transformation	Second transforma
i	G	H	Translate $\begin{pmatrix} 6 \\ 0 \end{pmatrix}$	
ii	G	H	Rotate 90° clockwise about (1, $^-2$)	
iii	G	H	Rotate 180° about (0, $^-1$)	

d What single transformation maps G on to H?

5 These transformations map A on to J, K and L.

Transformations			
Object	First	Second	Image
A	Enlarge SF 2 with centre (0, 1)	Reflect in $y = 1$	J
A	Reflect in $x = 1$	Enlarge SF 2 with centre (5, 0)	K
A	Enlarge SF 2 with centre (5, 0)	Reflect in $x = 4$	L

a On a new diagram, on axes with: $^-10 \leqslant x \leqslant 12$ and $^-10 \leqslant y \leqslant 10$ draw the images J, K and L

b What single transformation maps

  i J on to K   **ii** L on to K?

Probability B

In an experiment a Multilink cube was dropped on to a hard surface and its resting position was recorded.

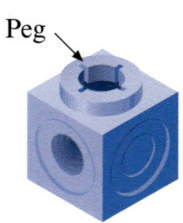

Peg

Position	Frequency
Peg up	7
Peg to side	51
Peg down	5
Peg tilting	2

1 Give the relative frequency to 2 dp of the cube coming to rest:

  a peg down
  b peg to side
  c peg tilting
  d peg up.

2 Why should all the unrounded answers to Questions **1a** to **1d** add up to 1?

3 What is the relative frequency of a cube resting:

  a without the peg down
  b with either a peg up or a peg down
  c with neither a peg to the side nor a peg up?

4 From the results do you think that each face of the cube is equally likely to be on the top. Explain your answer.

5 The same Multilink cube is dropped 150 times on the same surface. Estimate the number of times it will come to rest:

  a peg down
  b peg to the side
  c peg tilting.

6 Do an experiment yourself to find the relative frequency of different positions a Multilink cube can come to rest.

  a Compare your relative frequencies with the results of the experiment above.
  b How could you improve the accuracy of your experiment to find the relative frequencies?

7 The probability that Mike is late for work on a Monday is 0.4 and on a Tuesday it is 0.2

  a Draw a tree diagram to show the outcomes and probabilities.
  b Estimate the probability that Mike is late on Monday and Tuesday.
  c Estimate the probability that Mike is late on neither day.
  d Estimate the probability that he is late at least once in the two days.

8 Would you use theoretical probability or relative frequency to find the probability that:

  a a person in the UK is right- or left-handed
  b the next volcanic eruption occurs in May
  c the next person you meet in school is female
  d the next wine gum in the tube is black
  e a toothpaste tube will leak before it is finished
  f if you visited a foreign country at random you would be expected to drive on the left
  g your toast lands jam side down on the floor if you drop it
  h it rains when you have no waterproofs?

A biased spinner has seven sides. On the sides are the numbers 7, 15, 24, 25, 29, 30, 36.

The probability of the spinner landing on each number is given in this table.

Number	7	15	24	25	29	30	36
Probability	0.04	0.15	0.19	0.14	0.14	0.13	0.21

9 **a** What is the probability of the spinner landing on a number less than 20?
  b Why can you add the probabilities in this case?

10 **a** Complete the Venn diagram to show multiples of 3 and multiples of 5 for the spinner.

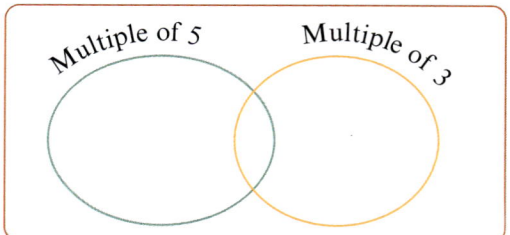

Multiple of 5   Multiple of 3

  b Why can you not add probabilities in this case?

11 Calculate the probability of getting in one spin:

  a a multiple of 5 and a multiple of 3
  b either a multiple of 3 or a multiple of 5
  c neither a multiple of 3 nor a multiple of 5
  d a multiple of 3 but not a multiple of 5.

12 Find the probability of getting either a square number or an even number.

Inequalities

1 Copy each of these and insert > or < in place of the box to make it correct.

 a 9 ☐ ⁻4   **b** ⁻8 ☐ ⁻3
 c 4 ☐ ⁻2   **d** ⁻2 ☐ 2
 e ⁻5.6 ☐ ⁻4.23   **f** ⁻8 ☐ ⁻11

2 Which of these numbers is not a possible value for t, where $t \leqslant \, ⁻7$?

 ⁻8, 4, 71, ⁻3.2, ⁻7.003, ⁻28.6, ⁻7, ⁻0.33

3 What integer values for k satisfy these inequalities? Show each answer on a number line.

 a $9 > k \geqslant 4$ **b** $⁻3 \leqslant k \leqslant \, ⁻1$
 c $⁻5 \leqslant k < 1$ **d** $⁻4 \geqslant k > \, ⁻7$
 e $⁻56 < k < \, ⁻57$ **f** $24 \leqslant k \leqslant 24$
 g $⁻14 \geqslant k > \, ⁻16$ **h** $6 > k > 0$

4 Write two other inequalities in h which describe the same integer values as $⁻2 < h < 2$.

5 Explain why these two inequalities are different types of inequality from each other.

> • This chair is suitable for people with weights given by $5 \leqslant w \leqslant 15$, where w is their weight in stones.

> • The waiters in a restaurant are given by $5 \leqslant w \leqslant 15$ where w is the number of waiters.

6 This diagram shows the inequality $⁻2 < x \leqslant 3$.

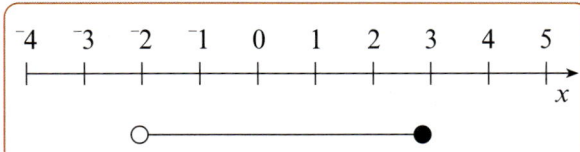

Draw similar diagrams to show these inequalities:

 a $7 \leqslant x \leqslant 10$ **b** $⁻4 \leqslant x < 1$
 c $⁻5 < x < \, ⁻2$ **d** $0 < x \leqslant 4$

7 Solve the following inequalities.

 a $3x \geqslant 27$ **b** $6t < \, ⁻42$
 c $42 \geqslant 4p$ **d** $k^2 \leqslant 121$
 e $1 + 3w < 7$ **f** $5q + 7 > 53.5$
 g $19 \leqslant 3t - 5$ **h** $7 - 2h \geqslant 3$
 i $4(3f - 2) < 28$ **j** $c^2 - 5 < 76$
 k $36 \leqslant 3(2g + 3)$ **l** $3x + 5 > x - 1$
 m $3j - 2 < 2j + 17$ **n** $5d + 20 \geqslant 6 - 2d$
 o $24 - 8v > 6 - 4v$ **p** $2(x + 3) \leqslant x - 7$
 q $5 - 3s > 5 + 3s$ **r** $9u + 6 \leqslant 7u$

8 The conditions on x are given by the inequality $⁻4 \leqslant x < 3$.
What is:

 a the smallest possible value of x^2
 b the largest possible value of x^2?

9 Sketch graphs to show the regions which satisfy the following conditions.

> Ⓐ   ***Anita Southgate Floral Blinds***
> Blinds fit windows between 1 metre and 2 metres wide and up to 1 metre 20 cm high.

> Ⓑ   **ANTOK SUPPLY SERVICES**
> Drivers should be between 21 and 55 with a clean driving licence. We need drivers who have had experience of working with at least four previous companies.

> Ⓒ   ***Clarkson Hi Fi Sale***
> Each rack holds up to 45 compact discs. We have different models with prices from £12.

> Ⓓ   **Opus 4123R Fax Machine**
> Will take fax rolls up to 214 millimetres wide and up to 50 metres long.

> Ⓔ   **REGENT CAR SALES**
> Our new models will do up to 50 miles per gallon at speeds between 50 and 60 miles per hour.

10 Sketch graphs and shade the regions which satisfy the following conditions.

 a $x > 4$ **b** $⁻3 < y < 4$
 c $y < 6$ **d** $4 > x > 1$
 e $x < 4$ and $y > 6$
 f $⁻1 < x < 2$ and $y < \, ⁻2$
 g $0 < x < 4$ and $0 > y > \, ⁻4$
 h $y > x - 2$
 i $x + y < 6$
 j $y < 2x + 2$

11 What two inequalities define the shaded region on this sketch?

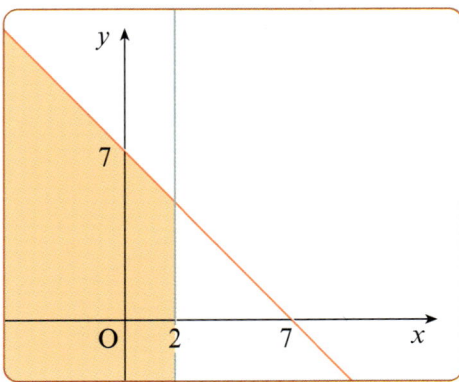

Working in 3D

1 Each of these solids is made from three cubes.

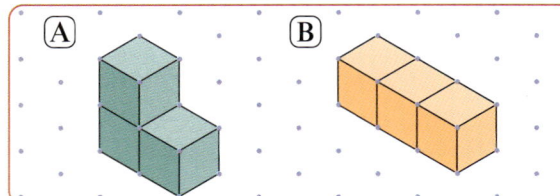

How many planes of symmetry has each solid?

2 a Draw a solid made from five cubes with 1 plane of symmetry.
b Draw a solid made from six cubes with 3 planes of symmetry.

3 How many planes of symmetry has:
a a cuboid with no square faces
b a cube?

4

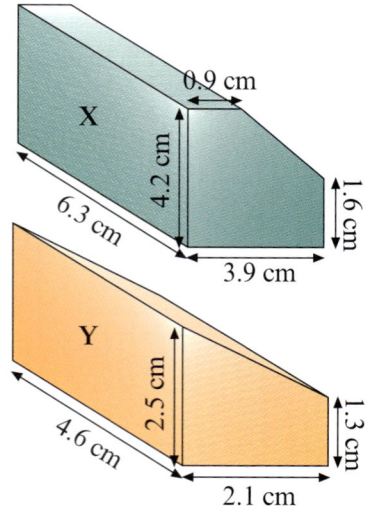

1 cm

a Copy and complete this diagram to make the net of a prism.
b What is the volume of the box that can be made from the net?
c What is the surface area of the box that can be made from the net?

5 Solids X and Y are prisms.

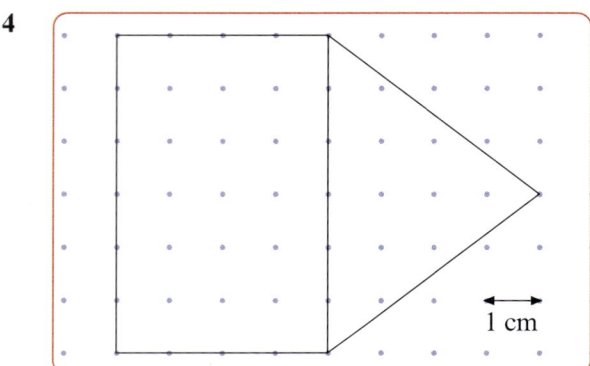

Calculate the volume of each solid.

6 Each prism P and Q has a volume of 1000 cm³.

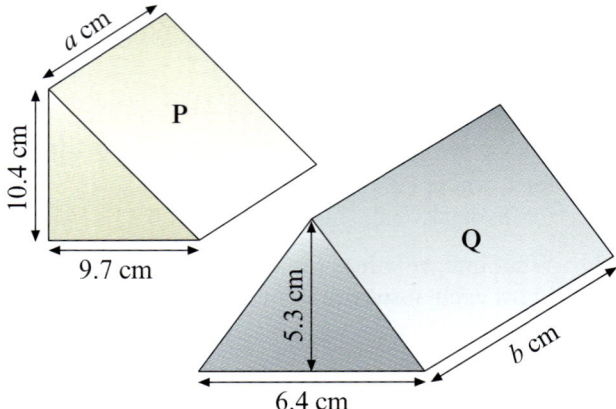

Find the values of a and b correct to 1 dp.

7 Each of these containers is in the shape of a cylinder.

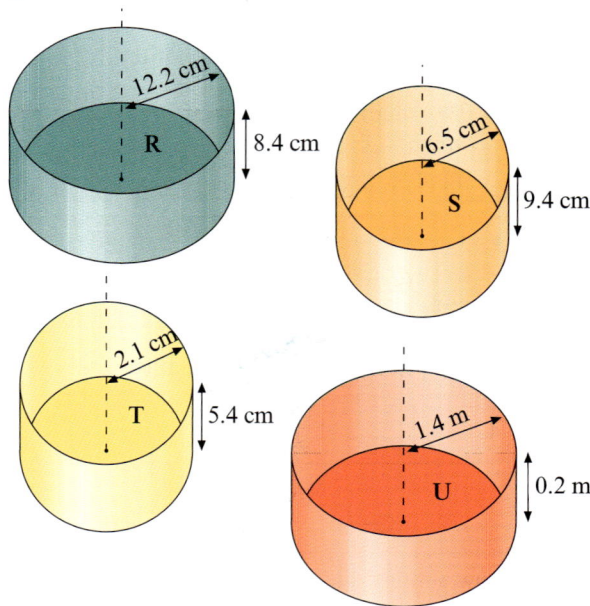

Find the volume of each container, to 1 dp:
a in ml   **b** in litres.

8 In the following expressions, r, h and l each represent a length.
Decide if each expression represents:
a length, area, volume or none of these.
Give reasons for each answer.

a $4h$ **b** h^2
c rh **d** rhl
e $2(h + l)$ **f** pr^2
g $hl + r$ **h** $\frac{1}{3}rh^2$
i $\sqrt{h^2 + l^2}$ **j** $prh - 3l^2$
k $r^3 + h^2 + rl$ **l** $h^3 + rl^2$
m $5(rh + hl + rl)$ **n** $prh(h + l)$
o $2h(r - l)$ **p** $\frac{4}{3}p(r + h)$

Quadratics

1 Factorise and solve these quadratic equations:

a $c^2 + 5c + 6 = 0$ b $g^2 + g - 20 = 0$
c $h^2 - 6h - 16 = 0$ d $t^2 + 3t - 54 = 0$
e $p^2 - 4p - 45 = 0$ f $k^2 + 3k - 40 = 0$
g $x^2 - 7x + 10 = 0$ h $v^2 + 9v - 36 = 0$
i $d^2 + 8d + 15 = 0$ j $h^2 - 9h + 18 = 0$
k $n^2 + 6n - 72 = 0$ l $y^2 - 7y - 30 = 0$
m $a^2 + 5a - 36 = 0$ n $x^2 + 8x - 48 = 0$
o $u^2 - 12u + 27 = 0$ p $c^2 + 18c + 77 = 0$
q $g^2 - g - 72 = 0$ r $y^2 + y - 42 = 0$
s $k^2 + 15k + 56 = 0$ t $x^2 - 7x - 60 = 0$

2 Find a value for x by trial and improvement.
(Give your answer correct to 1 dp.)

a $x^2 - x = 8$ b $x^3 + x = 5$
c $x^2 - 2x = 4$ d $x^3 - x = 4$
e $2x^2 - x = 5$ f $2x^3 - x = 6$

3 a Draw up a table of values for the graph of:
$$y = x^2 - 4x$$
with values of x from $^-2$ to $^+5$.
b From your table of values, draw and label the graph.

4 On a pair of axes **sketch**, and label these graphs:

a $y = x^2$ b $y = x^2 + 2$ c $y = x^2 - 3$

5 a Draw up a table of values for the graph of:
$$y = x^2 - 5x + 4$$
with values of x from 0 to $^+5$.
b From your table of values, draw and label the graph.

6 a Draw up a table of values for the graph of:
$$y = x^2 + 2x - 8$$
with values of x from $^-5$ to $^+3$.
b From your table of values draw and label the graph.
c Use your graph to solve the equation:
$$x^2 + 2x - 8 = 0$$
Explain your answers.

7 a With values of x from $^-7$ to $^+4$, draw up a table of values for the graph of:
$$y = x^2 + 3x - 18$$
b Use your graph to solve the equaton:
$$x^2 + 3x - 18 = 0$$
c Explain how you can check your answer.

8 a Draw the graph of $y = x^2 - 6x + 5$ for values of x from $^-1$ to $^+6$.
b Use your graph to solve the equation:
$$x^2 - 6x + 5 = 0$$
c On the same axes draw a graph of:
$$y = 2$$
d Use your graphs to solve the equation:
$$x^2 - 6x + 5 = 2$$
Give reasons for your answer.

9 a Draw the graph of $y = x^2 + 4x - 21$ for values of x from $^-8$ to $^+4$.
b Use your graph to solve the equation:
$$x^2 + 4x - 21 = 0$$
c On the same axes draw a graph of $y = ^-5$.
d Use your graphs to solve the equation:
$$x^2 + 4x - 21 = ^-5$$
e i Explain how you would solve the equation:
$$x^2 + 4x - 21 = 4$$
ii Solve the equation: $x^2 + 4x - 21 = 4$

10 Show how each of these equations can be made by rearranging $x^2 + 8x - 20 = 0$.

a $x^2 - 20 = ^-8x$ b $20 = x^2 + 8x$
c $x^2 = 20 - 8x$

11 Show how each of these equations can be made by rearranging $x^2 - 7x = 60$.

a $x^2 - 60 = 7x$ b $x^2 = 7x + 60$
c $x^2 - 7x - 60 = 0$

12 Rearrange the equation $x^2 - 55 = 6x$ to make three other quadratic equations.

13 Make three other quadratic equations by rearranging $60 - 11x = x^2$.

14 With values of x from $^-4$ to $^+3$:

a Draw the graph of $y = x^2 + 3$.
b On the same axes draw a graph of $y = 2x + 4$.
c Use your graphs to solve the equation:
$$x^2 + 3 = 2x + 4$$
d Either by drawing another graph or using these graphs solve the equation:
$$x^2 - 2x - 1 = 0$$

15 Sketch the graph of $y = x^3 + 1$.

16 For values of x from $^-3$ to $^+3$ draw a graph of $y = x^3 + 2x$.

17 Use trial and improvement to solve the equation:
$$x^5 = 20$$
(Give your answer correct to 3 sf.)

Processing data

1

Town Centre Survey

1 Do you come into the town centre often? ☐ Yes ☐ No

2 Do you agree that the town centre
should be pedestrianised? ☐ Yes ☐ No

3 What do you think about buses? _____

    a Criticise each of these questions.
    b Write an improved question for each one.

2

Body Matters
Right-handed people are more likely to
have a stronger left eye than right eye.

Design an experiment to test this hypothesis.

3   **a** Carry out your experiment.
    b Analyse the data you collect.

4   **a** Do you think the hypothesis is true or false?
    b Explain why.

5

Body Matters
Do right-handed people fold their arms
differently to left-handed people?

Design an experiment to answer this question.

6   **a** Carry out your experiment.
    b Analyse the data you collect.
    c Interpret your results to answer the question.

7

Motorbikes – Size of Engine & Price

Engine size (cc)	250	900	600	125	650	900	500	750
Price (£)	4500	8100	6700	2400	5100	6700	3400	9200

  a Draw axes: horizontal 0 to 1200
                  vertical 0 to 12 000
  b Plot the motorbike data on your diagram.
  c Draw a line of best fit.
  d Describe any correlation between size of engine
    and price.

8 Estimate the price of a motorbike with engine size:
  a 800 cc   **b** 450 cc

9 Estimate the engine size of a motorbike costing:
  a £4000   **b** £6500

10 Extend the line of best fit on your scatter
diagram to estimate:
  a the price of a motorbike with a 1000 cc engine
  b the engine size of a motorbike costing £10 000.

11

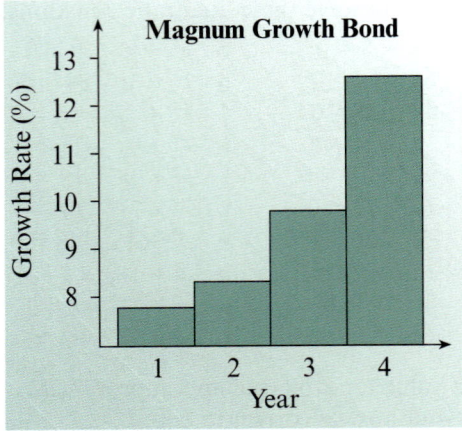

Explain why this diagram is misleading.

12

Explain why this diagram is misleading.

13 The graph was used under the headline
'Fantastic improvement in safety'.

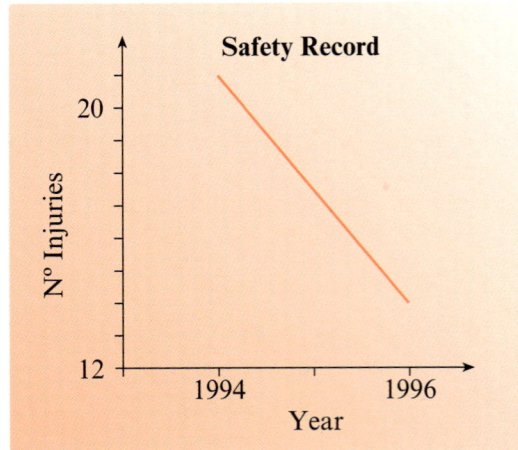

  a What data does the graph actually show?
  b Explain what was done to the graph to support
    the headline.
  c Draw a graph of the data showing how you
    think it should look.
  d What can you say about the number of injuries
    in 1995?

Trigonometry and Pythagoras' rule

1

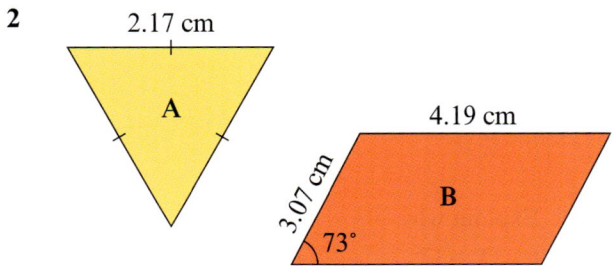

For each triangle calculate the side or angle marked with a letter, correct to 3 sf.

2

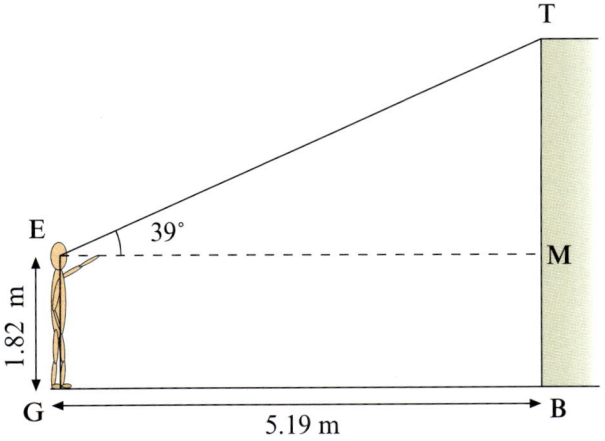

Calculate the area and perimeter of shapes A and B, correct to 2 dp.

3 This diagram shows the top of a building T and a person at G.
The horizontal distance GB is 5.19 m.
The point E is 1.82 m above the ground.
The angle of elevation from E to T is 39°.

a Calculate, to the nearest 0.01 m:
   i the vertical height MT
   ii the total height TB.
b Calculate the angle of depression from E to B to the nearest degree.

4 This diagram shows the position of three markers A, B and C on a map.
N is a point due north of A and due west of B.

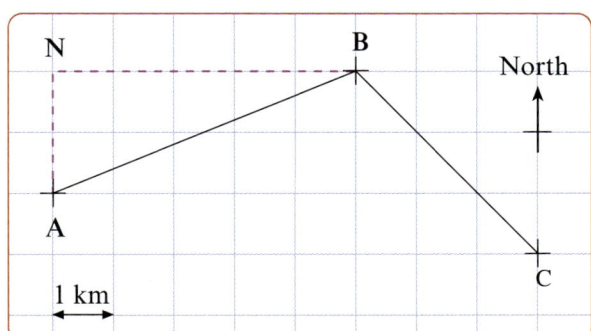

a How far is A:
   i west of B   **ii** south of B?
b Give the bearing of :
   i C from B   **ii** B from A   **iii** A from C
c In Δ ABN, calculate:
   i AB in km to 2 dp
   ii NB̂A to the nearest degree.
d Calculate the distance BC in km to 2 dp.
e Calculate the distance between A and C.

5 Double-decker buses are tested for stability for angles up to 28° from the horizontal.

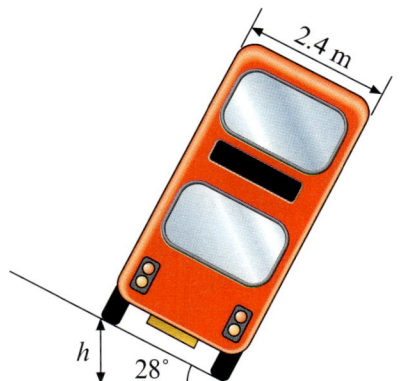

a Calculate the height *h* in metres, for a double-decker when it is tipped to the maximum test angle, correct to 1 dp.

b For a single-decker bus the test angle is smaller. A single decker bus 2.8 metres wide is tested and the height *h* is 1.31 metres.
Find the test angle, correct to 1 dp.

Circle properties

1 Draw any circle, and a chord AB.

    a Label:
      i the minor segment
      ii the major segment.

    A chord CD is such that the circle is divided into two segments that are the same size.

    b What can you say about the chord CD?

2

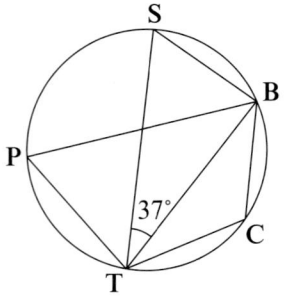

    a Name an angle in the minor segment.

    b Calculate the size of TBS when $T\hat{P}B = 75°$. Explain your answer.

3 In the diagram PT is a chord and O the centre of the circle.

    Calculate the size of $T\hat{P}O$. Give reasons for your answer.

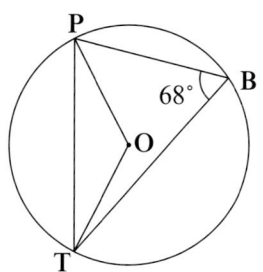

4 The circle has a centre O.

    a Calculate the size of $P\hat{T}B$. Explain your answer.

    b What is the size of $A\hat{C}B$? Explain your answer.

    c Explain why $A\hat{B}P$ is 76°.

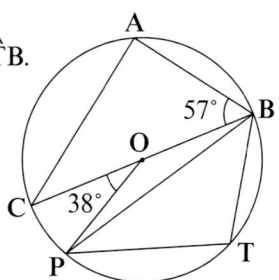

5 The circle has a centre O.

    a Explain why $P\hat{A}T$ is 42°.

    b Calculate $P\hat{C}T$. Explain your answer.

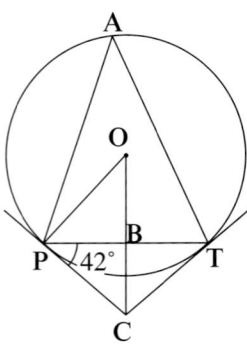

6 **a** Explain why $T\hat{B}P$ is 35°.

    b Calculate $B\hat{P}T$. Give reasons for your answer.

    c Explain why $A\hat{C}B$ is 73°.

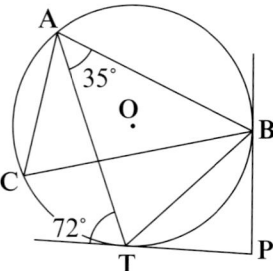

7 **a** Explain why $E\hat{C}D = 31°$.

    b Calculate the size of $A\hat{B}C$. Explain your answer.

    c $E\hat{A}B = 52°$ Give two different explanations to show this.

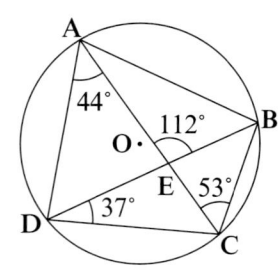

8 Show why an expression for $D\hat{A}B$ is given as:

    $180 - a$

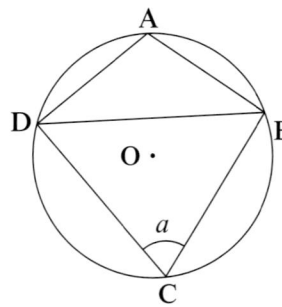

For your GCSE Maths you will need to produce two pieces of coursework:

1. Investigative
2. Statistical

Each piece is worth 10% of your exam mark.
This unit gives you guidance on approaching your coursework and will tell you how to get good marks for each piece.

Investigative coursework

In your investigative coursework you will need to:

> ◆ Say how you are going to carry out the task and provide a plan of action
>
> ◆ Collect results for the task and consider an appropriate way to represent your result
>
> ◆ Write down observations you make from your diagrams or calculations
>
> ◆ Look at your results and write down any observations, rules or patterns – try to make a general statement
>
> ◆ Check out your general statements by testing further data – say if your test works or not
>
> ◆ Develop the task by posing your own questions and provide a conclusion to the questions posed
>
> ◆ Extend the task by using techniques and calculations from the content of the higher tier
>
> ◆ Make your conclusions clear and link them to the original task giving comments on your methods.

Assessing the task

Before you start, it is helpful for you to know how your work will be marked. This way you can make sure you are familiar with the sort of things which examiners are looking for.

Investigative work is marked under three headings:

1. **Making and monitoring decisions to solve problems**
2. **Communicating mathematically**
3. **Developing skills of mathematical reasoning**

Each strand assesses a different aspect of your coursework. The criteria are:

1. Making and monitoring decisions to solve problems

This strand is about deciding what needs to be done then doing it. The strand requires you to select an appropriate approach, obtain information and introduce your own questions which develop the task further.

For the higher marks you need to analyse alternative mathematical approaches and, possibly, develop the task using work from the higher tier GCSE syllabus content.

For this strand you need to:

◆ solve the task by collecting information
◆ break down the task by solving it in a systematic way
◆ extend the task by introducing your own *relevant* questions
◆ extend the task by following through alternative approaches

♦ develop the task by including a number of mathematical features
♦ explore the task extensively using higher level mathematics

2. Communicating mathematically

This strand is about communicating what you are doing using words, tables, diagrams and symbols. You should make sure your chosen presentation is accurate and relevant.

For the higher marks you will need to use mathematical symbols accurately, concisely and efficiently in presenting a reasoned argument.

For this strand you need to:

♦ illustrate your information using diagrams and calculations
♦ interpret and explain your diagrams and calculations
♦ begin to use mathematical symbols to explain your work
♦ use mathematical symbols consistently to explain your work
♦ use mathematical symbols accurately to argue your case
♦ use mathematical symbols concisely and efficiently to argue your case

3. Developing skills of mathematical reasoning

This strand is about testing, explaining and justifying what you have done. It requires you to search for patterns and provide generalisations. You should test, justify and explain any generalisations.

For the higher marks you will need to provide a sophisticated and rigorous justification, argument or proof which demonstrates a mathematical insight into the problem.

For this strand you need to:

♦ make a general statement from your results
♦ confirm your general statement by further testing
♦ make a justification for your general statement
♦ develop your justification further
♦ provide a sophisticated justification
♦ provide a rigorous justification, argument or proof

This unit uses a series of investigative tasks to demonstrate how each of these strands can be achieved.

Task 1

TRIANGLES

Patterns of triangles are made as shown in the following diagrams:

Pattern 1

Pattern 2

Pattern 3

What do you notice about the pattern number and the number of triangles?

Investigate further.

Note: You are reminded that any coursework submitted for your GCSE examination must be your own. If you copy from someone else then you may be disqualified from the examination.

Planning your work

A straightforward approach to this task is to continue the pattern further. You can record your results in a table and then see if you can make any generalisations.

Collecting information

A good starting point for any investigation is to collect information about the task.

You can see that:

- Pattern 1 shows 1 triangle
- Pattern 2 shows 4 triangles
- Pattern 3 shows 9 triangles

You can then extend this by considering further patterns.

Pattern 4   Pattern 5

16 triangles

25 triangles

> **Moderator comment**
>
> It is a good idea not to collect too much data as this is time consuming ... aim to collect 4 or 5 items of data to start with.

Drawing up tables

The information is not easy to follow so it is **always** a good idea to illustrate it in a table.

A table to show the relationship between the pattern number and the number of triangles

> It is important to give your table a title and to make it quite clear what the table is showing.

Pattern number	1	2	3	4	5
No. of triangles	1	4	9	16	25

> **Now you try ...**
>
> Using the table:
>
> - Can you see anything special about the number of triangles?
> - Ask yourself: are they odd numbers, even numbers, square numbers, triangle numbers, are they multiples, do they get bigger, do they get smaller ...
> - What other relationships might you look for?

Being systematic

Another important aspect to your work is to be systematic.

This task shows you what that means.

Task 2

ARRANGE A LETTER

How many different ways can you arrange the letters ABCD?

Investigate further.

Note: You are reminded that any coursework submitted for your GCSE examination must be your own. If you copy from someone else then you may be disqualified from the examination.

You could haphazardly try out some different possibilities:

ABCD
BCDA
CDAB etc

You may or may not find all the arrangements this way. To be sure of finding them all you should list the arrangements systematically.

To be systematic you group letters starting with A:

Notice the system here:
A followed by B,
then A followed by C,
then A followed by D
and so on.

ABCD
ABDC
ACBD
ACDB
ADBC
ADCB

Now you try ...

Now see if you can systematically produce all of the arrangements starting with B (you should find six of them) then complete the other arrangements.

You should also be systematic about the way you collect your data. In task 2, you were asked to find the arrangements of 4 letters: A, B, C and D.

To approach the whole task systematically, you should start by finding the arrangements for:

- 1 letter: A
- 2 letters: A and B
- 3 letters: A, B and C

This will give you more information about the task.

For different numbers of letters you should find:

Arrangements of different numbers of letters				
Number of letters	1	2	3	4
Number of arrangements	1	2	6	24

You should now try and explain what your table tells you:

From my table I can . . .

Finding a relationship – making a generalisation

To make a generalisation, you need to find a relationship from your table of results. A useful method is to look at the differences between terms in the table.

Here is the table of results from task 1:

A table to show the relationship between the pattern number and the number of triangles

Pattern number	1	2	3	4	5
No of triangles	1	4	9	16	25
		+3	+5	+7	+9	+11
			+2	+2	+2	+2

In this table the first "differences" are going up in two's.

The "second differences" are all the same ... this tells you that the relationship is quadratic, so you should compare the numbers with the square numbers

Here are some generalisations you could make:

From my table I notice that the number of triangles are all square numbers.

From my table I notice that the number of triangles are the pattern numbers squared.

Moderator comment

Remember that the idea is to identify some relationship from your table or, graph so that there is some point in you using it in your work ... a table or graph without any comment on the findings is of little use.

A graph may help see the relationship:

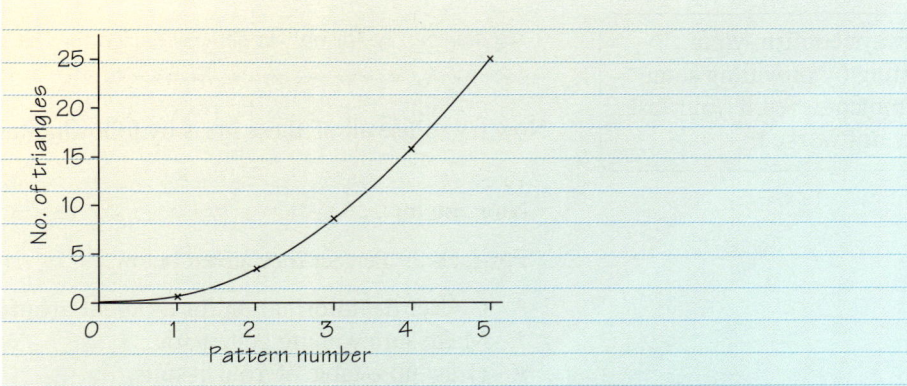

From my graph I notice that there is a quadratic relationship between the number of triangles and the pattern number.

Using algebra

You will gain marks if you can express your generalisation or rule using algebra.

The rule:

> I notice that the number of triangles are the pattern numbers squared.

can be written in algebra:

> $t = p^2$ where t is the number of triangles and p is the
> pattern number

> Remember that you must explain what t and p stand for.

Testing generalisations

You need to confirm your generalisation by further testing.

> From my table I notice that the number of triangles are the
> pattern numbers squared so that the sixth pattern will have $6^2 = 36$
> triangles.

You can confirm this with a diagram:

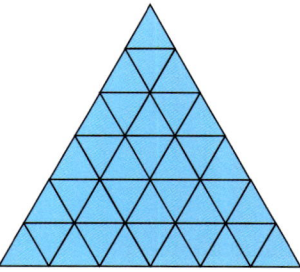

> My diagram confirms the generalisation for the sixth pattern.

> Always confirm your testing by providing some comment, even if your test has not worked!

Now try to use all of these ideas by following task 3.

> **Now you try ...**
>
> For task 3: 'Perfect tiles' given below:
>
> ◆ Collect the information for different arrangements
> ◆ Make sure you are systematic
> ◆ Draw up a table of your results.

Task 3

PERFECT TILES

Floor spacers are used to give a perfect finish when laying tiles on the kitchen floor.

Three different spacers are used including

     L spacer
     T spacer
     + spacer

Here is a 3 × 3 arrangement of tiles

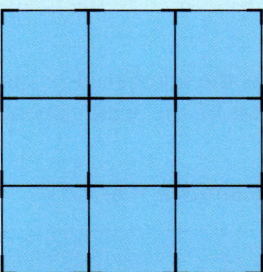

Draw picture of 3 × 3 arrangement

This uses   4 **L** spacers
            8 **T** spacers and
            4 **+** spacers

Investigate different arrangements of tiles.

Note: You are reminded that any coursework submitted for your GCSE examination must be your own. If you copy from someone else then you may be disqualified from the examination.

You should be able to produce this table:

Number of spacers for different arrangements of tiles

Size of square	1 × 1	2 × 2	3 × 3	4 × 4
Number of L spacers	4	4	4	4
Number of T spacers	0	4	8	12
Number of + spacers	0	1	4	9

Now you try ...

What patterns do you notice from the table?
What general statements can you make?
Now test your general statements.

Generalisations for the 'Perfect tiles' task might include:

$L = 4$ where L is the number of L spacers

$T = 4(n - 1)$ where T is the number of T spacers and
                        n is the size of the arrangement ($n × n$)

> The formula $T = 4(n - 1)$ can also be written as $T = 4n - 4$.

Test:

For a 5 × 5 arrangement (ie $n = 5$)

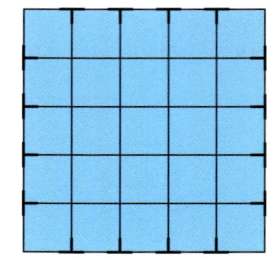

$T = 4(n - 1)$
$T = 4(5 - 1)$
$T = 4 \times 4$
$T = 16$

The number of T spacers is 16 so the formula works for a 5 × 5 arrangement.

$+ = (n - 1)^2$ where + is the number of + spacers and
          n is the size of the arrangement ($n \times n$)

Test:
For a 6 × 6 arrangement (ie $n = 6$)

$+ = (n - 1)^2$
$+ = (6 - 1)^2$
$+ = (5)^2$
$+ = 25$

The number of + spacers is 25 so the formula works for a 6 × 6 arrangement.

Making justifications

Once you have found the general statement and tested it then you need to justify it. You justify the statement by explaining WHY it works.

For example, here are possible justifications for the rules in task 3:

WHY is the number of L spacers always 4?

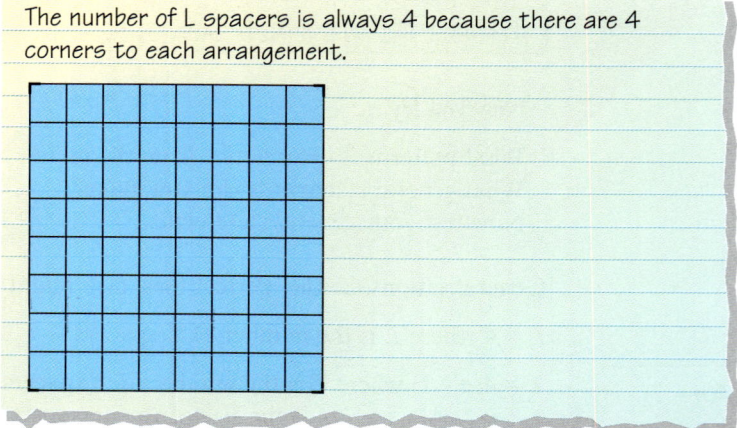

The number of L spacers is always 4 because there are 4 corners to each arrangement.

WHY are all of the T spacers multiples of 4?

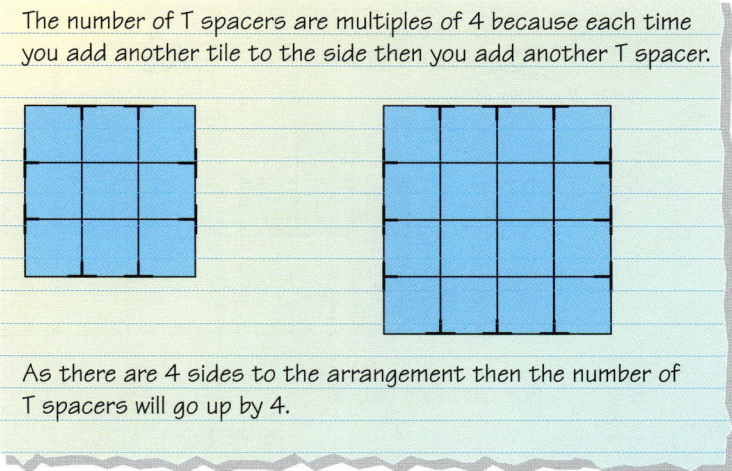

The number of T spacers are multiples of 4 because each time you add another tile to the side then you add another T spacer.

As there are 4 sides to the arrangement then the number of T spacers will go up by 4.

WHY are all of the + spacers square numbers?

Now you try ...

Explain why the number of + spacers are **always** square numbers.

Extending the problem – investigating further

Once you have understood and explained the basic task, you should extend your work to get better marks.

To extend a task you need to pose your own questions.
This means that you must think of different ways to extend the original task.

Extending task 1: The triangle problem

You could ask and investigate:

What about different patterns?

For example, triangles

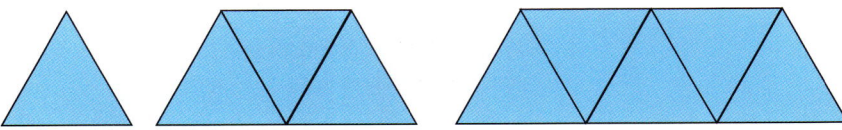

> What about patterns of different shapes?

For example, squares

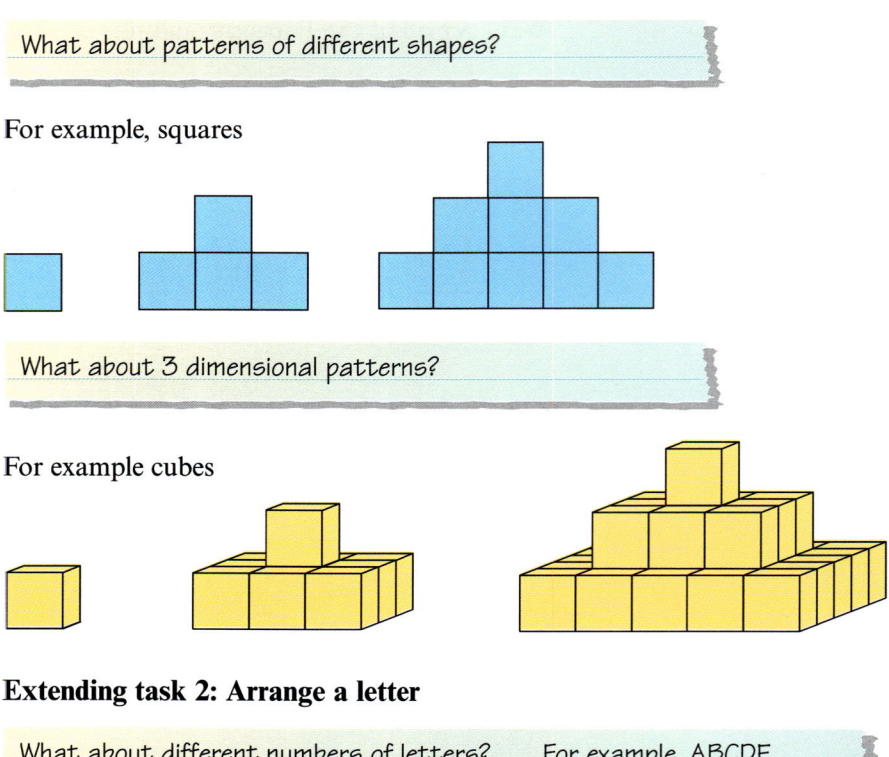

> What about 3 dimensional patterns?

For example cubes

Extending task 2: Arrange a letter

> What about different numbers of letters?　　For example, ABCDE

> What happens when:
> ◆ a letter is repeated?　　　　　　　　　For example, AABCD
> ◆ a letter is repeated more than once?　　For example, AAABC
> ◆ more than one letter is repeated?　　　For example, AAABB

Of course, you will have to do more than just pose a question – you will have to carry out the investigation and come to a conclusion.

> **Now you try ...**
>
> Think of some different ways to extend task 3: Perfect Tiles.
> What sort of questions might you ask?
> What other areas might you explore?

Extending task 3: Perfect Tiles

You could ask:

> What about rectangular arrangements of tiles?

> What about different shaped tiles?
> For example, triangles

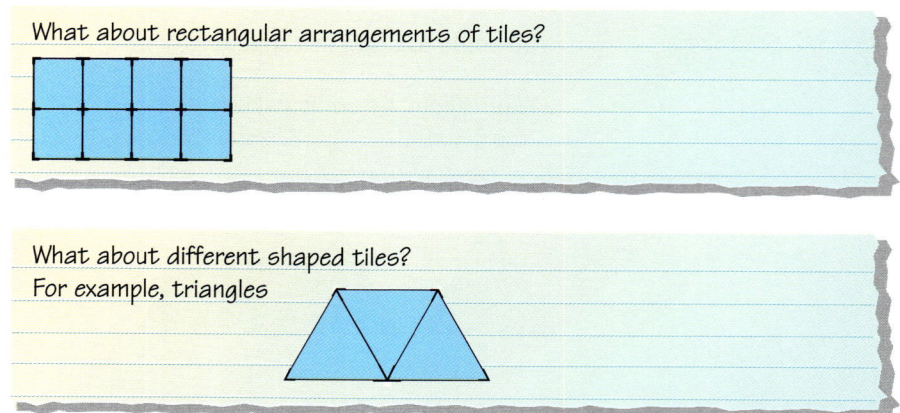

> Remember to work systematically.

Consider rectangular arrangements of tiles:

Number of spacers for different arrangements of tiles

Size of arrangement	2 × 1	2 × 2	2 × 3	2 × 4	2 × 5
Number of L spacers	4	4	4	4	4
Number of T spacers	2	4	6	8	10
Number of + spacers	0	1	2	3	4
Number of spacers	6	9	12	15	18

Number of spacers for different arrangements of tiles

Size of arrangement	3 × 1	3 × 2	3 × 3	3 × 4	3 × 5
Number of L spacers	4	4	4	4	4
Number of T spacers	4	6	8	10	12
Number of + spacers	0	2	4	6	8
Number of spacers	8	12	16	20	24

Rules for the 'Perfect tiles' task extension might include:

$L = 4$	where L is the number of L spacers
$T = 2(x - 1) + 2(y - 1)$	where T is the number of T spacers and x is the length of the arrangement and y is the width of the arrangement ($x \times y$)
$P = (x - 1)(y - 1)$	where P is the number of + spacers and x is the length of the arrangement and y is the width of the arrangement ($x \times y$)
$N = (x + 1)(y + 1)$	where N is the total number of spacers and x is the length of the arrangement and y is the width of the arrangement ($x \times y$)

> Note that the formula $T = 2(x - 1) + 2(y - 1)$ can also be written
> $T = 2(x - 1) + 2(y - 1)$
> $T = 2x - 2 + 2y - 2$
> $T = 2x + 2y - 4$

> It is sensible to replace + by P so that P stands for the number of + spacers.

Moderator comment

The use of higher order algebra can result in high marks awarded on the middle strand as well as an opportunity to provide a sophisticated justification

The total number of spacers should equal the number of L spacers plus the number of T spacers plus the number of + spacers

Proof:
$$N = L + T + P$$
$$= 4 + [2(x-1) + 2(y-1)] + [(x-1)(y-1)]$$
$$= 4 + 2x - 2 + 2y - 2 + xy - y - x + 1$$
$$= xy + x + y + 1$$
$$= (x + 1)(y + 1)$$

Since $N = (x + 1)(y + 1)$ is true, then my theory is proved

Task 4

SQUARE SEA SHELLS

Square sea shells are formed from squares whose pattern of growth is shown as follows:

Day 1

Day 2

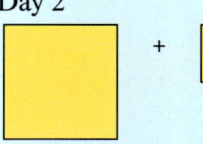

The length of the new square is half that of the previous day

Day 3

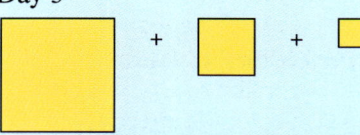

The length of the new square is half that of the previous day

Day 4

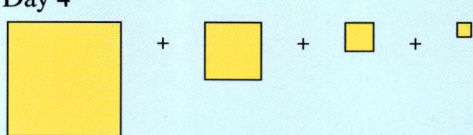

The length of the new square is half that of the previous day

This pattern of growth continues

Investigate the area covered by square sea shells on different days.

Investigate further.

For a square of side a

$$\text{Day 1} \quad \text{Area} = a^2$$

$$\text{Day 2} \quad \text{Area} = a^2 + \frac{a^2}{4}$$

$$\text{Day 3} \quad \text{Area} = a^2 + \frac{a^2}{4} + \frac{a^2}{16}$$

$$\text{Day 4} \quad \text{Area} = a^2 + \frac{a^2}{4} + \frac{a^2}{16} + \frac{a^2}{64}$$

The proof of the formula is beyond GCSE and the rest of the work is left as a challenge for the reader.

In this unit we have tried to give you some hints on approaching investigative coursework to gain your best possible mark.

This mathematics is often useful in investigative tasks:

- Creating tables
- Drawing and interpreting graphs
- Recognising square numbers
- Recognising triangular numbers
- Finding the nth term of a linear sequence

Summary

These are the grade criteria your coursework will be marked by:

Identifying information and making statements (grade E/F)

To achieve this level you must:
- solve the task by collecting information
- illustrate your information using diagrams and calculations
- make a general statement from your results

Testing general statements (grade D)

To achieve this level you must:
- break down the task by solving it in an orderly manner
- interpret and explain your diagrams and calculations
- confirm your general statement by further testing

Posing questions and justifying (grade C)

To achieve this level you must:
- extend the task by introducing your own relevant questions
- begin to use mathematical symbols to explain your work
- make a justification for your general statement
- provide a sophisticated justification

Making further progress (grade B)

To achieve this level you must:
- extend the task by following through alternative approaches
- use mathematical symbols consistently to explain your work
- develop your justification further

Justifying a number of mathematical features (grade A)

To achieve this level you must:
- develop the task by including a number of mathematical features
- use mathematical symbols accurately to argue your case
- provide a sophisticated justification

Statistical coursework

In your statistical coursework you will need to:
- Provide a well considered hypothesis and provide a plan of action to carry out the task
- Decide what data is needed and collect results for the task using an appropriate sample size and sampling method
- Consider the most appropriate way to represent your results and write down any observations you make
- Consider the most appropriate statistical calculations to use and interpret your findings in terms of the original hypothesis
- Develop the task by posing your own questions – you may need to collect further data to move the task on
- Extend the task by using techniques and calculations from the content of the higher tier
- Make your conclusions clear – always link them to the original hypothesis recognising limitations and suggesting improvements.

Assessing the task

It may be helpful for you to know how the work is marked. This way you can make sure you are familiar with the sort of thing that examiners are looking for.

Statistical work is marked under three headings:

1. Specifying the problem and planning
2. Collecting, processing and representing the data
3. Interpreting and discussing the results

Each strand assesses a different aspect of your coursework as follows:

1. Specifying the problem and planning

This strand is about choosing a problem and deciding what needs to be done then doing it. The strand requires you to provide clear aims, consider the collection of data, identify practical problems and explain how you might overcome them. For the higher marks you need to decide upon a suitable sampling method, explain what steps were taken to avoid possible bias and provide a well structured report.

2. Collecting, processing and representing the data

This strand is about collecting data and using appropriate statistical techniques and calculations to process and represent the data. Diagrams should be appropriate and calculations mostly correct. For the higher marks you will need to accurately use higher level statistical techniques and calculations from the higher tier GCSE syllabus content.

3. Interpreting and discussing the results

This strand is about commenting, summarising and interpreting your data. Your discussion should link back to the original problem and provide an evaluation of the work undertaken. For the higher marks you will need to provide sophisticated and rigorous interpretations of your data and provide an analysis of how significant your findings are.

This unit uses a series of statistical tasks to demonstrate how each of these strands can be achieved.

Planning your work

Statistical coursework requires careful planning if you are to gain good marks. Before undertaking any statistical investigation, it is important that you plan your work and decide exactly what you are going to investigate – do not be too ambitious!

Before you start you should:

- decide what your investigation is about and why you have chosen it
- decide how you are going to collect the information
- explain how you intend to ensure that your data is representative
- detail any presumptions which you are making

Getting started – setting up your hypothesis

A good starting point for any statistical task is to consider the best way to collect the data and then to write a clear hypothesis you can test.

Task 1

WHAT THE PAPERS SAY

Choose a passage from two different newspapers and investigate the similarities and differences between them.

Write down a hypothesis to test.

Design and carry out a statistical experiment to test the hypothesis.

Investigate further.

Note: You are reminded that any coursework submitted for your GCSE examination must be your own. If you copy from someone else then you may be disqualified from the examination.

First consider different ways to compare the newspapers:

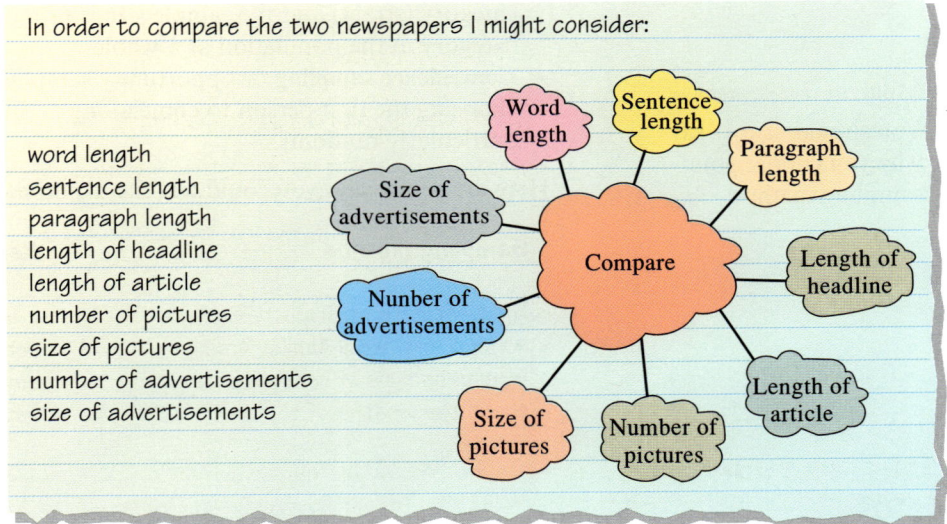

In order to compare the two newspapers I might consider:

word length
sentence length
paragraph length
length of headline
length of article
number of pictures
size of pictures
number of advertisements
size of advertisements

Now formulate your hypothesis.

Here are some possible hypotheses:

For my statistics coursework, I am going to investigate the hypothesis that 'tabloid' papers use shorter words than 'broadsheet' newspapers.

My hypothesis is that word lengths in the tabloid newspaper will be shorter than word lengths in the broadsheet newspaper.

My hypothesis is that sentence lengths in the tabloid newspaper will be shorter than sentence lengths in the broadsheet newspaper.

My hypothesis is that the number of advertisements in the tabloid newspaper will be greater than the number of advertisements in the broadsheet newspaper.

345

> **Remember:**
> it does not matter whether your hypotheses are true or false and you will still gain marks if your hypothesis turns out to be false.

> **Now you try ...**
>
> See if you can add some hypotheses of your own.
> Explain how you would proceed with the task.

Choosing the right sample

Moderator comment

It is important to choose an appropriate sample, give reasons for your choice and explain what steps you will take to avoid bias.

Once you have decided your aims and set up a hypothesis then it is important to consider how you will test your hypothesis.

Sampling techniques include:

- **Random sampling** is where each member of the population has an equally likely chance of being selected. An easy way to do this would be to give each person a number and then choose the numbers randomly.
- **Systematic sampling** is the same as random sampling except that there is some system involved such as numbering each person and then choosing every 20th number to form the sample.
- **Stratified sampling** is where each person is divided into some particular group or category (strata) and the sample size is proportional to the size of the group or category in the population as a whole.
- **Convenience sampling** or opportunity sampling is one which involves simply choosing the first person to come along ... although this method is not particularly random!

Moderator comment

You should always say why you choose your sampling method.

Here are some ways you could test the hypotheses for task 1:

Sampling method	Reason
To investigate my hypothesis I am going to choose a similar article from each type of newspaper and count the lengths of the first 100 words.	I decided to choose similar articles because the words will be describing similar information and so will be better to make comparisons.
To investigate my hypothesis I will choose every tenth page from each type of newspaper and calculate the percentage area covered by pictures.	I decided to choose every tenth page from each type of newspaper because the types of articles vary throughout the newspaper (for example headlines at the front and sports at the back of the paper).

Collecting primary data

Moderator comment

It is important to carefully consider the collection of reliable data. Appropriate methods of collecting primary data might include observation, interviewing, questionnaires or experiments.

If you are collecting primary data then remember:

- **Observation** involves collecting information by observation and might involve participant observation (where the observer takes part in the activity), or systematic observation (where the observation happens without anyone knowing).
- **Interviewing** involves a conversation between two or more people. Interviewing can be formal (where the questions follow a strict format) or informal (where they follow a general format but can be changed around to suit the questioning).
- **Questionnaires** are the most popular way of undertaking surveys. Good questionnaires are
  - simple, short, clear and precise,
  - attractively laid out and quick to complete

  and the questions are
  - not biased or offensive
  - written in a language which is easy to understand
  - relevant to the hypothesis being investigated
  - accompanied by clear instructions on how to answer the questions.

> **Moderator comment**
>
> It is always a good idea to undertake a small scale 'dry run' to check for problems. This 'dry run' is called a pilot survey and can be used to improve your questionnaire or survey before it is undertaken.

Avoiding bias

You must be very careful to avoid any possibility of bias in your work. For example, in making comparisons it is important to ensure that you are comparing like with like.

Jean undertook task 1: 'What the papers say'.
She collected this data from two newspapers by measuring (observation).

	Tabloid	Broadsheet
Number of advertisements	35	30
Number of pages	50	30
Area of each page	1000cm^2	2000cm^2

Her hypothesis is:

> The tabloid newspaper has more advertisements than the broadsheet newspaper.

A quick glance at the table may make her claim look true: the broadsheet has 30 adverts but the tabloid has 35, which is more.

However, the sizes of the newspapers are different so Jean is not really comparing like with like.
To ensure Jean compares like with like she should take account of:

- the area of the pages
- the number of pages

and so on.

Percentage coverage per page:
Tabloid $= 35 \div 50 \times 100\% = 70\%$
Broadsheet $= 30 \div 30 \times 100\% = 100\%$

This shows that:

> The broadsheet newspaper has more advertisements per
> page than the tabloid newspaper.

The total area of the pages is:
Tabloid $= 50 \times 1000\text{cm}^2 = 50\,000\text{cm}^2$
Broadsheet $= 30 \times 2000\text{cm}^2 = 60\,000\text{cm}^2$

So the percentage coverage is:
Tabloid $= 35 \div 50\,000 \times 100\% = 0.07\%$
Broadsheet $= 30 \div 60\,000 \times 100\% = 0.05\%$

> Note: this still doesn't take account of the size of the adverts!

This shows:

> The tabloid newspaper has more advertisements per area
> of coverage than the broadsheet newspaper.

Methods and calculations

> You can represent the data using statistical calculations such as the mean, median, mode, range and standard deviation.

Once you have collected your data, you need to use appropriate statistical methods and calculations to process and represent your data.

Task 2

> **GUESSING GAME**
>
> Dinesh asked a sample of people to estimate the length of a line and weight of a packet.
>
> Write down a hypothesis about estimating lengths and weights and carry out your own experiment to test your hypothesis.
>
> Investigate further

Note: You are reminded that any coursework submitted for your GCSE examination must be your own. If you copy from someone else then you may be disqualified from the examination.

Maurice and Angela decide to explore the hypothesis that:

> Students are better at estimating the length of a line
> than the weight of a package.

To test the hypothesis they collect data from 50 children, detailing their estimations of the length of a line and the weight of a parcel.

Here are their findings:

> **Moderator comment**
>
> It is important to consider whether information on all of these statistical calculations is essential.

A table to show the estimations for the length of a line and the weight of a parcel		
	Length (cm)	Weight (g)
Mean	15.9	105.2
Median	15.5	100
Mode	14	100
Range	8.6	28
SD	1.2	2.1

Note: the actual length of the line is 15cm and the weight of the parcel is 100g.

Now you try …

Using the table:

- What do you notice about the average of the length and weight?
- What do you notice about the spread of the length and weight?
- Does the information support the hypothesis?

Graphical representation

Other statistical representations might include pie charts, bar charts, scatter graphs and histograms.

Moderator comment

You should only use appropriate diagrams and graphs.

Remember to consider the possibility of bias in your data:

The percentage error for the lengths is
$0.9/15 \times 100\% = 6\%$

The percentage error for the weights is
$5.2/100 \times 100\% = 5.2\%$

The data for task 2 was sorted into different categories and represented as a table.

It may be useful to show your results using graphs and diagrams as sometimes it is easier to see trends.

Graphical representations might include stem and leaf diagrams and cumulative frequency graphs and box-and-whisker diagrams.

Once you have drawn a graph you should say what you notice from the graph:

From my representation, I can see that the estimations for the line are generally more accurate than the estimations for the weight because:

- the mean is closer to the actual value for the lengths and
- the standard deviation is smaller for the lengths

However, on closer inspection:

- the percentage error on the mean is smaller for the weights
- so the median and mode value are better averages to use for the weights

Using secondary data

Moderator comment

If you use secondary data there must be enough 'to allow sampling to take place' – about 50 pieces of data.

You may use secondary data in your coursework.

Secondary data is data that is already collected for you.

Task 3

GENDER DIFFERENCES IN EXAMINATIONS

Jade is conducting a survey in the GCSE examination results for Year 11 students at her school.

She has collected data on last year's results, and wants to compare the performance of boys and girls at the school.

Jade explores the hypothesis that:

> Year 11 girls do better in their GCSE examinations than boys

To test the hypothesis she decided to concentrate on the core subjects and her findings are shown in the table:

A table to show the performance of Year 11 girls and boys in their GCSE examinations

		%A*-C	%A*-G	APS
English	Girls	62	93	4.9
	Boys	46	89	4.3
	All	54	91	4.6
Mathematics	Girls	47	91	4.2
	Boys	45	89	4.2
	All	46	90	4.2
Science	Girls	48	91	4.4
	Boys	45	88	4.3
	All	47	90	4.4

Now you try ...

Using the table:

- What do you notice about the percentages of A*-C grades?
- What do you notice about the percentages of A*-G grades?
- What do you notice about the average point scores?
- Does the information support the hypothesis?

The data has been sorted into different categories and represented as a table.

Jade could use comparative bar charts to represent the data as it will allow her to make comparisons more easily.

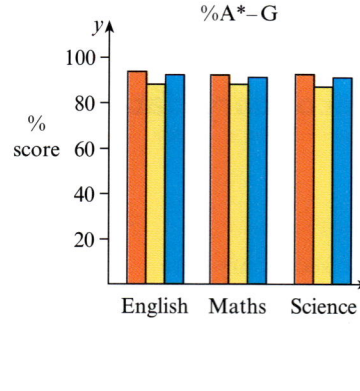

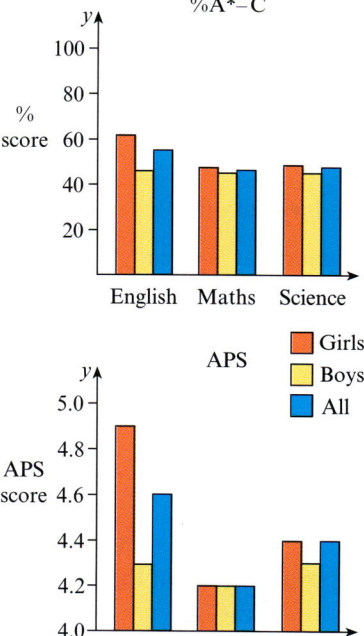

Summarising and interpreting data

Jade summarises her findings like this:

> From my graph, I can see that Year 11 girls do better in their GCSE examinations than boys in terms of A*-C grades, A*-G grades and average point scores.
>
> The performance of Year 11 girls is significantly better than boys in English although less so in mathematics.

Moderator comment

You should refer to your original hypothesis when you summarise your results

Hint:
You need to appreciate that the data is more secure if the sample size is 500 rather than 50.

Moderator comment

In your conclusion you should also suggest limitations to your investigation and explain how these might be overcome.

You may wish to discuss:
♦ sample size
♦ sampling methods
♦ biased data
♦ other difficulties

Extending the task

To gain better marks in your coursework you should extend the task in light of your findings.

In your extension you should:

♦ Give a clear hypothesis
♦ Collect further data if necessary
♦ Present your findings using charts and diagrams as appropriate
♦ Summarise your findings referring to your hypothesis

Extending task 3: Gender differences in examinations

Jade extends the task by looking at the performance of individual students in combinations of different subjects.

> I am now going to extend my task by looking at the performance of individual students in English and mathematics. My hypothesis is that there will be no correlation between the two subjects.

She presents the data on a scattergraph.

Note:
You should only draw a line of best fit on the diagram to show the correlation if you make some proper use of it (for example to calculate a students' likely English mark given their mathematics mark.)

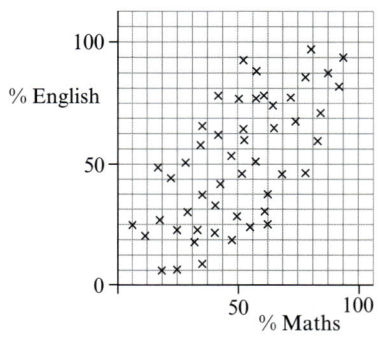

She summarises her findings:

> From my scattergraph, I can see that there is some correlation between the performance of individual students in English and mathematics.

She extends the task further:

> I shall now look at the performance of individual students in mathematics and science. My hypothesis is that there will be a correlation between the two subjects.

She presents the data on a scattergraph.

Note:
The strength of the correlation could be measured using higher level statistical techniques such as Spearman's Rank Correlation.

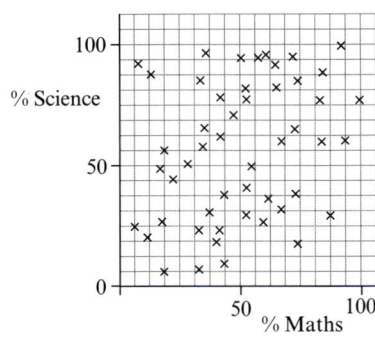

> From my scattergraph, I can see that there is no correlation between the performance of individual students in mathematics and science.

Extending task 2: Guessing game

Maurice extends the 'Guessing game' like this:

> I am now going to extend my task by looking to see whether people who are good at estimating lengths are also good at estimating weights. My hypothesis is that there will be a strong correlation between peoples' ability at estimating lengths and estimating weights.

He draws a scattergraph:

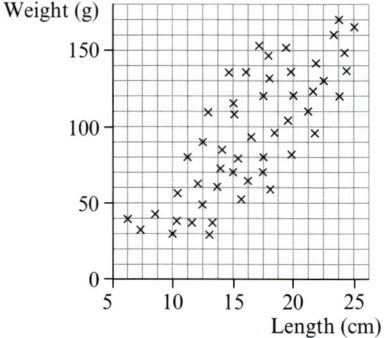

He summarises his findings:

> From my scattergraph I can see that there is a strong correlation between peoples' ability at estimating lengths and estimating weights.

The strength of the correlation can be measured using higher level statistical techniques such as Spearman's Rank Correlation.

50 people were in the survey so $n = 50$

I am now going to extend my investigation by using Spearman's Rank Correlation to calculate the rank correlation coefficient.

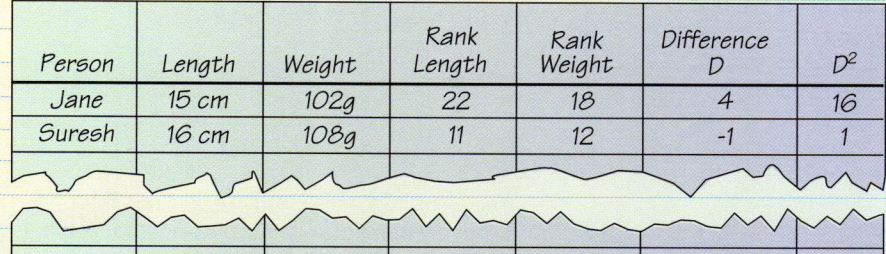

Person	Length	Weight	Rank Length	Rank Weight	Difference D	D^2
Jane	15 cm	102g	22	18	4	16
Suresh	16 cm	108g	11	12	-1	1
					Total	2250

$$r = 1 - \frac{6(\sum D^2)}{n(n^2 - 1)}$$

$$r = 1 - \frac{6(2250)}{50(50^2 - 1)}$$

$$r = 1 - \frac{13500}{50(2499)}$$

$$r = 1 - \frac{13500}{124950}$$

$$r = 1 - 0.108043\ldots$$

$$r = 0.891956\ldots$$

$$r = 0.89 \ (2dp)$$

The value for the rank correlation coefficient is quite close to 1 so that there is a strong positive correlation between my results.

This tells me that there is a strong correlation between peoples' ability at estimating lengths and estimating weights and confirms my original hypothesis.

Using a computer

It is quite acceptable that calculations and representations are generated by a computer, as long as any such work is accompanied by some analysis and interpretation.

Remember:
make sure that your computer generated scattergraph has labelled axes and a title to make it quite clear what it is showing

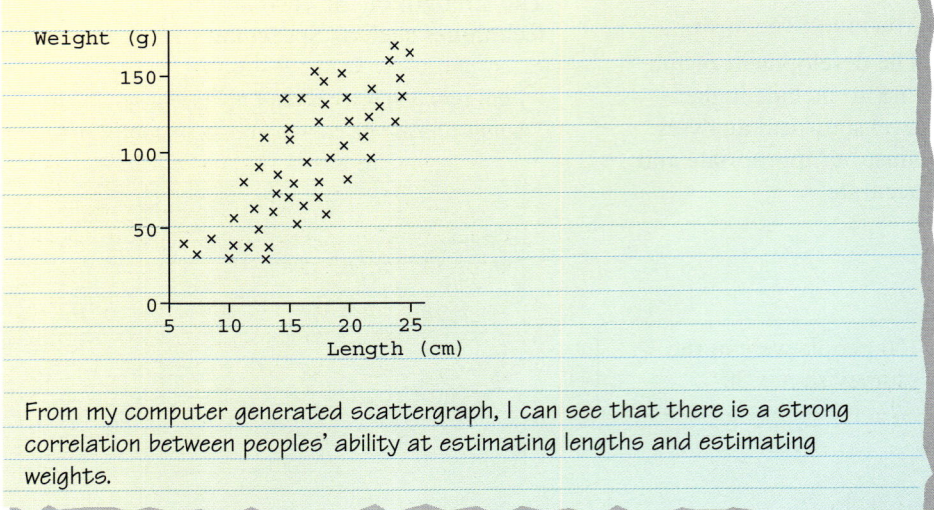

From my computer generated scattergraph, I can see that there is a strong correlation between peoples' ability at estimating lengths and estimating weights.

In this unit we have tried to give you some hints on approaching statistical coursework to gain your best possible mark.

This statistics is often useful in investigative tasks:

◆ Calculating averages (mean, median and mode)
◆ Finding the range
◆ Pie charts, bar charts, stem and leaf diagrams
◆ Constructing a cumulative frequency graph
◆ Finding the interquartile range
◆ Histograms
◆ Calculating the standard deviation
◆ Drawing a scatter graph and line of best fit
◆ Sampling techniques
◆ Discussing bias

Summary

These are the grade criteria your coursework will be marked by:

Foundation statistical task (grade E/F)

To achieve this level you must:
◆ set out reasonably clear aims and include a plan
◆ ensure that the sample size is of an appropriate size (about 25)
◆ collect data and make use of statistical techniques and calculations

For example: pie charts, bar charts, stem and leaf diagrams, mean, median, mode and scattergraphs

◆ summarise and interpret some of your diagrams and calculations

Intermediate statistical task (grade C)

To achieve this level you must:
◆ set out clear aims and include a plan designed to meet those aims
◆ ensure that the sample size is of an appropriate size (about 50)
◆ give reasons for your choice of sample
◆ collect data and make use of statistical techniques and calculations

For example: pie charts, bar charts, stem and leaf diagrams, mean, median, mode (of grouped data), scatter graphs and cumulative frequency

◆ summarise and correctly interpret your diagrams and calculations
◆ consider your strategies and how successful they were

Higher statistical task (grade A)

To achieve this level you must:

- set out clear aims for a more demanding problem
- include a plan which is specifically designed to meet those aims
- ensure that sample size is considered and limitations discussed
- collect relevant data and use statistical techniques and calculations

   For example: pie charts, bar charts, stem and leaf diagrams, mean, median, mode (of grouped data), scatter graphs, cumulative frequency, histograms and sampling techniques

- summarise and correctly interpret your diagrams and calculations
- use your results to respond to your original question
- consider your strategies, limitations and possible improvements

⊗ EXAM QUESTIONS CONTENTS

Past exam questions

Practice exam papers

Formula sheet

In the Edexcel GCSE examination you will be given a formula sheet like this one.
You will be required to memorise all other formulae, such as area of a circle, Pythagoras' theorem and trigonometry.
The formula sheet is the same for all Examining Groups.

Area of trapezium $= \frac{1}{2}(a + b)h$

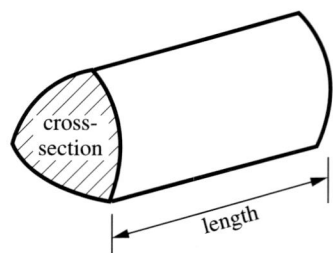

Volume of prism = Area of cross-section × Length

Number

N1 Mark has a market stall. Mark bought 25 melons for his stall.
He paid £16 for 25 melons.

a Work out the price Mark paid for each melon.
Do NOT use a calculator and show ALL your working.

Two of the melons were bad.
Mark sold the other 23 melons for 149p each.

b Work out the total amount for which Mark sold the melons.
Do NOT use a calculator and show ALL your working.

Mark usually sells oranges for 40p each.

He reduces the price to $\frac{7}{8}$ of this.

c Work out the new price of an orange.

Mark bought his potatoes for 30p for each kilogram.
He sold the potatoes and made a profit of 40%.

d At what price did Mark sell the potatoes?

[Edexcel]

N2 £500 is invested for 2 years at 6% per annum compound interest.

a Work out the total interest earned over the 2 years.

£250 is invested for 3 years at 7% per annum compound interest.

b By what single number must £250 be multiplied to obtain the total
amount at the end of the 3 years?

[Edexcel]

N3 Fred has a recipe for 30 biscuits.
Here is a list of ingredients for 30 biscuits.

Self-raising flour   :   230 g
Butter   :   150 g
Caster sugar   :   100 g
Eggs   :   2

Fred wants to make 45 biscuits.

a Complete his new list of ingredients for 45 biscuits.

Self-raising flour   :
Butter   :
Caster sugar   :
Eggs   :

The recipe gives the baking temperature as 350° Fahrenheit, *F*.
A modern oven shows baking temperature in Celsius, *C*.

b Use the formula $C = \dfrac{5(F-32)}{9}$

to change 350° Fahrenheit to Celsius.
Give your answer correct to the nearest degree.

Gill has only 1 kilogram of self-raising flour. She has plenty of the other
ingredients.

c Work out the maximum number of biscuits that Gill could bake.

[Edexcel]

N4   The star Sirius is 81 900 000 000 000 km from the Earth.

   a   Write 81 900 000 000 000 in standard form.

   Light travels 3×10^5 km in 1 second.

   b   Calculate the number of seconds that light takes to travel from Sirius to the Earth.
   Give your answer in standard form correct to 2 significant figures.

[Edexcel]

N5   Compact discs (CDs) are sold for £12 each.

   $\frac{2}{5}$ of the £12 goes to the record company.

   a   Work out the amount of money which goes to the record company.
   30% of the £12 goes to the shopkeeper.

   b   **i**   Write 30% as a decimal.
       ii   Work out the amount of money which goes to the shopkeeper.

   For each CD sold, £0.84 goes to the singer.

   c   Work out the percentage of the £12 which goes to the singer.

   For each CD sold, £0.24 goes to the song-writer.

   d   Work out the fraction of the £12 which goes to the song-writer.
   Give your fraction in its simplest form.

[Edexcel]

N6   Tom uses his calculator to multiply 17.8 by 0.97.
   His answer is 18.236.

   a   Without finding the exact value of 17.8×0.97, explain why his answer must be wrong.

   Sally estimates the value of $\dfrac{42.8 \times 63.7}{285}$ to be 8.

   b   Write down three numbers Sally could use to get her estimate.

[Edexcel]

N7   A shop is having a sale. Each day, prices are reduced by 20% of the price on the previous day.
   Before the start of the sale, the price of a television is £450.
   On the first day of the sale, the price is reduced by 20%.

   a   Work out the price of the television on
       i   the first day of the sale,
       ii   the third day of the sale.

   On the first day of the sale, the price of a cooker is £300.

   b   Work out the price of the cooker before the start of the sale.

[Edexcel]

N8   Shreena is placing an advertisement in a weekly newspaper.

> Single Column Advertisement:
> £3.25 for each line plus VAT at $17\frac{1}{2}$%.

Shreena wants to put a Single Column Advertisement in the paper.
The advertisement is 9 lines long.

   a   Work out the total cost, including VAT, of placing a Single Column Advertisement 9 lines long.

A 4-column advertisement is to be 8 centimetres high.

The cost, in £, of this advertisement is worked out using the formula:

$$\text{Cost} = \frac{\text{Height (cm)} \times \text{Number of columns} \times 1745}{100}$$

b Work out the cost, in £, of this advertisement.

[Edexcel]

N9 A series of patterns is drawn using dots.

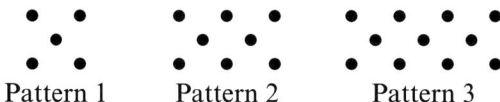

Pattern 1   Pattern 2   Pattern 3

a Draw patterns to show the number of dots needed for

   i Pattern 4   **ii** Pattern 5

The table shows the number of dots needed for different patterns.

Pattern	1	2	3	4	5	6	7
Number of dots	5	8	11				

b Complete the table.

c Explain how you would work out the number of dots needed for Pattern 12.

[Edexcel]

N10 A shop has a sale of jackets and shirts.
In the sale there are a total of 120 jackets and shirts.

Shirts are to be sold at a price of £8.00 plus VAT at $17\frac{1}{2}\%$.

a What is the cost of buying one shirt, including the $17\frac{1}{2}\%$ VAT?

The jackets and shirts are in the ratio 5 : 3.

b Work out the number of jackets.

c Calculate the percentage that are shirts.

[Edexcel]

N11 In this question you must NOT use a calculator.
You must show ALL your working.
Tom buys 67 cameras at £312 each.

a Work out the total cost.

b Write down two numbers you could use to get an approximate answer to your calculation.

[Edexcel]

N12 A school buys a trampoline.
The school is given a discount of $\frac{1}{8}$ of the price.

a Write $\frac{1}{8}$ as
   i a decimal,
   ii a percentage.

The price of the trampoline is £3218.

b Work out the amount the school actually has to pay.

[Edexcel]

N13 Janet goes on holiday to Spain.
The exchange rate is £1 = 230.6 pesetas.
She changes £150 into pesetas.

a How many pesetas should Janet get?

Janet comes back home.
She changes 550 pesetas back into pounds.
The exchange rate is the same.

b How much money should she get?
Give your answer to the nearest penny.

[Edexcel]

N14
$$F = \frac{ab}{a - b}$$

Imran uses this formula to calculate the value of F.
Imran estimates the value of F without using a calculator.
$a = 49.8$ and $b = 30.6$.

a **i** Write down approximate values for a and b that Imran could use
to estimate the value of F.

ii Work out the estimate for the value of F that these approximations
give.

iii Use your calculator to work out the accurate value for F.
Use $a = 49.8$ and $b = 30.6$.
Write down all the figures on your calculator display.

Imran works out the value of F with two new values for a and b.

b Calculate the value of F when
$a = 9.6 \times 10^{12}$ and $b = 4.7 \times 10^{11}$.

[Edexcel]

N15 Jack shares £180 between his two children Ruth and Ben.
The ratio of Ruth's share to Ben's share is 5 : 4.

a Work out how much each child is given.

Ben then gives 10% of his share to Ruth.

b Work out the percentage of the £180 that Ruth now has.

[Edexcel]

N16 At midday the temperature in Moscow was ⁻6 °C.
At midday the temperature in Norwich was 4 °C.

a How many degrees higher was the temperature in Norwich than the
temperature in Moscow?

At midnight the temperature in Norwich had fallen by 7 degrees from
4 °C.

b Work out the midnight temperature in Norwich.

[Edexcel]

N17 The table shows the value of each prize and the number of winners in a lottery.

Value of each prize	Number of winners
£1 000 000	1
£100 000	4
£50 000	10
£25 000	17
£10 000	44
£5000	87

a Work out the total number of winners.

b Work out the total amount of prize money won.

[Edexcel]

N18 A clothes shop has a sale.
All the original prices are reduced by 24% to give the sale price.
The sale price of a jacket is £36.86.
Work out the original price of the jacket.

[Edexcel]

N19 Judy and Anna share a flat.
The rent for the flat is £125 each week.

Judy pays $\frac{3}{5}$ of the £125.

a Work out how much rent Judy pays each week.

They have their washing machine repaired.
The repair costs £28 plus VAT at 17.5%.

b Work out how much VAT they pay.

They hire a television.
The hire cost is £24.50 each month.
Judy and Anna share the £24.50 in the ratio 3 : 4.

c Work out how much each of them pays for the television each month.

[Edexcel]

N20 In part (a) of this question you must NOT use a calculator.
You must show ALL your working.
Fatima buys 48 toys at £9.75 each.

a Work out the total amount that Fatima pays for the 48 toys.

You MAY use a calculator for this part of the question.
Fatima sells each toy at a profit of £3.90.

b Work out the percentage profit that Fatima makes on each toy.

[Edexcel]

N21 Ann wins £160. She gives

$\frac{1}{4}$ of £160 to Pat,

$\frac{3}{8}$ of £160 to John
and £28 to Peter.
What fraction of the £160 does Ann keep?
Give your fraction in its simplest form.

[Edexcel]

N22 Asif sells computers.
Each week Asif is paid £150 plus £20.50 for each computer he sells that week.
Last week Asif was paid £888.

a Work out how many computers Asif sold last week.

Jane is going to buy a computer for £480 + $17\frac{1}{2}$% VAT.

b Work out the total price, including VAT, that Jane will pay for the computer.

[Edexcel]

N23 It needs 20 litres of orange drink to fill 50 cups.

a Work out how many litres of orange drink are needed to fill 60 cups.

The total cost of orange drink for 50 cups is £7.50.
Each cup of drink is sold at a 20% profit.

b Work out the price at which each cup of orange drink is sold.

[Edexcel]

N24 Tracey and Wayne share £7200 in the ratio 5 : 4
Work out how much each of them receives.

[Edexcel]

N25 Tim paid £5.44 for 17 pencils.
Each pencil costs the same.
Work out the cost of each pencil.

[Edexcel]

N26 Write these numbers in order of size.
Start with the largest number.

$$0.8 \qquad 70\% \qquad \frac{7}{8} \qquad \frac{3}{4}$$

[Edexcel]

N27 Here are the first five terms of a sequence.

$$30, 29, 27, 24, 20, ..., ...$$

a Write down the next two terms in the sequence.

Here are the first five terms of a different sequence.

$$1, 5, 9, 13, 17, ...$$

b Find, in terms of n, an expression for the nth term of the sequence.

[Edexcel]

N28

| 3.629 | 3.738 | 3.662 | 3.649 | 3.751 | 3.747 |

a Write the numbers in order of size, smallest first.

b Write 3.629 correct to 3 significant figures.

[Edexcel]

N29 Mr Daley buys a car for £3400.
He later sells it and makes a loss of 20%.
Work out how much he sells it for.

[Edexcel]

N30 Work out

i $\frac{2}{5} + \frac{3}{10}$

ii $\frac{3}{4} - \frac{2}{5}$

iii $\frac{3}{4} \times \frac{3}{5}$

[Edexcel]

N31 Fred buys some apples.
They weigh 3.65 kilograms.
Work out the approximate weight of the apples in pounds.

[Edexcel]

N32 Browsers Bookshops has a shop in each of the five towns shown in the
diagram below.
The diagram shows the distances in miles between the towns.

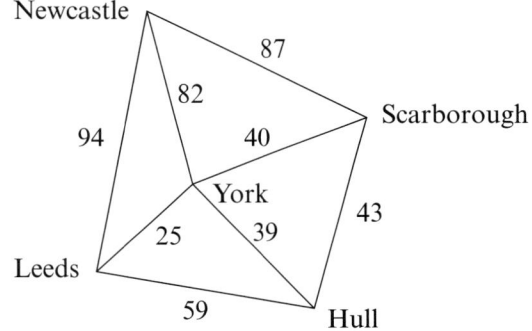

Shirley has to take books to each of the five towns.
She starts at York and finishes at York.
Write down the shortest route she can take.

[Edexcel]

N33 Write down the value of

a √25,

b the cube of 4.

[Edexcel]

N34 Robert used these ingredients to make 24 buns.

100 g of butter

80 g of sugar

2 eggs

90 g of flour

30 ml of milk

Robert wants to make 36 similar buns.
Write down how much of each ingredient he needs for 36 buns.

[Edexcel]

N35   There are 27 wall tiles in a pack.
Only full packs of tiles are sold.
Barry needs 200 tiles.

a   How many full packs of tiles must he buy?

Each tile is a rectangle 20 cm by 15 cm.

b   Work out the area of one tile.

Navdeep wants to tile a wall.
The wall is a rectangle 3 metres by 2.4 metres.

c   Work out the number of tiles she needs to cover the wall completely.

[Edexcel]

N36   Jo got 36 out of 80 in an English test.

a   Work out 36 out of 80 as a percentage.

Jo got 65% of the total number of marks in a French test.
Jo got 39 marks.

b   Work out the total number of marks for the French test.

[Edexcel]

N37   Anna, Beth and Cheryl share the total cost of a holiday in the ratio
$6 : 5 : 4$.
Anna pays £294.

a   Work out the total cost of the holiday.

b   Work out how much Cheryl pays.

[Edexcel]

N38   Write down two different fractions that lie between $\frac{1}{4}$ and $\frac{1}{2}$.

[Edexcel]

N39   Kylie went to Paris.
She changed £200 into French francs.
The exchange rate was £1 = 9.60 French francs.

a   Work out the number of French francs Kylie got.

Kylie brought 25 French francs back from Paris.
The exchange rate was now £1 = 10 French francs.

b   Work out how much Kylie got in pounds.

[Edexcel]

N40   In this question do NOT use a calculator and show ALL your working.
Calculate

a 256×37

b $954 \div 37$

[Edexcel]

N41   Fred won a prize of £12 000.
He put some of the money in a Building Society.
He put the rest of the money in the Post Office.
The money was put in the Building Society and Post Office in the ratio
$2 : 3$.

a   Calculate the amount of money put in the Building Society.

After a number of years the money put in the Building Society had
increased by 9%.

b Calculate the amount of money Fred then had in the Building Society.

After the same number of years the money Fred had put in the Post Office had increased by an eighth.

c Calculate the increase in the amount of money in the Post Office.

[Edexcel]

N42 Use your calculator to evaluate

$$\frac{2.8(3.75 - 1.53)}{17.74 - 3.96}$$

[Edexcel]

N43 A teacher is organising a trip to Pleasureland. There are 560 pupils going on the trip. The pupils will go by coach.
45% of the pupils are boys.

a Work out the number of boys going on the trip.

Do NOT use your calculator for part (b) of this question. Show all your working.
Each pupil pays £14 to go on the trip.

b Work out the total amount paid by the pupils.

[Edexcel]

N44 Mr McDonald is making sheep pens. He uses fences to make pens as shown in the diagram below. The pens are arranged in pairs in a row.

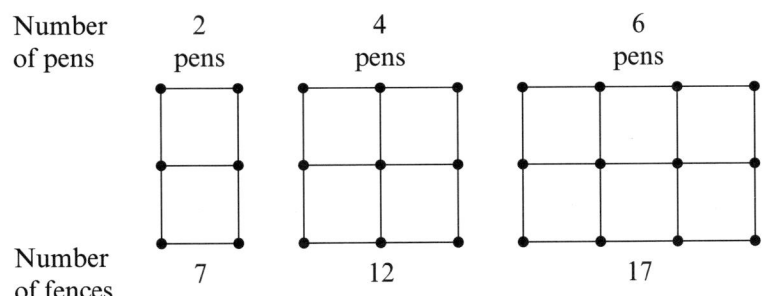

| Number of pens | 2 pens | 4 pens | 6 pens |

| Number of fences | 7 | 12 | 17 |

a Draw diagrams to show the number of fences needed for

i 8 pens,   **ii** 10 pens.

The table below shows the number of fences needed for different numbers of pens.

Number of pens	2	4	6	8	10	12	14	16
Number of fences	7	12	17					

b Copy and complete the table.

c Work out the number of fences needed for 30 pens.

[Edexcel]

N45 The mass of a neutron is 1.675×10^{-24} grams.
Calculate the total mass of 1500 neutrons.
Give your answer in standard form.

[Edexcel]

N46 Mortar is made by mixing 5 parts by weight of sand with 1 part by weight of cement
How much sand is needed to make 8400 kg of mortar?

[Edexcel]

N47 Chetna makes patterns using sticks.

Pattern Number 1   Pattern Number 2   Pattern Number 3

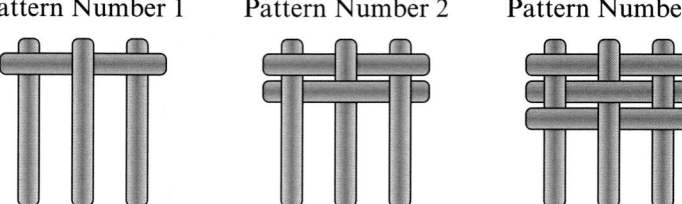

Complete the table for Pattern Number n.

Pattern Number	Number of lollipop sticks
1	4
2	5
3	6
4	7
n	

[Edexcel]

N48 **a** Work out an estimate for

$$\frac{3.08 \times 693.89}{0.47}$$

The length of a rod is 98 cm correct to the nearest centimetre.
b **i** Write down the maximum value that 98 cm could be.
   ii Write down the minimum value that 98 cm could be.

[Edexcel]

N49 Use your calculator to find the exact value of $\dfrac{19.72 + 4.29^2}{8.72 - 3.63}$.

[Edexcel]

N50 The ratio of Winston's age to his mother's age is 2 : 5.

a Write down the ratio of his mother's age to Winston's age.

Winston is 16 years old.

b Work out his mother's age.

The ratio of Flora's age to her father's age is 1 : 3.
Their total age is 56 years.

c Work out Flora's age.

[Edexcel]

Algebra

A1 A shop sells two types of lollipops.
The shop sells Big lollipops at 80p each and Small lollipops at 60p each.
Henry buys x Big lollipops

  a Write down an expression, in terms of x, for the cost of Henry's lollipops.

Lucy buys r Big lollipops and t Small lollipops

  b Write down an expression, in terms of r and t, for the total cost of Lucy's lollipops.

The cost of g Big lollipops and 2 Small lollipops is £10.80

  c Write this as an equation in terms of g.

  d Use your equation to find the value of g.

[Edexcel]

A2

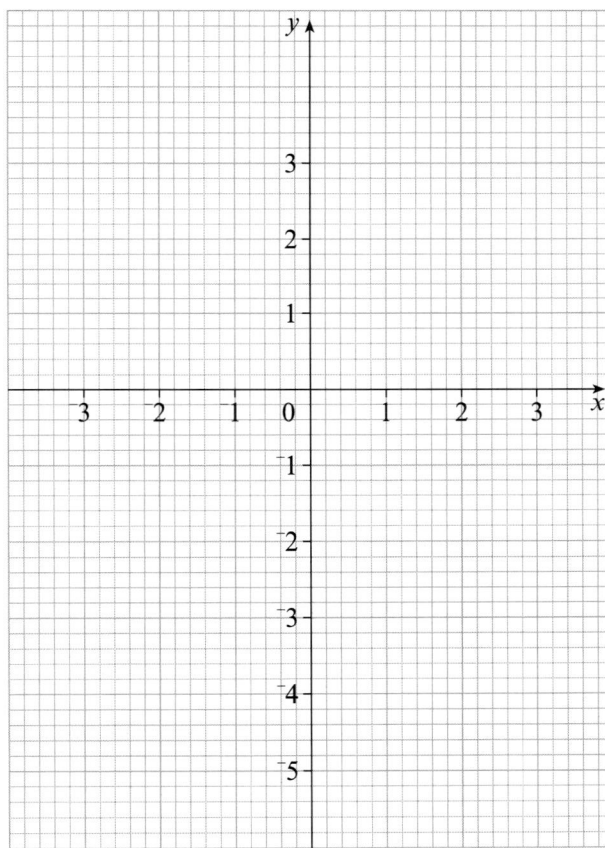

  a On a copy of the grid, draw the graph of $y = x^2 - x - 4$
     Use values of x between ⁻2 and +3

  b Use your graph to write down an estimate for

    i the minimum value of y.

    ii the solutions of the equation $x^2 - x - 4 = 0$

[Edexcel]

A3

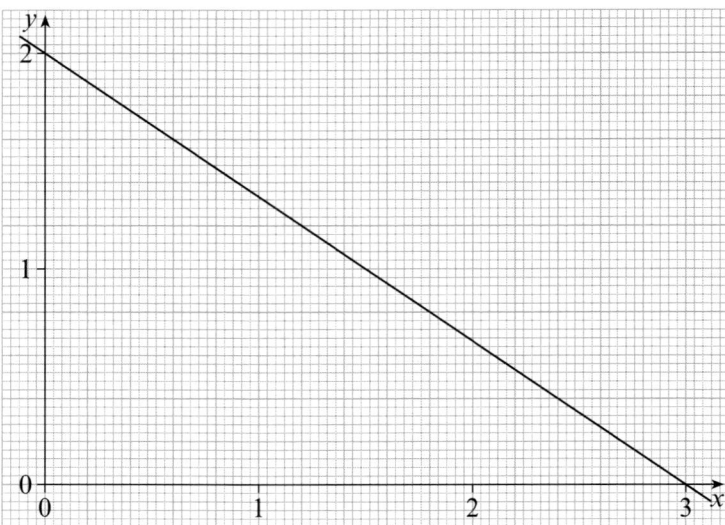

The line with equation $3y = -2x + 6$ has been drawn on the grid.

a Draw the graph of $y = 2x - 2$ on a copy of the same grid.

b Use the graphs to find the solution of the simultaneous equations

$$3y = -2x + 6$$
$$y = 2x - 2$$

A line is drawn parallel to $3y = -2x + 6$ through the point (2, 1).

c Find the equation of this line.

[Edexcel]

A4 **a** Solve $\qquad 7 - \dfrac{3x}{2} = 11$

b **i** Factorise $\qquad x^2 + 4x - 12$

Hence, or otherwise:
   ii Solve

$$x^2 + 4x - 12 = 0$$

[Edexcel]

A5 Here are the first five terms of a sequence.

17, 14, 11, 8, 5.

a **i** Write down the next two terms of the sequence.
   ii Explain how you worked out your answers.

b Find, in terms of n, an expression for the nth term of the sequence.

c Find the 50th term of the sequence.

[Edexcel]

A6 y is an integer and $-2 < y \leqslant 2$.

a Write down all the possible values of y.

b **i** Solve the inequality $3n > -8$.
   ii Write down the smallest integer which satisfies the inequality $3n > -8$.

[Edexcel]

A7 **a** Complete the table of values for $y = 2x - 1$.

x	-2	-1	0	1	2	3
y						

Make a copy of this grid. Label and graduate each axis.

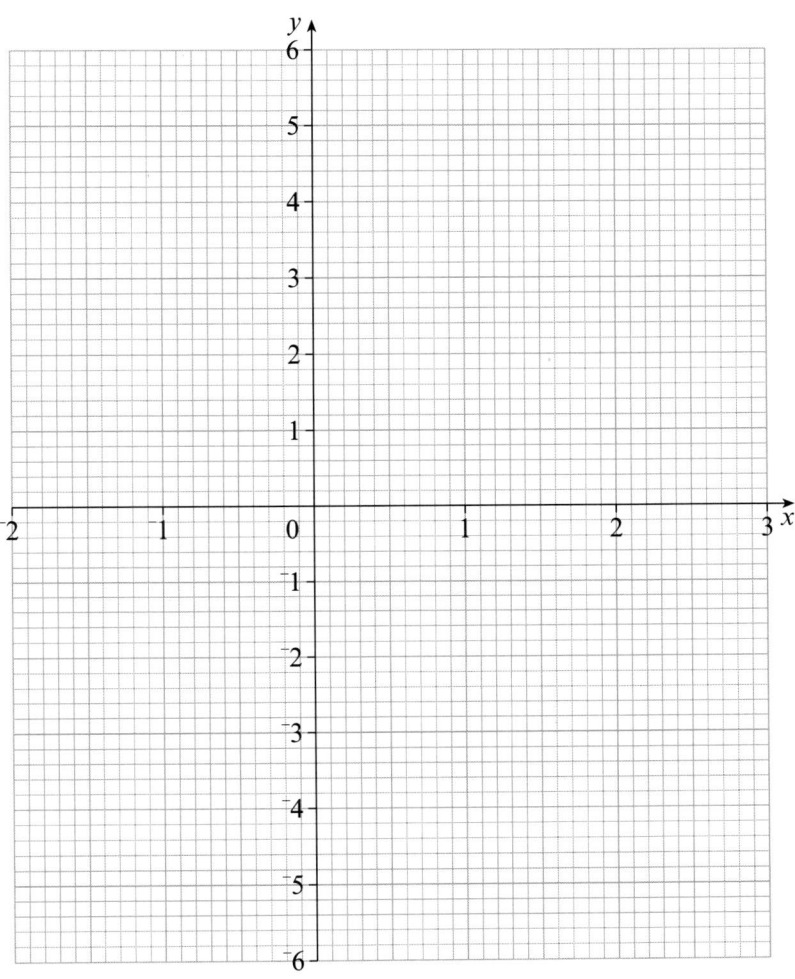

b Draw the graph of $y = 2x - 1$.

c Use your graph to find
i the value of y when $x = -1.4$
ii the value of x when $y = 3.8$

[Edexcel]

A8 Use the method of trial and improvement to solve the equation

$$x^3 - 2x = 37$$

Give your answer correct to two decimal places. You must show ALL your working.

[Edexcel]

A9 $C = 180R + 2000$

The formula gives the capacity, C litres, of a tank needed to supply water to R hotel rooms.

$R = 5$.

a Work out the value of C.

$C = 3440$

b Work out the value of R.

A water tank has a capacity of 3200 litres.

c Work out the greatest number of hotel rooms it could supply.

d Make R the subject of the formula

$$C = 180R + 2000.$$

[Edexcel]

A10 **a** Complete this table of values for $y = 2x + 3$.

x	−3	−2	−1	0	1	2
y		−1				

b Draw the graph of $y = 2x + 3$ on a copy of this grid.

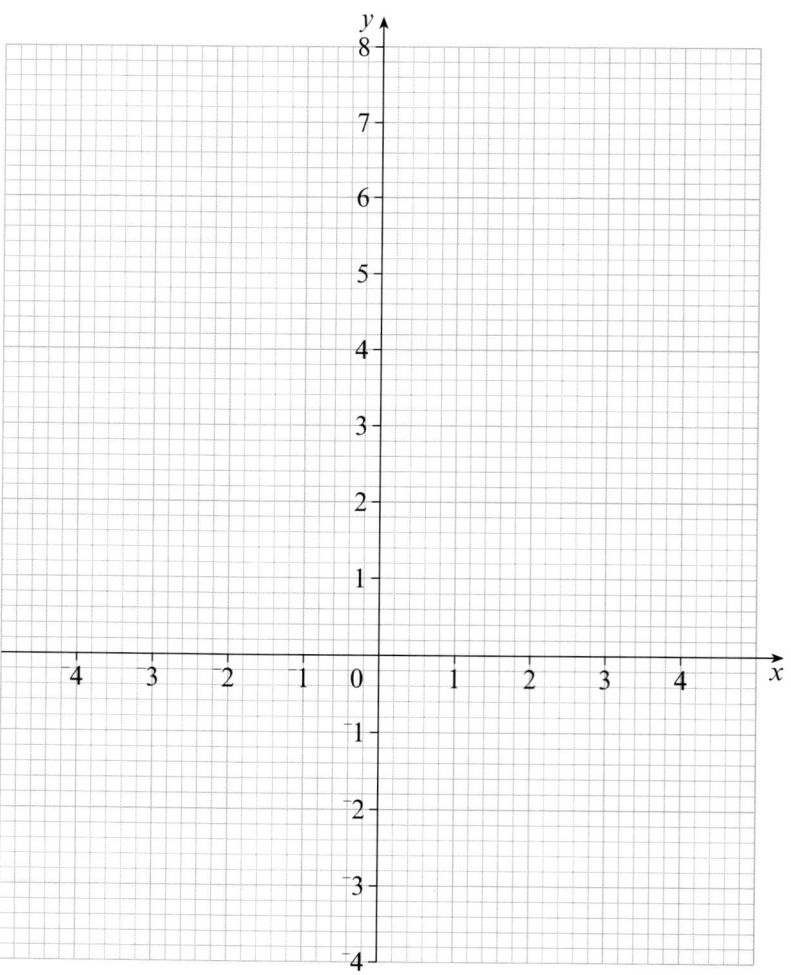

c Use your graph to find
   i the value of y when $x = 1.5$,
   ii the value of x when $y = -0.5$.

[Edexcel]

A11   **a**   Solve the equation
$$2x = 10.$$
       b   Solve the equation
$$6y + 1 = 25.$$
       c   Solve the equation
$$8p - 3 = 3p + 13.$$
       d   Solve the equation
$$4x + 3 = 2(x - 3).$$
       e   Solve the inequality
$$2x + 3 \leqslant 8.$$

[Edexcel]

A12   **a**   Expand and simplify
$$(2x - 5)(x + 3).$$
       b   **i**   Factorise
$$x^2 + 6x - 7.$$
           ii   Solve the equation
$$x^2 + 6x - 7 = 0.$$

[Edexcel]

A13   The cost, C pounds, of a coat rack with h hooks can be worked out using the formula
$$C = 3h + 7.$$
       a   Work out the cost of a coat rack with four hooks.

       Another coat rack costs £43.

       b   Use the same formula to work out the number of hooks this coat rack has.

       c   Make h the subject of the formula
$$C = 3h + 7.$$

[Edexcel]

A14   Sharon earns b pounds an hour.
       She worked for h hours.
       She also earned a bonus of c pounds.
       Write down a formula for her total earnings, P pounds.

[Edexcel]

A15 The diagram shows a rectangular pond and a path.
The outside edges of the path form the rectangle *ABCD*.

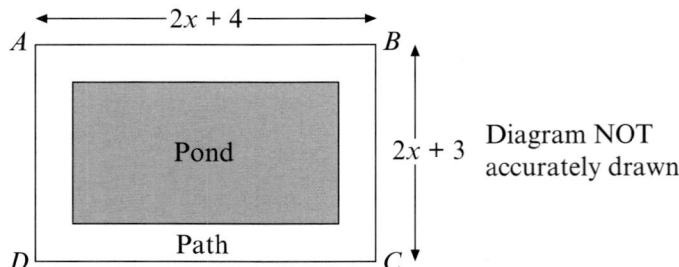

The length, in metres, of *AB* is $2x + 4$.
The length, in metres, of *BC* is $2x + 3$.

a Write down, in terms of x, an expression for the perimeter of the rectangle *ABCD*.
Write your expression in its simplest form.

The area of the pond is 12 m².

b Show that the area, in m², of the path is $4x^2 + 14x$.

$x = 1.2$

c Use the expression

$$4x^2 + 14x$$

to find the area of the path when $x = 1.2$.

[Edexcel]

A16 **a** Solve $4p + 6 = 26$.
b Solve $5(2q + 6) = 25$.
c Solve $18y - 27 = 10y - 25$.

[Edexcel]

A17 The diagrams show 3 shapes made with sticks.

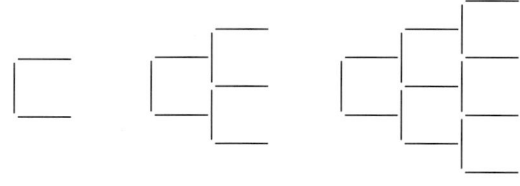

Shape Number   1   2   3

a Draw Shape Number 4.
b Copy and complete this table.

Shape Number	1	2	3	4	5	6
Number of sticks	3	8	15	24		

c Find an expression, in terms of n, for the number of sticks in Shape Number n.

[Edexcel]

A18 $v^2 = \dfrac{GM}{R}$

$G = 6.6 \times 10^{-11}$
$M = 6 \times 10^{24}$
$R = 6\ 800\ 000$

   a Calculate the value of v. Give your answer in standard form, correct to 2 significant figures.

   b Rearrange the formula $v^2 = \dfrac{GM}{R}$ to make M the subject.

[Edexcel]

A19   **a** Simplify

$$2x - 1 + x + 4 + x$$

   b Solve the equation

$$5(2p - 3) = 50$$

   c Solve the equation

$$\frac{16 - q}{3} = 3$$

[Edexcel]

A20   **a** Expand and simplify

$$4(x + 3) + 3(2x - 3)$$

   b Expand and simplify

$$(2x - y)(3x + 4y)$$

[Edexcel]

A21   Solve the simultaneous equations

$$2x + 5y = -1$$
$$6x - y = 5$$

[Edexcel]

A22   **a** Simplify

$$x^3 \times x^5$$

   b Simplify

$$y^6 \div y^2$$

   c Simplify

$$\frac{8w^7}{2w^2 \times w^3}$$

[Edexcel]

A23   **a** Expand and simplify

$$(x + 5)(x - 3)$$

   b Factorise completely

$$6a^2 - 9ab$$

[Edexcel]

A24 The equation

$$x^3 - 5x = 38$$

has a solution between 3 and 4.
Use a trial and improvement method to find this solution.
Give your answer correct to 1 decimal place.
You must show ALL your working.

[Edexcel]

A25 **a** Multiply out

$$t^2(t^3 - t^4)$$

b Multiply out and simplify

$$3(2a + 6) - 2(3a - 6)$$

c Simplify

$$\frac{12a^2b}{4ab}$$

[Edexcel]

A26 **a** Simplify

$$3x \times 4y$$

b Multiply out

$$5(3h + 2)$$

[Edexcel]

A27 Solve the equations
a $3p + 7 = 34$
b $3(2q - 5) = 36$
c $5r + 6 = 2r - 15$

[Edexcel]

A28 Use a trial and improvement method to solve the equation

$$x^3 + 3x = 60.$$

Show ALL your working and give your answer correct to 2 decimal places.

[Edexcel]

A29 Solve
a $5x - 1 = 2x + 5$
b $4(y + 3) = 30$

A30
$$P = \frac{mv}{t}$$

$m = 324$, $v = 76.8$ and $t = 0.413$.

a Work out an estimate for the value of P.

b Make v the subject of the formula $P = \dfrac{mv}{t}$.

[Edexcel]

A31 **a** Expand and simplify $(x + 6)(x - 4)$.
b Solve $x^2 - 5x + 6 = 0$.

[Edexcel]

A32 This formula is used to change degrees Fahrenheit to degrees Celsius.

$$C = \frac{5}{9}(F - 32)$$

C represents the temperature in degrees Celsius and F represents the temperature in degrees Fahrenheit

a Calculate the value of C when $F = 60$.

b Make F the subject of the formula.

[Edexcel]

A33 Solve the simultaneous equations

$$2p - 3q = 7$$
$$p + q = 1$$

[Edexcel]

A34 Solve the following equations:

a $4x - 7 = 20$

b $3(y + 5) = 42$

[Edexcel]

A35

$$v^2 = u^2 + 2as$$

a Calculate the value of v when $u = -6$, $a = 5$, and $s = 0.8$. Give your answer to one significant figure.

b Make u the subject of the formula $v^2 = u^2 + 2as$.

[Edexcel]

A36

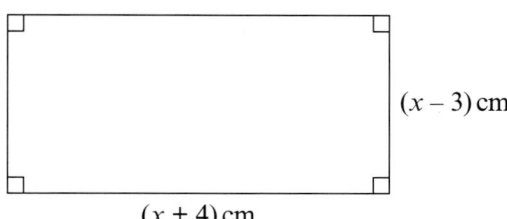

Diagram NOT accurately drawn.

$(x - 3)$ cm

$(x + 4)$ cm

The length of a rectangle is $(x + 4)$ cm.
The width is $(x - 3)$ cm.
The area of the rectangle is 78 cm².

a Use this information to write down an equation in terms of x.

b i Show that your equation in part **a** can be written as

$$x^2 + x - 90 = 0$$

ii Find the values of x which are the solutions of the equation

$$x^2 + x - 90 = 0$$

iii Write down the length and the width of the rectangle.

[Edexcel]

A37 The cost C, in pounds, of a fence with n panels is given by the formula

$$C = 11n + 4(n + 1)$$

a Expand the brackets and express the formula as simply as possible.

b Make n the subject of the formula.

[Edexcel]

A38   Mrs Rogers bought 3 blouses and 2 scarfs.
She paid £26.
Miss Summers bought 4 blouses and 1 scarf.
She paid £28.
The cost of a blouse was x pounds.
The cost of a scarf was y pounds.

   a   Use the information to write down two equations in x and y.

   b   Solve these equations to find the cost of one blouse.

                                  [Edexcel]

A39   A coach has x passengers upstairs and y passengers downstairs.

   a   Write down an expression, in terms of x and y, for the total number of passengers on the coach.

The tickets for the journey on the coach cost £5 each.

   b   Write down an expression, in terms of x and y, for the total amount of money paid by the passengers on the coach.

                                    [Edexcel]

A40   The diagram shows part of a sequence of shapes.
The shapes are made from rectangles with a dot at each corner and a dot in the centre.

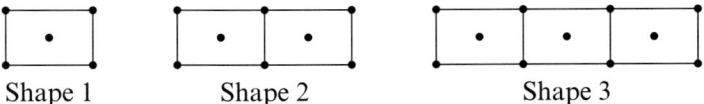

Shape 1   Shape 2   Shape 3

   a   Copy and complete the table to show the number of dots in each of the first 5 shapes in the sequence.

Shape number (n)	1	2	3	4	5
Number of sticks (d)	5	8	11	14	. . . . .

   b   Write down a formula which can be used to calculate the number of dots, d, in terms of the shape number, n.

                                    [Edexcel]

A41   **a**   Complete the table of values for $y = x^2 - 5$.

x	-3	-2	-1	0	1	2	3
y			-4				4

   b   On a grid, draw the graph of $y = x^2 - 5$.

                                    [Edexcel]

A42   **a**   Complete this table of values for $y = 3x - 1$.

x	-2	-1	0	1	2	3
y			-1			

   b   Draw the graph of $y = 3x - 1$ on a grid.

   c   Use your graph to find

      i   the value of x when $y = 3.5$

      ii   the value of y when $x = -1.5$

                                    [Edexcel]

Shape and Space

S1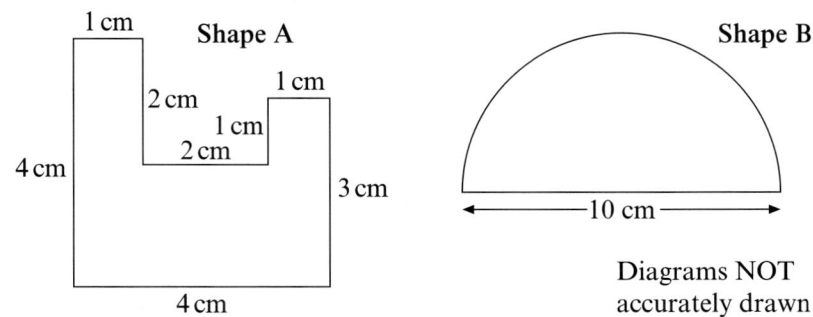

Diagrams NOT accurately drawn

a Work out the area of Shape A.

b i Work out the perimeter of the semicircle, Shape B.
ii Work out the area of the semicircle, Shape B.

[Edexcel]

S2 Simone made a scale model of a 'hot rod' car on a scale of 1 to 12.5.
The height of the model car is 10 cm.

a Work out the height of the real car.

The length of the real car is 5 m.

b Work out the length of the model car. Give your answer in centimetres.

The angle the windscreen made with the bonnet on the real car is 140°.

c What is the angle the windscreen makes with the bonnet on the model car?

The width of the windscreen in the real car is 119 cm correct to the nearest centimetre.

d Write down the smallest length this measurement could be.

[Edexcel]

S3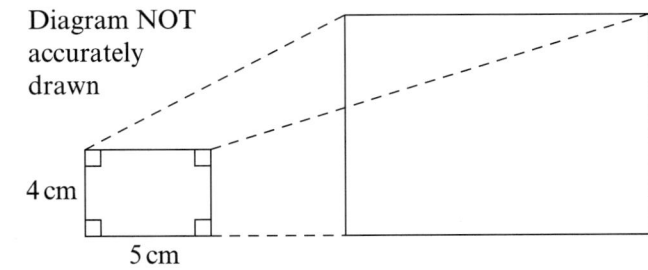

Diagram NOT accurately drawn

The diagram represents two photographs.

a Work out the area of the small photograph.

State the units of your answer.

The photograph is to be enlarged by scale factor 3.

b Write down the measurements of the enlarged photograph.

c How many times bigger is the area of the enlarged photograph than the area of the small photograph?

[Edexcel]

S4

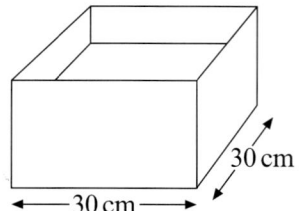

Water is stored in a tank in the shape of a cuboid with a square base. The sides of the base are 30 cm long. The depth of the water is 20 cm.

a Work out the volume of the water.

More water is put in the tank. The depth of the water rises to 21.6 cm.

b Calculate the percentage increase in the volume of the water in the tank.

[Edexcel]

S5

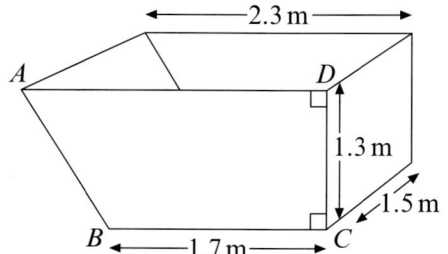

Diagram Not accurately drawn

A skip is in the shape of a prism with cross-section *ABCD*.
AD = 2.3 m, *DC* = 1.3 m and *BC* = 1.7 m.
The width of the skip is 1.5 m.

a Calculate the area of the shape *ABCD*.

b Calculate the volume of the skip.

The weight of an empty skip is 650 kg.
The skip is full to the top with sand.
1 m³ of sand weighs 4300 kg.

c Calculate the total weight of the skip and the sand.

[Edexcel]

S6

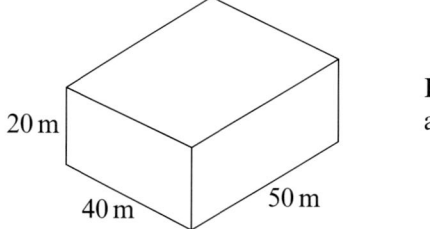

Diagram NOT accurately drawn

The diagram shows a cuboid.

Work out the volume, in cm³, of the cuboid.

[Edexcel]

S7 Jon cycled a distance of 18 km from Guildford to Cranleigh.
The graph shows Jon's cycle ride.
On the way, Jon stopped to buy a drink at a shop.

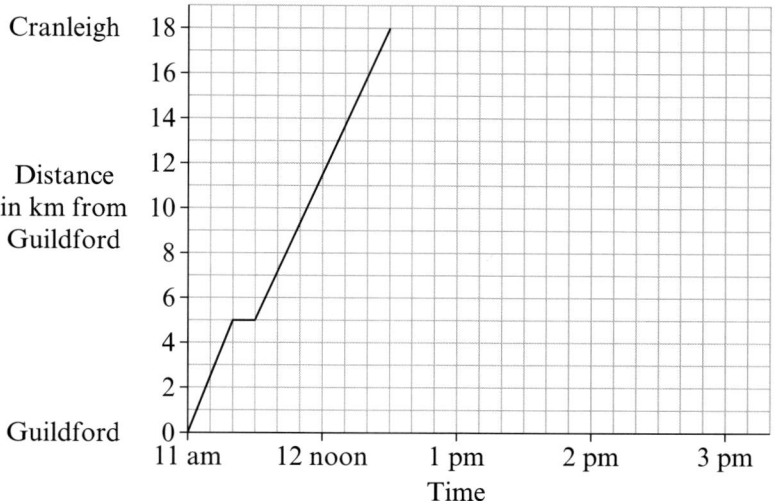

a i Write down the distance of the shop from Guildford.
ii Write down the time at which Jon stopped.
iii For how long did he stop?

Jon stayed in Cranleigh for lunch
He left Cranleigh at 1.30 pm.
He cycled back to Guildford at a steady speed.
He reached Guildford at 3 pm.

b Copy the grid and complete the graph of Jon's journey.

c Work out the steady speed at which he cycled back to Guildford.

[Edexcel]

S8

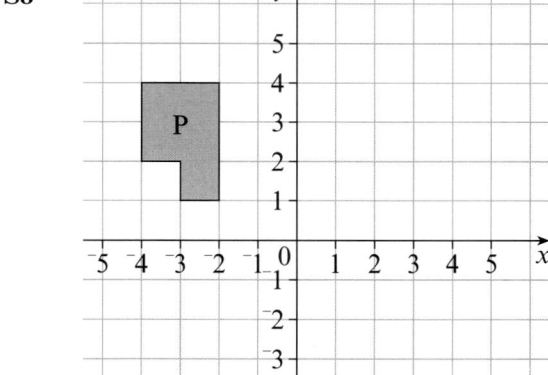

The shape **P** has been drawn on the grid. Copy the grid.

a Reflect the shape, **P** in the y axis.
Label the image **Q**.

b Rotate the shape **Q** through 180° about (0, 0).
Label this image **R**.

c Describe fully the single transformation which maps the shape **P** to
the shape **R**.

[Edexcel]

S9 A treasure chest is buried on an island.
P and Q are two trees on this island.

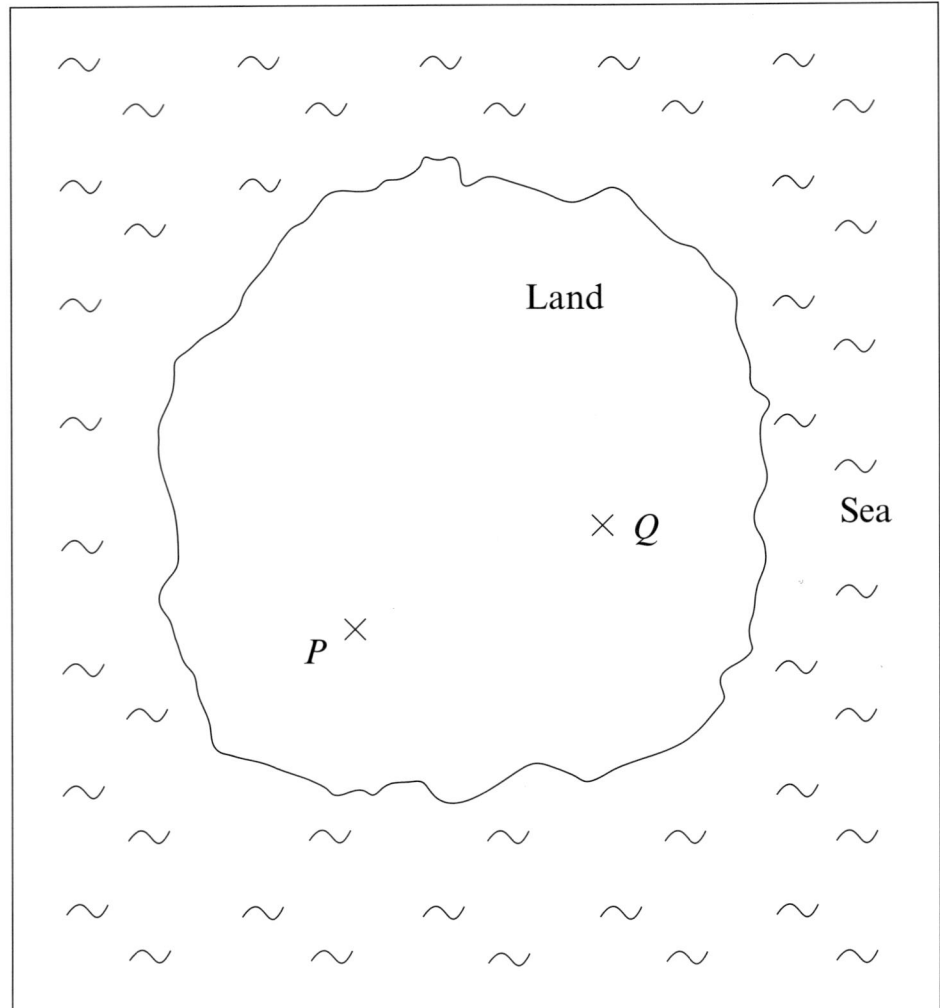

Scale:
1 cm represents 5 m

The treasure chest is buried the same distance from P as it is from Q.

a Trace the diagram, then draw accurately the locus of points which are the same distance from P as they are from Q.

On the diagram, 1 centimetre represents 5 metres.
The treasure chest is buried 20 metres from P.

b On your diagram, draw accurately the locus which represents all the points which are 20 metres from P.

c Find the point where the treasure chest is buried.
On your diagram, mark the point clearly with a T.

[Edexcel]

S10

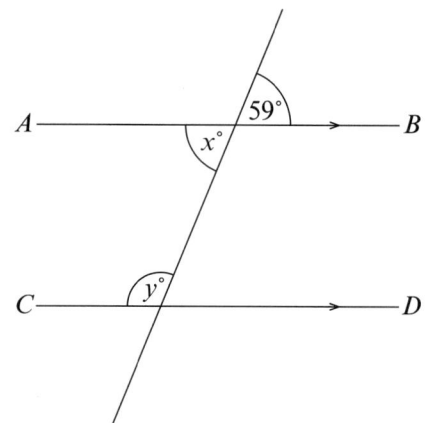

Diagram NOT
accurately drawn

AB is parallel to *CD*.

a i Write down the size of the angle marked $x°$.
   ii Give a reason for your answer.

b i Work out the size of the angle marked $y°$.
   ii Explain how you worked out your answer.

[Edexcel]

S11

THE SENATE

Petrol consumption:
9 litres per 100 km

1 km = $\frac{5}{8}$ mile. 1 gallon = 4.54 litre.
Change 9 litres per 100 km into miles per gallon.

[Edexcel]

S12

1 m^3 = 220 gallons.
1 m^3 = 10^6 cm^3.

a How many m^3 are equal to one gallon?
   Write your answer in standard form correct to 3 significant figures.
The petrol tank of a small car holds 6 gallons when it is 80% full.

b What is the capacity of the petrol tank in cm^3?

[Edexcel]

S13

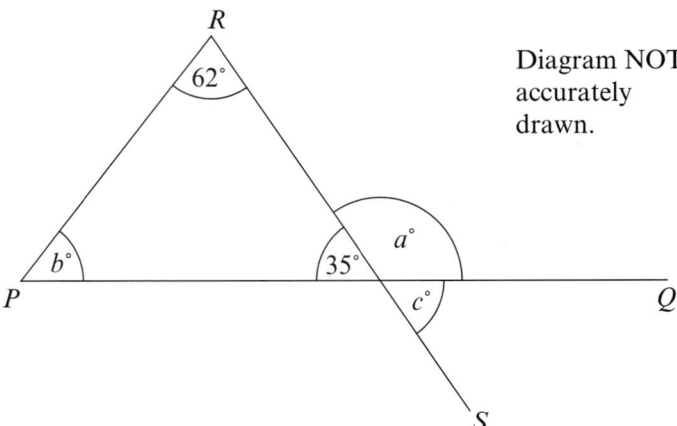

Diagram NOT
accurately
drawn.

In the diagram *PQ* and *RS* are straight lines.

a i Work out the value of *a*.
  ii Give a reason for your answer.
b i Work out the value of *b*.
  ii Give a reason for your answer.
c i Work out the value of *c*.
  ii Give a reason for your answer.

[Edexcel]

S14

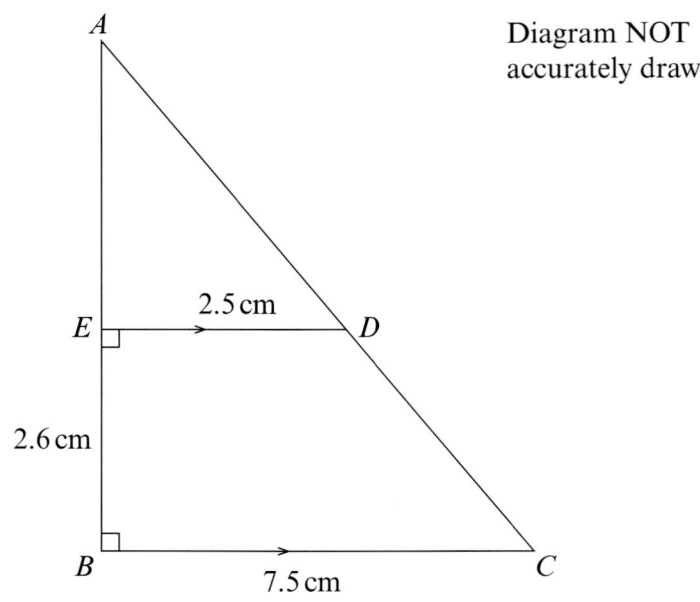

Diagram NOT
accurately drawn.

ABC is a right-angled triangle.
ED is parallel to *BC*.
BC = 7.5 cm,   *ED* = 2.5 cm,   *EB* = 2.6 cm.
Calculate the area of trapezium *BCDE*.

[Edexcel]

S15

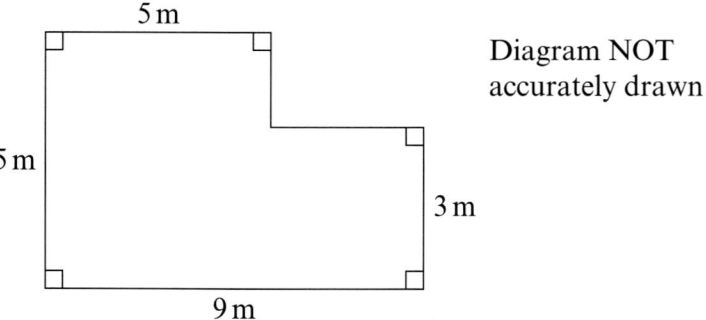

Diagram NOT
accurately drawn

This diagram shows the floor plan of a room.
Work out the area of the floor.
Give the units with your answer.

[Edexcel]

S16

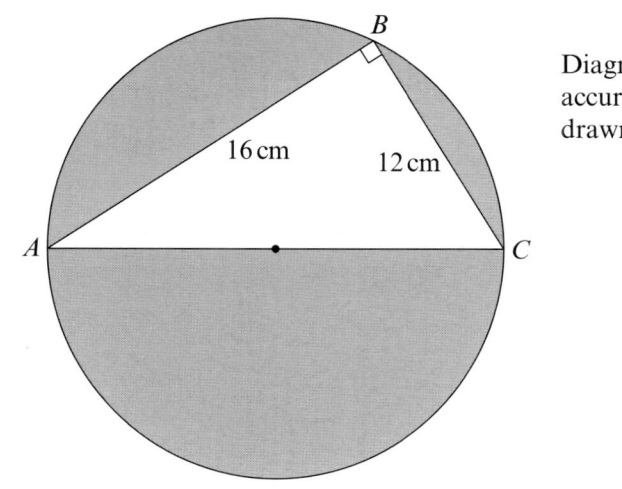

Diagram NOT
accurately
drawn

The diagram shows a right-angled triangle *ABC* and a circle.
A, *B* and *C* are points on the circumference of the circle.
AC is a diameter of the circle.
The radius of the circle is 10 cm.
AB = 16 cm and *BC* = 12 cm.

Work out the area of the shaded part of the circle.
Give your answer correct to the nearest cm^2.

[Edexcel]

S17

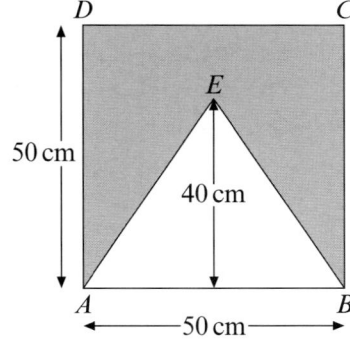

Diagram NOT
accurately drawn

ABCD is a square of side 50 cm.
E is a point inside the square.
E is 40 cm from the line *AB*.
Work out the area of the shaded region.
State the units with your answer.

[Edexcel]

S18

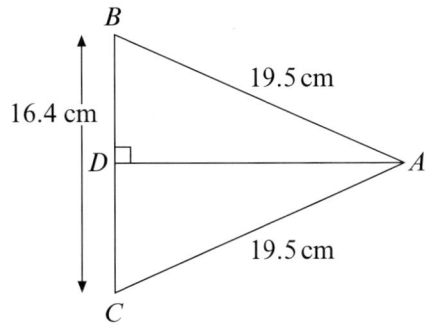

Diagram NOT
accurately drawn

$AB = 19.5$ cm, $AC = 19.5$ cm and $BC = 16.4$ cm.

Angle $ADB = 90°$.

BDC is a straight line.

Calculate the length of AD.

Give your answer in centimetres, correct to 1 decimal place.

[Edexcel]

S19

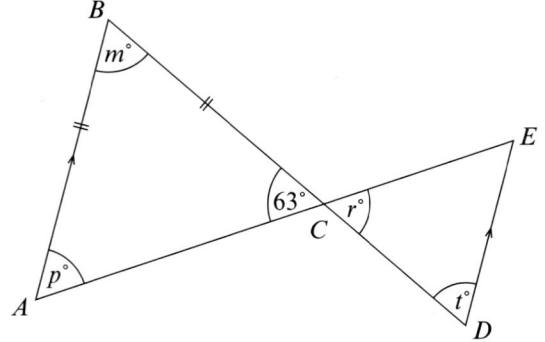

Diagram NOT accurately drawn

$AB = BC$.

Angle $ACB = 63°$.

ACE and BCD are straight lines.

a i Find the size of the angle marked $p°$.

  ii Give a reason for your answer.

b Work out the size of

  i the angle marked $m°$,

  ii the angle marked $r°$.

AB is parallel to DE.

c i Find the size of the angle marked $t°$.

  ii Explain how you worked out your answer.

[Edexcel]

S20

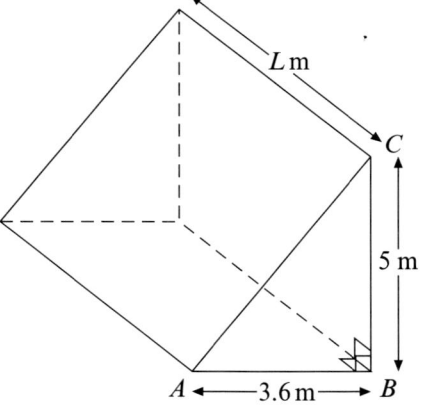

Diagram NOT accurately drawn

The diagram shows a triangular prism.

A cross-section of the prism is the triangle ABC.

The width AB of the triangle is 3.6 m.

The height BC of the triangle is 5 m.

Angle $ABC = 90°$.

The length of the prism is L metres.

The volume of the prism is 22.5 m³.

a Calculate the value of *L*.

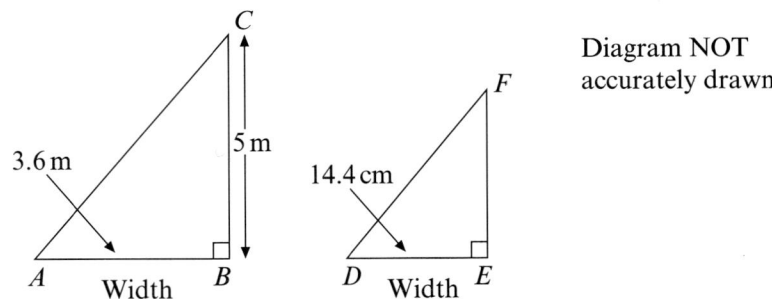

Diagram NOT accurately drawn

The cross-section of a similar prism is the triangle *DEF*.
The width, *DE*, of the prism is 14.4 cm.

b Calculate the height, *EF*, of the prism. Give your answer in centimetres.

[Edexcel]

S21 On Monday, Gareth drove from Swindon to Newcastle.
The distance was 325 miles.

He left Swindon at 08 00.
He arrived in Newcastle at 14 30.

a Work out Gareth's average speed.

On Tuesday, Gareth left Newcastle at 10 00 to drive back to Swindon.
He drove for 160 miles at an average speed of 64 miles per hour.

He stopped at a Service Station for one hour, before completing the journey.

He arrived in Swindon at 16 30.

b Calculate Gareth's average speed from the Service Station to Swindon.

[Edexcel]

S22

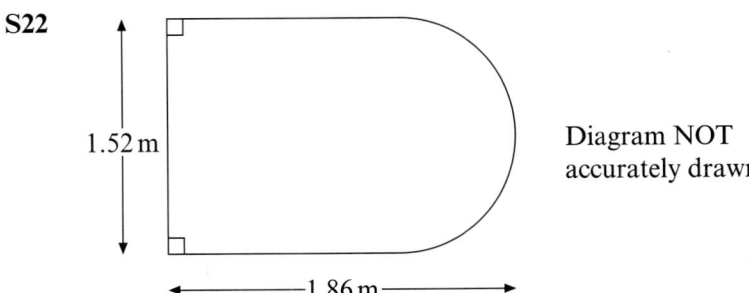

Diagram NOT accurately drawn

A mat is made in the shape of a rectangle with a semicircle added at one end.

The width of the mat is 1.52 metres.

The length of the mat is 1.86 metres.

Calculate the area of the mat.
Give your answer in square metres, correct to 2 decimal places.

[Edexcel]

S23   A circle has a radius of 32 cm.

Work out the circumference of the circle.
Give your answer correct to the neatest centimetre.

[Edexcel]

S24   **a**

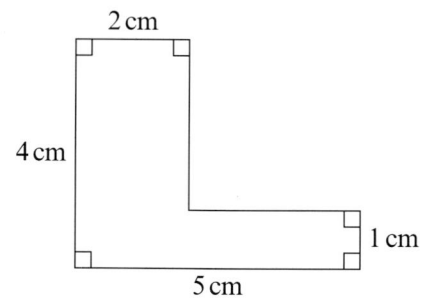

Diagram NOT
accurately drawn.

Work out the perimeter of the shape above.

b

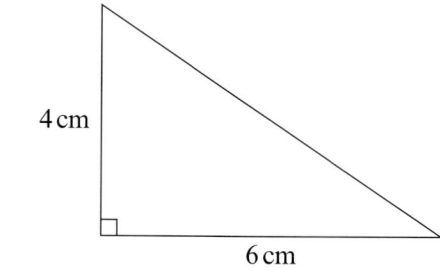

Diagram NOT
accurately drawn.

Work out the area of the triangle.
State the units with your answer.

[Edexcel]

S25

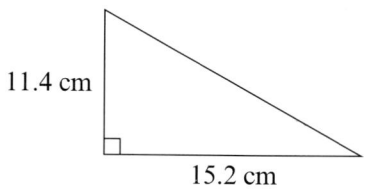

Diagram NOT
accurately drawn.

The diagram shows a sketch of a triangle.

a   Work out the area of the triangle.
State the units of your answer.

b   Work out the perimeter of the triangle.

[Edexcel]

S26   The radius of a circular table is 52 cm.

a   Work out the circumference of the circular table.
Give your answer to an appropriate degree of accuracy.

b   Work out the area of the circular table.

[Edexcel]

S27

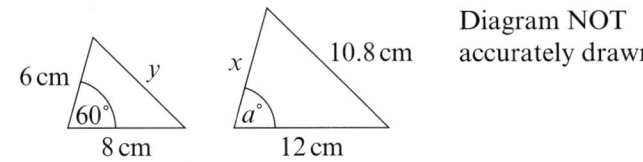

Diagram NOT accurately drawn

The triangles shown in the diagram are similar to each other.

a Write down the size of the angle marked $a°$.

b Work out the lengths of the sides marked

 i x,

 ii y.

[Edexcel]

S28

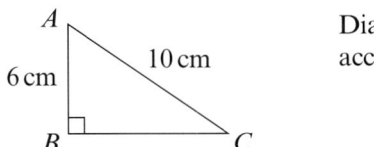

Diagram NOT accurately drawn

The diagram shows the cross-section of a triangular prism.
Angle $B = 90°$. $AB = 6$ cm. $AC = 10$ cm.
The length of the prism is 5 cm.
Work out the volume of the prism.

[Edexcel]

S29 Here is a diagram of a company logo.

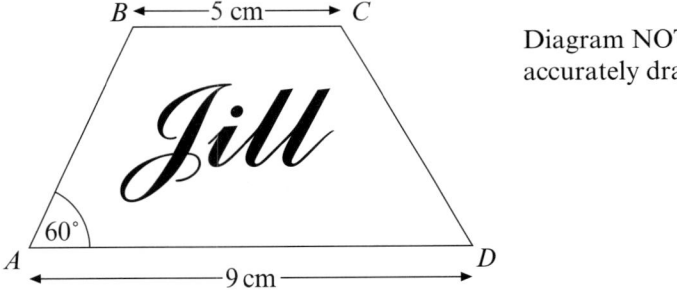

Diagram NOT accurately drawn

The diagram is enlarged so that the length BC becomes 7.5 cm.

a Work out the length of the enlarged side AD.

The enlarged side AB is 6 cm.

b Work out the length AB on the original diagram.

c What is the size of angle A in the enlarged diagram?

[Edexcel]

S30

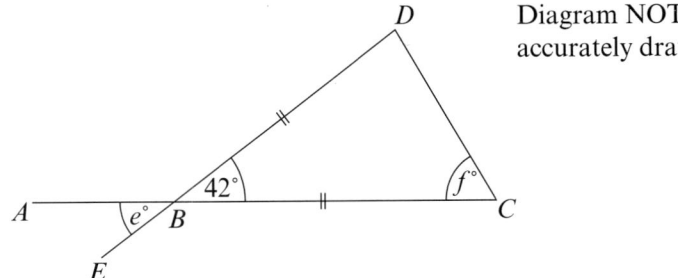

Diagram NOT accurately drawn

ABC and *EBD* are straight lines.

BD = *BC*.

Angle *CBD* = 42°.

a Write down the size of the angle marked *e*°.

b Work out the size of the angle marked *f*°.

[Edexcel]

S31

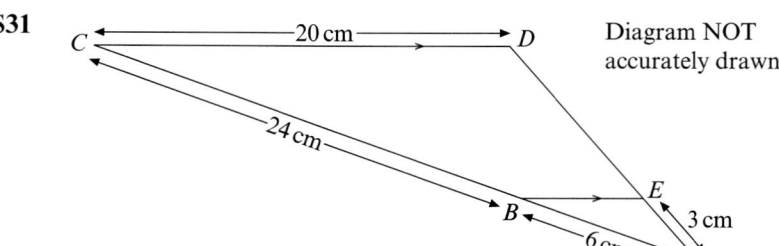

BE is parallel to *CD*.

ABC and *AED* are straight lines.

AB = 6 cm, *BC* = 24 cm, *CD* = 20 cm, *AE* = 3 cm.

a Calculate the length of *BE*.

b Calculate the length of *DE*.

[Edexcel]

S32

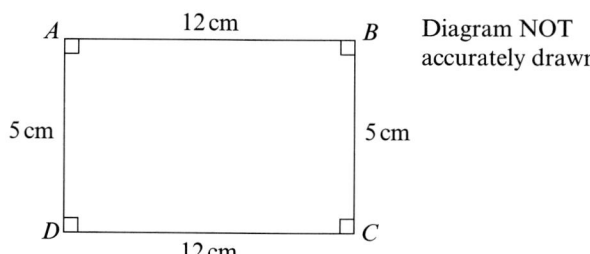

Calculate the length of the diagonal *AC* of the rectangle *ABCD* which has length 12 centimetres and width 5 centimetres.

[Edexcel]

S33

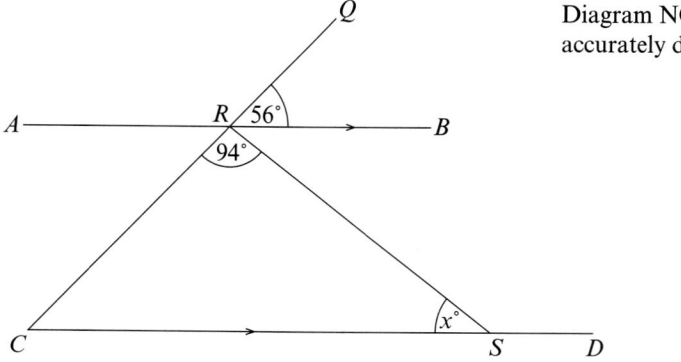

In the diagram, the lines *AB* and *CD* are parallel.

CRQ is a straight line.

Angle *CRS* = 94°.

Angle *QRB* = 56°.

Angle *RSC* = *x*°.

Find the value of *x*.

[Edexcel]

S34

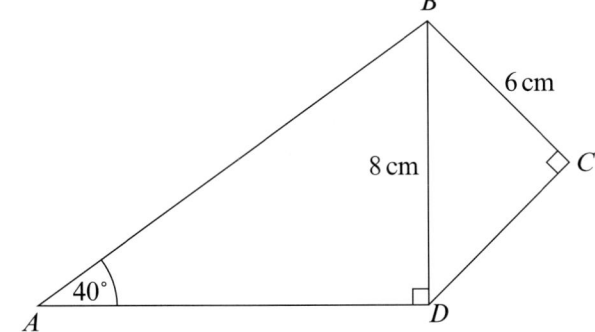

Diagram NOT accurately drawn.

ABCD is a quadrilateral.
Angle *BDA* = 90°, angle *BCD* = 90°, angle *BAD* = 40°.
BC = 6 cm, *BD* = 8 cm.

a Calculate the length of *DC*. Give your answer correct to 3 significant figures.

b Calculate the size of angle *DBC*.
Give your answer correct to 3 significant figures.

c Calculate the length of *AB*. Give your answer correct to 3 significant figures

[Edexcel]

S35

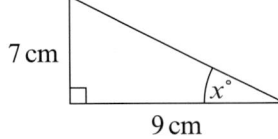

Diagram NOT accurately drawn.

Work out the size of the angle marked *x*°.
Give your answer correct to 1 decimal place.

[Edexcel]

S36

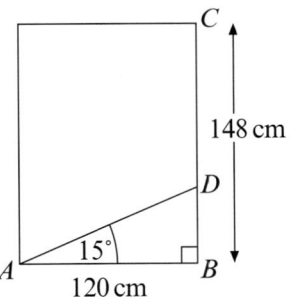

Diagram NOT accurately drawn

AB and *BC* are two sides of a rectangle.
AB = 120 cm and *BC* = 148 cm.
D is a point on *BC1*.
Angle *BAD* = 15°.

Work out the length of *CD*.
Give your answer correct to the nearest centimetre.

[Edexcel]

S37

Diagram NOT
accurately drawn

PQRS is a parallelogram.

angle *QSP* = 47°
angle *QSR* = 24°
PST is a straight line.

a **i** Find the size of the angle marked *x*.
   ii Give a reason for your answer.

b **i** Work out the size of angle *PQS*.
   ii Give a reason for your answer.

[Edexcel]

S38

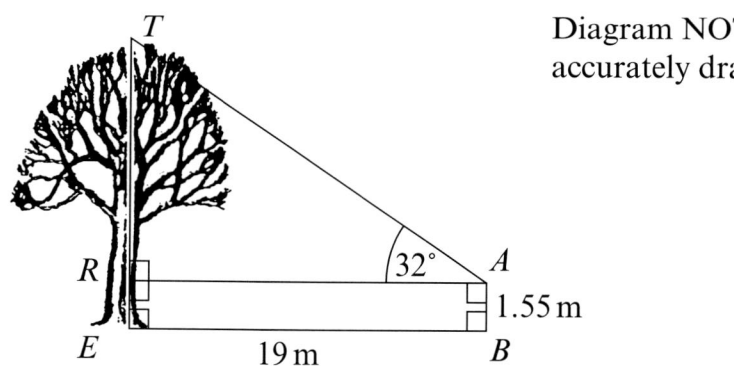

Diagram NOT
accurately drawn

Abbi is standing on level ground, at *B*, a distance of 19 metres away from
the foot *E* of a tree *TE*.

She measures the angle of elevation of the top of the tree at a height of
1.55 metres above the ground as 32°.

Calculate the height *TE* of the tree. Give your answer correct to 3
significant figures.

[Edexcel]

S39

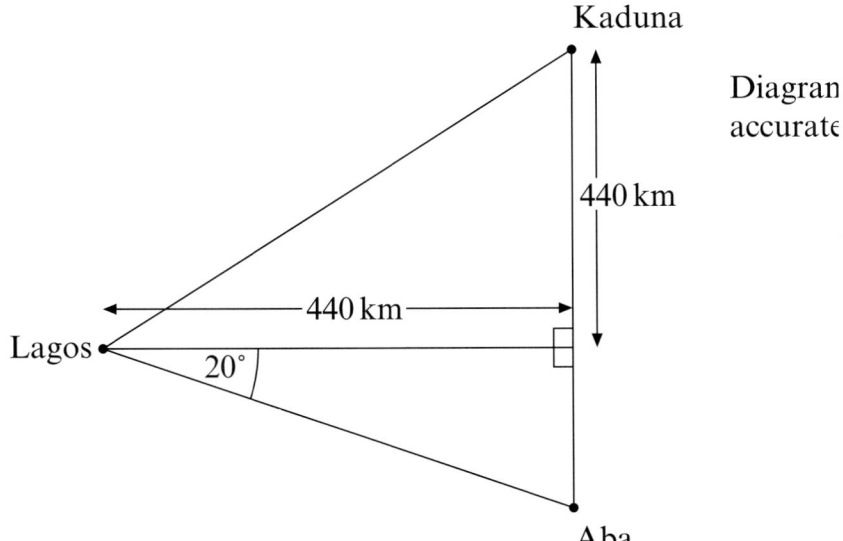

Kaduna

Diagran
accurate

440 km

440 km

Lagos

20°

Aba

The diagram is part of a map showing the positions of three Nigerian towns.

Kaduna is due North of Aba.

a Calculate the direct distance between Lagos and Kaduna.

b Calculate the distance between Kaduna and Aba.

Give your answer to the nearest kilometre.

[Edexcel]

S40

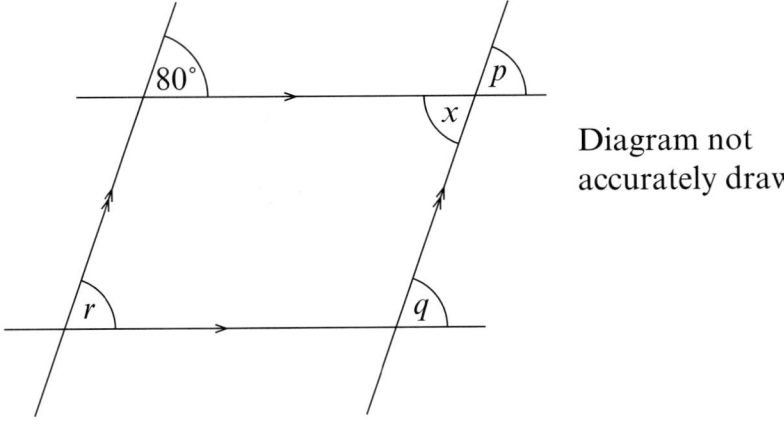

80°

p

x

r

q

Diagram not
accurately drawn

The diagram has two pairs of parallel lines.

Angles marked *p* and *q* are equal.

a What geometrical name is given to this type of equal angles?

b Write down the size of angle *r*.

c **i** Write down the size of angle *x*.

 ii What geometrical name is given to the pair of angles *x* and *q*?

[Edexcel]

S41

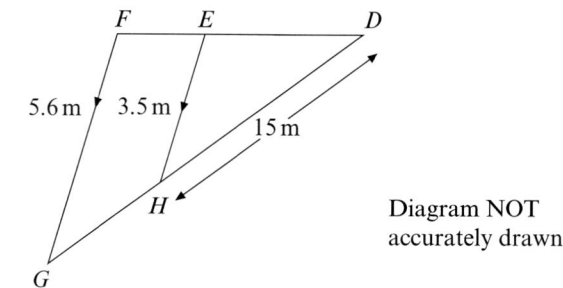

Two ports, P and Q, are shown on the map.

Measure and write down the bearing of P from Q.

[Edexcel]

S42

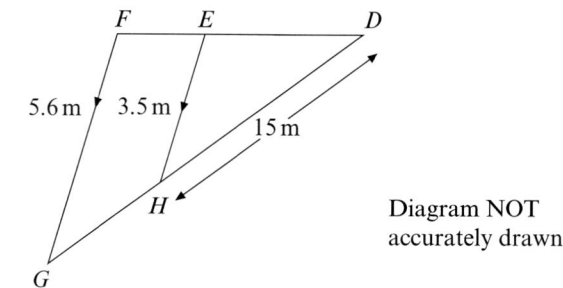

In the diagram *FG* = 5.6 metres, *EH* = 3.5 metres and *DH* = 15 metres.
EH is parallel to *FG*. *FED* and *DHG* are straight lines.

Calculate the length of *DG*.

[Edexcel]

S43

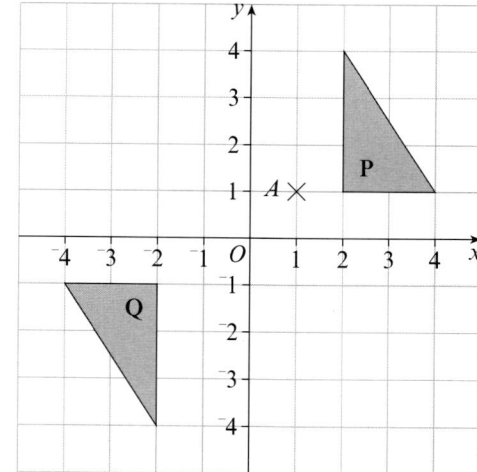

Copy the grid as shown.

a Describe fully the single transformation that maps shape **P** on to shape **Q**.

b Rotate shape **P** 90° anticlockwise about the point *A*(1, 1).

Give the coordinates of each vertex of the image.

[Edexcel]

S44

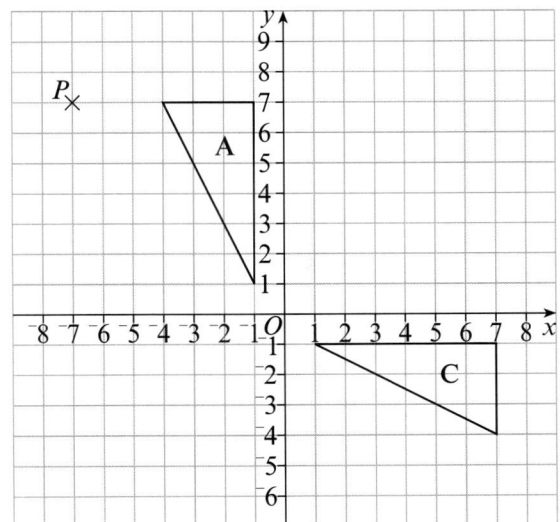

Copy the grid as shown.

a Enlarge triangle **A** by
the scale factor $\frac{1}{3}$
with centre the point $P(-7, 7)$.

b Describe fully the single transformation which maps triangle **A** onto
triangle *C*.

[Edexcel]

S45 Elizabeth went for a cycle ride.
The distance-time graph shows her ride.

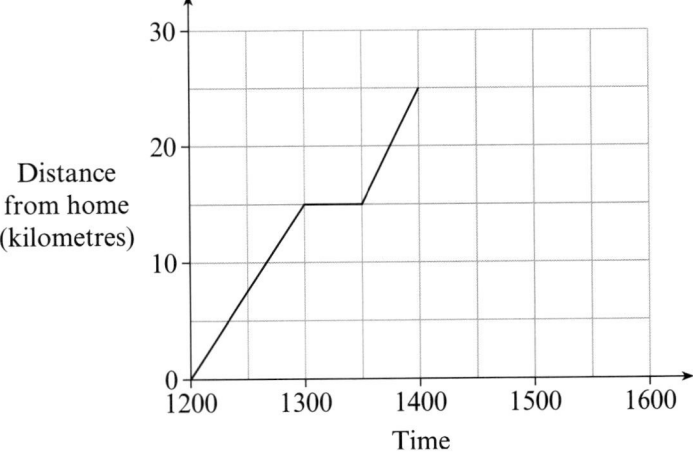

She set off from home at 1200 and had a flat tyre at 1400.
During her ride, she stopped for a rest.

a i At what time did she stop for a rest?
ii At what speed did she travel after her rest?

It took Elizabeth 15 minutes to repair the flat tyre.
She then cycled home at 25 kilometres per hour.

b Copy and complete the distance-time graph to show this information.

[Edexcel]

S46

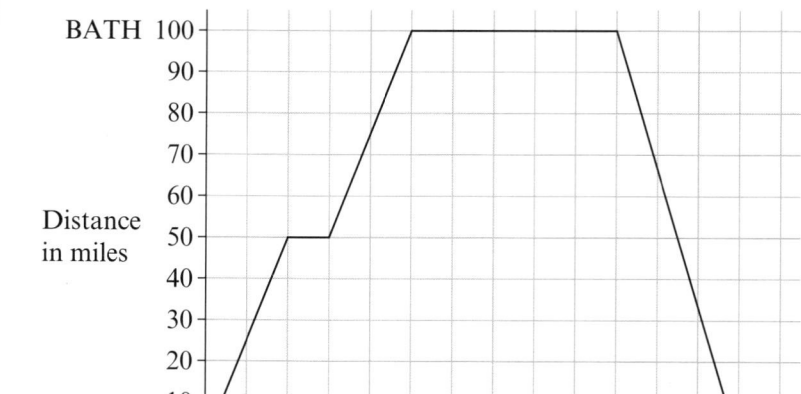

Ester took a coach trip from London to Bath and back again.
The diagram shows a distance-time graph of her trip.

a i At what time did the coach first stop to let some passengers off?
 ii For how long did it stop?

b Write down the distance between London and Bath.

c Work out the average speed of the coach on the return journey from Bath to London. Give your answer correct to the nearest whole number.

[Edexcel]

S47 The diagram shows a conversion graph between Pounds (£) and German Deutschmarks (DM).

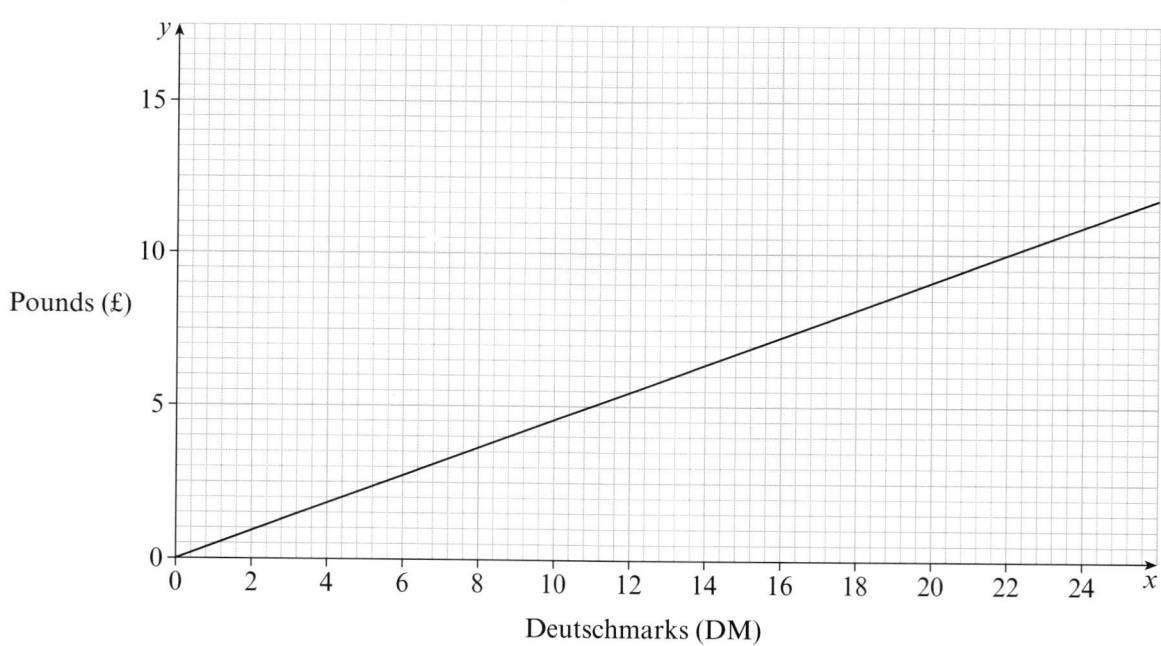

Use the graph to write down how many

 i Deutschmarks can be exchanged for £10.
 ii Pounds can be exchanged for DM10.

[Edexcel]

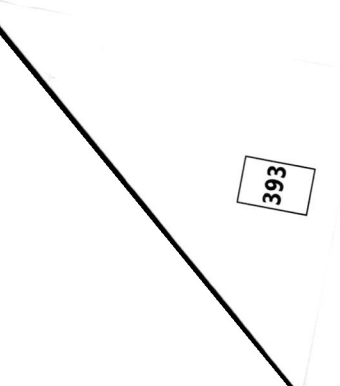

S48   There are 2.54 centimetres in 1 inch.
There are 12 inches in 1 foot.
There are 3 feet in 1 yard.
i   Calculate the number of yards in 10 metres.
ii   Calculate the number of metres in 10 yards.

[Edexcel]

S49

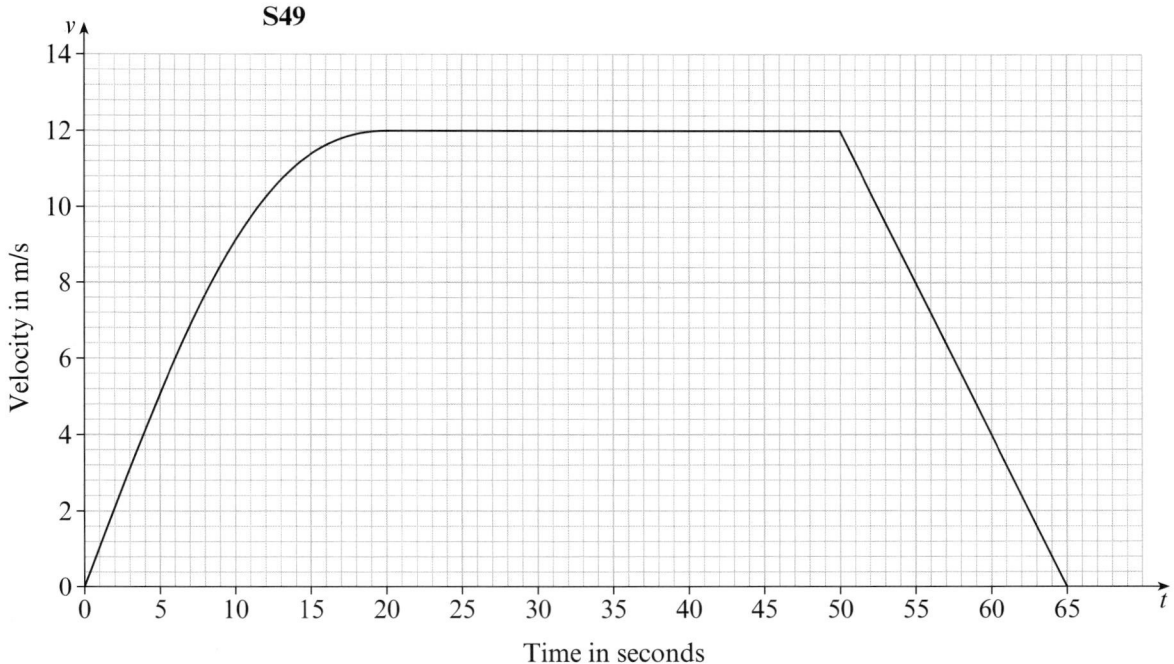

Time in seconds

A car travels between two sets of traffic lights.
The diagram is the velocity-time graph of the car.

The car leaves the first set of traffic lights.

a   Use the graph to find the velocity of the car after 15 seconds.

b   Describe fully the journey of the car between the two sets of traffic
lights.

[Edexcel]

S50   Here are 3 expressions.

Expression	Length	Area	Volume	None of these
$3rl$				
$\dfrac{2(r+l)^2}{h}$				
$\dfrac{4\pi r^4}{3l}$				

r, l and h are lengths.
π, 2, 3 and 4 are numbers and have no dimension.

Put a tick in the correct column to show whether the expression can ⌐
used for length, area, volume or none of these.

Handling Data

D1 Lauren and Yasmina each try to score a goal.
They each have one attempt.
The probability that Lauren will score a goal is 0.85.
The probability that Yasmina will score a goal is 0.6.

   a Work out the probability that **both** Lauren **and** Yasmina will score a goal.

   b Work out the probability that Lauren **will** score a goal **and** Yasmina **will not** score a goal.

[Edexcel]

D2 Martin bought a packet of mixed flower seeds.
The seeds produce flowers that are Red or Blue or White or Yellow.
The probability of a flower seed producing a flower of a particular colour is:

Colour	Red	Blue	White	Yellow
Probability	0.6	0.15		0.15

   a Write down the most common colour of a flower.

Martin chooses a flower seed at random from the packet.

   b **i** Work out the probability that the flower produced will be White.
      ii Write down the probability that the flower produced will be Orange.

[Edexcel]

D3 A lorry contains 232 boxes of crisps.
Each box has either plain crisps or cheese and onion flavour crisps.
The probability that a box selected at random holds plain crisps is $\frac{1}{3}$ of the probability that the box holds cheese and onion crisps.

   a Calculate the number of boxes of plain crisps.

Each box holds 48 packets of crisps.
One in every 8 packets of plain crisps has a prize in it. One in every 16 packets of cheese and onion flavour crisps has a prize in it.
A packet is to be selected at random from the lorry.

   b Calculate the probability that the packet will have a prize in it.

[Edexcel]

D4 Helen tries to win a coconut at the fair.
She throws a ball at a coconut.
If she knocks a coconut off its stand, she wins the coconut.
Helen has two throws.
The probability that she will win a coconut with her first throw is 0.2.
The probability that she will win a coconut with her second throw is 0.3.
Work out the probability that, with her two throws, Helen will win

   i 2 coconuts,

   ii exactly 1 coconut.

[Edexcel]

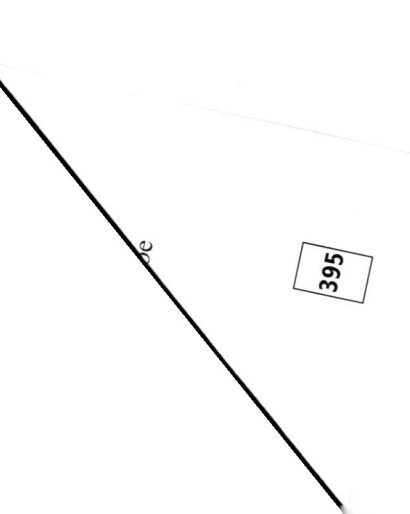

D5 The probability of a person having brown eyes is $\frac{1}{4}$.
The probability of a person having blue eyes is $\frac{1}{3}$.
Two people are chosen at random.
Work out the probability that

  i both people will have brown eyes

  ii one person will have blue eyes and the other person will have brown eyes.

[Edexcel]

D6 Here is a 5-sided spinner.

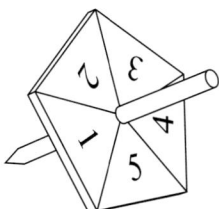

Its sides are labelled 1, 2, 3, 4, 5.
Alan spins the spinner and throws a coin.
One possible outcome is (3, Heads).

  a List all possible outcomes.

The spinner is biased.
The probability that the spinner will land on each of the numbers 1 to 4 is given in the table.

Number	1	2	3	4	5
Probability	0.36	0.1	0.25	0.15	

Alan spins the spinner once.

  b i Work out the probability that the spinner will land on 5.
   ii Write down the probability that the spinner will land on 6.
   iii Write down the number that the spinner is most likely to land on.
   iv Work out the probability that the spinner will land on an even number.

Alan spins the spinner and throws a fair coin.

  c Work out the probability that the spinner will land on 3 and the coin will show Heads.

[Edexcel]

D7 Sharon has 12 computer discs.
Five of the discs are red.
Seven of the discs are black.
She keeps all the discs in a box.
Sharon removes one disc at random. She records its colour and replaces it in the box.
Sharon removes a second disc at random, and again records its colour.

a Copy and complete the tree diagram.

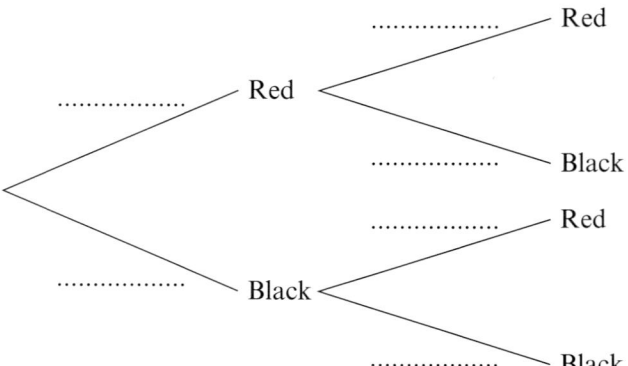

b Calculate the probability that the two discs removed
  i will both be red,
  ii will be different colours.

[Edexcel]

D8 Ben has some coloured cubes in a bag.
The table shows the number of cubes of each colour.

Red	Blue	Yellow	Brown
7	4	8	6

Ben is going to take one cube at random from the bag.
Write down the probability that Ben
i will take a yellow cube,
ii will not take a brown cube.

[Edexcel]

D9 A and B are two fair spinners.

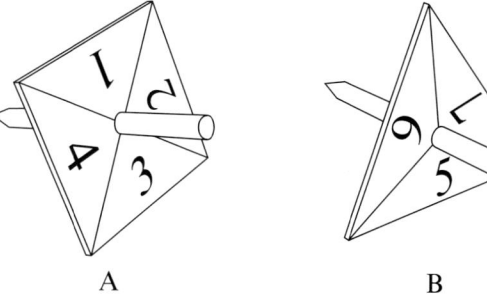

A   B

Jane spins the two spinners together once.

a Construct a table to show all the possible results and the total scores.

b Use your table to find the probability that Jane will get a total score of 7.

Tom spins the two spinners together 60 times.

c Work out the number of times you would expect Tom to get a total score of 7.

[Edexcel]

D10 Twenty five people took part in a competition.
The points scored are grouped in the frequency table below.

Points scored	Number of people		
1 to 5	1		
6 to 10	2		
11 to 15	5		
16 to 20	7		
21 to 25	8		
26 to 30	2		

a Work out the class interval which contains the median.

b Work out an estimate for the mean number of points scored.

c Copy and complete the table below to show the cumulative frequency for this data.

Points scored	Cumulative frequency
1 to 5	
1 to 10	
1 to 15	
1 to 20	
1 to 25	
1 to 30	

d Draw a cumulative frequency graph for this data.

[Edexcel]

D11 A class took a test. The mean mark of the 20 boys in the class was 17.4.
The mean mark of the 10 girls in the class was 13.8.

a Calculate the mean mark for the whole class.

5 pupils in another class took the test.
Their marks, written in order, were 1 2, 3, 4 and x.
The mean of these 5 marks is equal to twice the median of these 5 marks.

b Calculate the value of x.

[Edexcel]

D12   Mrs Chowdery gives her class a maths test.
Here are the test marks for the girls.

$$7, 5, 8, 5, 2, 8, 7, 4, 7, 10, 3, 7, 4, 3, 6$$

a   Work out the mode.

b   Work out the median.

The median mark for the boys was 7 and the range of the marks of the boys was 4.
The range of the girls' marks was 8.

c   By comparing the results explain whether the boys or the girls did better in the test.

[Edexcel]

D13   Some students took a mental arithmetic test.
Information about their marks is shown in the frequency table.

Mark	Frequency
4	2
5	1
6	2
7	4
8	7
9	10
10	3

a   Work out how many students took the test.

b   Write down the modal mark.

24 students had a higher mark than Caroline.

c   Work out Caroline's mark.

d   Find the median mark.

e   Work out the range of the marks.

[Edexcel]

D14 The following is a record of the lengths, in centimetres, of 40 guinea pigs.

21	22	11	16	22	13	11	25	9	17
21	24	27	25	12	14	8	12	6	17
23	7	12	26	14	8	12	26	17	19
23	29	21	19	26	26	18	21	13	9

a Complete the frequency table.

Length (l) cm	Tally	Frequency
$5 \leqslant l < 10$		
$10 \leqslant l < 15$		
$15 \leqslant l < 20$		
$20 \leqslant l < 25$		
$25 \leqslant l < 30$		

b Draw a frequency diagram for this information, on a copy of the axes shown.

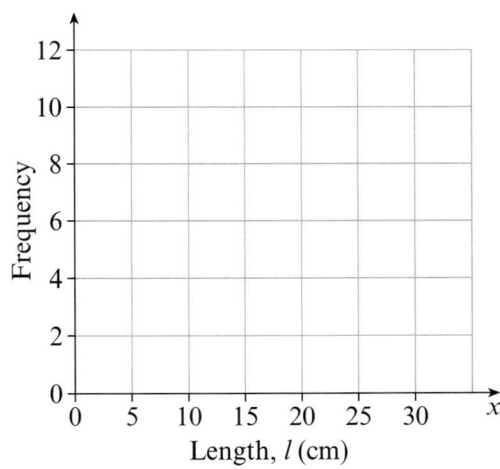

c How many guinea pigs were under 15 cm in length?

d Write down the modal class interval of the lengths.

[Edexcel]

D15   A set of 25 times in seconds is recorded.

12.9	10.0	4.2	16.0	5.6	18.1	8.3	14.0	11.5	21.7
22.2	6.0	13.6	3.1	11.5	10.8	15.7	3.7	9.4	8.0
6.4	17.0	7.3	12.8	13.5

a   Complete the frequency table below, using intervals of 5 seconds.

Time (t) seconds	Tally	Frequency
$0 \leqslant t < 5$		

b   Write down the modal class interval.

[Edexcel]

D16   Ben asked 50 people how much they paid for a new computer.
The results are shown in this frequency table.

Price (£P)	Number of computers		
$0 < P \leqslant 500$	7		
$500 < P \leqslant 1000$	20		
$1000 < P \leqslant 1500$	11		
$1500 < P \leqslant 2000$	9		
$2000 < P \leqslant 2500$	3		

a   Calculate an estimate for the mean price paid for a new computer.

By the end of each year, the value of a computer falls by 15% of its value at the start of that year.
A new computer has a value of £1200.

b   Calculate its value by the end of the third year.

[Edexcel]

D17 The names and prices of four second-hand cars are shown in the table

Name	Nippy sports	Tuff hatchback	Ace supermini	Mega estate
Price	£12 000	£4000	£6000	£18 000

On the diagram each letter represents one of the cars.

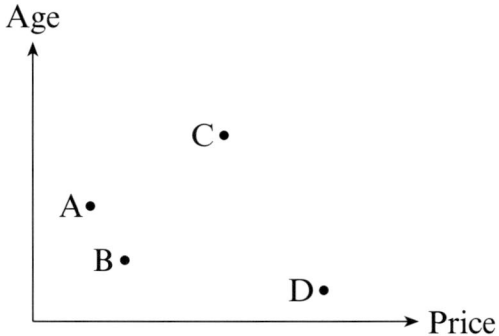

i   Use the information shown in the table and in the diagram to write down the letter which represents each car.

Nippy sports   Letter . . . .

Tuff hatchback   Letter . . . .

Ace supermini   Letter . . . .

Mega estate   Letter . . . .

ii Write down the name of the oldest car.

[Edexcel]

D18 Derek asked 45 students to name their favourite flavour of yoghurt. His results are shown in the table below.

Flavour of Yoghurt	Number of students	Angle in degrees
Strawberry	16	
Orange	5	
Peach	9	
Blackberry	4	
Other	11	

Draw an accurate pie chart to show this information.

[Edexcel]

D19 Fred carried out a survey of the time, in seconds, between one car and the next car on a road.

His results are shown in the cumulative frequency graph on the grid opposite.

a How many cars were there in the survey?

b Use the graph to estimate the median time.

c Use the graph to estimate the percentage of times that were greater than 25 seconds.

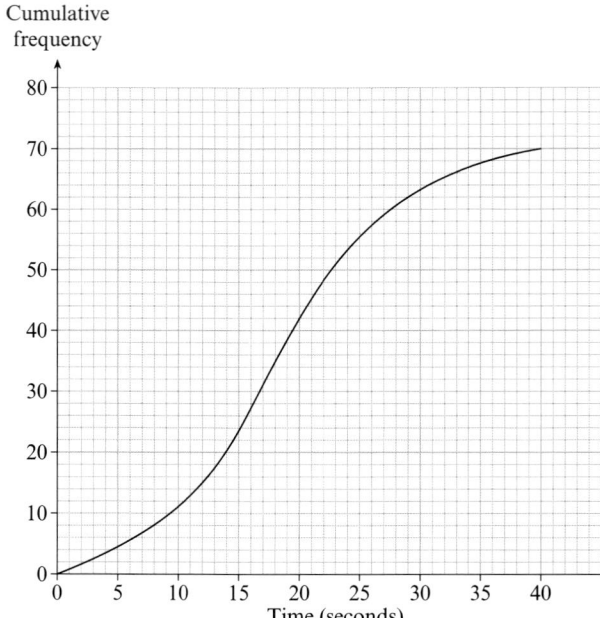

[Edexcel]

D20 The cumulative frequency table gives information about the heights, in centimetres, of some students at Bax School.

Heights (h) of students in cm	Cumulative frequency
$h < 135$	0
$135 \leqslant h < 140$	2
$135 \leqslant h < 145$	10
$135 \leqslant h < 150$	29
$135 \leqslant h < 155$	70
$135 \leqslant h < 160$	115
$135 \leqslant h < 165$	143
$135 \leqslant h < 170$	156
$135 \leqslant h < 175$	163

a Draw a cumulative frequency graph for the data in the table.

b Use your cumulative frequency graph to estimate

  i the median height,

  ii the number of students with height less than 151 cm.

[Edexcel]

D21 Inside Box *A* there are five coins,

one 1p coin,
one 5p coin,
one 10p coin,
one 20p coin,
one 50p coin.

Inside Box *B* there are four coins,

one 1p coin,
one 2p coin,
one 10p coin,
one 20p coin.

Jack picks one coin from Box *A* and one coin from Box *B*.
Complete this table to show all the possible totals, in pence, for the two coins.

		Box *A*				
		1p	5p	10p	20p	50p
Box	1p	2	6			
B	2p	3	7			
	10p					
	20p					

[Edexcel]

D22 Helen and Joan are going to take a swimming test.
The probability that Helen will pass the swimming test is 0.95
The probability that Joan will pass the swimming test is 0.8
The two events are independent.

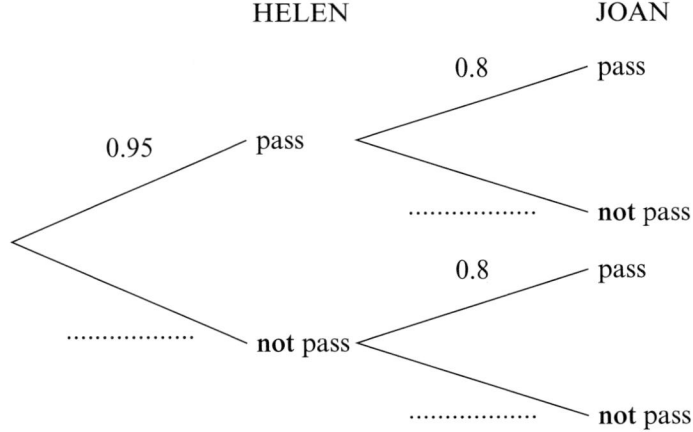

a Copy and complete the probability tree diagram.

b Work out the probability that both Helen and Joan will pass the swimming test.

c Work out the probability that one of them will pass the swimming test and the other one will not pass the swimming test.

[Edexcel]

D23 Mr Hulme chose 10 boys and 10 girls at random from his school.
He counted the numbers of different vowels in their first names.
This table shows the results.

Number of different vowels in first name	One	Two	Three	Four	Five
Number of boys	3	4	2	1	0
Number of girls	2	3	4	0	1

There are 1000 pupils in the school.
There are 480 boys and 520 girls.
Estimate the number of pupils in the school who have exactly three
different vowels in their first names.

[Edexcel]

D24 A bag contains counters which are white or green or red or yellow.
The probability of taking a counter of a particular colour at random is

Colour	White	Green	Red	Yellow
Probability	0.15	0.25		0.4

Laura is going to take a counter at random and then put it back in the
bag.

a i Work out the probability that Laura will take a red counter.
  ii Write down the probability that Laura will take a blue counter.

Laura is going to take a counter from the bag at random 100 times.
Each time she will put the counter back in the bag.

b Work out an estimate for the number of times that Laura will take a
yellow counter.

[Edexcel]

D25 Chris is going to roll a biased dice.
The probability that he will get a six is 0.09

a Work out the probability that he will not get a six.

Chris is going to roll the dice 30 times.

b Work out an estimate for the number of sixes he will get.

Tina is going to roll the same biased dice twice.

c Work out the probability that she will get
  i two sixes,
  ii exactly one six.

[Edexcel]

D26 Jack has two fair dice.
One of the dice has 6 faces numbered from 1 to 6.
The other dice has 4 faces numbered from 1 to 4.
Jack is going to throw the two dice.
He will add the scores on the two dice to get the total.
Work out the probability that he will get

i a total of 7,
ii a total of less than 5.

[Edexcel]

D27 | M | A | T | H | E | M | A | T | I | C | S |

Each of the eleven letters in the word 'mathematics' is written on a card.
Sita picks a card without looking.
Find the probability that she will pick

i an E,

ii an M,

iii one of the first five letters of the alphabet.

[Edexcel]

D28 There are four blood groups A, B, AB and O.
The probability that a person picked at random will be in blood group A,
B or AB is shown in the table.

Blood group	A	B	AB	O
Probability	0.42	0.09	0.03	

a Work out the probability that a person chosen at random will be in
blood group O.

Mathstown High School has 1000 students.

b Work out an estimate for the number of students who will be in blood
group A.

[Edexcel]

D29 Donna has a bag of marbles. Each marble is red or blue or green.
Without looking, Donna picks a marble from the bag.
She records the colour of the marble and puts it back in the bag.
She carries out this experiment 200 times.
The table shows her results.

Colour	red	blue	green
Frequency	86	62	52

She then picks another marble from the bag.

a Work out an estimate for the probability that it will be
  i red,
  ii red or blue.
There are 50 marbles in the bag.

b Work out an estimate for the number of green marbles in the bag.

[Edexcel]

D30 A packet contains only yellow counters and green counters.
There are 8 yellow counters and 5 green counters.
A counter is to be taken from the packet at random.

Write down the probability that
i a yellow counter will be taken,
ii a yellow counter will not be taken.

A second counter is to be taken from the packet.
Write down all the possible outcomes of taking out two counters.

[Edexcel]

D31 The probability that a GCSE candidate, chosen at random, will obtain a grade C or higher in English is 0.55.
Calculate the probability that a GCSE candidate, chosen at random, will obtain a grade lower than grade C in English.

[Edexcel]

D32 A box contains only blue pencils and red pencils.
6 of the pencils are blue and 5 are red.
A pencil is to be taken at random from the box.
Write down the probability that

a a blue pencil will be taken,

b a blue pencil will not be taken.

[Edexcel]

D33 A shop employs 8 men and 2 women.
The mean weekly wage of the 10 employees is £396.
The mean weekly wage of the 8 men is £400.
Calculate the mean weekly wage of the 2 women.

[Edexcel]

D34 20 students took part in a competition.
The frequency table shows information about the points they scored.

Points scored	1	2	3
Frequency	9	4	7

Work out the total number of points scored by the 20 students.

[Edexcel]

D35 The table gives some information about the favourite Mathematics topics of 90 students.

Mathematics Topic	Frequency	Angle
Number	18	
Algebra	36	
Shape, Space and Measure		
Handling Data		
Total	90	

a   Use the information in the table to copy and complete the pie chart.

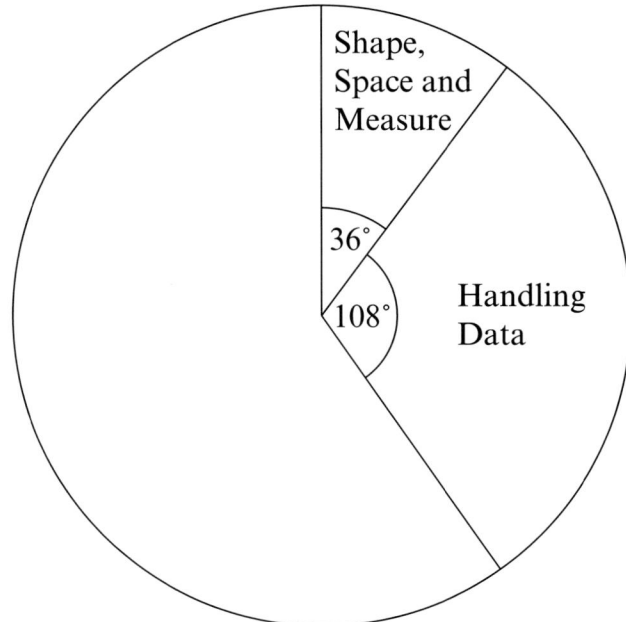

b   Use the information in the pie chart to copy and complete the table.

c   Work out the fraction of the 90 students whose favourite topic is Algebra.
Give your answer in its simplest form.

[Edexcel]

D36   In a town 1800 cars were stolen in a year. The table shows information about the times of day when they were stolen.

Time	Number of cars
Midnight to 6 am	700
6 am to midday	80
Midday to 6 pm	280
6 pm to midnight	470
Time unknown	270

This information can be shown in a pie chart.

a   Work out the angle of each sector of the pie chart.

b   Construct a pie chart to show the data.

c   What fraction of the number of cars was stolen between Midday and 6 pm? Write your fraction in its simplest form.

[Edexcel]

D37

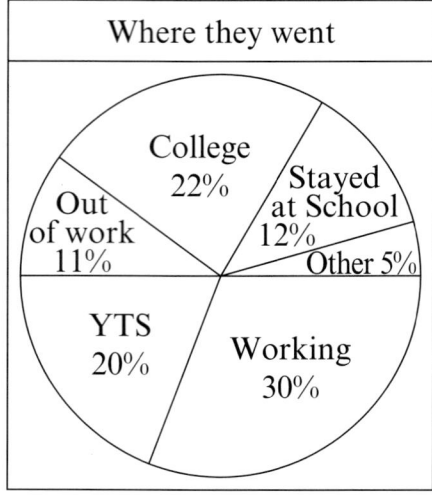

Where they went

College 22%
Stayed at School 12%
Out of work 11%
Other 5%
YTS 20%
Working 30%

Diagram not accurately drawn

300 young people were asked what they did after completing Year 11 at school.
The pie chart shows the results of the survey.

a How many of the young people were working?

Gwen made an accurate drawing of the pie chart.
She first drew the sector representing the young people out of work.

b Calculate the size of the angle of this sector.
Give your answer correct to the nearest degree.

c Change to a decimal the percentage going to college.

d What fraction of the young people stayed at school?
Give your answer in its simplest form.

[Edexcel]

D38 The table gives information about the times 60 teachers take to travel to school.

Time in minutes	Frequency	Angle
Less than 10	5	
Between 10 and 15	15	
Between 15 and 20	12	
More than 20	28	

Draw a pie chart to show this information.

[Edexcel]

D39 The diagram shows a target.

Billy fires at the target ten times.
The frequency table gives information about his scores.

Score	Frequency
5	2
6	1
7	3
8	1
9	2
10	1

a Work out his total score.

[Edexcel]

D40 Sybil weighed some pieces of cheese.
The table gives information about her results.

Weight (w) grams	Frequency		
$90 < w \leqslant 94$	1		
$94 < w \leqslant 98$	2		
$98 < w \leqslant 102$	6		
$102 < w \leqslant 106$	1		

Work out an estimate of the mean weight.

[Edexcel]

D41 Geoff made 24 telephone calls last week.
He recorded the time, in minutes, each call lasted.
He used the information to complete this frequency table.

Time in minutes	Frequency
$0 < t \leqslant 10$	14
$10 < t \leqslant 20$	6
$20 < t \leqslant 30$	3
$30 < t \leqslant 40$	1

a Write down the modal class interval.

b On a grid draw a frequency polygon to show this information.

[Edexcel]

D42 A survey was carried out to find out how much time was needed by a group of pupils to complete homework set on a particular Monday evening. The results are shown in the table below.

Time, t hours, spent on homework	Number of pupils		
0	3		
$0 < t \leqslant 1$	14		
$1 < t \leqslant 2$	17		
$2 < t \leqslant 3$	5		
$3 < t \leqslant 4$	1		

Copy the table and calculate an estimate for the mean time spent on homework by the pupils in the group.

[Edexcel]

D43 The grouped frequency table gives information about the weekly rainfall (d) in millimetres at Heathrow airport in 1995.

Weekly rainfall (d) in mm	Number of weeks		
$0 \leqslant d < 10$	20		
$10 \leqslant d < 20$	18		
$20 \leqslant d < 30$	6		
$30 \leqslant d < 40$	4		
$40 \leqslant d < 50$	2		
$50 \leqslant d < 60$	2		

a Copy the table and hence calculate an estimate for the mean weekly rainfall.

b Write down the probability that the rainfall in any week in 1995, chosen at random, was greater than or equal to 20 mm and less than 40 mm.

c Copy and complete this cumulative frequency table for the data.

Weekly rainfall (d) in mm	Cumulative frequency
$0 \leqslant d < 10$	
$0 \leqslant d < 20$	
$0 \leqslant d < 30$	
$0 \leqslant d < 40$	
$0 \leqslant d < 50$	
$0 \leqslant d < 60$	

d Draw a cumulative frequency graph for this data.

e Use your cumulative frequency graph to estimate the median weekly rainfall.
You must show your method clearly.

[Edexcel]

Do NOT use a calculator for this exam paper.

1   A window cleaner charges £1.50 plus 25p per window.

   a   Roy has seven windows cleaned.
      How much is the total charge?

                                                                        (3 marks)

   b   Hayley has x windows cleaned.
      Write down an expression for the amount she pays.

                                                                        (2 marks)

2   **a**   Here is a number machine.
      Copy and complete the answer box.

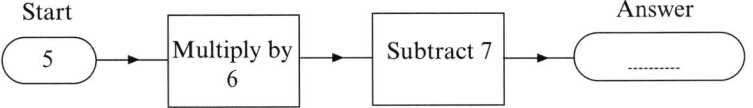

                                                                       (1 mark)

   b   Here is an algebra machine.
      Copy and complete the answer box.

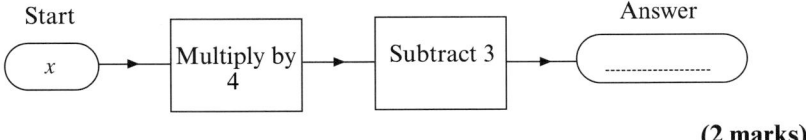

                                                                       (2 marks)

3   **a**   On a copy of the grid, reflect the shape in the dotted line.

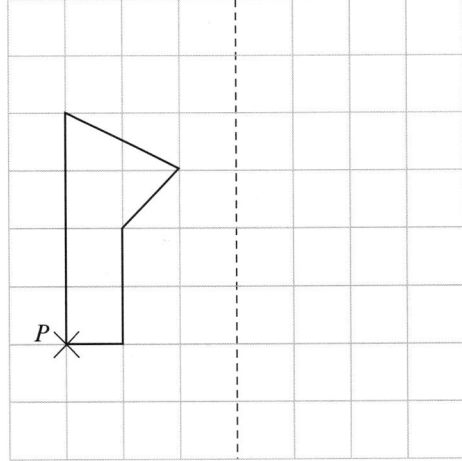

                                                                       (2 marks)

   b   Rotate your original shape through 180° about the point P.

                                                                       (2 marks)

4 120 customers in a bread shop were asked which type of bread they prefer.
Here are the results.

Type of bread	White	Brown	Rolls	other
Number of customers	25	43	36	16

a Draw a pie chart to show the results.

(4 marks)

b Calculate the percentage of customers who prefer rolls.

(2 marks)

5 The triangle ABC is isosceles.

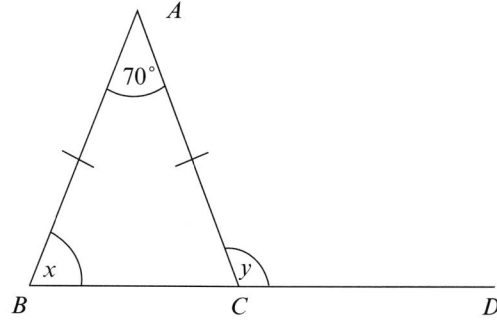

a Work out the value of angle x.

(2 marks)

b Work out the value of angle y.

(2 marks)

6 **a** Find the value of $2x + 3y$ when $x = \frac{1}{2}$ and $y = 4$

(2 marks)

b Solve the equation
$$4x + 5 = 29$$

(2 marks)

c Solve the equation
$$3(x - 2) = x + 7$$

(3 marks)

7 A motorist travelled 100 miles in 3 hours.

a Calculate his average speed.
Give your answer to a suitable degree of accuracy.

(4 marks)

b He then travels a further 60 miles at 30 miles per hour.
Calculate how long the whole journey takes.

(3 marks)

8 A group of 15 pupils decide to record how long it takes to complete a Mathematics homework. Here are the results to the nearest minute.

| 11 | 15 | 20 | 23 | 35 | 38 | 31 | 25 |

| 17 | 15 | 12 | 23 | 15 | 21 | 22 |

  a Draw a stem and leaf diagram to show this information.

                                    (3 marks)

  b Find the median of the recorded times.

                                    (3 marks)

  c The homework was expected to take 20 minutes to complete.
Do you think that this was reasonable?
Explain your answer.

                                    (1 mark)

9 The weights of two parcels differ by 500 grams.
The combined weights of the two parcels is 4.6 kg.
Calculate the weight of the heavier parcel.

                                    (4 marks)

10 The diagram shows a triangle ABC.

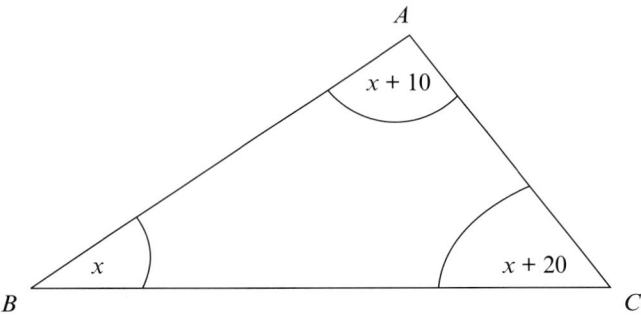

Work out the size of x.

                                    (4 marks)

11 **a** Express 72 as a product of its prime factors.

                                    (3 marks)

   b What is the highest common factor of 16 and 72?

                                    (1 mark)

   c What is the least common multiple of 6 and 8?

                                    (2 marks)

12 Measure this angle and copy it onto paper. Using ruler and compasses only construct the bisector of the angle.

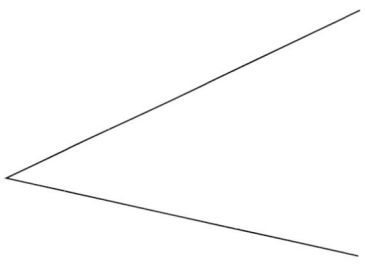

                                    (3 marks)

13 Simplify

a $x^2 \times x^3$

(1 mark)

b $\dfrac{x^7}{x^5}$

(1 mark)

c $\dfrac{x^2 \times x^4}{x^3}$

(2 marks)

14 The grid shows a line segment AB.

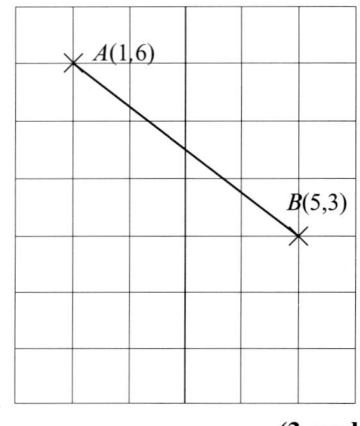

Find the length of the line segment AB.

(3 marks)

15

Which savings scheme makes the most money after two years?
Explain your answer.

(4 marks)

16 A logo is made up of a rectangle and a semicircle as shown.
Find the perimeter of the logo.
Give your answer in terms of π.

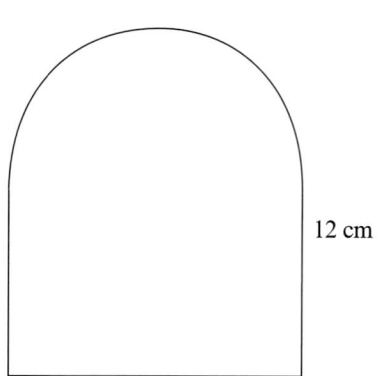

12 cm

20 cm

(4 marks)

17 Clare tiles a wall with 180 tiles.
There are plain tiles and patterned tiles.
The ratio of plain tiles to patterned tiles is 7 : 2
How many plain tiles does she use?

(3 marks)

18 Solve the simultaneous equations
$$2x + y = 7$$
$$x - 3y = 0$$

(4 marks)

19 The graph shows a straight line passing through the points A (0, 2) and B (6, 5).

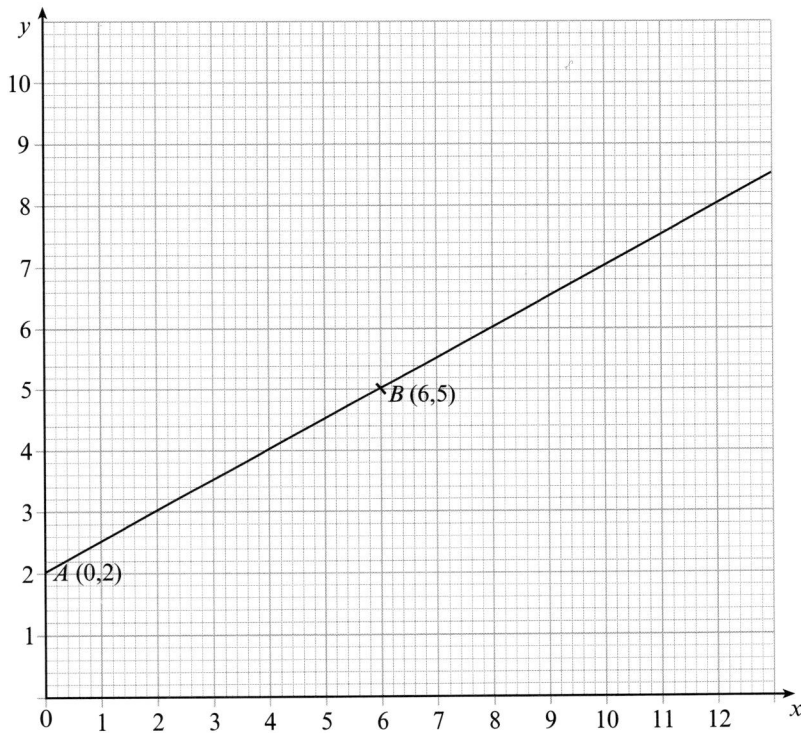

Find the equation of the line.

(4 marks)

20 When man first walked on the moon 600 million people worldwide watched it on television.

a Write 600 million as a number in standard form.

(2 marks)

b 600 million was $\frac{1}{5}$ of the world's population.
Calculate the world population.
Give your answer in standard form.

(3 marks)

21   The box plot shows the results of a Science examination.

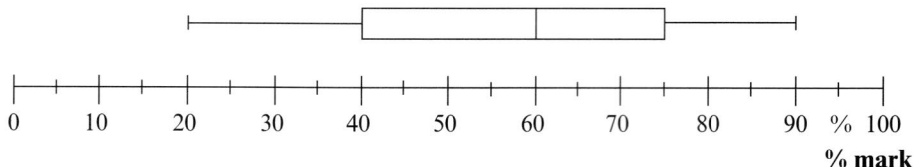

The following data is collected for an English examination.

Lowest score	30
Lower quartile	48
Median	60
Upper quartile	72
Highest score	80

a Draw the box plot for the English data.

(3 marks)

b Make two comparisons between the Science data and the English data.

(2 marks)

22   The diagram shows a circle.

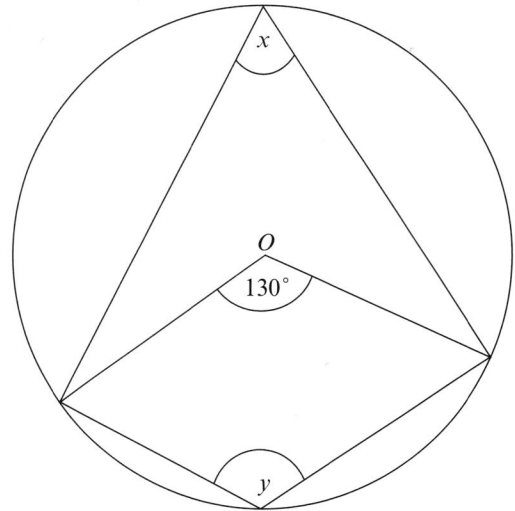

O is the centre of the circle.

a Work out the value of x.

(2 marks)

b Work out the value of y.

(2 marks)

100 marks total

1   A magazine costs £2.40
     The price is increased by 15%.
     Calculate the new cost.

                                                                    (3 marks)

2   **a**   Simplify

                     $3a + 2b - 2a + 4b$

                                                                    (2 marks)

     b $x = 5$ and $y = -2$
             Find the value of $3x + 4y$

                                                                    (2 marks)

     c   Solve

                     $9x + 1 = 4x + 25$

                                                                    (3 marks)

3

Country	£1 is equal to
France	9.90 Francs
Germany	2.96 Marks
Greece	511 Drachma
Spain	250.7 Pesetas
U.S.A.	1.43 Dollars

     a   Belinda goes to Germany.
             How many marks should she get if she changes £150?

                                                                    (2 marks)

     b   Jonathan comes home from the U.S.A. with 200 dollars.
             How much money should he get when he changes his money back
             into pounds?

                                                                    (2 marks)

4   A garage has x cars.
     Each car has 4 wheels.

     a   Write down an expression, in terms of x, for the total number of
             wheels.

                                                                    (1 mark)

     b   Each wheel uses 4 nuts.
             Write down an expression, in terms of x, for the total number of
             nuts.

                                                                    (1 mark)

     c   The total number of nuts used is 144.
             How many cars are there?

                                                                    (2 marks)

5

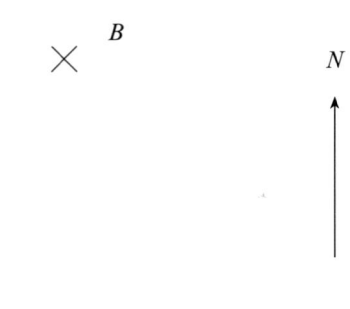

a Using tracing paper copy the diagram.
Measure the bearing of *B* from *A*.

(2 marks)

b *C* is due east of *A*.
C is also on a bearing of 160° from *B*.
Mark accurately the position of *C* on the diagram.

(3 marks)

6 A teacher says that anyone who scores 60% or more in a test will pass.
Matthew scores 38 out of 60.
Did he score enough to pass?
Give a reason for your answer.

(4 marks)

7 A bus leaves the station at 9.00 a.m.
The graph shows the journey.

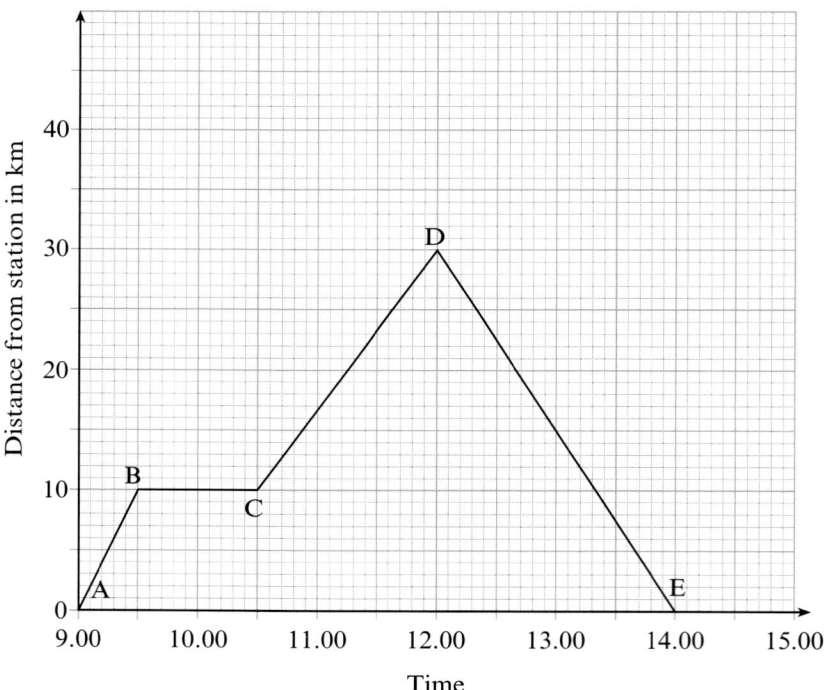

a Describe each stage of the journey.

(3 marks)

b What was the average speed of the bus, in km/h, when travelling from *C* to *D*?

(2 marks)

8 The table shows the scores awarded by 10 judges for a diving competition between two competitors.

Competitor 1	6	8	7	9	9	7	8	8.5	6.5	7
Competitor 1	6.5	4	7	9.5	8.5	6.5	7	9.5	6.5	8

a Plot the scores on a scatter diagram.

(2 marks)

b What do you notice about the correlation between the scores awarded to the two competitors?

(1 mark)

c i Draw a line of best fit.

(1 mark)

   ii Use your graph to estimate the score awarded to Competitor 2 if Competitor 1 is awarded 7.5.

(1 mark)

9 A circle has a radius of 6.4 cm.
Calculate the area of the circle.
Give your answer to a suitable degree of accuracy.

(3 marks)

10 Here is a number pattern

    Line 1: $1 \times 2 = 2^2 - 2$
    Line 2: $2 \times 3 = 3^2 - 3$
    Line 3: $3 \times 4 = 4^2 - 4$
    Line 4: $4 \times 5 = 5^2 - 5$

a Write down the fifth line of the pattern.

(2 marks)

b Write down the *nth* line of the pattern.

(2 marks)

11 Use your calculator to work out

a $\dfrac{(0.3 + 0.51)}{(0.27 + 0.34)}$

(2 marks)

b $\sqrt{\dfrac{8}{0.4}}$

(2 marks)

12 The times taken for a group of pupils to get to school one day is shown.

Time taken T minutes	Number of people
$0 < T \leqslant 5$	10
$5 < T \leqslant 10$	18
$10 < T \leqslant 15$	15
$15 < T \leqslant 20$	7
$20 < T \leqslant 25$	6
$25 < T \leqslant 30$	4
	Total = 60

a Write down the modal class.

(1 mark)

b Estimate the mean time taken to get to school.

(4 marks)

13 Estimate the value of

$$\frac{593 \times 4.98}{0.52}$$

(3 marks)

14 Solve the equation

$$\frac{24 - x}{3} = 5$$

(3 marks)

15 Use trial and improvement to solve the equation
$$x^3 - 11x = 12$$
Give your answer to 1 decimal place.

(3 marks)

16 The diagram shows an airport car park.

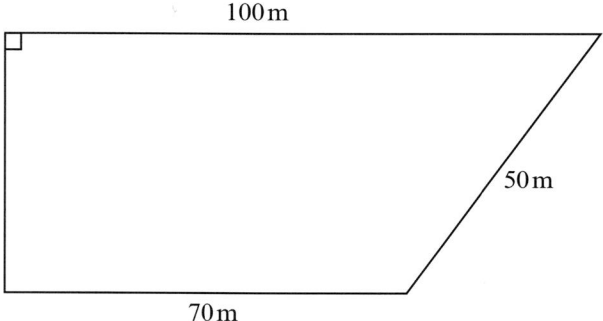

Calculate the area of the car park.

(7 marks)

17 Work out
   a $\sqrt{25}$ **b** 4^0 **c** $\sqrt[3]{27}$ **d** 2^{-3}

   (4 marks)

18 The diagram shows a circle.

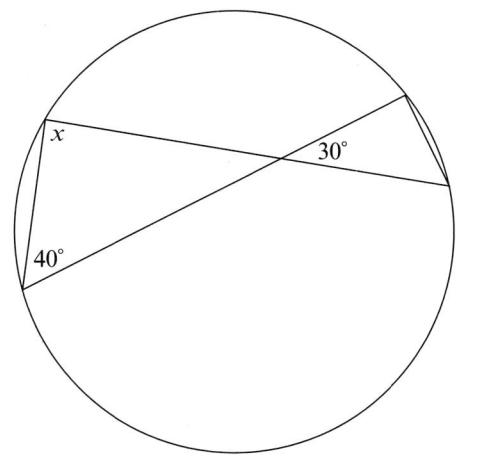

Not to scale

Calculate the value of x.
Show your working.

   (3 marks)

19 In the diagram $AB = 7.2$ cm, $CD = 2.3$ cm and angle $B = 48°$

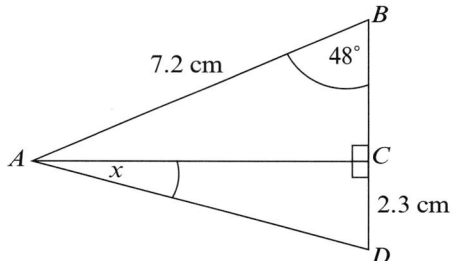

Calculate the size of the angle marked x.
Give your answer to 1 decimal place.

   (5 marks)

20 A rectangle is drawn so that the length is 4 cm more than the width as shown.

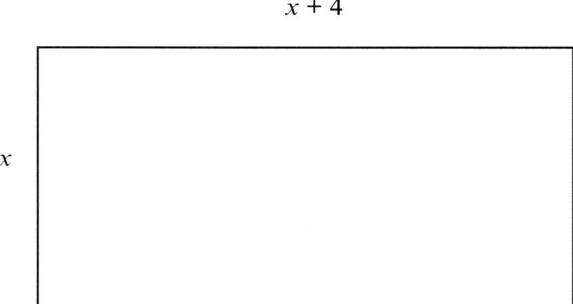

The area of the rectangle is 45 cm².

   a Form an expression, in terms of x, for the area of the rectangle.

   (1 mark)

b Show that the equation can be simplified to
$$x^2 + 4x - 45 = 0$$

(2 marks)

c Solve the equation $x^2 + 4x - 45 = 0$ and hence write down the perimeter of the rectangle.

(3 marks)

20 **d** A different rectangle has length 11 cm and width 9 cm.
All measurements are to the nearest centimetre.
What is the minimum value of the area of the rectangle?

(3 marks)

21 The probability that the school bus is late on a Monday morning is 0.4.

a Write down the probability that the school bus is not late on a Monday morning.

(1 mark)

b The probability that the bus is late on Tuesday morning is 0.3.
   i Copy and complete the tree diagram.

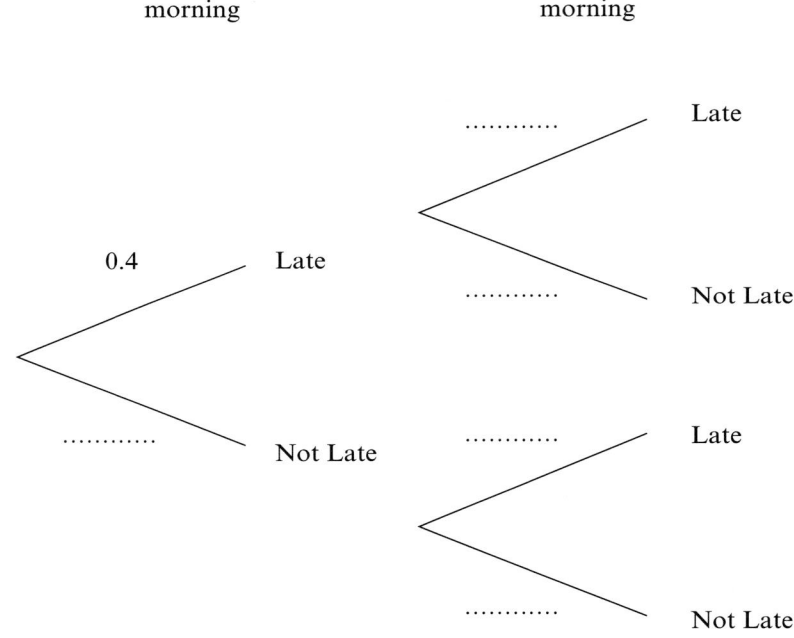

(2 marks)

   ii Calculate the probability that the bus is late on both Monday and Tuesday mornings.

(2 marks)

22 **a** Given that $2^n = 64$
Work out the value of n.

(1 mark)

b $3^p = 3^2 \times 3^4$
Write down the value of p.

(1 mark)

23 Rearrange the formula to make x the subject.
$$ax = bx + c$$

(3 marks)

100 marks total

Section 1 Number bites

Starting points

A1   a $^-2$ b $^-7$ c 4   d 10   e $^-4$ f 5

A2   a $^-2$ b $^-2$ c 9   d $^-9$ e $^-3$ f 2   g $^-4$ h $^-5$ i $^-4$ j $^-6$

B1   a 5, 10, 15, 20, 25, 30, 35 for example
     b 8, 16, 24, 32, 40, 48, 56 for example
     c 12, 24, 36, 48, 60, 72, 84 for example

B2   a 40, 80, 120 for example   b 24, 48, 72 for example

B3   a 1, 2, 3, 4, 6, 8, 12, 24   b 1, 2, 3, 5, 6, 10, 15, 30   c 1, 5, 25   d 1, 17

B4   a 1, 2, 3, 6   b 1, 5

B5   It has only one factor

B6   a 2, 3, 5, 7, 11, 13, 17, 19

C1   a 5^2 b 2^5

C2   a 64   b 81   c 2.25   d 1.61051

D1   a 2, 3, 7   b 2, 7, 11

D2   a $2^3 \times 3 \times 5$ b $2 \times 5^2 \times 7$

E2   a $1\frac{1}{3}$ b $2\frac{1}{6}$ c $5\frac{3}{4}$

E3   a $\frac{7}{3}$ b $\frac{10}{3}$ c $\frac{5}{1}$ d $\frac{37}{13}$ e $\frac{11}{1}$

F1   a $\frac{4}{10}, \frac{6}{15}, \frac{8}{20}$ b $\frac{1}{3}, \frac{2}{6}, \frac{8}{24}$ c $\frac{2}{3}, \frac{4}{6}, \frac{8}{12}$

G1   a 0.4375   b $0.\dot{5}$ c $0.6\dot{3}$ d $0.4\dot{2}5\dot{9}$

H1   £90   b 92 litres   c 99 m

Exercise 1.1

1   A, B, D, F, H, I

2   a $3 \times {^-5} = {^-15}$ b $^-2 \times 6 = {^-12}$ c $5 \times {^-4} = {^-20}$ d $^-3 \times {^-6} = 18$
     e $^-12 \div {^-3} = 4$ f $^-8 \div 4 = {^-2}$ g $24 \div {^-6} = {^-4}$ h $^-40 \div 5 = {^-8}$

3   a $4 \times {^-7} = {^-28}$ b $^-15 \div {^-5} = 3$ c $^-4 \times 8 = {^-32}$ d $^-3 \times {^-6} = 18$
     e $^-35 \div {^-7} = 5$ f $36 \div {^-6} = {^-6}$ g $^-3 \times {^-12} = 36$ h $14 \div 7 = 2$

4   a 9.43   b 9.43   c $^-9.43$ d $^-7.2$ e 7.2   f $^-7.2$

5   If you multiply an odd number of negative numbers you get a negative number. If you multiply an even number of negative numbers you get a positive number.

Exercise 1.2

2   a 5.5, $^-5.5$ b 8.6, $^-8.6$ c 3.5, $^-3.5$
     d 6.7, $^-6.7$ e 7.7, $^-7.7$ f 9.7, $^-9.7$

3   a 6.93, $^-6.93$ b 8.49, $^-8.49$ c 6.25, $^-6.25$
     d 5.18, $^-5.18$ e 30.00, $^-30.00$ f 0.1, $^-0.1$

4   a 13, $^-13$ b 14, $^-14$ c 10, $^-10$
     d 9, $^-9$ e 15, $^-15$

Exercise 1.3

2   a 729   b 27 000   c 42.875   d $^-729$ e $^-1.331$ f 0.064

4   a 2   b 4   c 10   d 3   e 9   f 7   g 2.5   h 20   i 5   j 0.7

5   a $^-125, ^-64, ^-27, ^-8, ^-1, 0, 1, 8, 27, 64, 125$

Exercise 1.4

1   a 0.25   b 10   c 0.4   d $0.\dot{3}$ e 1.5   f 1.2   g Not defined

2   a $\frac{5}{3}$ b $\frac{4}{7}$ c $\frac{2}{3}$ d $\frac{3}{10}$ e $\frac{5}{6}$

3   Graph passing through $(1, 1)$, $(2, \frac{1}{2})$, $(4, \frac{1}{4})$ and $(10, \frac{1}{10})$.

Exercise 1.5

1   a 3.8×10^5 b 4.51×10^7 c 9.2×10^{-4} d 2.62×10^{-5}

2   a 94 200   b 0.0025   c 741 400 000   d 0.000 000 627

3   a **P:** 13 is greater than 10.   **Q:** 1000 isn't written as a power of 10.
     R: Should be a multiplication not a division.
     S: 0.58 is less than 1.
     b **P:** 1.3×10^5,   **Q:** 5.7×10^3,   **R:** 6.42×10^{-3},   **S:** 58×10^{-5}

5   a 8.76×10^6 b 1.7×10^5 c 1.98×10^4
     d 7.88×10^{10} e 7.68×10^{-15} f 3.972×10^4

Exercise 1.6

1   a 3^7 b 4^2 c 2^3 d 7^{-6} e 6^{-2} f 7^{-6} g 5^0 h 2^5 i 6^{-1} j 7^{-3}

2   a $2^3 \times 2^4 = 2^7$ b $4^5 \times 4^{-2} = 4^3$ c $3^5 \div 3^3 = 3^2$ d $7^7 \times 7^{-3} = 7^4$
     e $8^4 \div 8^6 = 8^{-2}$ f $5^4 \div 5^2 = 5^2$ g $2^{-6} \times 2^5 = 2^{-1}$ h $3^4 \div 3^1 = 3^3$

3   a 3^8 b 5^9 c 4^{-10} d 4^{-10} e 2^{-16} f 7^0

4   a $2^3 \times 2^2 = 32$ b $3^6 \times 3^{-3} = 27$ c $2^8 \div 2^2 = 64$
     d $4^2 \div 4^{-1} = 64$ e $(3^2)^2 = 81$ f $(5^3)^1 = 125$
     g $(2^{-3})^1 = 0.125$ h $(5^{-1})^2 = 0.04$

Exercise 1.7

1   a 2.4×10^4 b 2×10^{-6} c 2×10^3 d 6×10^{-11}

Exercise 1.8

1   a 36   b 420   c 420   d 3150   e 630   f 1890   g 14 850

2   45, 90, 315 or 630

Exercise 1.9

1   a $\frac{5}{6}$ b $\frac{13}{20}$ c $\frac{1}{6}$ d $\frac{11}{20}$ e $\frac{31}{40}$ f $\frac{3}{4}$ g $\frac{1}{2}$ h $\frac{17}{20}$

2   a $1\frac{1}{6}$ b $1\frac{1}{20}$ c $1\frac{1}{12}$ d $1\frac{5}{8}$

3   a $2\frac{3}{10}$ b $1\frac{3}{10}$ c $3\frac{5}{6}$ d $\frac{9}{10}$ e $3\frac{23}{24}$ f $\frac{13}{20}$

Thinking ahead to ...

A   a 8   b 8   c 12.8   d 12.8

B   Same as multiplying by its reciprocal

Exercise 1.10

1   a $\frac{5}{12}$ b $\frac{1}{2}$ c $\frac{12}{35}$ d $\frac{3}{7}$ e $\frac{2}{5}$ f $\frac{7}{9}$ g $\frac{8}{15}$ h $\frac{5}{12}$

2   a $2\frac{2}{5}$ b 6   c 12   d $10\frac{1}{2}$ e 21   f $31\frac{1}{2}$ g $4\frac{2}{7}$ h $4\frac{3}{13}$

3   a $4\frac{1}{2}$ b $2\frac{11}{12}$ c $2\frac{4}{7}$ d $7\frac{1}{3}$ e 1   f $1\frac{1}{9}$ g 6   h $\frac{20}{33}$

Exercise 1.11

1   a 36   b 15   c 77   d 3

End points

A1   a $7 \times {^-2} = {^-14}$ b $^-3 \times {^-5} = 15$ c $^-12 \div 4 = {^-3}$ d $20 \div {^-5} = {^-4}$
     e $6 \times {^-3} = {^-18}$ f $^-16 \div {^-2} = 8$ g $^-4 \times {^-9} = 36$ h $^-21 \div 7 = {^-3}$

B1   a 1.2   b 35.937   c 6   d $^-1.5$

B2   a $\frac{1}{8}$ b 5   c 1.8   d $2\frac{1}{2}$ e $\frac{4}{5}$

C1   a 3^7 b 4^3 c 7^6

C2   a $2^3 \times 2^4 = 128$ b $3^5 \times 3^{-2} = 27$ c $(4^2)^{-1} = 0.0625$

C3   a 2^2 b 3^{-9} c 5^{10} d 4^{-6}

C4   a $3^4 \times 3^6 = 3^{10}$ b $4^3 \div 4^{-4} = 4^7$ c $7^{-2} \times 7^7 = 7^5$ d $(6^{-2})^0 = 6^0$

D1   a 7.062×10^7 b 3.75×10^{-6}

D2   a 0.000 000 069   b 10 300 000 000

D3   a 9.984×10^{14} b 9.4×10^{-8} c 4.8×10^{27}

D4   a 4.2×10^{-3} b 2×10^4

E1   a $\frac{13}{15}$ b $\frac{13}{30}$ c $\frac{5}{8}$ d $3\frac{37}{42}$

E2   a $\frac{3}{10}$ b $\frac{1}{4}$ c 6   d $1\frac{3}{7}$

F1   a 84   b 360

F2   a 7   b 35

Section 2 Sequences

Starting points

A1   a 21   b 3   c 0   d 18

A2   a 19   b 36   c 4   d 12

B1

B2   1, 9, 25, 49 for example

B3   1, 3, 6, 10, 15, 21

B4   5, 25, 125, 625

B5   34

B6 29, 47, 76, 123

C1 33, 39

C2 31

C3 **a** 31, 42, 55 **b** 32, 38, 44 **c** 47, 65, 86 **d** 57, 83, 114

D1 **a**

b i 16 **ii** 26

c

Pattern number (n)	Number of matches (m)
1	6
2	11
3	16
4	21
5	26

d $m = 5n + 1$

Exercise 2.1

1

Pattern number		Number of matches
1	→	3
2	→	4
3	→	5
4	→	6
20	→	22
n	→	$n + 2$

2 **a**

Pattern number		Number of matches
1	→	4
2	→	8
3	→	12
4	→	16
5	→	20
6	→	24

b $n → 4n$ **c** 400

Thinking ahead to ...

A 301 matches; $3n + 1$

Exercise 2.2

1 **a** $m = 2n + 1$ **b** 17

2 **a** $m = 4n + 1$ **b** 161 **c** 32nd

3 **a** $m = 6n - 2$ **b** 598

4 **b** $m = 4n + 2$

5 **a** $50 → 249$, $n → 5n - 1$ **b** $36 → 293$, $p → 8n + 5$
    c $2 → 5$, $3 → 9$, $4 → 13$, $25 → 97$

Thinking ahead to ...

A **a** 19, 21 **b** 29 **c** 105

Exercise 2.3

1 **a** A: $3n + 3$, B: $5n - 4$, C: $10n + 3$, D: $8n - 6$
    b A: 153, B: 246, C: 503, D: 394

2 **a** He should have had $3n$ instead of $n + 3$. **b** $3n + 2$

3 $n + 7$

4 $22 - 2n$

Exercise 2.4

1 **a** 5th number, $40 = 5 × 8$ **b** 130

2 **a** 5th triangle number, $15 = \frac{(5 × 6)}{2}$ **b** 78 **c** $\frac{\{n × (n + 1)\}}{2}$

3 **a** 5th power, $32 = 2 × 2 × 2 × 2 × 2 = 2^5$ **b** 256 **c** 2^n

Exercise 2.5

1 **c** 204 **d** 80th

2 **a** $2n + 6$, $2(n + 3)$

End points

A1 **a** $m = 2n + 3$ **b** 203

B1 **a**

1	→	2
2	→	5
3	→	8
4	→	11
10	→	29
p	→	$3p - 1$

b

1	→	7
2	→	12
3	→	17
4	→	22
20	→	102
n	→	$5n + 2$

C1 **a** A: $3n + 4$, B: $5n - 3$, C: $4n + 1$, D: $n + 5$, E: $3n + 6$, F: $7n + 4$
    b A: 64, B: 97, C: 81, D: 25, E: 66, F: 144

SECTION 3 Properties of shapes

Starting points

A1 **a i** reflex **ii** acute **iii** right angle
    b i acute **ii** right angle **c** Triangles AED, BAD

B1 **a** 60° **b** 300° **c** 45

B2 **a** 35°, 72.5°
    b The three angles in a triangle add up to 180°, so if one angle is more than 90°, the other two must be less than 90°.

C1 116°, 116°, 64°, 64°

D1 trapezium; rectangle or square; square or rhombus; rectangle, square, parallelogram or rhombus

E1 1080°

E2 142°

E3 129°

E4 **a** 1800° **b** 150°

Exercise 3.1

1 **b** 74°, 53°, 53°, 74°, 53°, 53°

2 62°, 120°, 117°

Exercise 3.2

1 **a** The corresponding angles are: any two out of a, c and e; any two out of b, d and f; any two out of g, i and k; any two out of h, j and l
    b The alternate angles are: b and i or k, d and k, h and c or e, j and e

2 72°, 135°, 45°

3 **a i** The corresponding angles are:
    any pair out of angles ADC, DHG, HLK and LPO;
    any pair out of angles EDH, IHL, MLP and QPS;
    any pair out of angles ADE, DHI, HLM and LPQ;
    any pair out of angles CDH, GHL, KLP and OPS
    ii The corresponding angles are:
    any pair out of angles BDC, DGF, GKJ and KON;
    any pair out of angles BDE, DGH, GKL and KOP;
    any pair out of angles CDG, FGK, JKO and NOR;
    any pair out of angles EDG, HGK, LKO and POR
    b BR and AS are not parallel

c The alternate angles are:
angles CDG and DGH, GKL or KOP;
angles FGK and GKL or KOP;
angles JKO and KOP;
angles EDH and DHG, HLK or LPO;
angles IHL and HLK or LPO;
angles MLP and LPO;
angles EDG and DGF, GKJ or KON;
angles HGK and GKJ or KON;
angles LKO and KON;
angles CDH and DHI, HLM or LPQ;
angles GHL and HLM or LPQ;
angles KLP and LPQ

d i 65° **ii** 54° **iii** 61° **iv** 54° **v** 126° **vi** 61°

4 42°, 77°, 61°, 72°, 72°, 72°, 68.5°, 68.5°, 68.5°

Exercise 3.3

1 a 56°, 42°, 69°, 104°, 89°

2 a The 12 exterior angles are 12 equal turns which are equivalent to one full turn, 360°.
 b 30° **c** 150°

3 40°, 140°, 70°, 40°

Exercise 3.4

1 a AMD is isosceles, so angle AMD = 180° − 2 × 22.5° = 135°
 b i 67.5°, 157.5°

2 a i 67.5° at A, 67.5° at C, 45° at F
 ii 112.5° at A, 112.5° at C, 135° at F
 b 5 **c** 45°, 67.5°, 90°, 112.5°, 135°, 157.5°

End points

A1 42°, 67°, 109°

B1 C is regular, B and E have only one obtuse angle

C1 a B 120°, 30°, 30°; C 60°, 60°, 60°; D 90°, 90°, 90°, 90°; E 60°, 90°, 120°, 90°
 b B 60°, 150°, 150°; C 120°, 120°, 120°; D 90°, 90°, 90°, 90°; E 120°, 90°, 60°, 90°

C2 45°

SECTION 4 Linear graphs

Starting points

A1 d Straight line passing through (¯3, 2), and (3, 2), and another straight line passing through (¯1, ¯3) and (¯1, 3).
 e (¯1, 2)

A2 a (0, 2)
 b No, the lines are parallel.
 c The *x*-coordinates will always be 0.

A3 a horizontal **b** vertical **c** vertical
 d horizontal **e** horizontal **f** vertical

B1 a

x	¯2	¯1	0	1	2	3
y	¯7	¯4	¯1	2	5	8

b

x	¯2	¯1	0	1	2	3
y	0	1	2	3	4	5

c

x	¯2	¯1	0	1	2	3
y	¯2	¯1	0	1	2	3

d

x	¯2	¯1	0	1	2	3
y	¯5	¯4	¯3	¯2	¯1	0

e

x	¯2	¯1	0	1	2	3
y	¯2	¯0.5	1	2.5	4	4.5

B2 a Straight line passing through (¯1, ¯4) and (1, 2).
 b Straight line passing through (¯2, 0) and (0, 2).
 c Straight line passing through (¯1, ¯1) and (1, 1).
 d Straight line passing through (¯1, ¯4) and (3, 0).
 e Straight line passing through (¯2, ¯2) and (1, 2).

B3 (¯2, ¯7), (2, 1), (3, 3) and (¯0.25, ¯3.5) all lie on the line $y = 2x − 3$

Exercise 4.1

1 a 30p **b** 10p

2 a more **b** 26p

3 34p

4 20 seconds at low call charge is cheaper

5 Low call charge

6 26p

7 1 minute 40 seconds

8 c Peak charge **d** Weekend, Low call, Infotel, Peak
 e The higher the charge rate the greater the steepness.

9 16p

10 a $c = t \div 2$ **b** 45p

Exercise 4.2

1 a 3 **b** 2 **c** 1 **d** 2 **e** 2 **f** 1 **g** 3 **h** 1

3 The higher the gradient the greater the steepness.

Exercise 4.3

1 a i $\frac{3}{2}, \frac{4}{3}, \frac{4}{5}, \frac{5}{4}, \frac{3}{2}, \frac{5}{2}, \frac{1}{2}$ **ii** 1.5, 1.3, 0.8, 1.25, 1.5, 2.5, 0.5
 b f = 2.5, a = 1.5, e = 1.5, b = 1.3, d = 1.25, c = 0.8, g = 0.5

2 Lines that are parallel have the same gradient.

Exercise 4.4

1 a a, c, e, f, h
 b i $\frac{-1}{2}, \frac{5}{2}, \frac{-3}{2}, \frac{3}{2}, \frac{-3}{4}, \frac{-1}{2}, \frac{1}{2}, -1$ **ii** ¯0.5, 2.5, ¯1.5, 1.5, ¯0.75, ¯0.5, 0.5, ¯1

3 Positive gradient

4 A

5 G

6 F

7 ¯1

8 C

9 gradient of E = $\frac{-3}{2}$

10 gradient of D = $\frac{2}{3}$

11 a–d Straight line passing through (¯5, 5) and (3, 3), and another passing through (¯3, ¯5) and (3, 3).
 e (3, 3) **f** ¯5 **g** (¯1, 4)

Exercise 4.5

1 2

2 $\frac{-1}{2}$

3 h

4 a 3 **b** $\frac{-1}{3}$ **c** ¯1 **d** yes

5 Their gradients multiply to give ¯1

6 1 and g

7 yes

8 d

9 e

10 gradient of g × gradient of f = 3 × $\frac{-1}{3}$ = ¯1

Exercise 4.6

1 **a** Multiply the *x*-coordinate by 3 then add 2 to get the *y*-coordinate.

   b

x	$^-2$	$^-1$	0	1	2	3
y	$^-4$	$^-1$	2	5	8	11

   c $(^-2, ^-4)$, $(^-1, ^-1)$, $(0, 2)$, $(1, 5)$, $(2, 8)$, $(3, 11)$
   d & e Straight line passing through $(^-1, ^-1)$ and $(1, 5)$.

2 **a** Multiply the *x*-coordinate by 2 then subtract 3 to get the *y*-coordinate.

   b

x	$^-1$	0	1	2	3
y	$^-5$	$^-3$	$^-1$	1	3

   c Straight line passing through $(^-1, ^-5)$ and $3, 3)$

3 **a**

x	$^-2$	$^-1$	0	1	2	3
y	$^-3$	$^-1$	1	3	5	7

   b Straight line passing through $(^-2, ^-3)$ and $(1, 3)$.

4 Straight line passing through $(^-2, ^-9)$ and $(1, 3)$.

5 **a** Straight line passing through $(^-2, ^-5)$ and $(1, 1)$ and another passing through $(^-1, ^-6)$ and $(2, 3)$.
   b $(2, 3)$

Exercise 4.7

1 **a** 2 **b** $^-5$

2 **a** 3 **b** $^-1$

3 $y = 4x + 2$ and $y = 2 + 4x$

4 **a** $3, ^-2$ **b** $1, 2$ **c** $4, 0$ **d** $5, ^-4$ **e** $2, 3$ **f** $3, ^-\frac{1}{2}$ **g** $5, 1$ **h** $\frac{1}{2}, 0$
   i $^-5, 5$ **j** $1, 0$ **k** $1, ^-2$ **l** $^-4, 0$

5 K: $y = 3x + 5$, M: $y = \frac{1}{2}x - 1$, P: $y = 2x + 1.5$, R: $y = -3x + 2$, T: $y = x$

Exercise 4.8

1 **a** $y = 4x - 3$ passes through $(0, ^-3)$ and $(1, 1)$.
   $y = x - 1$ passes through $(0, ^-1)$ and $(1, 0)$.
   $y = 2x + 1$ passes through $(^-1, ^-1)$ and $(0, 1)$.

2 **a** $y = 3x$ passes through $(0, 0)$ and $(1, 3)$.
   $y = x + 2$ passes through $(0, 2)$ and $(1, 3)$.
   $y = 3x - 3$ passes through $(0, ^-3)$ and $(1, 0)$.
   b $y = 3x$, $y = x + 2$
   c $y = x + 2$

3 **a** $^-3, 2$ **b** Straight line passing through $(0, 2)$ and $(1, ^-1)$.

Exercise 4.9

1 **a** 2 **b** $^-4$ **c** $y = 2x - 4$

2 **a** $^-3$ **b** $y = 3x - 3$

3 C: $y = 3x + 1$, D: $y = 2x + 3$, E: $y = ^-3x + 1$

Exercise 4.10

1 **a** 4 **b** $y = 2x - 5$

2 **a** 2 **b** $y = \frac{3}{2}x + 2$

3 **a** $y = 2x + 4$ **b** $y = \frac{4}{3}x - 3$ **c** $y = x - 2$ **d** $y = \frac{3}{5}x$
   e $y = \frac{2}{5}x + 1$ **f** $y = 3x - 2$ **g** $y = \frac{3}{2} - 2x$ **h** $y = \frac{1}{4}x$
   i $y = 1 - x$ **j** $y = 2x + 2$ **k** $y = \frac{1}{5}x + \frac{1}{5}$ **l** $y = \frac{2}{3}x$

4 $3y = 6x - 9$

Exercise 4.11

1 **a** $y = \frac{3}{2}x + 1$ **b** $\frac{3}{2}$ **c** 1
   d Straight line passing through $(0, 1)$ and $(2, 4)$.

2 **a** $\frac{2}{3}$ **b** -1
   c Straight line passing through $(0, -1)$ and $(3, 1)$.

3 Straight line passing through $(0, -3)$ and $(1, 0)$.

4 **c** $(4, 2)$.

5 Straight line passing through $(-1, -2)$ and $(2, 2)$.

6 Yes, $2 \times 3 = (5 \times 2) - 4$

Exercise 4.12

1 **a** $(2.5, 3)$ **b** $5y = 2x + 10$, $y = 2x - 2$

2 **b** $(2, 3)$

3 $(1.2, 2.2)$

End points

B1 1

B2 **a** $\frac{7}{4}$ **b** 1.75

B3 $\frac{2}{3}, ^-1$

C1

x	$^-1$	0	1	2	3
y	$^-7$	$^-4$	$^-1$	2	5

C2 Straight line passing through $(-1, -7)$ and $(1, 0)$.

D1 $y = \frac{3}{4}x + \frac{3}{2}$

E1 $2, ^-5$

E2 Straight line passing through $(0, 1)$ and $(2, 2)$.

E3 Straight line passing through $(0, ^-2)$ and $(3, 3)$.

F1 $3y = 2x + 6$, $y = 2x - 2$

SECTION 5 Comparing Data

Starting points

B1

Number of children in car	Tally	Number of cars
0	ＩＩＩＩ ＩＩＩＩ Ｉ	11
1	ＩＩＩＩ ＩＩ	7
2	ＩＩＩＩ	4
3	Ｉ	1
4	Ｉ	1

B2

Colour of car	Frequency
Red	5
Blue	6
Green	2
White	3
Silver	2
Black	4
Brown	2

C1 bars with heights 5, 6, 2, 3, 2, 4, 2.

C2 bars with heights 11, 7, 4, 1, 1.

D1

Colour of car	Percentage
Red	20.8%
Blue	25%
Green	8.3%
White	12.5%
Silver	8.3%
Black	16.7%
Brown	8.3%

E1 **a** 15 and 42 **b** 39.5 **c** 42.5 **d** 69

E2 **a** 25 **b** 38 **c** 48.8 **d** 93

E3 2, 3, 3, 5, 7, 7, 10, 11 for example

F1 blue

F2 **a** 0 **b** 4

G1 **a** £340 **b** £340 **c** £314.89
   d The mean is not sensible, because there is an extreme value, £112.

H1 **a** 54.4% **b** 34.8% **c** 10.8%

H2 **a** 38.8% **b** 33.0% **c** 28.2%

Exercise 5.1

1 **a** 72, 48, 100 **b** 5°, 7.5°, 3.6°
   c

	Caring for Children	Child Action	Children in Crisis
Child Care	306°	252°	252°
Fund raising	32°	66°	54°
Administration	8°	18°	36°
Other costs	14°	24°	18°

Thinking ahead to ...

A **a** 7 **b** 4

Exercise 5.2

1 **a** 2 **b** 5, 5, 5, 5, 6, 6, 7, 8, 8, 9 **c** 6
2 **a** 15 **b** 5, 5, 5, 6, 6, 7, 7, 7, 7, 9, 9, 9, 10, 10, 10 **c** 7
3 **a** Peta fired 18 arrows, so the median is halfway between the 9th and 10th values.
   b 5
4 For example

Ring score	2	3	4	5	6	7	8	9	10
Frequency	6	0	0	0	1	1	0	0	6

Thinking ahead to ...

A **a** 24 points

Exercise 5.3

1 **a** Jodie 84, Geeta 156 **b** Jodie 16, Geeta 20
   c Jodie 5.25, Geeta 7.8
2 **a** They have not multiplied the frequency by the ring score to find the total score.
   b 7.4

Exercise 5.4

1 **a** Javed 6, Lisa 5.5 **b** Javed 6, Lisa 3
   c Javed's scores are higher on average. Lisa is more consistent.
2 **a** Amy's mean = 5.8, range = 4; Paul's mean = 5.7, range = 5
   b Amy's scores are slightly higher on average. She is also more consistent.
3 **a** Imran's mode = 5, range = 8; Viv's mode = 7, range = 8
   b Most of her scores are between 4 and 7, but the range is distorted by a couple of extreme values.

Exercise 5.5

1 **a** William: 2, 4, 4, 4, 5, 5, 6, 6, 6, 6, 6, 7, 7, 7, 8, 8
      Bryony: 4, 4, 5, 5, 5, 5, 6, 6, 6, 7, 7, 7, 9, 9
      Daniel: 2, 3, 4, 4, 5, 5, 5, 6, 6, 6, 6, 6, 7, 7, 7, 9, 9, 9, 10, 10
   b William's lower quartile = 4.5, upper quartile = 7
      Bryony's lower quartile = 5, upper quartile = 7
      Daniel's lower quartile = 5, upper quartile = 8
   c William's interquartile range = 2.5, Bryony's = 2, Daniel's = 3
2 Sheera's interquartile range = 2
   Dave's interquartile range = 3.5
   Sheera is more consistent

Thinking ahead to ...

A **a** 5 **b** 8 **b** 8

Exercise 5.6

1

Ring score	Cumulative Frequency
≤5	2
≤6	7
≤7	19
≤8	26
≤9	31

2

Ring score	2	3	4	5	6	7	8
Frequency	3	7	12	19	13	9	4
Cumulative frequency	3	10	22	41	54	63	67

3 **a** 15 **b** 36 **c** 64

4

Age	Cumulative frequency
10 and under	8
11 and under	15
12 and under	25
13 and under	36
14 and under	44
15 and under	53
16 and under	64

5

Distance to target in metres	10	15	20	25	30	35	40
Number of arrows hitting target	20	19	17	16	15	13	11
Cumulative frequency	20	39	56	72	87	100	111

6

	Ring score									
	≤1	≤2	≤3	≤4	≤5	≤6	≤7	≤8	≤9	≤10
April	4	11	18	26	31	37	39	42	47	48
May	6	11	15	24	29	32	40	42	44	48
June	3	9	15	19	24	28	36	40	42	48

b She is improving: she is getting more high scores and fewer low ones.

Thinking ahead to ...

A **b** 8

Exercise 5.7

1 **a** Lee

Ring score	3	4	5	6	7	8
Cumulative Frequency	7	17	22	23	29	40

Faith

Ring score	4	5	6	7	8
Cumulative Frequency	11	18	26	35	45

Aqib

Ring score	1	2	3	4	5	6	7	8	9
Cumulative Frequency	2	6	8	11	12	16	23	32	39

Tegan

Ring score	1	2	3	4	5	6	7	8	9	10
Cumulative Frequency	6	15	20	24	26	31	37	41	44	48

   c Lee's median score is 5
      Faith's median score is 6
      Aqib's median score is 7
      Tegan's median score is 4.5
2 **a** Median is between 40th and 41st score i.e. halfway between 7 and 8
   b Upper quartile – lower quartile = 9 – 6 = 3

Exercise 5.8

1 **a**
```
4 | 0 6
3 | 2 3 8
2 | 0 3 4 6
1 | 2 2 4 9
0 | 7
    _____
    1 2 3 4 5 6
```
   b
```
4 | 0
3 | 0 1 2
2 | 0 2 4 4 4 6 6
1 | 0 2 5
0 |
    _____
    1 2 3 4 5 6 7
```
   c
```
4 | 1 2 4
3 | 0 6 0
2 | 4 5 5
1 | 6 6 8
0 | 6 6 7 9
    _____
    1 2 3 4 5 6
```
   d
```
5 | 6
4 | 0 0 6
3 | 0 2 7 8
2 | 2 4 8 8 8
1 | 4 8
0 |
    _____
    1 2 3 4 5 6 7
```

2 **a** 6 | 2
5 | 0 0 0 1 2 2 3 4 4 6 7 9
4 | 0 0 0 1 2 4 4 7
3 | 0 4 8 8
2 |
1 |
0 |

1 2 3 4 5 6 7 8 9 10 11 12

b 50

Thinking ahead to ...
A **b i** X 38   Y 36   Z 38
ii X 3.5   Y 4   Z 2.5

End points
A1 Pie chart with angles: child care 230°, fundraising 85°, administration 25°, other 20°.
B1 **a** 5.5
b i

Ring score	1	2	3	4	5	6	7	8
Cumulative Frequency	2	3	8	16	23	28	31	33

ii 5
C1 **a** 5.3   **b** 4.6
D1 Manoj's mean is higher with a smaller range.
D2 Kim's mode is higher with a smaller range.
E1 **a** LQ3 UQ 7   **b** LQ5.5 UQ 8

SECTION 6 Working in 2D

Starting points
A1 23.2 cm
A2 57.5 cm
A3 44.4 cm
A4 **a** 16.3 cm   **b** 2.2 cm   **c** 15.1 cm
B1 121.6 cm²
B2 56.3 cm²
B3 15.1 cm²
B4 **a** 58.1 cm²   **b** 176.7 cm²   **c** 483.1 cm²
C1 **a** 128.6 cm²   **b** 53.0 cm²
D1 4.1 cm, 9.7 cm, 8.7 cm

Exercise 6.1
1 **a** 102.5 cm²   **b** 5.6 cm²
2 **a** 92 cm²   **b** 120 cm²   **c** 238 cm²   **d** 299 cm²
e 344 cm²   **f** 22.5 cm²   **g** 592 cm²   **h** 222 cm   **i** 2450 mm²
j 558 cm²
3 **a** 101 cm²   **b** 303 cm²
4 **b** trapezium   **c** 12 797 cm²
5 **b** 675.8 cm²
6 **b** 2827.5 cm²

Exercise 6.2
1 **a** 162.9 cm²   **b** 120.8 cm²   **c** radius = 1.5, area = $1.5^2\pi = 2.25\pi$
d 8π **e** $\sqrt{53}$
2 **a** 18 mm   **b** 1017.9 mm²   **c** 180 mm   **d** 3010.6 mm²
3 120 424.8 mm²
4 **a** triangle   **b** 61.4 cm²   **c** 52.4 cm²
5 **a** 183.7 cm²   **b** less   **c** 85.4 cm
6 **a** 154 mm   **b** 136 mm   **c** 504 mm²   **d** 928 mm²   **e** 16 288 mm²
f 4656 mm²   **g** between $\frac{1}{5}$ and $\frac{1}{4}$
7 **a** 18 mm   **b** $\frac{1}{4}$ **c** 254.5 mm²   **d** 763.4 mm²   **e** 278.1 mm²

Exercise 6.3
1 **a** 2.9 cm   **b** 4.7 cm   **c** 13.0 cm²
2 30.0 cm²
3 **a** 6.5 cm   **b** 21.5 cm²
4 **a** 9.5 cm²   **b** 21.6 cm²
5 10.4 cm
6 62.4 cm²
7 8.6 cm²
8 49.8 cm²
9 262.6 cm²
10 413.4 cm²
11 73.4 cm

Exercise 6.4
1 **a** 59.7 cm   **b** 77.0 cm   **c** $2 \times \text{radius} \times \pi = 2 \times 4.8 \times \pi = 9.67\pi$
d 7.5 cm   **e** 55.3 cm
2 **a** 99.0 m   **b** 33.0 m
3 32.5
4 23 cm
5 6240 tins

Exercise 6.5
3 **a** triangle   **b** rectangle   **c** rectangle
5 9600 mm²
7 **b** 18

End points
A1 **a** 27.1 cm²   **b** 29.4 cm²   **c** 46.3 cm²
B1 Composite shapes are made up from more than one shape.
B2 **a** 109 cm²   **b** 147 cm²
B3 41.4 cm²
C1 18.3 m

SKILLS BREAK 1A

1 In 2001 it is 44 years.
2 9.5 hours
3 £2.70
4 Malnourished
5 12.3 kg
6 1000 kg
7 0.55 kg
8 50%
9 7319
10 In 2001, 26 years.
11 35 minutes
12 $\frac{1}{3}$
13 60%
14 8
15 $\frac{1}{3}$
16 5%
17 2
18 2
19 0.85 kg
20 770 g
21 14.25 kg
22 Tony

23 24.5%

24 a 19 kg b 18.5 kg c 19 kg

25 23 kg, 17 kg, 12 kg, 24 kg, 19 kg

26 65 000 g, 56 000 g, 66 000 g, 98 000 g, 85 000 g

27 150 kg

28 Outer Hebrides

29 2100, 9500, 7000, 900, 900

30 2000, 10 000, 7000, 1000, 1000

32 80 000

33 a 40 km b $1\frac{3}{4}$ hours c 3 minutes d $\frac{1}{3}$ km

34 24 000

35 45 minutes

36 a $\frac{1}{6}$ b $\frac{1}{6}$

37 a 46p b £1.44 c 81p

SKILLS BREAK 1B

1 17 miles

2 4 square miles

3 32 square miles

4 20 square miles

5 right

6 Silchurch

7 a Robridge b 286°

8 a 382 594 b 345 585 c 362 589, 378 593 d 354 585 e 422 580

10 a 48 km b 88 km c 128 km d 40 km e 208 km f 456 km

11 a 34 miles b 60 miles c 25 miles d 75 miles
e 147 miles f 234 miles

12 $\frac{5}{8}$

13 27 km

14 43 minutes

15 a 25 minutes b 50 minutes

16 2 hours 5 minutes

17 13 minutes

18 a 61 days b 183

19 a 15 625 b 9375

20 No, some of the adults may be senior citizens and pay a cheaper fare so ticket sales will be less.

21 410 passengers

22 5469

23 No, the running costs will be nearly £120 000.

24 Adult £5.80, Child £3.20, Senior Citizen £3.70

25

	Frequency
Adult	67
Child	19
Senior Citizen	93

26 £1844.50

27 £10.30

29 2.2 tons

30 500 gallons of water

31 58.7 tons

32 $p + n = 8$

33 a $27p + 37n$ b $55 + 27p + 37n$

SECTION 7 Solving Equations

Starting points

A1 a 17 b 6 c 24 d $^-3$

A2 a 13 b 7 c 49 d $^-11$

B1 C

B2 a $9t$ b $10s + 9k$ c $8b + 3c$ d $11x + 2$ e $7v + 16$

C1 C

C2 a $z = 1.5$ b $y = 3$ c $x = ^-2$ d $w = 4$ e $v = 8$ f $t = ^-1$ g $s = 2$

D1 A and C

D2 a

x	$^-2$	$^-1$	0	1	2
y	1	3	5	7	9

b

x	$^-2$	$^-1$	0	1	2
y	8	7	6	5	4

Thinking ahead to ...

A P and U, Q and S, R and T

Exercise 7.1

1 a $8t$ b $5x$ c $3h + 3$ d a e $1 - 5y$ f $6 - 3f$ g $x + 4y - 5$
h $2x - 5y$ i $2x + 2y - 4$ j $6ax + 3x$ k $2xy + 2x + y$ l $4x - 2y - 10$
m $4x + 5y - 3$ n $3y - 4x$ o $4 - x - 7y$ p $12 - 5x - 15y$
q $2y - x - 3$ r $4ax + ay + y$

2 b It is a magic square

13	1	10
5	8	11
6	15	3

c $6x - 3y$

Thinking ahead to ...

A A and F, B and E, C and G, D and H

Exercise 7.2

1 16

2 0

3 $2(a + 4)$ and $2a + 8$, $8a - 12$ and $4(2a - 3)$, $2(a + 2)$ and $2a + 4$, $2a - 12$ and $2(a - 6)$

4 b $12 + 3x$

5 $5(2n - 1) = (5 \times 2n) - (5 \times 1) = 10n - 5$

6 a $4n + 4$ b $5m - 15$ c $6c - 54$ d $6p + 4$ e $16s - 40$
f $12 + 4t$ g $15 - 3k$ h $14n - 10$ i $60f - 390$ j $900 + 540h$
k $30 - 70q$ l $10y + 15z$

7 $2(n + 2) + 2n = 2n + 4 + 2n = 4n + 4$

8 a $8p + 10$ b $5q - 4$ c $26 - x$ d $54r + 30$ e $6s + 2t$ f $7x - 1$
g $^-2x - 2$ h $5x - 9$ i $^-2x - 23$

Exercise 7.3

1 a $x = 4$ b $y = 6$ c $x = 7$ d $w = 18$ e $v = 4$ f $u = 1.2$ g $t = 7.25$
h $s = 5.375$ i $r = 9$ j $q = 1.8$ k $p = 2$ l $n = 1.5$ m $m = 4$
n $l = 3.6$ o $k = 2.5$ p $j = 10$ q $h = 5$ r $g = 3.5$

2 a $z = ^-5$ b $y = ^-1$ c $x = -2$ d $w = ^-3$ e $v = ^-4$ f $u = ^-6$
g $t = ^-0.5$ h $s = ^-1.2$ i $r = ^-3.5$

3 a $z = \frac{1}{3}$ b $y = \frac{5}{6}$ c $x = 3.5$ d $w = 5$ e $v = 2$ f $u = \frac{2}{5}$

Exercise 7.4

1 a 5 cm, 10 cm, 11 cm b 20 cm c $(6x - 4)$ cm
d 18 cm, 36 cm, 50 cm

2 **a** square 80 m, rectangle 50 m **b** $8p$ m **c** $(4p + 10)$ m
  d 20 m, 25 m **e** $p = 2.5$ m

3 $t = 1$ cm

4 **a** $x = 11$
  b

22	9	32
31	21	11
10	33	20

  c $x = 20$
  d **i**

26	11	38
37	25	13
12	39	24

  e Sum of corners = $2x + 3x - 1 + x - 1 + 2x - 2 = 8x - 4 = 4(2x - 1)$
    = 4 × centre number

Exercise 7.5
1 A

2 A $3n + 5 = 13 - n, n = 2$
  B $4(n - 2) = 5(n - 3), n = 7$
  C $3(n - 1) = 4(8 - n), n = 5$
  D $10 - 6n = 7 - 3n, n = 1$
  E $2n - 11 = 15 - 2n, n = 6.5$

3 **b** $n = 7$

Exercise 7.6
1 **a** $t + 2b = 70, 3t + b = 130$
  b $t + 2b = 70$

t	0	10	20	30	40
b	35	30	25	20	15

  $3t + b = 130$

t	0	10	20	30	40
b	130	100	70	40	10

  d tea = 38p, biscuit = 16p

2 **a** 2, ⁻3 **d** $x = 1.5, y = 4.5$

3 **a** 2,1 **d** $x = ⁻1, y = 3$

4 **c** A: $x = 3, y = ⁻2$, B: $x = ⁻2.5, y = 1$, C: $x = 1.8, y = 3.6$

Exercise 7.7
1 A: £9.20, C: £2.30

2 **a** £3.04 **b** £1.24 **c** £4 **d** 96p **e** 28p **f** 68p

3 The woman is 160 cm tall, each twin is 120 cm tall.

Thinking ahead to ...
A **a** **i** 33 **ii** $4q$ **iii** 34 **iv** 12 **v** 6
  b $p = 1, q = 5$

Exercise 7.8
1 **a** $x = 6, y = 9$ **b** $x = 8, y = 1$ **c** $x = 1.5, y = 3.5$

2 A false B true C false

3 $m = ⁻1, n = 10$

4 **a** $v = 2, w = 12$ **b** $v = ⁻1, w = 3$ **c** $v = 2.5, w = ⁻0.5$

Exercise 7.9
1 **a** $a = 10, b = 2$ **b** $a = 2, b = 5$ **c** $a = 5.5, b = 1.5$
  d $a = ⁻2, b = 8$ **e** $a = 3, b = 1$ **f** $a = 6, b = ⁻3$

2 **a** $m = 5$ **b** $n = 10$

3 $p = 3, q = 1$

4 $m = 6, n = 1$

5 **a** $x = 4, y = 2$ **b** $x = 3, y = 5$ **c** $x = 2, y = 1$
  d $x = 2.5, y = 1.5$ **e** $x = 5, y = ⁻1$ **f** $x = 7, y = ⁻4$

Exercise 7.10
1 38 and 35

2 26 large, 63 small

End points
A1 **a** $2k + 9m$ **b** $2p + 8$

B1 **a** $z = 6.5$ **b** $y = 3$ **c** $x = 1.5$
  d $w = 1$ **e** $v = ⁻3$ **f** $t = 1.5$

B2 **a** $3x + 2$ **b** $3x + 2 = 56$, 18 cm, 12 cm, 26 cm

B3 $2n - 1 = 14 - 4n, n = 2.5$

C1 $x = 0.5, y = ⁻2$

D1 **a** $m = 1.5, n = 0.25$ **b** $m = ⁻2, n = 3$ **c** $m = 12, n = 2$

D2 Fifty-seven 50p coins, eight-three 20p coins

SECTION 8 Estimation and Approximation

Starting points
A1 **a** 2000 **b** 2200 **c** 2180 **d** 2176

A2 **a** 45.64 **b** 45.6

A3 **a** 34.60 **b** 3 **c** 38.5 **d** 3500

A4 c, d, e, f

B1 **a** 177.753 **b** 43.463 **c** 3.984 **d** 3442.8895

B2 **a** 27.98 **b** 11.17 **c** 149.86 **d** 232.236

B3 23.4 has been added instead of 2.34.

B4 The hundredths and thousandths digits of 2.278 have been added instead of subtracted.

Exercise 8.1
1 **a** Round up, in case it costs more than expected.
  b Do not round at all, or you will sit in someone else's seat.
  c Round up, in case you make any mistakes when cutting the paper.
  d Do not round at all, or you will visit someone else.
  e Round down, in case you do not earn as much as you expect.

3 **a** Round down to 95 cm to make sure it fits.
  b Round down to 67 500 miles, because lower mileage cars are more saleable.
  c Round up to 80, just in case a few more people come.
  d Round up to 15.5 metres, in case there is a slight error in your calculations.

Exercise 8.2
1 **a** 10:43 should be 11:00, 71 934 should be 70 000, 63 479.6 cm should be 635 m, 19.6 cm should be 20 cm, 1452.56 metres² should be 1500 metres², 56 minutes should be 1 hour, 492 or so should be about 500
  b 10.84 sec, because the record will only have been broken by a fraction of a second; 100 m and 2003.
  c 19.6, because it is a bigger percentage of the number.

3 1000 km or 600 miles

4 **a** 10 243 **b** 24.6

5 At least to the nearest metre.

Exercise 8.3
1 **a** 1450 **b** 21 700 **c** 143 **d** 2 130 000 **e** 150 000

2 **a** 70 000 **b** 72 000 **c** 71 900 **d** 71 930

3 Because the unit 9 is rounded up.

4 3200, 3210 or 3150 for example

5

Number	45 287	2395	302 604.32	14.823
to 2 sf	45 000	2400	300 000	15
to 3 sf	45 300	2400	303 000	14.8
to 4 sf	45 290	2395	302 600	14.82

6 The length measurements are only given to 3 sf, so the area measurement should not be given more accurately than that. Round to 1500 m².

7 419 500 for example

8 3 to 6 sf

Exercise 8.4

1. 0.008 cm

2. The last two zeros indicate that the number has been rounded to 4 sf.

3.

to 1 sf	to 2 sf	to 3 sf
0.03	0.025	0.0254
0.009	0.0091	0.00914
0.04	0.039	0.0394
0.6	0.62	0.621
0.002	0.0018	0.00176
0.06	0.061	0.0610
0.03	0.028	0.0283
0.005	0.0045	0.00454

4. Japan

5. Countries with similar areas, for example Bangladesh and Cambodia, could end up with very different looking areas. Other countries with different areas, for example Cambodia and United Kingdom, would end up with exactly the same area.

6. **a** 244 020 **b** 460 **c** 370 000 **d** 9 561 000 **e** 2

7. Congo and Bangladesh

Exercise 8.5

1. **a** $30 \times 40 = 1200$ **b** $700 \times 4 = 2800$ **c** $900 \times 80 = 72\,000$
   d $500 \times 50 = 25\,000$ **e** $60\,000 \div 80 = 750$ **f** $60\,000 \div 500 = 120$
   g $500 \div 2 = 250$ **h** $30\,000 \div 1 = 30\,000$

2. **a** £18 000 **b** Both numbers get rounded up.

3. 120 people per km^2

4. In the first calculation both numbers are rounded down a lot, in the second they are both rounded up a lot.

5. **a** too large **b** too small **c** about right **d** too small **e** too small
   f too large **g** about right **h** too small

6. The size of the numbers is too different. You would be rounding the first number to the nearest hundred-thousand, then adding on a multiple of one hundred.

Thinking ahead to ...

A 9.234, smaller

B 12 666.67, larger

C **a** it gets smaller **b** it gets bigger

Exercise 8.6

1. **a** 17 784 **b** 17.784

2. estimate 0.1, actually 0.1035

3. **a** 7500, 8657.5 **b** 18, 15.738 **c** 0.18, 0.17524 **d** 1000, 1360
   e 4, 4.2558 **f** 1 000 000, 1 189 915.556

4. **a** 0.6 m^2, 70 m^2 **b** 0.822 m^2, 97.35042 m^2

5. 40 m × 0.5 m, 50 000 m × 0.04 m

6. 36 000 cm^3, 20 000 cm^3, 84 cm^3

7. 0.5 cm

Exercise 8.7

1. **a** The £5 has reduced in size each time except in 1971 when the length increased.

   b

Year	Method 1	Method 2
1925	286	294
1957	142	144
1963	118	112
2971	114	120
1994	93	98

   d Method 1

End points

A1 Alez tuned into her favourite station. Channel 162 on the infrawave. She knew the slot lasted for about 90 minutes so she would need a compulsory meal before the end. She dined on 27 of her favourite food pills with 785 ml of ice-cold isophoric delight. Jeq materialised in 11 minutes raving about some antique maths book with about 400 pages that he'd found and dated as 1997.

A2 **a** Round up, to make sure you get enough paint.
   b Round down, just in case the maximum is inaccurate.

B1 **a** 346 **b** 345.68 **c** 300

B2 **a** 34.7 **b** 1.97 **c** 195 000 **d** 0.00319 **e** 144 **f** 6.99

C1 **a** 10 m^3 **b** 8.90 m^3

Section 9 Probability A

Starting points

A1 **a** $\frac{1}{4}$ **b** $\frac{1}{12}$ **c** $\frac{1}{12}$ **d** $\frac{1}{6}$ **e** $\frac{1}{12}$

A2 **a** $\frac{5}{12}$ **b** $\frac{1}{3}$ **c** $\frac{2}{3}$ **d** $\frac{1}{6}$ **e** 1 **f** 0

A4 **a** $\frac{1}{6}$ **b** $\frac{1}{2}$ **c** $\frac{1}{3}$

B1 $\frac{1}{5}$

C1 **a** $\frac{3}{16}$ **b** $\frac{5}{16}$ **c** $\frac{2}{7}$ **d** $\frac{3}{32}$

D1

		X	X1	X2	X3	X4
Outcome of		Y	Y1	Y2	Y3	Y4
Spinner A		Z	Z1	Z2	Z3	Z4
			1	2	3	4

Outcome of Spinner B

D2 **a** $\frac{1}{12}$ **b** $\frac{1}{6}$

E2 $\frac{1}{8}$

Exercise 9.1

1. **a** $\frac{1}{18}$ **b** $\frac{1}{36}$ **c** $\frac{1}{6}$ **d** $\frac{1}{6}$ **e** $\frac{5}{6}$ **f** $\frac{1}{12}$ **g** $\frac{5}{18}$ **h** $\frac{1}{4}$ **i** $\frac{5}{12}$

2.

Number on red dice	9	9,0	9,1	9,2	9,3	9,4	9,5	9,6	9,7	9,8	9,9
	8	8,0	8,1	8,2	8,3	8,4	8,5	8,6	8,7	8,8	8,9
	7	7,0	7,1	7,2	7,3	7,4	7,5	7,6	7,7	7,8	7,9
	6	6,0	6,1	6,2	6,3	6,4	6,5	6,6	6,7	6,8	6,9
	5	5,0	5,1	5,2	5,3	5,4	5,5	5,6	5,7	5,8	5,9
	4	4,0	4,1	4,2	4,3	4,4	4,5	4,6	4,7	4,8	4,9
	3	3,0	3,1	3,2	3,3	3,4	3,5	3,6	3,7	3,8	3,9
	2	2,0	2,1	2,2	2,3	2,4	2,5	2,6	2,7	2,8	2,9
	1	1,0	1,1	1,2	1,3	1,4	1,5	1,6	1,7	1,8	1,9
	0	0,0	0,1	0,2	0,3	0,4	0,5	0,6	0,7	0,8	0,9
		0	1	2	3	4	5	6	7	8	9

Number on blue dice

3. **a** $\frac{1}{10}$ **b** $\frac{3}{20}$ **c** $\frac{1}{4}$

Exercise 9.2

1. **a** $\frac{2}{9}$ **b** $\frac{7}{27}$ **c** $\frac{2}{9}$ **d** $\frac{2}{3}$ **e** $\frac{19}{27}$ **f** $\frac{1}{27}$ **g** $\frac{1}{9}$ **h** $\frac{2}{3}$

3. **a** $\frac{1}{8}$ **b** $\frac{1}{4}$ **c** $\frac{7}{8}$ **d** $\frac{3}{4}$ **e** 0 **f** $\frac{3}{8}$

Exercise 9.3

1. **a**

EHNW	HEWN	NEHW	WEHN
EHWN	HENW	NEWH	WENH
ENHW	HWEN	NHEW	WHEN
ENWH	HWNE	NHWE	WHNE
EWNH	HNEW	NWEH	WNEH
EWHN	HNWE	NWHE	WNHE

   c $\frac{1}{24}$

2. NOT, NTO, ONT, OTN, TON, TNO

3. 2

4. **a** and **c**

No. of letters	1	2	3	4	5	6
No. of arrangements	1	2	6	24	120	720

   b 2, 3, 4; the number you multiply by increases by 1 each time.

5 **a** NNN, NNO, NON, ONN, NNT, NTN, TNN, OOO, OON, ONO, NOO, OOT, OTO, TOO, TTT, TTO, TOT, OTT, TTN, TNT, NTT, NOT, NTO, ONT, OTN, TON, TNO

b 27

c By showing that there is a choice of 3 letters for each of the 3 spaces in the word, giving $3 \times 3 \times 3 = 27$ possibilities.

Exercise 9.4

1

Rolls of dice	Number of sixes	Relative frequency
20	7	0.34
50	17	0.34
70	24	0.34

2 **a** 0.23 **b** 2 **c** 0.77

3 **a** 0.25 **b** 0.75 **c** 150

4 0.26

5 0.49

End points

A1

		R	RR	RY	RB	RP
Outcome of	Y		YR	YY	YB	YP
Spinner A	B		BR	BY	BB	BP
			R	Y	B	P

Outcome of Spinner B

A2 **a** $\frac{1}{4}$ **b** $\frac{1}{12}$ **c** $\frac{1}{12}$ **d** 0

A4 **a** $\frac{1}{8}$ **b** $\frac{3}{4}$ **c** $\frac{1}{4}$ **d** $\frac{3}{8}$

B1 **a** 24 **b** $\frac{1}{2}$

C1 0.36

SECTION 10 Using algebra A

Starting points

A1 **a** $3a + 12$ **b** $2b - 12$ **c** $6c + 15$ **d** $8d - 8$ **e** $32 - 8e$ **f** $15 + 30f$

B1 **a** $5a + 4b - 6$ **b** $4p + 6q - 10$

B2 **a** $8a + 10$ **b** $14a + 10$

C1 **a** $p = 3$ **b** $q = 10$ **c** $t = 0.75$ **d** $x = 18.75$ **e** $x = {}^-4.5$

D1 A and E, B and F, C and D

E1 **a** 31.28 **b** 12.6 **c** 4.12 **d** 27.92 **e** 29.76 **f** 45.08

Exercise 10.1

1 **a i** 16.3 m **ii** 3.4 m **iii** 11.9 m
b 54.1 m **c** 8.2 m × 10.5 m **d** 26.9 m **e** 108 m² **f** 306 m²

2 **a** 27.9 m **b** 19.4 m **c** 21.4 m² **d** 86.12m²

Exercise 10.2

1 **b i** $13.4 + w$ **ii** $16.2 - 2w$ **d i** $32.4 - 2w$ **ii** 28.4 m
e ii $131.22 - 16.2w$ **f** 2.7 m

2 **a i** $46 + p$ **ii** $24 - p$ **iii** $138.24 - 9.6p$ **iv** $24p$
b 48 m **c** 2.4

Exercise 10.3

1 **a** $2a + 8$ **b** $3b - 15$ **c** $8c + 12$ **d** $d^2 + 7d$ **e** $6e - e^2$ **f** $4f + f^2$
g $2p^2 + 9p$ **h** $4r^2 - 2r$

2 **a** $a(a + 5)$, $b(b - 2)$, $c(c - 6)$, $d(12 - d)$, $4(e + 8)$, $7(8 - f)$
b $a^2 + 5a$, $b^2 - 2b$, $c^2 - 6c$, $12d - d^2$, $4e + 32$, $56 - 7f$

3 **a** $5(2p + 2)$, $4(3q - 2)$, $r(2r + 2)$ **b** $10p + 10$, $12q - 8$, $2r^2 + 2r$

4 $2a, b, c$

5 $d + 3, e + 2, 5 + f, 2g + 3$

6 **a** $4(x + 3)$ **b** $2(3p - 2)$ **c** $2(a + 4)$ **d** $5(2 - q)$ **e** $b(b - 6)$
f $y(y + 4)$ **g** $r(2r + 3)$ **h** $d(3d - 5)$ **i** $t(3t + 4)$ **j** $s(6s + 7)$

Exercise 10.4

1 **a** $(a + 8)$ cm **b** $(4a + 16)$ cm **c** 209 cm²

2 **a** $5d$ cm **b** 23.2 cm

3 225 m²

4 84.5 cm²

Exercise 10.5

1 **a** $6ab$ **b** $3pq$ **c** $20xy$ **d** $30pq$ **e** $2x^2$ **f** a^2b
g $2xy^2$ **h** $6a^2b$ **i** $6p^2q$ **j** a^5 **k** $6b^3$ **l** $10bc^3$

2 A and E, B and F, C and H, G and I

3 **a** $ab + 4a$ **b** $xy + xz$ **c** $2mn + 3mp$ **d** $6xy + 4xz$ **e** $ac + c^2$
f $p^2 - pq$ **g** $3bc + 3b^2$ **h** $4a^2 - 4ab$ **i** $3p^2 - 4p$ **j** $2ab - 4ac$
k $6a^2 + 8ab$ **l** $8pq - 12p^2$ **m** $6p^2q + 4pq^2$ **n** $4x^2y - 8xy^2$
o $3abx^2 - 3aby^2$

4 **a** $9a - 22b$ **b** $2x^2 + x$ **c** $4a + 5ab$ **d** $x^2 + x^3 - 2x$ **e** $11a + 2b$
f $6x^2 + 4xy$ **g** $4p^2 + 6q$ **h** $6a$

5 **a** $9a + 26b$ **b** $10x + 5y$ **c** $2x^2 + x$ **d** $10x^2 + 3xy$ **e** $2a^2b$
f $3a^2b + 3ab + 2ab^2 - 2b^2$

6 **a** $6a^2 + 6ab + 12b^2$ **b** $16x^2y + 20xy^2 - 6xy$ **c** $8p^2q - 7pq^2$
d $14m^2n - 8mn^2 - 12m^3$

Exercise 10.6

1 **a** $5(ax + 2y)$ **b** $3(2ay - 5bc)$ **c** $3(3xy - 4a)$ **d** $2x(2y + 5)$
e $7(3xy + 5c)$ **f** $9(4ab - 5y)$ **g** $3(2ax + 4y + 5x)$
h $4(2ac + 5cy - 6ax)$ **i** $3(3x + 5ay - 7xy)$ **j** $6(5kx + 3y + 7ac)$
k $14(2xy + 4y - 3c)$ **l** $12(12ac - 5y + 9x)$

2 **a** $a(3x - 5y)$ **b** $x(6y + 7a)$ **c** $y(9x - 5a)$ **d** $x(6a - 17y)$
e $a(3bc + 5x)$ **f** $b(9a + 4)$ **g** $y(7x - 9)$ **h** $ax(6 + 7y)$ **i** $c(7ab - 8)$
j $x(3a + 5y - 7c)$ **k** $y(5x - 7a + 6)$ **l** $y(12x - 19a + 15)$
m $c(6ab + 5b + 7)$ **n** $x(8 + 7ay + 9by)$ **o** $ab(7c + 1)$
p $y(16x + 25 - 13ax)$ **q** $3bx(7a - 5c)$

3 **a** $2a(x - 4y)$ **b** $5x(5y + 3a)$ **c** $4y(4a + 7b)$ **d** $8x(4y - 3a)$
e $7b(6ac - 5x)$ **f** $14b(4a + 3y)$ **g** $3a(2b - 3y + 4)$
h $2y(2x + 3a - 5c)$ **i** $3x(7y - 5 + 4a)$ **j** $5xy(3k + 5a)$
k $3x(6c + 4ay + 5)$ **l** $7ac(3b + 8x)$ **m** $3x(5ay + 6a - 10y)$
n $bc(a + 5 - 15x)$ **o** $8x(3y - 2x + a)$ **p** $2x(x + 2 - 3y)$
q $5x(1 - 2y - 5x)$ **r** $3ab(2 - 3a + c)$

End points

A1 53.8 m

A2 16.7 m

A3 **a** 15.9 m **b** 796 m²

B1 **a** $20f - 16$ **b** $m^2 + mn$ **c** $2pr - 2p^2$ **d** $a^2b + ab^2$
e $2m^2n + 3mn^2$ **f** $6xy^2 - 10x^2y$

B2 **a** $22b + 16$ **b** $26b + 14$ **c** $10ab^2 + 4a^2b$ **d** $3x^2y + 7xy^2 + 19xy$

C1 **a** **i** $(p - 8.4)$ cm **ii** $(p - 6.1)$ cm **b i** $4p$ cm **ii** 14 cm

C2 **a** $(a + 5)$ cm **b** $(4a + 10)$ cm **c** $a(a + 5) = a^2 + 5a$
d 11.5 cm **e** 51.51 cm²

D1 **a** $4(2p + q)$ **b** $6(a - 2b)$ **c** $a(a + b)$ **d** $m(7 + 3n)$ **e** $2y(4x + 5)$
f $2a(2b + 3a)$ **g** $xy(3y - 5x)$ **h** $2pq(2 - 3p)$ **i** $h(2g^2 - 5h)$
j $2x(15y + 12x - 8y^2)$ **k** $5ab(5a - 3b)$ **l** $7x(2ay - 3x + 5xy)$
m $50y(2x + ay - 5x^2)$

SECTION 11 Constructions and loci

Starting points

A1 **a** AI **b** EL **c** EF, EL or EC **d** triangle ECL
e triangle AGI **f** triangle HIJ **g** AK or CL **h** GD **i** FC

B1 F

Exercise 11.1

2 Rydon and Barton

3 Rydon

7 about 20 miles

Exercise 11.2

2 isosceles

3 **a** right-angled **b** scalene **c** equilateral **d** scalene

4 This triangle does not exist. The sum of the lengths of the two shorter sides must always be greater than the length of the longest side.

6 104 miles

Exercise 11.3

2 angle SQR = 77°

3 Triangles LMN and PQR look identical.

Thinking ahead to ...

E **b** The locus is a line perpendicular to the line AB.

Exercise 11.5

1 **c** It is always the same distance from A and B.

3 **c** The three bisectors intersect at one point.
d yes

Thinking ahead to ...

A A, C and H

Exercise 11.6

2 Each angle is 60°. When you bisect one of the angles you make an angle of 30°.

Exercise 11.7

1 **a** 2 cm **b** 20 m

End points

B1 The gradients are the same.

C2 The bisectors intersect at one point.

Section 12 Ratio

Starting points

A1 **a** 3 : 1 **b** 1 : 3

B1 **a** 12 : 3 **b** 1 : $2\frac{1}{2}$: 2 **c** 3 : 4, 6 : 8

B2 2 : 5, 1 : $2\frac{1}{2}$ and 10 : 25, 1 : 3 : 2 and 3 : 9 : 6, 10 : 20 and $2\frac{1}{2}$: 5

B3 **a** 3 : 2 **b** 2 : 3 **c** 2 : 4 : 1 **d** 10 : 1 **e** 4 : 3 : 6

C1 **a** 0.625 **b** 1.8 **c** 0.58$\dot{3}$ **d** 1.3$\dot{6}$

C2 **a** 61.$\dot{1}$% **b** 50% **c** 70.8$\dot{3}$% **d** 87.5%

D1 **a** 1, 3, 4, 7, 11, 18, 29, 47 **b** 2, 4, 6, 10, 16, 26, 42
c 1, 5, 6, 11, 17, 28, 45, 73 **d** 3, 6, 9, 15, 24, 39, 63, 102

E1 **a** GI **b** GHI **c** HGI and GIH **d** GH

F3 JKL and PQR

Exercise 12.1

1 **a** 480 ml **b** 1200 ml

2 **a** 315 ml **b** 525 ml

3 **a** 50 ml **b** 175 ml

4 **a** 250 kg **b i** 2 : 1 : 9 **ii** 2400 kg

5 **a** 8 m **b** 9 cm

6 **a** 15 cm **b** 2.25 m

7 **a** 6.3 cm **b** 1 : 5000

Exercise 12.2

1 Sarah gets 20, Denzil gets 15

2 £3, £6, £9

3 £75, £45, £30

4 **a** $1\frac{1}{4}$ hours **b** $1\frac{1}{2}$ hours

Exercise 12.3

1 345 g caster sugar, 450 ml water, 75 g cocoa, 1800 ml chilled milk

2 **a** 60 g **b** 75 ml
c 300 g white sugar, 100 ml milk, 100 ml water, 50 g butter, 20 g cocoa

3 **a** 7 drops **b** 1.75, which is not practical

Exercise 12.4

1 **a** 2 : 1 **b** $\frac{2}{1}$

2 2 : 3, $\frac{2}{3}$

3 **a** $\frac{5}{3}$ **b** $\frac{3}{5}$

4 **a** $\frac{1}{5}$ **b** $\frac{4}{5}$

5 **a** $\frac{2}{3}$ **b** That would be height as a fraction of height and diameter.

Exercise 12.5

1 **a** 1.5 **b** 7.5 cm

2 **a** 2.5 **b** 12.75 cm

3 **a** 0.67 **b** 7.2 cm

End points

A1 **a** 450 ml **b** 630 ml

B1 £160, £128, £112

C1 **a i** 375 g **ii** 10 tablespoons **b** 3 eggs

D1 **a** 3 : 4 **b** $\frac{3}{4}$

D2 $\frac{2}{5}$

Skills break 2a

1 £1.27, £1.69, £2.54, £4.24

2 8 litre

3 12 litre

4 yes

5 2 gallon

6 5105 g

7 **a** 0.315 kg **b** Less, it will weigh closer to 7 kg.

8 **a** 33 g **b** $2\frac{1}{4}$ scoops

9 **a** 66 gallons **b** 10 g **c i** $\frac{1}{100}$ **ii** 0.01 **iii** 1%

10 Yes, 300 litres is about the same as 66 gallons.

11 £3502

12 **a** 24.75 m² **b** 15p

13 4913 cm²

14 30 metre

15 **a** three 50 m rolls and one 30 m roll, or six 30 m rolls, or for 28p per metre
b six 30 m rolls, which costs £44.94

16 2.4 m²

17 **a** rectangle **b** 5.5 m by 1.9 m

18 6.3 m²

19 34.4 m²

20 No, nearly 9 litres will be required.

21 76.0°

22 **a** rectangle **b** 2.9 m **c** 16.5 m² **d** 72.4 kg

23 **a** £180 **b** £14.40 **c** £10.80

24 £24, £26.67, £33.33

Skills break 2b

1 12

2 12

3 40.8 m

4 132.7 m²

5 **a** octagon **b** 45° **c** 135°

6 696 cm²

7 4

8 1110 tons

9 50 m

10 16 cm

11 370

12 a 1.01×10^4 b 5.0×10^5

13 a 1650   b 1700   c 2000

14 c

15 a 225°   b 45°

16 £7.43

17 a 308m FF   b £41m

18 a 2 200 000   b 2 190 000   c 2 189 000   d 2 188 900

19

Vowel	French	English
a	41	42
e	73	51
i	26	35
o	21	36
u	27	12

21 a EP, EV, EM, PV, PM, VM   b $\frac{1}{2}$ c $\frac{1}{6}$ d i $\frac{1}{3}$ ii $\frac{1}{2}$ e loss

22 a £851   b £99   c £89   d £277

SECTION 13 Distance, speed and time

Starting points

A1 a 5000 m   b 3.4 m

A2 a 4800 m   b 1.5 m   c 1.8 m

A3 a 4 cm   b 400 000 cm   c 390 cm   d 18 cm

A4 a 5.13 km   b 11.36 km

A5 12.5 miles

B1 a 15.4 pounds   b 8.4 pounds

B2 a 7.3 kg   b 2.3 kg   c 3.6 kg   d 2.1 kg

B3 2.2 pounds

B4 a

Gallons	1	2	3	4	5	6	7	8
Litres	4.55	9.1	13.65	18.2	22.75	27.3	31.85	36.4

     c i 11.4 litres   ii 3.3 gallons

C1 a 9:30 am   b 2:21 pm

C2 a 02:23   b 17:25

C3 75 minutes

C4 6 hours 40 minutes

D1 13:00

D2 0625: 4 hours 40 minutes, 0825: 4 hours 45 minutes,
0955: 4 hours 45 minutes, 1145: 4 hours 50 minutes,
1345: 4 hours 55 minutes, 1700: 4 hours 45 minutes

D3 The 1400 bus from Torquay take the longest to get to Exeter.

D4 5:15 pm

D5 3 hours 12 minutes

D6 9:45 pm

Exercise 13.1

1 a i 5.1 m   ii 25.5 m   iii 11.48 m   b i 153 m   ii 9180 m
    c i 413 seconds   ii 6 minutes 53 seconds

2 b Alperton: 29.8 seconds, The Angel: 80 seconds, Chancery Lane:
15.2 seconds: Kentish Town: 66.8 seconds   c 2:02 pm

Exercise 13.2

1 520 mph

2 a 1 hour and 50 minutes is not the same as 1.5 hours   b 608.2 mph

3 a 10.33 miles per minute   b 103.3 miles

4 a 6.72 miles per minute   b 17.9 minutes

5 a 35.5 minutes   b 3487.5 miles

Exercise 13.3

1 b 175 m   c 2.7 seconds   d 120 m
    f i 20 m/s, 10 m/s   ii red   iii red has the steepest graph
    g i 3.5 seconds   ii It is the point where their graphs cross.

2 a Purple 90 m, black 100 m, blue 118 m, red 140 m
    b Black   c i 4 seconds   ii 3.3 seconds
    d Purple 20 m/s, black 40 m/s, blue 30 m/s, red 0 m/s
    e It broke down!

3 b 200 m

Exercise 13.4

1 a 30 miles   b 50 minutes   c 18.5 miles
    d i 9 minutes   ii 21 minutes   e 11 miles   f A3, B2, C4, D5, E1
    g It is a graph of Geeta's distance from home, not a diagram of the
     roads.

Exercise 13.5

1 a 0.2 miles per minute   b 12 miles per hour

2 a 0.875 miles per minute   b 52.5 miles per hour

3 a 58.9 mph

4 1:40 pm

5 16 miles

6 a i 20 minutes   ii 33 miles per hour
    b 25.44 mph, 29.4 mph   c 45 miles   d 1:46 pm

Exercise 13.6

1 a 1500 m   b A   c 10 minutes
    d A: 1500 m, B: 1200 m, C: 0 m, D: 700 m   e 18 km/h
    f i A is Sharon, B is Jane, C is Jason, D is Lorna
     ii 500 m   iii Walked, she was the slowest.
    g i A: 150 m/min, B: 100 m/min, C: 300 m/min, D: 60 m/min
     ii A: 9000 m/h, B: 6000 m/h, C: 18 000 m/h, D: 3600 m/h
     iii A: 9 km/h, B: 6 km/h, C: 18 km/h, D: 3.6 km/h

2 a i 13:40   ii 20 minutes   b i 14:00 to 14:16   ii 45 mph
    c The motorbike passed the car in the opposite direction.
    e i The motorbike travelled 60 miles, the car travelled 30 miles
     ii motorbike 33 mph, car 18 mph

Exercise 13.7

1 a ii 11.1 km/h   b ii 3:16 pm   iii 2:44 pm

2 a ii 9 mph   iii 9.2 mph   b i 10 miles   iii 12.8 mph
    iv Alice was buying cakes as Emily passed   c ii 20 mph

Exercise 13.8

1 a A: 5.6 gallons, B: 3.6 gallons   b 35 miles   c twice
    d 6.4 gallons   e 3 gallons, 2.8 gallons
    f i 9.2 gallons, 12.6 gallons   ii 43.5 mpg, 31.7 mpg

2 a 0.89 cars per household   b 55 cars per kilometre   c No

3 a 0.45 cars per household

End points

A1 225.25 m

A2 4 minutes 49 seconds

A3 45.6 mph

A4 16:32

B1 a 2.5 miles   b 26 minutes   c twice   d 10.3 mph
    e i 20:40 and 21:08   ii steepest part   f 50 minutes   g 60 mph
    h i 21:26   ii 15.5 miles

C1 b 7.7 mph

D1 3000 people

D2 0.51 litres per second

D3 £1.95 per person

SECTION 14 Trigonometry

Starting points

A1 **a** GI **b** angle GHI **c** angle GIH **d** GH

B1 A and C, B and E

B2 **a** XZ **b** FG

Thinking ahead to ...

A **b** 0.55 **c** they are the same

B **a** 1.83 and 1.83 **c** they are the same

Exercise 14.1

1 **a i** BC/AC, ED/EF **ii** 1.18
   b i AC/BC, EF/ED **ii** 0.85
   c AB/AC = DF/EF = HI/GI = 0.81, AC/AB = EF/DF = GI/HI = 1.22
   AB/BC = DF/ED = HI/GH = 0.69,
   BC/AB = ED/DF = GH/HI = 1.44

2 **a** All corresponding angles are equal.
   b 0.75, 1.33, 0.6, 1.67, 0.8, 1.25

3 6

Exercise 14.2

1

adj	opp	hyp	adj/hyp	adj/opp	opp/hyp	opp/adj	hyp/adj	hyp/opp
40	23	46	0.87	1.74	0.5	0.56	1.15	2

Exercise 14.4

1 **b** A: $\sin x = \frac{21.6}{22.5} = 0.96$, $\cos x = \frac{6.3}{22.5} = 0.28$, $\tan x = \frac{21.6}{6.3} = 3.43$
   B: $\sin x = \frac{3}{5} = 0.6$, $\cos x = \frac{4}{5} = 0.8$, $\tan x = \frac{3}{4} = 0.75$
   C: $\sin x = \frac{4.5}{11.7} = 0.38$, $\cos x = \frac{10.8}{11.7} = 0.92$, $\tan x = \frac{4.5}{10.8} = 0.42$
   D: $\sin x = \frac{1.05}{1.75} = 0.6$, $\cos x = \frac{1.4}{1.75} = 0.8$, $\tan x = \frac{1.05}{1.4} = 0.75$
   E: $\sin x = \frac{5}{13} = 0.38$, $\cos x = \frac{12}{13} = 0.92$, $\tan x = \frac{5}{12} = 0.42$
   F: $\sin x = \frac{8}{10} = 0.8$, $\cos x = \frac{6}{10} = 0.6$, $\tan x = \frac{8}{6} = 1.33$
   G: $\sin x = \frac{12}{12.5} = 0.96$, $\cos x = \frac{3.5}{12.5} = 0.28$, $\tan x = \frac{12}{3.5} = 3.43$
   H: $\sin x = \frac{9.6}{10.4} = 0.92$, $\cos x = \frac{4}{10.4} = 0.38$, $\tan x = \frac{9.6}{4} = 2.4$

Exercise 14.5

2 EF = 3.9 cm, GH = 5.7 cm, JK = 3.5 cm, MN = 20.7 cm

3 74.9 cm

4 AB = 6.9 cm, DE = 6.5 cm, KL = 3.8 cm, GI = 7.1 cm

5 JK = 3.1 cm, NO = 8.1 cm, AC = 3.7 cm, PR = 11.5 cm

Exercise 14.6

2 **a** 21° **b** 46° **c** 69°

3 **a** 21° **b** 65° **c** 59°

4 28°

5 $a = 42°$, $b = 56°$, $c = 46°$

6 $a = 37°$, $b = 52°$, $c = 30°$, $d = 41°$, $e = 61°$,
   $f = 60°$, $g = 38°$, $h = 44°$, $i = 60°$, $j = 51°$

7 **a** Wayne must be incorrect
   b The cosine of an angle cannot be greater than 1.

Thinking ahead to ...

A **a** less than
   b RS is opposite the smallest angle

C 5.41 cm

Exercise 14.7

1 **a** A: greater than hyp, B: isosceles triangle, C: opposite 50°, but almost as large as hyp
   D: opposite smallest side but greater than 45°, E: smaller than opp, F: isosceles triangle
   b 3.79 cm, 4.32 cm, 9.46 cm, 42.1°, 10.2 cm, 45°

2 42.0°, 1.33 cm, 13.2 cm, 31.3°, 72.4°, 22.6°

Exercise 14.8

1 **a** 75.52° **b** 4.84 m

2 **a** 1.39 m **b** 1.09 m **c** 4.28 m

3

	72°	76°
4 m	3.80 m	3.88 m
6 m	5.71 m	5.82 m
8 m	7.61 m	7.76 m

4 **a** 312 mm **b** 140 mm **d** 24.2°

5 **a** 0.42 m **b** 8.03 m **c** 8.83 m **d** 8.01 m **e** 10.41 m

6 **a** 2.86°
   b A: 3.33°, unsafe B: 2.01°, safe C: 1.51°, safe D: 1.06°, safe

7 2.80°

8 **a** 32.74° **b** 21.66°

End points

A1 ABC and MNO

B1 48.59°, 35.54°, 59.74°

C1 10.7 cm, 3.75 m, 6.63 cm

SECTION 15 Using algebra B

Starting points

A1 48 cm²

A2 **a** 19 **b** 2.5 **c** 49 **d** 20

B1 **a** 12 **b** 39

C1 **a** $n = 20$ **b** $n = 2$

C2 **a** $7n + 9 = 65$ **b** 8

D1 **a** $5x + 8$ **b** $9a + 2b$ **c** $5x^2 + 2x$ **d** $5a^2 + 6a - 7$
   e $3t^2$ **f** $10cd$ **g** $12k^2$ **h** $7w^2$

Thinking ahead to ...

A **a** 3 hours 20 minutes **b** 3 lb

Exercise 15.1

1 **a** $p = t - 1$ **b** $p = \frac{s}{7}$ **c** $p = \frac{(y + 2)}{5}$ **d** $p = 10 - v$ **e** $p = \frac{(12 - h)}{9}$
   f $p = r - q$ **g** $p = \frac{t}{q}$ **h** $p = \frac{(x - f)}{2}$ **i** $p = \frac{(k + m)}{l}$ **j** $p = 9g$
   k $p = mn$ **l** $p = 2(s - 1)$

2 **a** £10.25 **b** $n = \frac{(c - 50)}{15}$ **c** 129 **d** 63

3 **a** $t = 5d$ **b** 7.5 seconds

4 **a** 720° **b** $n = \frac{S}{180} + 2$ **c** 20
   d Putting 600° into the formula does not give a whole number of sides.

5 **a** 58 **b** 3 **c** $d = \frac{(100 - n)}{3}$ **d** 30 days

6 **a** 550 **b** $x = \frac{(800 - n)}{5}$ **c** 60p
   d The formula would give a negative value for the number of ice-creams she would sell.

7 **a** 94 miles **b** 600 mph **c** $t \rightarrow \times s \rightarrow d$
   d $d \rightarrow \div s \rightarrow t$, $t = \frac{d}{s}$ **e** $2\frac{1}{2}$ hours

Exercise 15.2

1 **a** $w = \frac{5k}{3}$ **b** $w = \frac{4x}{3}$ **c** $w = \frac{5(k - 1)}{3}$ **d** $w = 2(3x - 4)$ **e** $w = 3(3y - 3)$
   f $w = 3(2k - 4)$ **g** $w = \frac{5(4 - x)}{3}$ **h** $w = 3k - 1$ **i** $w = 6k + 2$
   j $w = \frac{3}{4}k + 1\frac{1}{2}$ **k** $w = -\frac{4k}{3}$ **l** $w = k + 4$

2 **a** 17.6 lb **b** $k = \frac{5}{11}p$ **c** 1.4 kg

3 **a** $h = \frac{2A}{b}$ **b** 80 cm

4 **a** 20 °C
   b $F \rightarrow [-32] \rightarrow F - 32 \rightarrow [\times 5] \rightarrow 5(F - 32) \rightarrow [\div 9] \rightarrow \frac{5(F - 32)}{9} = C$
   c $C \rightarrow [\times 9] \rightarrow 9C \rightarrow [\div 5] \rightarrow \frac{9C}{5} \rightarrow [+32] \rightarrow \frac{9C}{5} + 32 = F$
   d 75.2 °F

5 **a** $y = \frac{4z}{3}$ **b** $x = \frac{3m}{x}$ **c** $y = 2(d - x)$ **d** $y = \frac{7(k + h)}{4}$

6 **a** 36.14 km/h **b** $d = \frac{5st}{18}$ **c** $t = \frac{18d}{5s}$

Exercise 15.3

1 **a** 403.44 cm² **b** $x\sqrt{\left(\frac{5}{6}\right)}$ **c** 4.47 cm

2 **a** 7.07 cm² **b** B **c** 7.98 cm

3 **a** 207.03 m **b** $t = \sqrt{\left(\frac{10d}{49}\right)}$ **c** 9.51 seconds

4 **a** $k = \sqrt{(m-1)}$ **b** $k = \sqrt{\left(\frac{m-1}{3}\right)}$ **c** $k = \sqrt{(m-n)}$

5 **a** $y = \sqrt{\left(\frac{k-2}{4}\right)}$ **b** $y = \sqrt{\left(\frac{1-3x}{2}\right)}$ **c** $y = \sqrt{\left(\frac{ax}{5}\right)}$ **d** $y = \sqrt{\left(\frac{ax+4}{3}\right)}$

  e $y = \sqrt{(2(x+4))}$ **f** $y = x\sqrt{\left(\frac{a}{b}\right)}$ **g** $y = \sqrt{\left(\frac{a^2-3x}{5}\right)}$

  h $y = \sqrt{(3-a^2-x^2)}$ **i** $\sqrt{\left(\frac{x^2-4a^2+3x)}{2}\right)}$ **j** $y = \sqrt{\left(\frac{4}{a-3}\right)}$ **k** $y = \sqrt{\left(\frac{2w}{3-x}\right)}$

  l $y = \sqrt{\left(\frac{4}{5+a}\right)}$ **m** $y = x\sqrt{\left(\frac{(b)}{(c-a)}\right)}$ **n** $y = \sqrt{\left(\frac{(3-x^2)}{(a-5)}\right)}$ **o** $y = \sqrt{\left(\frac{(2w+3x^2)}{x}\right)}$

  p $y = \sqrt{\left(\frac{4x}{1+4a}\right)}$ **q** $y = \sqrt{(3-2x^2)}$ **r** $y = \sqrt{\left(\frac{(4a^2)}{x^2}\right)}$

Exercise 15.4

1 **a** 42, 42, 42, 42 **b** 16, 20, 20, 12 **c** 6, 12, 12, 0
  d 10.75, 15.75, 15.75, 5.75

2 82.5 m

3 **a** 121.5 cm³ **b** 64 cm³ **c** $x = 2$

4 87.29 m/s

5 **a** 31.11 m/s **b** 21.14 m/s

Exercise 15.5

1 **a** 1500 cm², 180 cm², 150 cm², 18 cm² **b** 1848 cm²

2 **b** 378 cm²

3 **a**

×	40	+5
60	2400	300
+1	40	5

So 45 × 61 = 2745

  b

×	20	+5
20	400	100
+7	140	35

So 25 × 27 = 675

  c

×	30	+2
810	2400	160
+4	120	8

So 32 × 84 = 2688

Thinking ahead to ...

A **a** $n^2, 3n, 2n, 6$ **b** $n^2 + 5n + 6$

Exercise 15.6

1 **a** $a^2 + 4a + 3$ **b** $b^2 + 6b + 8$ **c** $c^2 + 13c + 40$ **d** $d^2 + 14d + 45$
  e $e^2 + 8e + 7$ **f** $f^2 + 13f + 22$ **g** $x^2 + 7x + 12$ **h** $x^2 + 11x + 24$
  i $x^2 + 18x + 81$ **j** $x^2 + 9x + 14$ **k** $x^2 + 24x + 135$ **l** $x^2 + 26x + 160$
  m $x^2 + 4x + 3$ **n** $x^2 + 15x + 56$ **o** $x^2 + 27x + 126$

2 **a** **i** $x^2 + 2x + 1$ **ii** $x^2 + 4x + 4$ **iii** $x^2 + 6x + 9$

3

×	2x	+4
3x	6x²	12x
+12	2x	4

So $(2x+4)(3x+1) = 6x^2 + 14x + 4$

4 **a** $2a^2 + 9a + 4$ **b** $3b^2 + 14b + 15$ **c** $6c^2 + 25c + 14$
  d $12d^2 + 25d + 12$ **e** $4e^2 + 12e + 9$ **f** $25f^2 + 10f + 1$
  g $10x^2 + 13x + 4$ **h** $12x^2 + 43x + 35$ **i** $9x^2 + 12x + 4$
  j $56x^2 + 43x + 5$ **k** $20x^2 + 9x + 1$ **l** $16x^2 + 20x + 6$

5 B

6 **a** $5 \times 8 = 40, 6 \times 7 - 2 = 40$
  b $(n+1)(n+2) - 2 = n^2 + 3n + 2 - 2 = n^2 + 3n = n(n+3)$

7 $(2x+1)(2x+3) - 3 = 4x^2 + 8x + 3 - 3 = 4x^2 + 8x = 4x(x+2)$

Exercise 15.7

1 **a** $x^2 - x - 12$ **b** $x^2 + x - 20$ **c** $x^2 - 5x - 14$ **d** $x^2 - 2x - 24$
  e $x^2 - x - 56$ **f** $x^2 + 4x - 32$ **g** $x^2 - x - 42$ **h** $x^2 + 7x - 18$
  i $x^2 + 4x - 21$ **j** $x^2 + 4x - 96$ **k** $x^2 - 3x - 54$ **l** $x^2 + 7x - 8$

2 **a** $x^2 + x - 6$ **b** $x^2 + x - 30$ **c** $x^2 - 4x - 32$ **d** $x^2 - 2x - 35$
  e $x^2 + x - 72$ **f** $x^2 - 5x - 14$ **g** $x^2 - 5x - 36$ **h** $x^2 - 3x - 28$
  i $x^2 - 4x - 45$ **j** $x^2 - 2x - 80$ **k** $x^2 - 7x - 60$ **l** $x^2 - x - 56$

3 **a** $x^2 - 11x + 24$ **b** $x^2 - 10x + 24$ **c** $x^2 - 13x + 40$ **d** $x^2 - 10x + 9$
  e $x^2 - 10x + 21$ **f** $x^2 - 13x + 36$ **g** $x^2 - 19x + 84$ **h** $x^2 - 19x + 90$
  i $x^2 - 11x + 30$ **j** $x^2 - 27x + 180$ **k** $x^2 - 19x + 84$ **l** $x^2 - 29x + 100$

4 **a** $6x^2 + 11x + 4$ **b** $12x^2 + 23x + 10$ **c** $30x^2 + 39x + 12$ **d** $8x^2 + 2x - 3$
  e $15x^2 + 8x - 12$ **f** $54x^2 - 3x - 12$ **g** $18x^2 - 2$ **h** $28x^2 - 13x - 6$
  i $32x^2 - 8x - 12$ **j** $42x^2 - 46x + 12$ **k** $18x^2 - 60x + 18$
  l $20x^2 - 41x + 20$

Thinking ahead to ...

A $(x+2)(x+6)$

B **a**

×	p	−2
p	p²	−2p
+7	7p	−14

  b $(p-2)(p+7)$

Exercise 15.8

1 **a** $(x+1)(x+7)$ **b** $(x+1)(x+2)$ **c** $(x+1)(x+3)$
  d $(x+2)(x+6)$ **e** $(x+3)(x+3)$ **f** $(x+4)(x+9)$

2 **a** $(x+1)(x+5)$ **b** $(x+8)(x-6)$ **c** $(x+7)(x-2)$
  d $(x+7)(x-3)$ **e** $(x+9)(x-3)$ **f** $(x+9)(x-2)$
  g $(x+10)(x-1)$ **h** $(x+12)(x-5)$ **i** $(x+10)(x-2)$

3 **a** $(x+1)(x-7)$ **b** $(x+5)(x-9)$ **c** $(x+5)(x-8)$
  d $(x+6)(x-10)$ **e** $(x+1)(x-5)$ **f** $(x+5)(x-12)$
  g $(x+7)(x-9)$ **h** $(x+2)(x-7)$ **i** $(x+3)(x-12)$

4 **a** $(x-2)(x-7)$ **b** $(x-5)(x-4)$ **c** $(x-2)(x-4)$
  d $(x-2)(x-5)$ **e** $(x-4)(x-7)$ **f** $(x-5)(x-9)$
  g $(x-6)(x-5)$ **h** $(x-7)(x-8)$ **i** $(x-7)(x-13)$

End points

A1 **a** 100 m **b** 175 m **c** 100 m

A2 52.4 cm³

B1 **a** $n = \frac{(C-20)}{30}$ **b** 15 days

B2 **a** $m = 5(C-10)$ **b** 100 miles

B3 **a** $k = \sqrt{\left(\frac{h}{5}\right)}$ **b** $k = 5m - n$ **c** $k = \sqrt{(A-9)}$

C1 C

C2 **a** $a^2 + 9a + 8$ **b** $b^2 + 8b + 16$ **c** $5c^2 + 17c + 6$
  d $d^2 + d - 2$ **e** $6e^2 - 5e - 50$ **f** $3f^2 - 23f + 14$

D1 $(k-2)(k+7)$

D2 **a** $(x+1)(x+11)$ **b** $(x+2)(x+7)$ **c** $(x+5)(x-1)$
  d $(x+3)(x-2)$ **e** $(x+4)(x-5)$ **f** $(x-3)(x-2)$

SECTION 16 Grouped data

Starting points

A1 2.305 m

A2 2.25 m

B1 1.97 m

B2 1.91 m

C1 **a** 14 cm **b** 12 cm

C2 **a** 7 cm **b** 6 cm

E1 **a** 6 years **b** 4 years

E2 **a** bars with heights 8, 13, 10, 3
  b bars with heights 3, 17, 14, 8, 5, 1

Exercise 16.1

1　a 70–72　b 70–72

2　a 1st round

　b The frequency polygon for the 1st round scores is higher in the low score section of the graph than the frequency polygon for the 4th round.

3　a 17, 18, 19, 20, 21, 22, 23, 24, 25, 26, 27　b 22　c 22, 33, 44, 55

5　No, you need roughly the same number to compare.

Thinking ahead to ...

A　a 64, 65, 66　b 79, 80, 81

B　17, 16, 15, 14, 13

C　15, the middle value and happens most often

Exercise 16.2

1　a 5　b 69, 74, 79, 84　c 15

2　a 4　b 68.5, 72.5, 76.5, 80.5　c 12

3　a 128, 129, 130, 131, 132　b 130

　c The mid-class value is the best estimate of the score in a class, so the mid-class value multiplied by the frequency is the best estimate of the total.

Exercise 16.3

1　43 900

2　252 750

3　8153.2

4　a 3089 m　b 2837 m

Exercise 16.4

1　63.0 m

2　About the same as the mean throw is the same

3　18.7 m

4　20.1 m

5　The men threw further than the women

Exercise 16.5

Reaction time	0.120–	0.140–	0.160–	0.180–	0.200–	0.220–	0.240–	0.260–	Total
Frequency	5	8	23	5	4	2	0	0	47

1

2　a i 0.14 seconds　ii 0.1 seconds
　b i 0.1738 seconds　ii 0.1704 seconds

3　true, the mean is lower and the range is narrower

Reaction time	0.120–	0.140–	0.160–	0.180–	0.200–	0.220–	0.240–	0.260–	Total
Frequency	10	16	46	10	8	4	0	0	94

4　a

　d No, the finalists may have reacted exactly the same in the semifinals, the athletes that didn't make it to the finals perhaps had the slowest reaction times in the semifinals.

Thinking ahead to ...

A　12.73, 12.68, 12.749

Exercise 16.6

1　a 9.835 to 9.845 seconds　b 7.115 to 7.125 metres
　c 0.815 to 0.825 metres　d 0.1735 to 0.1745 seconds

2　a 12.375 to 12.385 seconds　b 9.15 to 9.25 metres
　c 7.25 to 7.35 seconds　d 7.295 to 7.305 seconds

Exercise 16.7

1　10.93 to 10.94 seconds

2　7.00 to 7.01 metres

3　So that the time/distance recorded isn't better than they performed.

Exercise 16.8

1　a

Points	< 700	< 800	< 900	< 1000	< 1100
Cumulative frequency	0	8	31	47	50

2　a

Distance (metres)	< 76	< 78	< 80	< 82	< 84	< 86	< 88	< 90
Cumulative Frequency	0	2	6	20	33	41	47	48

Exercise 16.9

1　a

Distance (metres)	< 72	< 74	< 76	< 78	< 80	< 82
Cumulative Frequency	0	1	12	35	48	52

2　a 77.2　b 2.3

3　The javelin was, on average, thrown further but the hammer was throw more consistently.

4　b 840　c 180

5　a 875　b 115

6　More points were scored on day 1 with a smaller spread.

7　a

Distance (metres)	< 56	< 58	< 60	< 62	< 64	< 66	< 68
Cumulative Frequency	0	7	18	28	38	45	46

8　a at 11.5, 23, 34.5　b 61, 4.5

Exercise 16.10

1　5

2　23　3 b 15

Exercise 16.11

1　a 12, 10.7, 9, 12, 8, 11, 9, 10, 14
　b 12, 13, 10, 9, 6, 9, 5, 12, 9, 15
　c 9, 12, 10, 11, 6, 15, 14, 17.3, 7, 10

2　a 23, 20.8, 27.2, 27.8, 25.8, 25, 25.2, 16
　b 12.8, 12, 12.6, 11, 14.6, 21.4, 21.2
　c 19.8, 17.4, 14.2, 11.8, 12.6, 15.2
　d 12.4, 10.8, 10.6, 18, 17, 16.4, 18.8

3　a 16.25, 15, 10.25, 11.25, 13.25, 20.25, 23.25, 20.5, 18
　b 20.5, 19.75, 18.25, 19, 14.75, 19.5, 16.5, 24.75
　c 10, 20.75, 27.75, 32, 38.25, 29.75, 32.75, 44
　d 31.5, 19, 17.25, 16.5, 8, 17, 29.5, 40.25, 48

4　a 8, 7.5, 6, 5, 6, 6.75, 9, 8, 8
　c Sales of boots are quite steady over each 4 week period.

5　a 20.25, 16.75, 15.5, 14, 13.5, 16.5, 15.5, 14.25, 15.25, 14.75, 10.25, 11.75, 10.5, 7.75
　c Watch sales declined over the period

End points

A1　a bars with heights 7, 22, 23, 10, 3, 2, 2, 1
　b bars with heights 5, 10, 30, 17, 5, 3, 1

A2　a 0.16–　b 0.16–

B1　a

Reaction Time (s)	< 0.12	< 0.14	< 0.16	< 0.18	< 0.20	< 0.22	< 0.24	< 0.26	< 0.28
Cumulative Frequency	0	7	29	52	62	65	67	69	70

　c i 0.165 seconds　ii 0.03 seconds

C1　a 0.12 seconds　b 0.14 seconds

C2   **a**   0.175 seconds   **b**   0.170 seconds

D1   The men were, on average, quicker but the women were more consistent

E1   **b i** 8   **ii** 15

F1   **a**   7, 9, 10, 11, 7, 9, 10, 12, 7, 10, 12, 13
    c   The number of lates rises then falls in cycles.

SECTION 17   Working with percentages

Starting points

A1   **a**   0.75, 75%   **b**   0.625, 62.5%   **c**   1.6, 160%   **d**   1.25, 125%
    e   1.2, 120%   **f**   0.65, 65%   **g**   0.25, 25%   **h**   1.5, 150%
    i   0.4375, 43.75%   **j**   2.2857, 228.57%   **k**   1.125, 112.5%   **l**   1.8, 180%

B1   **a**   325 miles   **b**   £300   **c**   2 marks   **d**   2484   **e**   10 marks   **f**   171 kg
    g   £240.70   **h**   360 g   **i**   18p   **j**   16.2 m   **k**   £1.71   **l**   1.8 cm   **m**   £6.25
    n   2.4 miles   **o**   10 km   **p**   £480   **q**   110.7 miles   **r**   9 cm   **s**   8.4 m
    t   438 m

C1   **a**   £13.60   **b**   5.6 mm   **c**   1540 g   **d**   20p   **e i**   50.7 mm   **ii**   5.07 cm
    f i   9p   **ii**   £0.09   **g i**   1.4 tonnes   **ii**   1400 kg

Exercise 17.1

1   50%

2   400%

3   a little more

4   **i**   8   **ii** $80 \times \frac{10}{100}$

5   50%

6   **a**   16   **b**   21

7   **a**   9   **b**   His number is a little more than 8 but not as much as 10.

8   25%, 25%, 20%, 25%, 75%, 10%

9   **a**   December and March   **b**   January   **c**   62.5%

Exercise 17.2

1   **a**   44.44%   **b**   65.00%   **c**   29.17%   **d**   46.20%
    e   80.00%   **f**   12.50%   **g**   3.75%   **h**   133.33%
    i   166.67%   **j**   187.50%

2   **a**   69.09%   **b**   Pretty accurate, rounded to the nearest 5%.

3   14%

4   **a**   170%

Thinking ahead to ...

A   45 ml, less   **B** $\frac{3}{20}$ **C**   345 ml

Exercise 17.3

1   **a**   110%   **b**   484 g

2   **a**   420 kg   **b**   514.05 km   **c**   5.49 m   **d**   £37 800

3   206 000 000

4   26g

Exercise 17.4

1   **a**   266 kg   **b**   312 km   **c**   20.9 m   **d**   £287 300
    e   £18.36   **f**   16 588 tonnes   **g**   4 717 000   **h**   1.2 cm

2   1466 kg

3   **a**   1125
    b   Round down to the whole number below, because fractions of cars are irrelevant.

4   **a**   26 650   **b**   14 350

5   7644 kg

6   **a**   £19.60   **b**   £4.90

Thinking ahead to ...

A   Divide by 10, halve the answer, halve the answer again, and add the three amounts.

B   £235

C   **i**   £14.69   **ii**   Round to the nearest penny.

Exercise 17.5

1   **a**   £18.80   **b**   £51.99   **c**   £217.38   **d**   £22.80
    e   85p   **f**   £432.89   **g**   £307.85   **h**   £28.85   **i**   £10.99

2   **a**   £287.88   **b**   £42.88

Exercise 17.6

1   £387.60

2   **a**   £155   **b**   £181.35   **c**   £301.35

3   **a**   £2612.50   **b**   £112.50

4   £89 979

Exercise 17.7

1   **a**   £324   **b**   £67.20   **c**   £1350   **d**   £10 200   **e**   £90

2   **a**   £5950   **b**   £9450

3   £231

4   £66 440 000

Exercise 17.8

1   **a**   £103.26   **b**   £175.71   **c**   £1217.61
    d   £1529.58   **e**   £1030.32   **f**   £8967.60

2   **a**   £2346.64   **b**   £7346.64   **c i**   £7346.64   **ii**   Round each year.

3   **a**   £1900.16   **b**   £7542.60   **c**   £9881.86   **d**   £10 438.16

4   £5898.58

5   £30 653.86

Exercise 17.9

1   **a**   £115.32   **b**   £29.78   **c**   £63.82   **d**   £56.00
    e   £158.94   **f**   £12.33   **g**   £42.54   **h**   £680.83   **i**   £84.25

2   **a**   £56.25   **b**   £11.25

3   £26

4   444 g

End points

A1   25%

A2   300%

A3   28.13%

A4   7.7%

B1   556.8 km

B2   426 ml

C1   £52.45

C2   39 203 breakdowns

D1   17.5%

D2   **a**   £40.83   **b**   £186.12

E1   £221.27

F1   £115.20

F2   **a**   £180   **b**   £680

G1   **a**   £809.89   **b**   £2898.19   **c**   £4926.85   **d**   £9067.93
    e   £4810.70   **f**   £330.91

SECTION 18   Transformations

Starting points

A1 $x = 3, y = {}^{-}1$

B1   **b i**   Triangle with coordinates (3, 2), (3, 0), (2, 0).
      ii   Triangle with coordinates ($^{-}2$, 2), ($^{-}2$, 3), ($^{-}4$, 3).

C1   **a**   90°   **b**   (0, 1)

D1   U: (2, 2), (4, 3), (2, 3).
    V: ($^{-}1$, 0), (1, 0), ($^{-}1$, $^{-}1$)

E1   **a**   triangle: (1, 0), (1, 2), (5, 2)
    b   triangle: (0, 3.5), (0, 4), (1, 4)

Exercise 18.1

1   **a i**   0   **ii**   3   **b i**   1   **ii**   3

2   **a**   rotation of 60° clockwise   **b**   translation, reflection
    c   A and F   **d**   Z, N, V   **e i**   C, R, X   **ii**   E, J, T

Exercise 18.2

1 a i triangle 8 **ii** triangle 1 **iii** triangle 3 **b i** B **ii** 90°
   c i rotation of 90° about D **ii** rotation of 180° about P
   d i BD **ii** HD
   e triangle 7 **f** triangle 4 **g** rotation of $^-$90° about F
   h reflection in AE or rotation of 180° about P
   i rotation of 360° about any point, or reflection in GP

Exercise 18.3

1 a $\begin{pmatrix} 0 \\ -4 \end{pmatrix}$ **b** $(^-2, ^-1)$ **c** $y = ^-1$

2 a i C and E, D and H, F and G
   ii

Object	Mirror line	Image
C	$y = x$	E
D	$y = 0$	H
F	$x = ^-1$	G

b i C and F, D and G, E and H
   ii

Object	Vector	Image
C	$\begin{pmatrix} -4 \\ -4 \end{pmatrix}$	F
D	$\begin{pmatrix} 2 \\ -2 \end{pmatrix}$	G
E	$\begin{pmatrix} -4 \\ -2 \end{pmatrix}$	H

3 a $\begin{pmatrix} 1 \\ -2 \end{pmatrix}$ **b** $\begin{pmatrix} 4 \\ -1 \end{pmatrix}$ **c** $\begin{pmatrix} 4 \\ 2 \end{pmatrix}$ **d** $\begin{pmatrix} -3 \\ -1 \end{pmatrix}$ **e** $\begin{pmatrix} 4 \\ -3 \end{pmatrix}$ **f** $\begin{pmatrix} 5 \\ 3 \end{pmatrix}$ **g** $\begin{pmatrix} -6 \\ 3 \end{pmatrix}$ **h** $\begin{pmatrix} -5 \\ 5 \end{pmatrix}$ **i** $\begin{pmatrix} -5 \\ -3 \end{pmatrix}$

4 H

5 H

6 a B **b** $\begin{pmatrix} -4 \\ 3 \end{pmatrix}$

7 a $\begin{pmatrix} 1 \\ -3 \end{pmatrix}, \begin{pmatrix} -1 \\ -2 \end{pmatrix}, \begin{pmatrix} -4 \\ -1 \end{pmatrix}, \begin{pmatrix} 0 \\ 4 \end{pmatrix}, \begin{pmatrix} 3 \\ 1 \end{pmatrix}$ **b** $\begin{pmatrix} -1 \\ -1 \end{pmatrix}$

8 $\begin{pmatrix} 6 \\ 0 \end{pmatrix}$

Exercise 18.4

1 B: (1, 2), (2, 2), (1, 4)
   C: (0, 1), (0, 2), ($^-$2, 1).
   D: ($^-$2, $^-$1), ($^-$4, $^-$1), ($^-$2, $^-$2)
   E: (2, $^-$1), (3, $^-$1), (2, $^-$3).
   c Rotation 90° anticlockwise about (1, 1)
   d i Rotation 90° anticlockwise about (0, 1)
      ii Reflection in $y = x - 1$

2 B: (2, 2), (2, 3), (3, 4), (3, 2)
   C: (0, 3), (0, 4), ($^-$1, 4), ($^-$2, 3)
   D: (0, $^-$1), (0, 0), ($^-$1, 1), ($^-$1, $^-$1).

3 B: ($^-$1, 3), ($^-$1, 4), ($^-$3, 4), ($^-$2, 3), ($^-$2, 2)
   C: (2, 0), (2, $^-$1), (3, $^-$2), (1, $^-$2), (1, $^-$1)
   D: (2, 2), (3, 3), (4, 3), (4, 1), (3, 2)
   E: (0, 0), ($^-$1, $^-$1), ($^-$1, $^-$2), ($^-$2, $^-$1), ($^-$2, 0)

4 B: (0, 5), ($^-$1, 5), ($^-$2, 4), ($^-$1, 3), (0, 3), ($^-$1, 4)
   C: (1, 2), (2, 1), (3, 2), (3, 1), (2, 0), (1, 1)
   D: (0, $^-$1), (1, $^-$2), (1, $^-$3), (0, $^-$2), ($^-$1, $^-$3), ($^-$1, $^-$2).
   b They are the same
   d They are the same
   f They are the same

5 28 cm², 23 cm

6 6 cm², 12 cm

7 a area = 4 units², perimeter = 10 units
   d area = 4 units², perimeter = 10 units
   g They all have the same area and perimeter

Exercise 18.5

1 F: (2, 2), (3, 3), (4, 2), (4, 1)
   H: (4, 1), (5, 1), (6, 2), (5, 3)

2 B: ($^-$1, 0), ($^-$2, $^-$1), ($^-$3, $^-$1), ($^-$3, $^-$2), ($^-$2, $^-$2), ($^-$1, $^-$3)
   C: (2, $^-$3), (3, $^-$3), (3, $^-$4), (4, $^-$5), (1, $^-$5), (2, $^-$4)
   c The area and perimeter are all the same

3 b i translation [vector($^-$2, $^-$2)] **ii** 180° rotation about (2, 0)
      iii 180° rotation about (3, 1)

Note: there are a number of answers to the following questions. These are examples.

4 a rotation 90° clockwise about (2, 0) **b** reflection in $x = 5$.

5 A rotation of 90° clockwise about (0, 0) followed by a reflection in $x = 3$.

6 a [vector (6 0)] **b** reflection in $y = x - 8$

7 a reflection in $y = 2$ **b** rotation 90° anticlockwise about (7, 3)

8 A reflection in the x axis and a reflection in $x = 2$, in either order.

Exercise 18.6

1 b B
   c

Object	Image	SF	Centre
S	Y	2	(4, 8)
Y	S	$\frac{1}{2}$	(4, 8)
X	S	$\frac{1}{2}$	(8, 8)
X	U	$\frac{1}{2}$	(12, 8)

   d They are not in the same orientation.
   e i Rotate $^-$90° about (8, 8) **ii** Enlarge SF $\frac{1}{2}$ with centre (8, 12)
      iii Rotate 180° about (6, 8)

Exercise 18.7

1 A₁: (5, 4), (5, 2), (1, 4)
   A₂: (7, 4), (7, 0), (5, 0)
   A₃: (5, 6), (9, 6), (9, 4)
   A₄: (3, 4), (3, 8), (5, 8)

2 a B₁: $(\frac{1}{2}, \frac{1}{2}), (2, \frac{1}{2}), (2, 1\frac{1}{2}), (\frac{1}{2}, 2)$
   b i The lengths in B₁ are half those in B.
      ii 7.5 units²
      iii 1.875 units²
      iv lengths are divided by 2, areas are divided by 4

3 b i the lengths are divided by 4
      ii the area is divided 16
      iii the angles are the same

4 b i $\frac{1}{3}$ **ii** $\frac{2}{3}$ **iii** $\frac{1}{2}$ **c** 9 units² **d** 9 times
   e The perimeter of C₁ is 3 times the perimeter of A.
   f The angles are all the same.

5 b i (0, 0) **ii** ($^-$0.5, $^-$0.5) **c** $\frac{2}{3}$ **d** 27 units² **e** 6.75 units² **f** 12 units

End points

A1 a C, D **b** B, C, D **c** B, C, D

A2 b F: ($^-$1, $^-$1), ($^-$1, $^-$3), ($^-$2, $^-$3), ($^-$2, $^-$2), ($^-$4, $^-$2), ($^-$4, $^-$1).
   G: ($^-$4, 1), ($^-$4, 2), ($^-$2, 2), ($^-$2, 3), ($^-$1, 3), ($^-$1, 1).
   H: (3, 1), (3, $^-$1), (6, $^-$1), (6, 0), (4, 0), (4, 1)
   c i rotation 180° about (0, 0)
      ii reflection in $x = 0$
      iii translation $\begin{pmatrix} -2 \\ 2 \end{pmatrix}$
   d i reflection in $y = 0$
      ii rotation 180° about (1, $^-$1)

A3 B: (1, 1), (1, 3), (2, 2), (3, 2), (3, 1)
   C: (1, $^-$7), (1, $^-$3), (5, $^-$3), (5, $^-$5), (3, $^-$5)
   D: ($^-$5, $^-$2), ($^-$3, $^-$4), ($^-$1, $^-$4), ($^-$1, $^-$6), ($^-$5, $^-$6)
   b 10 units² **c** 2.5 units² **d** $\frac{1}{4}$ **e** lengths in B are half those in A
   g same area, perimeter, but mirror images **j** $\frac{1}{9}$
   k lengths in E are a third of those in A

B1 a K: ($^-$1, $^-$3), ($^-$4, $^-$3), ($^-$4, $^-$4), ($^-$2, $^-$4), ($^-$2, $^-$5), ($^-$1, $^-$5)
   L: ($^-$1, 1), ($^-$3, 1), ($^-$3, 2), ($^-$2, 2), ($^-$2, 4), ($^-$1, 4)
   M: (1, $^-$1), (2, $^-$1), (2, $^-$2), (4, $^-$2), (4, $^-$3), (1, $^-$3)
   b i rotation 180° about (0, $^-$1)
      ii rotation 90° anticlockwise about (0, 0)
      iii translation $\begin{pmatrix} 0 \\ -4 \end{pmatrix}$
   c i translation $\begin{pmatrix} 0 \\ 4 \end{pmatrix}$ **ii** reflect in $y = ^-3$
      iii rotation 180° about (1, 0)
   d i rotation 90° clockwise about (1, $^-$1)
      ii rotation 180° about (0, $^-$3)
      iii rotation 180° about (0, $^-$1)

ANSWERS

SKILLS BREAK 3A

1   **a** 22 : 40   **b** 02 : 20

2   3 hour 40 minutes

3   **a** 1045 crew   **b** 1600 people

4   61 people per boat

5   £287 970

6   £294 930

7   700 feet

8   **a** 14 knots   **b** 3 hours 19 minutes   **c** $1\frac{1}{2}$ hours

9   **a** $C = \frac{5(F-32)}{9}$ **b** $^-2.2\,°C$

10   **a** 1251   **b** 3800

11   **a** 0.61   **b** 0.43   **c** 0.25

12   38%

13   70 000 tons

14   £(3.0×10^7)

15   1°

16   200.97 ft

17   58°, because the cable is parallel to the tower.

18   14°, because angles in a triangle sum to 180°.

19   **a** 2 000 000   **b** 2 180 000   **c** 2 180 000   **d** 2 179 600

20   11 people

SKILLS BREAK 3B

1   3.53 m

2   5.88×102^{21}

3   **a** 40 080 km   **b** 40 100 km

4   12 756 km

5   **a** 0.2 m   **b** 3600 mm   **c** 7.9 cm

6   2561 days

7   0.6 m

8   307.7 m²

9   185 m³

10   **a** 14.8 mph   **b** 23.7 km/h

11   18 minutes

12   5.5×10^3

13   2200 cm³

14   **a** 20.2 cm   **b** 26 800 cm³   **c** 27 kg

15   1.75 m

16   1 : 3

17   **a** **i** 107 inches   **ii** 267.5 cm   **iii** 2.68 m

18   **a** 192   **b** 2277   **c** 8 digits per second

19   56 days

20   **a** 51.3°   **b** 242 m   **c i** 5 ms   **ii** 18 km/h

21   54 seconds

22   **a** 451.6 cm³   **b** 2.2 m³

23   **a i** 4 cans   **ii** 9 cans   **b** 576

24   after nearly 10 years

25   **a** 67 666   **b** 2 235 962

26   Once, 'a basin plant' is an anagram of 'banana split'.

SECTION 19 Probability B

Starting points

A1   0.23

A2   **a** 0.77   **b** 0.16   **c** 0.55   **d** 0.44

B1 $\frac{1}{6}$

B2 $\frac{1}{3}$

B3   **b** $\frac{1}{4}$

Thinking ahead to ...

A   **a** $\frac{5}{9}$ **b** $\frac{1}{9}$ **c** $\frac{4}{9}$ **d** $\frac{1}{3}$

Exercise 19.1

1   **a** 0.11   **b** 0.50

2   0.89

3   **a** 8   **b** 2   **c** 10

4   **a** 0.17

    b you don't know how many red, green or blue cubes there are.

Exercise 19.3

1   $1 - 0.24 - 0.36 = 0.4$

3   **a** 0.11   **b** 0.09   **c** 0.15   **d** 0.17

4   **a** 0.12   **b** 0.10   **c** 0.22

5   **a** 195   **b** 147   **c** 86   **d** 53   **e** 39   **f** 76

Exercise 19.4

1   **a** A   **b** A   **c** B   **d** B   **e** D   **f** C   **g** A   **h** B   **i** A   **j** D

2   **a** The outcomes may not be equally likely, perhaps more people are born in the summer, for instance.

    b Conduct a survey of people's birth dates.

3   c

4   **a** the sample is too small

    b people leaving a butcher's shop are unlikely to be vegetarian

    c not representative of the population

    d not representative of the population

    e wouldn't be able to answer

5   **a** not everybody plays soccer

6   True

Exercise 19.5

1   **a** $\frac{1}{4}$ **b** $\frac{7}{16}$ **c** $\frac{11}{16}$

2   c is the sum of a and b

3   **a** $\frac{5}{16}$ **b** $\frac{7}{16}$

4   **b** $\frac{1}{2}$ **c** Owen has a moustache and an earring

5   **a** $\frac{1}{4}$ **b** $\frac{1}{4}$ **c** $\frac{3}{8}$

Exercise 19.6

1   **b** No, some criminals have fair hair and a necklace.

2   **a** $\frac{11}{16}$ **b** $\frac{5}{16}$ **c** $\frac{5}{16}$ **d** $\frac{1}{8}$ **e** $\frac{1}{4}$

3   **a** $\frac{7}{16}$ **b** $\frac{3}{4}$ **c** $\frac{3}{4}$ **d** $\frac{9}{16}$

4   **a** No, some girls could play hockey and basketball.

    b Yes, boys and girls don't overlap.

    c Yes, you can't be a driver and a passenger.

    d No, tall women can drive buses.

    e No, some mechanics can dance.

Exercise 19.7

1   **a** 0.15   **b** 0.19   **c** 0.53

2   0.44

3   **a** a number can't be less than 5 and greater than 7

    b 0.55   **c** 0.45

End points

A1   **a** 0.17   **b** 0.07

A2   0.75

B1   **b** 0.42, below

C1   **a** theoretical   **b** relative   **c** relative

D1   **a** 0.36   **b** a tree cannot be an oak and a lime

D2   Some trees may have canker and deer damage.

SECTION 20 Inequalities

Starting points

A1 **a** $34 > 12$ **b** $2 < 15$ **c** $^-12 > ^-56$ **d** $5 > ^-2$ **e** $^-2 < 0$ **f** $^-17 < ^-5$

A2 **a** $56 > 28 > 9 > 5 > ^-4 > ^-24 > ^-30$ **b** $^-30 < ^-24 < ^-4 < 5 < 9 < 28 < 56$

A3 12, 5, 6.2

A4 **a** $^-5, ^-4, ^-3, ^-2, ^-1, 0, 1, 2$ **b** $^-4, ^-3, ^-2$

A5 There are no integers between $^-2$ and $^-1$

B1 **a** Straight line passing through $(0, ^-2)$ and $(3, 4)$.
b Straight line passing through $(0, 12)$ and $(4, 8)$.
c Straight line passing through $(0, 0)$ and $(4, 12)$.
d Straight line passing through $(0, ^-4)$ and $(4, ^-4)$.

Exercise 20.1

1 $45 \leqslant T \leqslant 80$

2 **a** $17 \leqslant P \leqslant 52$ **b** $10 < P < 64$ **c** $4 < P < 164$

3 **a** $5 \leqslant A \leqslant 65$ **b** $5 < A \leqslant 65$
c 5 year olds can use Fermatol but not Hypaticain.

4 $^-25 < T < ^-18$

Exercise 20.2

1 B, C

2 **b** 2, 3, 4

3 **b** $^-1, 0, 1, 2, 3$

4 **a** 5, 6, 7, 8, 9 **b** $^-2, ^-1, 0, 1, 2, 3$
c $^-3, ^-2, ^-1, 0, 1, 2, 3, 4, 5$ **d** 68, 69, 70
e $^-11, ^-10, ^-9, ^-8, ^-7, ^-6, ^-5, ^-4$ **f** $^-5, ^-4, ^-3, ^-2, ^-1, 0$
g $^-3, ^-2, ^-1, 0, 1, 2, 3, 4, 5, 6$ **h** 2164, 2165, 2166
i $^-2, ^-1, 0, 1, 2, 3$

5 $^-4 < h < 6$

6 $^-4 < g < 2$, $^-4 < g \leqslant 1$, $^-3 \leqslant g < 2$, $^-3 \leqslant g \leqslant 1$

7 **a** 6, 7 **b** $^-4, ^-3, ^-2, ^-1, 0, 1$ **c** 13, 14 **d** $^-1, 0$
e $^-2, ^-1, 0, 1, 2, 3, 4$ **f** $^-2, ^-1$

Exercise 20.3

1 **a** A – too large, E – too heavy **c** B

Exercise 20.4

2 **a** $x < 7$ **b** $5 < y < 12$ **c** $^-1 < x < 9$

4 $^-2 < x < 6$ and $1 < y < 5$

Exercise 20.5

1 **a** $x > 4$ **b** $x \leqslant 4$ **c** $x \leqslant 3$ **d** $x \geqslant ^-4$ **e** $x \geqslant \frac{1}{2}$ **f** $x \geqslant 9$ **g** $x \leqslant \frac{1}{2}$
h $x < 5$ **i** $x \leqslant 16$ **j** $x \geqslant 15$ **k** $x > 7$ **l** $x \leqslant 4$

2 **a** $y < 7$ **b** $s \leqslant 8$ **c** $k < 29$ **d** $p \geqslant 6$ **e** $b > \frac{7}{2}$ **f** $x > 3$ **g** $a \leqslant ^-8$
h $x > 8$ **i** $x < 5$ **j** $t \geqslant 4$ **k** $a \geqslant 7$ **l** $c \geqslant 8$ **m** $d < ^-5$ **n** $a \leqslant ^-13$
o $x > 6$

3 **a** $a \geqslant 11$ **b** $k < ^-2$ **c** $x > ^-9$ **d** $n \geqslant \frac{7}{2}$

End points

A1 **a** $^-1$ **b** $^-5, ^-4, ^-3, ^-2, ^-1, 0, 1, 2, 3$ **c** 4, 5, 6, 7
d $^-56, ^-57, ^-58, ^-59$ **e** $^-76, ^-75, ^-74, ^-73, ^-72$ **f** $^-2, ^-1, 0, 1, 2, 3$

B4 **a** $2 < x < 7$ **b** $y < 2, x > 4$

C1 **a** $x \leqslant 6$ **b** $f > 3$ **c** $a \geqslant 12$ **d** $k > ^-5$ **e** $a \geqslant 6$ **f** $d > 1.5$

SECTION 21 Working in 3D

Starting points

A1 **a** cube **b** sphere **c** cylinder **d** cuboid **e** cylinder

A3 They are easy to stack.

B1 **a** **ii** A, C, F **iii** B, D, E
b A cuboid, B tetrahedron, C cylinder, D pentagonal pyramid,
E octagonal pyramid, F hexagonal prism

C1 **a** 5, 7 **b** 8, 12 **c** 5, 7

D1 64 cm²

D2 85.5 cm²

D3 **a** 1 cm³ **b** 200 cm³

D4 2 m³

D5 13.5 cm³

Exercise 21.1

1 View 1 is from B, view 2 from A and view 3 from C.

2 B, C, D, G, J

3 A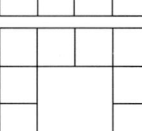
B
C

4 **a** 7
b A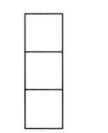
B
C

5 **a** 7
b A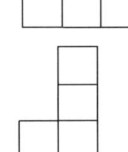
B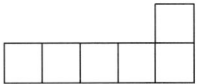
C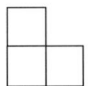

6 A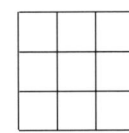
B
C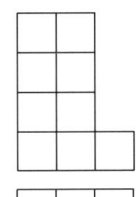

Exercise 21.2

1 **a** B and C **b** 1 **c** A

6

Shape	Number of faces	Number of edges	Number of vertices
P	5	9	6
Q	6	12	8
R	7	15	10
S	8	18	12

7 102 faces, 300 edges, 200 vertices, 101 planes of symmetry

8 $n + 2$ faces, $3n$ edges, $2n$ vertices, $n + 1$ planes of symmetry

9 **a i** C, D **ii** A, B
   b A is a octagonal pyramid, B is a square pyramid, C is a pentagonal prism, D is a triangular prism.

10 **b** X4, Y2, Z1

11 **a** There is only one more possible net, with the 4 triangles joined together in a line.
    b 3

Exercise 21.3

1 **a** 192 cm^3 **b** 216 cm^3 **c** 396 cm^3 **d** 1323 cm^3 **e** 490 cm^3
   f 871 cm^3 **g** 180 cm^3 **h** 558 cm^3 **i** 2070 cm^3

2 **a** 200 cm^3 **b** 87.5 cm^3

4 **b** $5 \text{ cm} \times 5 \text{ cm} \times 4 \text{ cm}$

5 **a** 63 cm^2 **b** $a = 14, b = 9$

6 **a** 23 cm^2 **b** $\sqrt{(23 \div \pi)}$

Exercise 21.4

1 16.5 cm

3 **c** With a radius of 6 cm and a height of 6 cm, the maximum volume is 679 cm^3.

Exercise 21.5

1 **a** 17.0 m^3 **b** 12.7 m^3 **c** 15.3 tonnes

2 **a** 27.2 m^3 **b i** 390 bags **ii** £1996.80

3 **a** 46.25 m^2 **b** 555 m^3 **c** 435 m^3 **d** 435 cm^3

Exercise 21.6

1 **a** 348 cm^2 **b** 120 cm^2 **c** 344 cm^2 **d** 526 cm^3 **e** 2974.5 cm^3

2 7 cm

3 The surface area of B will be 4 times that of A.

Exercise 21.7

1 **a** 300 cm^3 **b** 75 ml **c** 170 g

2 about 75 ml

3 B

4 1.37 litres

5 **a** $21\,714\,688 \text{ cm}^3$ **b** 21 715 litres

6 $217 \text{ cm}^3, 1.07 \text{ g/cm}^3; 649 \text{ cm}^3, 0.952 \text{ g/cm}^3$

7 A 1500 cm^3 B 720 cm^3 C 990 cm^3 **d** 490 cm^3 E 440 cm^3

8 By comparing densities: P contains crackers, Q contains Fruit and Fibre, R contains Fruit and Fibre, S contains Trichoc, T contains drinking chocolate

Exercise 21.8

1 **a** length **b** volume **c** area

2 **a** $\frac{1}{2}lz(x + y)$ **b** It is the only expression that represents a volume.

3 **a** $\pi(a + b)$ **b** πab

4 **a** area **b** length **c** length **d** volume **e** volume **f** area

5 The expressions xy and $y(x - z)$ represent areas, but xyz represents a volume.

End points

A1 4

A2 8 faces, 18 edges, 12 vertices

B1 110 cm^3

C1 **a** $6\,433\,982 \text{ cm}^3$ **b** 6434 litres

C2 19.3g/cm^3

D1 **a** $\pi r(l + r)$ **b** $\frac{1}{3}\pi r^2 h$

SECTION 22 Quadratics

Starting points

A1 **a** $c^2 + 3c - 28$ **b** $a^2 + 8a + 15$ **c** $y^2 - 3y - 40$ **d** $h^2 - 5h - 6$
   e $t^2 - 25$ **f** $y^2 - 6y + 9$ **g** $p^2 + 10p + 25$ **h** $k^2 - 11k + 30$

B1 **a** $(n - 1)(n + 5)$ **b** $(y + 2)(y - 5)$ **c** $(k - 1)(k - 6)$ **d** $(h + 4)(h + 5)$
   e $(w - 4)(w - 5)$ **f** $(v + 7)(v - 4)$ **g** $(d - 2)(d - 5)$ **h** $(a + 13)(a - 1)$

B2 **a** $(p + 3)^2$ **b** $(m - 5)^2$ **c** $(s - 2)^2$ **d** $(g + 4)^2$

Exercise 22.1

1 **a** $(n + 14)(n - 1) = 0, n = {}^-14 \text{ or } 1$ **b** $(n + 2)(n - 7) = 0, n = {}^-2 \text{ or } 7$
   c $(n + 1)(n - 14) = 0, n = {}^-1 \text{ or } 14$ **d** $(b + 5)(b - 3) = 0, b = {}^-5 \text{ or } 3$
   e $(b + 15)(b - 1) = 0, b = {}^-15 \text{ or } 1$ **f** $(b + 3)(b - 5) = 0, b = {}^-3 \text{ or } 5$
   g $(a + 10)(a - 3) = 0, a = {}^-10 \text{ or } 3$ **h** $(a + 3)(a + 10) = 0, a = {}^-3 \text{ or } {}^-10$
   i $(a + 6)(a - 5) = 0, a = {}^-6 \text{ or } 5$ **j** $(x + 5)(x - 1) = 0, x = {}^-5 \text{ or } 1$
   k $(x + 1)(x + 5) = 0, x = {}^-1 \text{ or } {}^-5$ **l** $(x - 1)(x - 5) = 0, x = 1 \text{ or } 5$
   m $(y - 3)(y - 5) = 0, y = 3 \text{ or } 5$ **n** $(y - 1)(y - 15) = 0, y = 1 \text{ or } 15$
   o $(y + 1)(y - 15) = 0, y = {}^-1 \text{ or } 15$ **p** $(k + 9)(k - 7) = 0, k = {}^-9 \text{ or } 7$
   q $(k - 7)(k - 9) = 0, k = 7 \text{ or } 9$ **r** $(k + 63)(k - 1) = 0, k = {}^-63 \text{ or } 1$

2 **a** $(d - 2)(d + 5), d = 2 \text{ or } {}^-5$ **b** $(d + 1)(d - 6), d = {}^-1 \text{ or } 6$
   c $(d + 8)(d - 5), d = {}^-8 \text{ or } 5$ **d** $(d - 11)(d + 3), d = 11 \text{ or } {}^-3$
   e $(d - 11)(d - 3), d = 11 \text{ or } 3$

3 **a** $(x - 4)^2$ **b** The two factors are the same.

4 **a** $c^2 + 5c - 6 = 0, (c + 6)(c - 1) = 0, c = {}^-6 \text{ or } 1$
   b $y^2 + 6y - 7 = 0, (y + 7)(y - 1) = 0, y = {}^-7 \text{ or } 1$
   c $p^2 - 6p + 9 = 0, (p - 3)(p - 3) = 0, p = 3 \text{ (twice)}$
   d $x^2 - 3x - 10 = 0, (x - 5)(x + 2) = 0, x = 5 \text{ or } {}^-2$
   e $w^2 - 4w + 4 = 0, (w - 2)(w - 2) = 0, w = 2 \text{ (twice)}$
   f $b^2 - 15b + 14 = 0, (b - 14)(b - 1) = 0, b = 14 \text{ or } 1$
   g $k^2 - k - 70 = 0, (k + 7)(k - 10) = 0, k = {}^-7 \text{ or } 10$
   h $v^2 + 6v + 9 = 0, (v + 3)(v + 3) = 0, v = {}^-3 \text{ (twice)}$
   i $t^2 - 3t - 28 = 0, (t - 7)(t + 4) = 0, t = 7 \text{ or } {}^-4$
   j $y^2 - 7y - 18 = 0, (y - 9)(y + 2) = 0, y = 9 \text{ or } {}^-2$
   k $g^2 - 8g + 16 = 0, (g - 4)(g - 4) = 0, g = 4 \text{ (twice)}$
   l $a^2 + 4a - 21 = 0, (a + 7)(a - 3) = 0, a = {}^-7 \text{ or } 3$
   m $p^2 - 3p - 4 = 0, (p - 4)(p + 1) = 0, p = 4 \text{ or } {}^-1$
   n $w^2 - 2w - 24 = 0, (w - 6)(w + 4) = 0, w = 6 \text{ or } {}^-4$
   o $u^2 + 7u + 10 = 0, (u + 2)(u + 5) = 0, u = {}^-2 \text{ or } {}^-5$

Exercise 22.2

1 **a** $^-3.4 \text{ or } 1.4$ **b** $^-1.6 \text{ or } 3.6$ **c** $^-4.2 \text{ or } 1.2$ **d** $^-4.4 \text{ or } 0.4$
   e $^-0.3 \text{ or } 3.3$ **f** $^-5.5 \text{ or } 0.5$ **g** $^-1.3 \text{ or } 2.3$ **h** $^-6.3 \text{ or } 0.3$
   i $^-5.5 \text{ or } 1.5$

2 $^-5.3 \text{ or } 1.3$

3 $^-0.4 \text{ or } 5.4$

4 **a** $^-1.35 \text{ or } 1.85$ **b** $^-1.7 \text{ or } 1.4$ **c** $^-0.85 \text{ or } 2.35$

5 It can be solved by factorisation

Exercise 22.3

1 **a**

x	$^-5$	$^-4$	$^-3$	$^-2$	$^-1$	0	1	2	3	4	5
y	25	16	9	4	1	0	1	4	9	16	25

b 25, 0

2 **a**

x	$^-4$	$^-3$	$^-2$	$^-1$	0	1	2	3	4
x^2	16	9	4	1	0	1	4	9	16
$+1$	1	1	1	1	1	1	1	1	1
y	17	10	5	2	1	2	5	10	17

c Same shape but $x^2 + 1 = 0$ is 1 unit higher.

4 a $y = x^2 + 1.5$

5 a

x	-3	-2	-1	0	1	2	3
x^2	9	4	1	0	1	4	9
$+x$	-3	-2	-1	0	1	2	3
y	6	2	0	0	2	6	12

6 a

x	-3	-2	-1	0	1	2	3
x^2	9	4	1	0	1	4	9
$+2x$	-6	-4	-2	0	2	4	6
y	3	0	-1	0	3	8	15

7 a $x = 0$ **b** $x = -1$ **c** $x = 0$ **d** $x = -1.5$

8 a

x	-4	-3	-2	-1	0	1	2	3
x^2	16	9	4	1	0	1	4	9
$+2x$	-8	-6	-4	-2	0	2	4	6
-3	-3	-3	-3	-3	-3	-3	-3	-3
y	5	0	-3	-4	-3	0	5	12

9 a

x	-4	-3	-2	-1	0	1	2	3
x^2	16	9	4	1	0	1	4	9
$-2x$	8	6	4	2	0	-2	-4	-6
-3	-3	-3	-3	-3	-3	-3	-3	-3
y	21	12	5	0	-3	-4	-3	0

10 a

x	-4	-3	-2	-1	0	1	2	3	4
x^2	16	9	4	1	0	1	4	9	16
$+3x$	-12	-9	-6	-3	0	3	6	9	12
-4	-4	-4	-4	-4	-4	-4	-4	-4	-4
y	0	-4	-6	-6	-4	0	6	14	24

d $(-4, 0), (1, 0)$

11 a

x	-4	-3	-2	-1	0	1	2	3	4
x^2	16	9	4	1	0	1	4	9	16
$-3x$	12	9	6	3	0	-3	-6	-9	-12
-4	-4	-4	-4	-4	-4	-4	-4	-4	-4
y	24	14	6	0	-4	-6	6	4	0

c $(-1, 0), (4, 0)$

12 a

x	-5	-4	-3	-2	-1	0	1	2	3	4	5
x^2	25	16	9	4	1	0	1	4	9	16	25
$-3x$	15	12	9	6	3	0	-3	-6	-9	-12	-15
-10	-10	-10	-10	-10	-10	-10	-10	-10	-10	-10	-10
y	30	18	8	0	-6	-10	-12	-12	-10	-6	0

b $(-2, 0), (5, 0)$

13 a

x	-5	-4	-3	-2	-1	0	1	2	3	4	5
x^2	25	16	9	4	1	0	1	4	9	16	25
$+2x$	-10	-8	-6	-4	-2	0	2	4	6	8	10
-8	-8	-8	-8	-8	-8	-8	-8	-8	-8	-8	-8
y	7	0	-5	-8	-9	-8	-5	0	7	16	27

c $(-4, 0), (2, 0)$

Exercise 22.4

1 a $y = 10$
b The solutions will be where they intersect
c $x = 2.2$ and $x = -7.2$
3 a Not enough of the graph has been plotted
b $x = 3.4$ **c** less than
4 a $x = 2$ and $x = 3$ **c** $x = 0.4$ and $x = 4.6$ **d** $x = -0.7$ and $x = 5.7$
5 a **i** $x = -1$ and $x = 5$ **ii** $x = -2.5$ and $x = 6.5$ **b** $x = 2$
6 b $x = 1$ and $x = 3$
7 c $x = 1$ and $x = 4$
8 b $x = -1$ and $x = 4$

Exercise 22.5

1 b $x = -1.6$
2 b $x = 1.3$
3 b from top to bottom: $y = x^3 + 10$, $y = x^3 + 6$, $y = x^3$, $y = x^3 - 5$
4 c $x = 0.7$

Exercise 22.6

1 a 2.62 **b** 3.42 **c** 2.67 **d** 2.24 **e** 2.88 **f** 3.32
g 1.16 **h** 1.51 **i** 4.94 **j** 2.76 **k** 2.17 **l** 9.46

End points

A1 A quadratic is an equation with one variable and where the highest power of the variable is two.
B1 a $(x + 5)(x - 2) = 0$, $x = -5$, $x = 2$
b $(x + 4)(x + 5) = 0$, $x = -4$, $x = -5$
c $(x - 2)(x - 8) = 0$, $x = 2$, $x = 8$
d $(x - 3)(x - 3) = 0$, $x = 3$ (twice)
C1 a $x^2 + 5x - 24 = 0$ **b** $p^2 - 3p - 2 = 0$
c $c^2 - 5c + 2 = 0$ **d** $y^2 - 6y - 2 = 0$
D1 a $v = -0.8, 3.8$ **b** $b = 0.6, 3.4$
c $p = -3.6, -1.4$ **d** $w = -7.7, 7.7$

E1

x	-4	-3	-2	-1	0	1	2	3	4
x^2	16	9	4	1	0	1	4	9	16
+2	+2	+2	+2	+2	+2	+2	+2	+2	+2
y	18	11	6	3	2	3	6	11	18

E2 There is symmetry about $x = 0$

F1

x	-4	-3	-2	-1	0	1	2	3
x^2	16	9	4	1	0	1	4	9
$+2x$	-8	-6	-4	-2	0	2	4	6
-3	-3	-3	-3	-3	-3	-3	-3	-3
y	5	0	-3	-4	-3	0	5	12

G1 $y = x^2 + 1$ doesn't cross the x-axis
H1 $x = 1, -5$
H2 $x = -6.4, 2.4$
H3 $x^2 + 4x - 5 = -8$
H4 $y = -9$ only touches the graph in one place.

SECTION 23 Processing data

Starting points

D1 **a** Q **b** R **c** P

Exercise 23.1

1 **a** not clear, needs specific times

2 **a** The question seems to expect the answer 'yes'.

3 **a** It has more than one meaning

4 **a** 1: Leading
   2: Not clear
   3: Ambiguous
   4: There is no option for between £1.50 and £2.50.

Exercise 23.3

1 **c** Negative

2 **c** Positive

Exercise 23.4

1 Moderate negative correlation

2 Strong negative correlation

3 Weak positive correlation

Exercise 23.5

1 **a** 50 miles **b** 80 miles

2 **a** £75 000 **b** £85 000

3 **a** 15 miles **b** £110 000

5 **b** Moderate positive correlation **c i** 272 days **ii** 3.8 kg

6 No, pregnancy doesn't usually last much longer than 9 months.

Exercise 23.6

1 **a** The second drop is a quarter of the area of the first one.

   b The vertical scale doesn't start at zero.

2 The tall rocket isn't three times the height of the small rocket.

3 **a** The vertical scale doesn't start at zero.

   b To make them look like they are growing faster than they are.

4 **a** The area of coin B will be 4 times that of coin A.

End points

A1 **a** Not clear **b** Ambiguous **c** Leading

C1 **d** Strong negative correlation

D1 **a** 33 mpg **b** 2 litres

D2 **a** 18 mpg **b** 4.8 litres

E1 A: The vertical scale doesn't start at zero.

   B: The areas of the flames are misleading.

SECTION 24 Trigonometry and Pythagoras' rule

Starting points

A1 **a i** 060° **ii** 240° **iii** 240° **b** 205° **c** 20 km

B1 **a ii** 90° **iii** 20 km **iv** 30 km **b** 36.1 km

C1 56°

C2 **a** 26 km **b** 15 km

Exercise 24.1

1 **a** trapezium **b i** 1 m **ii** 3.16 m

2 23.2 m

3 **a** 25.5 m² **b** £1.22

4 **a** isosceles **b i** 72° **ii** 54°

5 **a** 54° **b ii** 1.5 m **c i** 2.55 m **i** 2.06 m

6 **a** 3.10 m² **b** 15.5 m²

7 £15.20

8 **a** 5.83 m **b** A

Exercise 24.2

1 **a** 38.21°, 3.64 cm, 8.27 cm, 5.07 cm, 4.52 cm, 46.86°

Exercise 24.3

1 **a** 6.30 cm **b** 9.85 cm **c** 31.0 cm² **d** 25.9 cm

2 **a i** 30° **ii** 6.93 cm **b i** 26.6° **ii** 8.94 cm **c** 23.9 cm **d** 29.9 cm²

3 **a i** 9.80 cm **ii** 10.6 cm **b i** 24.1° **ii** 22.2°

   c i They get longer. **ii** They get smaller.

Exercise 24.4

1 **a i** 15 m **ii** 11 m **b i** 7 m **ii** 26 m **c i** 22 m **ii** 37 m **iii** 43 m

Exercise 24.5

1 **b** 252° **c i** 278° **ii** 082° **d** 9.4 km

2 **b** 258° **c** 5.7 km **d** 10.3 km **e** 10.5 km

3 **b** 7.6 km **c** 301° **d** 7.5 km **e** 5.3 km

4 **b** 255° **c** 4.9 km **d** 7.5 km **e** 9.9 km **f** 7.1 km

5 **b** 302° **c i** 277 km **d i** 90° **ii** 277 km **e i** 41° **ii** 073°

6 **a i** 178 km **ii** 111 km **b** 32° **c i** 95 km **ii** 153 km **d** 83 km

Exercise 24.6

1 **a i** 2 km **ii** 3 km **b i** 3.61 km **ii** 056° **c i** 3 km **ii** 3 km
   d i 4.24 km **ii** 135°

2 **b i** 140° **ii** 090° **iii** 180° **c i** 59.75 km **ii** 50.14 km

3 **a i** 14.44 km **ii** 19.17 km **b** 37.17 km
   c i 39.88 km **ii** 159° **d** 339°

4 **b i** 025° **ii** 18.79 km

Exercise 24.7

1 **a** 40.2° **b** 2.79 m **c** 3.82 m **d** 3.01 m **e** 5.79 m

2 **a** 2.63 m **b** 1.83 m **c** 4.28 m **d** 55.0° **e** 47° **f** 1.40 m

End points

A1 A 19.7 cm², 19.5 cm, B 6.0 cm², 11.6 cm

B1 **a i** 38° **ii** 256 m **b i** 93.0 m **ii** 39.2°

C1 **a** 6.9 km **b** 6.1km **c** 6.8 km **d** 65° **e** 3.3 km **f** 2.7 km

C2 **a** 027° **b i** 15.1 km **ii** 7.7 km **c** 27.3 km **d i** 056° **ii** 236°

D1 **a** 5.6 m **b** 16.5 m **c** 4.8 m

D2 **a** 130 feet **b** 22.6°

SECTION 25 Circle theorems

Starting points

A1 **b** The distances are the same. **c** yes **d** both 90°
   e both right angles **f** yes **h** both the same **i** yes **j** yes

A2 **b** angle AOB = 120°

A3 **b** OA = OB = OC = 3.6 cm

Exercise 25.1

1 **a** angles ADC and AEC **b** angles CAD and CED
   c 58° **d** 41° **e** 41°

2 **a** 74° **b** 37° **c** 90° **d** 53° **e** 37°

3 **a** angle EFD **b** 284° **c** 38° **d** 52°

4 **a** 41° **b** 49°
   c If a tangent (FC) and a chord (EF) meet, the angle between them (EFC) is always equal to the angle in the opposite segment (FGE).
   d 61°

Exercise 25.2

1 Opposite angles are each 90° and so add up to 180°.

2 Not unless it is a rectangle. One pair of opposite angles will sum to more than 180°, the other pair to less than 180°.

3 square

4 **a** $v + x = 180° = 12 \times 15°$ **b** 78°, 40°, 102°, 140°

6 102°

End points

A1   **a**   54°   **b**   61°

A2   **a**   angle EBD   **b**   angle CAB

B1   **a**   126°   **b**   54°

C1   62°

D1   54°

D2   72°

D3   36°

E1   The four vertices of the quadrilateral lie on the circumference of the same circle.
The angles at opposite vertices are supplementary.

E2   **a** $3k$ must be opposite $6k$, $4k$ must be opposite $5k$.
    b $9k = 180$, $k = 20°$

SKILLS BREAK 4A

1   UK side

2   **a**   2.69 m   **b**   4.18 m

3   November

4   73 minutes

5   No

6   9 minutes

7   160 km/h

8   5:27 pm

9   **a**   1:57 pm   **b**   Calais-Frethun, Paris Nord

10   11:43 am

11   12:44 pm

12   a 1:50 pm   **b**   6 : 34 pm

13   London

19   July 1995

20   23 383

22   **a** 1.5×10^5 **b**   23513

23   27355

24 5.2×10^6

25   35

26   No, the number of second-class seats cannot be divided exactly by the number of second-class carriages.

27   50

28   **a**   140 km/h   **b**   115 km/h

29   105 : 292

30   0.788%

31   19 minutes

32   **a**   6%   **b**   18%

33   883

SKILLS BREAK 4B

1   £9.59

2   £49.41

3   **a**   60 tiles (least amount of boxes)   **b**   £18.73

4   **a**   6   **b**   2   **c**   £20.97

5   28p

6   **b**   1120   **c**   1176

8   **a**   hexagon   **b**   120°   **c**   10 cm

9   **a**   He assumes that the tiles will fit exactly.
    b $18 \times 11 = 198$ **c**   £332.48

10   **a**   3.5 m   **b**   2.25 m

11   2.5 litre

12   **a**   4.11 m, 3.61 m   **b**   £43.47

13   £7.45

14   Forret

15   **a**   £16.59   **b**   £2.90

16   £1.65

17   £29.96

18   Floral design £6.31 per roll, £5.72 each for 12 rolls or more, Antique embossed £9.19 per roll.

IN FOCUS 1

1   B, C, D, E, H

2   **a** $^-6 \times 3 = ^-18$ **b** $^-4 \times ^-2 = ^-8$ **c** $^-14 \div 7 = ^-2$ **d** $^-18 \div 6 = ^-3$
    e $5 \times ^-4 = ^-20$ **f** $^-5 \times ^-7 = 35$ **g** $10 \div ^-2 = ^-5$ **h** $8 \times ^-3 = ^-24$
    i $^-15 \div ^-3 = 5$ **j** $^-4 \times 7 = ^-28$ **k** $^-16 \div ^-2 = 8$ **l** $^-36 \div 9 = ^-4$

3   **a**   8.54   **b**   9.71   **c**   80.40   **d**   3.62

4   **a**   512   **b**   19.683   **c**   0.125   **d** $^-64$ **e** $^-5.832$ **f** $^-0.027$
    g   30   **h**   9   **i** $^-1$ **j** $^-1.2$

5   **a**   0.5   **b**   0.1   **c**   2.5   **d** $1.\dot{6}$ **e**   3   **f**   6.25   **g**   100   **h**   0.2   **i** $0.08\dot{3}$
    j   4.5   **k** $0.0\dot{9}$ **l** $3.\dot{6}$

6   **a** $1\frac{1}{2}$ **b** $1\frac{1}{4}$ **c** $\frac{6}{11}$ **d** $\frac{8}{13}$ **e** $\frac{3}{4}$ **f** $\frac{5}{12}$ **g** $\frac{5}{7}$ **h** $\frac{5}{8}$

7

5^{-3}	5^{-2}	5^{-1}	5^0	5^1	5^2	5^3
$\frac{1}{125}$	$\frac{1}{25}$	$\frac{1}{5}$	1	5	25	125

8

6^{-3}	6^{-2}	6^{-1}	6^0	6^1	6^2	6^3
$\frac{1}{216}$	$\frac{1}{36}$	$\frac{1}{6}$	1	6	36	216

9 $7^0 = 1$

10   **a** $\frac{1}{4}$ **b** $\frac{1}{8}$ **c** $\frac{1}{9}$ **d** $\frac{1}{36}$ **e** $\frac{1}{1000}$ **f** $\frac{1}{10000}$
    g   1   **h** $\frac{1}{49}$ **i** $\frac{1}{64}$ **j** $\frac{1}{64}$ **k** $\frac{1}{81}$ **l**   1

11   **a**   33   **b**   12.5   **c**   15.75   **d**   0.027   **e**   104   **f**   180   **g**   72.9   **h**   6200

12   **a**   0.2   **b**   0.25   **c**   0.125   **d**   0.01   **e**   0.015625   **f** $0.\dot{1}$
    g   0.015626   **h**   0.008   **i**   0.001   **j**   0.01   **k** $0.02\dot{7}$ **l**   0.00001

13   **a** 6.17×10^6 **b** 9.2×10^{10} **c** 3.07×10^5
    d 2.5×10^{-5} **e** 2.603×10^{-4} **f** 1.0×10^{-7}

14   **a**   450 000 000   **b**   3 606 000 000 000   **c**   0.000 000 001 24
    d   70 000 000   **e**   0.000 000 000 051   **f**   0.000 006 104 7

15   **a** 1.305×10^{11} **b** 1.2384×10^{-3} **c** 8.7×10^{-8} **d** 8.55×10^{10}
    e 1.904×10^7 **f** 7.3×10^{-14} **g** 3.47×10^{15} **h** 4.1×10^9

16   **a** 4^{-4} **b** 3^6 **c** 5^4 **d** 2^8 **e** 5^7 **f** 7^{-6} **g** 4^{-9} **h** 6^5

17   **a** $3^8 \times 3^{-2} = 3^6$ **b** $6^5 \div 6^3 = 6^2$ **c** $2^{-4} \div 2^3 = 2^{-7}$ **d** $8^5 \times 8^{-3} = 8^2$
    e $3^{-2} \div 3^{-7} = 3^5$ **f** $7^{-5} \times 7^{-3} = 7^{-8}$

18   **a** 2×10^3 **b** 4×10^5 **c** 7×10^{-8} **d** 5×10^{-7}

19   **a** $\frac{11}{15}$ **b** $1\frac{7}{30}$ **c** $1\frac{7}{12}$ **d** $\frac{1}{15}$ **e** $\frac{7}{20}$ **f** $\frac{1}{2}$ **g** $4\frac{5}{6}$ **h** $2\frac{1}{2}$ **i** $4\frac{3}{10}$ **j** $\frac{2}{15}$

20   **a** $\frac{2}{15}$ **b** $\frac{1}{4}$ **c** $\frac{1}{3}$ **d** $\frac{8}{15}$ **e** $\frac{2}{5}$ **f** $1\frac{1}{9}$ **g** $\frac{18}{25}$
    h $1\frac{7}{18}$ **i** $4\frac{1}{2}$ **j** $\frac{3}{10}$ **k** $1\frac{9}{10}$ **l** $3\frac{4}{5}$

21   **a** $2\frac{1}{4}$ **b** $6\frac{2}{3}$ **c**   4   **c**   24   **e**   16   **f** $7\frac{1}{2}$
    g $3\frac{1}{3}$ **h**   38   **i** $7\frac{1}{5}$ **j**   3   **k**   20   **l** $6\frac{2}{5}$

22   **a**   36   **b**   24   **c**   240   **d**   6300   **e**   700

23   **a**   9   **b**   14   **c**   5   **d**   12   **e**   7

IN FOCUS 2

1   **b** $m = 2n + 1$, $m = 4n$, $m = 12n - 8$
    d   21 matches, 40 matches, 112 matches

2   **a** $20 \rightarrow 23$
       $n \rightarrow n + 3$
    b $40 \rightarrow 79$
       $t \rightarrow 2t - 1$
    c $2 \rightarrow 11$
       $3 \rightarrow 15$
       $4 \rightarrow 19$
       $25 \rightarrow 103$
    d $100 \rightarrow 902$ $p \rightarrow 9p + 2$

3 **a** $n \rightarrow 9n - 5$

4 $25 \rightarrow 629$ $n \rightarrow n^2 + 4$

5 **a** **A** $3n - 2$ **B** $3n + 2$ **C** $6n + 1$ **D** $5n - 1$ **E** $5n + 2$
   F $4n - 2$ **G** $6n - 3$ **H** $8n + 1$
  b **A** 88 **b** 92 **c** 181 **d** 149 **e** 152 **f** 118 **g** 177 **h** 241

6 **a** $5p + 10$ **b** $8n - 24$ **c** $4c + 20$ **d** $6r + 2$ **e** $35 + 5x$ **f** $14s - 21$
  g $3 - 15t$ **h** $6m - 12$ **i** $100k - 20$ **j** $32 - 8z$ **k** $50 - 20t$ **l** $4a + 4b$
  m $6g + 3h$ **n** $5x - 5y$ **o** $21c - 14d$

IN FOCUS 3

1 **a** 47°, 133°, 47°, 133° **b** 51.5°, 128.5°, 51.5°, 128.5°
  c 11°, 11°, 169°, 11°

2 **a** 34°, 73°, 73° **b** isosceles
  c **i** 34°, 146°, 34°, 146° **ii** 73°, 107°, 73°, 107°
  d All sides are of equal length; opposite sides are parallel.
  e Only pairs of opposite sides are equal in length.
  f **i** 90° **ii** 73° **iii** 73° **iv** 17° **v** 17°

3 **a** For an n-sided polygon, the sum of the exterior angles is $360° \div n$.

4 $360° \div 8 = 45°$

5 $a = 73°, b = 46°, c = 92°, d = 120°, e = 17°, f = 72°, g = 198°, h = 275°,$
  $i = 120°, j = 60°, k = 60°, l = 60°, m = 36°$

IN FOCUS 4

1 A: 4, B: $\frac{3}{2}$, C: $^-2$, D: 1

3 **a** $y = 2x + 1$

x	$^-1$	0	1	2	3	4
y	$^-1$	1	3	5	7	9

$y = 3x - 2$

x	$^-1$	0	1	2	3	4
y	$^-5$	$^-2$	1	4	7	10

$y = x - 2$

x	$^-1$	0	1	2	3	4
y	$^-3$	$^-2$	$^-1$	0	1	2

$y = 2x - 3$

x	$^-1$	0	1	2	3	4
y	$^-5$	$^-3$	$^-1$	1	3	5

$y = x + 2$

x	$^-1$	0	1	2	3	4
y	1	2	3	4	5	6

$y = x$

x	$^-1$	0	1	2	3	4
y	$^-1$	0	1	2	3	4

4 **a** 4, $^-3$ **b** 2, 1 **c** $^-3$, 2 **d** 1, 0 **e** $^-1$, 0 **f** 5, 2 **g** $^-2$, $^-1$ **h** 1, 1
  i 1.5, 2 **j** 1, 0.5 **k** $^-0.5$, 3 **l** 0, 1   **m** $\frac{1}{2}$, 0 **n** $\frac{3}{4}$, $^-1$ **o** 1, $\frac{^-3}{5}$
  p $^-1$, 1.6   **q** 1, 1.2 **r** $^-3$, 0

5 A: $y = 2x - 8$, B: $y = 3x + 4$, C: $y = \frac{3}{4}x - 1$,
  D: $y = 3 - 5x$, E: $y = ^-7x$

6 **a** $^-\frac{1}{2}$ **b** $^-\frac{1}{3}$ **c** $\frac{1}{3}$ **d** $^-1$ **e** 2 **f** $^-\frac{1}{5}$

7 **a** $y = 2x - 4$ **b** $y = 2x + 3$ **c** $y = 2x - 3$ **d** $y = 2x + \frac{1}{2}$
  e $y = \frac{2}{3}x + 2$ **f** $y = \frac{5}{2}x + \frac{7}{2}$ **g** $y = \frac{3}{4}x + \frac{1}{4}$ **h** $y = 2x - 1$

11 A: $y = \frac{4}{3}x + 1$, B: $y = 3x - 2$, C: $y = \frac{1}{2}x + 2$,
  D: $y = \frac{2}{3}x$, E: $y = \frac{1}{2}x - 1$, F: $y = ^-2$

12 a, b, c, d, e, f, g

IN FOCUS 5

1 Children in Care: 222°, 84°, 36°, 18°
  Child Safety: 234°, 72°, 36°, 18°

2 Chris: 8, Dani: 7, Sam: 4

3 **b** Bella: 6, Emily: 7, Gavin: 7.5

4 Bella: 5.4, Emily: 6.7, Gavin: 7.2

5 Asif: mode = 8, range = 6, Suki: mode = 5, range = 4

6 Kurt: median = 5, range = 6, Jean: median = 7, range = 5

7 Sally: mean = 5.9, range = 7, Gavin: mean = 6.4, range = 7

8 Peggy: mean = 5.7, median = 6, range = 6
  Toby: mean = 5.5, median = 5.5, range = 6
  Sue: mean = 6.2, median = 7, range = 7

9 **a** P: 36, Q: 38, R: 38.5
  b P: 35, 38, Q: 35.5, 39, R: 37, 40

IN FOCUS 6

1 39.4 cm², 96.4 cm², 12.6 cm², 77.2 cm², 109.6 cm²

2 **a** 3000 mm² **b** 154 905 mm² **c** 161 345 mm²

3 10.2 cm

4 **a** 20.4 cm, 33.2 cm² **b** 30.2 cm, 72.4 cm² **c** 31.9 cm, 60.4 cm²

5 26.25π

6 **a** 73.5 cm² **b** 35.8 cm

7 **a** 126.6 mm **b** 12 655 mm² **c** 395 **d** 5 251 275 mm²

IN FOCUS 7

1 **a** $6k$ **b** $6l + m$ **c** $5n - 7p$ **d** $4q + 6r$ **e** s **f** $9 + 5u + 7v$
  g $6w + 2x$ **h** $11 - 3y$ **i** $5z + 2$

2 **a & b**

31	18	35
32	28	24
21	41	25

  c no **e** Change bottom middle to $5g + h$

3 **a** $x = 5$ **b** $x = \frac{9}{2}$ **c** $x = \frac{19}{5}$ **d** $x = ^-2$ **e** $x = 4$ **f** $x = ^-3$ **g** $x = 2$
  h $x = \frac{3}{2}$ **i** $x = \frac{7}{2}$ **j** $x = ^-1$ **k** $x = 6$ **l** $x = 5$ **m** $x = -\frac{5}{2}$ **n** $x = 9$
  o $x = ^-1$ **p** $x = 2$

4 **a** A: 56 cm, B: 54 cm, C: 106 cm, D: 56 cm
  b A: $8p$ cm, B: $(6p + 12)$ cm, C: $(16p - 6)$ cm, D: $8p$ cm
  c 3 cm **d** 2.5 cm **e** 1.5 cm **f** 6 cm
  g 0.75 cm **h** 1.8 cm **i** They are both $8p$ cm

5 A $2n - 3 = 12 - n$, $n = 5$
  B $2(n + 1) = 3(n - 5)$, $n = 17$

6 A $x = 7, y = 1$ B $x = 1.5, y = 3$
  C $x = ^-2, y = 5$ D $x = 2.8, y = 6.8$

7 A $x = 1.5, y = 3$ B $x = 12, y = 1$
  C $x = 10, y = ^-2$ D $x = ^-1, y = 2.5$
  E $x = 2, y = 12$ F $x = 4.5, y = 2$
  G $x = 5, y = ^-1$ H $x = 1, y = 1$
  I $x = 3, y = 2.5$ J $x = 1, y = ^-5$

IN FOCUS 8

1 **a** 5.67 **b** 12.7 **c** 2140 **d** 534.69 **e** 35 **f** 3000 **g** 13 468.28
  h 0.07 **i** 60 **j** 60.0 **k** 9000 **l** 68 **m** 56.290

2 **a** 56.8 **b** 16 400 **c** 2.5 **d** 50 **e** 15.78 **f** 730 000 **g** 90
  h 93 748 000 **i** 564.2 **j** 200 **k** 6.8 **l** 15.68 **m** 673 500 **n** 0.036
  o 0.0280 **p** 3.00

3 **a** 40 000 **b** 200 **c** 54 000 **d** 250 000 **e** 350 **f** 7.5 **g** 120 000
  h 120 **i** 5000 **j** 150 **k** 350 **l** 3500 **m** 350

4 **a** 1200, 1216.9 **b** 20 000, 18 218.8 **c** 2000, 2381.721
  d 1000, 172.232 **e** 40 000, 32 421.57 **f** 0.15, 0.1863
  g 120, 110.88 **h** 1000, 965.0772

5 **a** 12 000 m² (400×30) **b** 200 m² (20×10)
  c 2 100 000 m² (7000×300)

IN FOCUS 9

1 **a** $\frac{1}{6}$ **b** $\frac{1}{2}$ **c** $\frac{1}{3}$ **d** $\frac{1}{2}$

2 **a** $\frac{2}{5}$ **b** $\frac{3}{10}$ **c** $\frac{1}{2}$ **d** $\frac{1}{10}$ **e** $\frac{9}{10}$ **f** 0 **g** $\frac{7}{10}$

3 probability of not B, for example

4

	1	2	3	4	5	6	7	8
1	1, 1	1, 2	1, 3	1, 4	1, 5	1, 6	1, 7	1, 8
2	2, 1	2, 2	2, 3	2, 4	2, 5	2, 6	2, 7	2, 8
3	3, 1	3, 2	3, 3	3, 4	3, 5	3, 6	3, 7	3, 8
4	4, 1	4, 2	4, 3	4, 4	4, 5	4, 6	4, 7	4, 8

5 **a** $\frac{11}{32}$ **b** $\frac{1}{8}$ **c** $\frac{1}{8}$ **d** $\frac{1}{32}$ **e** $\frac{1}{16}$ **f** $\frac{1}{4}$ **g** $\frac{1}{4}$ **h** $\frac{5}{16}$ **i** $\frac{9}{16}$ **j** $\frac{11}{32}$ **k** $\frac{1}{8}$ **l** $\frac{7}{8}$

7 **a** $\frac{1}{8}$ **b** $\frac{3}{4}$ **c** $\frac{1}{2}$ **d** $\frac{1}{8}$ **e** $\frac{1}{4}$ **f** 0

8 **a** $\frac{7}{15}$ **b** $\frac{1}{30}$ **c** $\frac{2}{15}$ **d** $\frac{7}{30}$ **e** $\frac{1}{15}$ **f** $\frac{1}{15}$

9 **a** PQRS, PQSR, PRQS, PRSQ, PSQR, PSRQ, QPRS, QPSR, QRPS, QRSP, QSPR, QSRP, RPQS, RPSQ, RQPS, RQSP, RSPQ, RSQP, SPQR, SPRQ, SQPR, SQRP, SRPQ, SRQP

10 **a** PQ, PR, PS, PT, QR, QS, QT, RS, RT, ST

IN FOCUS 10

1 **a** $(2b + 3.8)$m **b** $(2a + 4b + 17.4)$m **c** 37.4 m

2 **a** $(8p + 20)$cm² **b** 71.2 cm² **c** 4.25 cm

3 A and D, B and I, C and H

4 **a** $3a + 18$ **b** $5z + 30$ **c** $3n - 12$ **d** $6n + 12$ **e** $x^2 + 2x$ **f** $4p^2 - 12p$ **g** $3b^2 + 9b$ **h** $2a^2 - 8a$ **i** $12p^2 - 16p$

5 **a** $(w + 8)$cm **b** $(4w + 16)$cm **c** 14 cm

6 300 cm²

7 **a** $48pq$ **b** $9y^2$ **c** m^2n^2 **d** $4p^2q$ **e** $8m^2n$ **f** $18a^2b$ **g** a^3g^3 **h** $20p^3$ **i** $28a^3b$

8 **a** $24a - 18b$ **b** $pn - p^2$ **c** $s^2 + st$ **d** $6mn - 6m^2$ **e** $5x^2 + 5xy$ **f** $4u^2 + 3u$ **g** $5pq + 3pr$ **h** $12mn + 21n^2$ **i** $27ab + 9a^2$

9 D

10 **a** $7m + 2n$ **b** $21a^2 - a$ **c** $2b$ **d** $2xy + 8x - 2y$ **e** $4p + 3q$ **f** $8p^2 + 5q^2 + 2pq$ **g** $4m^2 + 2mn - 6n$ **h** $2x - 2xy$

11 **a** $56a - 5b$ **b** $19x + 13y$ **c** $16x^2 + 4x$ **d** $30x^2 + 13xy$

12 A: $2c + 3$, B: $3 + 4b$

13 **a** $p(7 + 3q)$ **b** $m(1 + 3n)$ **c** $4q(2p + 1)$ **d** $3y(1 - 4x)$ **e** $2b(b + 5)$ **f** $6b(3a + 4c)$ **g** $xy(x - 2)$ **h** $3ab(3a + 2b)$ **i** $3mn(4m - 3)$

IN FOCUS 12

1

Dye	Blue	Yellow	Total
Grass	120 ml	360 ml	480 ml
Lime	450 ml	1050 ml	1500 ml
Pea	200 ml	360 ml	560 ml
Lime	75 ml	175 ml	250 ml
Pea	600 ml	1080 ml	1680 ml
Grass	170 ml	510 ml	680 ml
Lime	645 ml	1505 ml	2150 ml
Pea	275 ml	495 ml	770 ml
Grass	240 ml	720 ml	960 ml
Lime	360 ml	840 ml	1200 ml
Grass	215 ml	645 ml	860 ml
Pea	250 ml	450 ml	700 ml
Grass	320 ml	960 ml	1280 ml
Lime	267 ml	623 ml	890 ml
Pea	375 ml	675 ml	1050 ml

2 **a** £24, £60 **b** £155, £93 **c** 135 g, 60 g **d** £40, £20, £60 **e** 75 g, 100 g, 25 g **f** 80 ml, 120 ml, 160 ml **g** 120 g, 180 g, 120 g **h** £55, £55, £385 **i** 36 cm, 45 cm, 18 cm **j** 1.25 m, 0.5 m, 0.25 m **k** 0.2 kg, 0.4 kg, 0.8 kg **l** 0.3 cm, 1.8 cm, 0.6 cm

3 **a** 90 g icing sugar, 405 g margarine, 180 g plain chocolate, 405 g flour **b** 20 g icing sugar, 90 g margarine, 40 g plain chocolate, 90 g flour

4 $1 : 2, 2 : 3, 2 : 1$

5 $\frac{2}{7}, \frac{3}{4}, \frac{4}{7}, \frac{1}{4}, \frac{3}{8}$

6 **a** 1.4 **b** 7.7 cm

7 **a** 0.75 **b** 5.7 cm

8 **a** £52 770, £79 155, £105 540, £158 310 **b** They should add up to the amount won

9 **a** 10.5 cm **b** 17.5 cm **c** 183.75 cm²

10 **a** $5 : 11 : 4$ **b** 281.25 ml, 618.75 ml **c** 437.5 ml, 962.5 ml, 350 ml

IN FOCUS 13

1 **a** 5.2 m **b** 13 m **c** 24.18 m **d** 156 m **e** 566.8 m

2 604 mph

3 43.7 seconds

4 0.7

5 **a** 2.15 m/s **b** 2000 m/s **c** 0.12 m/s **d** 5.6 m/s

6 **a** 120 km/h **b** 40 km/h **c** 65.1 km/h **d** 21.6 km/h

7 **a** 1 mile per minute **b** 0.5 miles per minute **c** 1.5 miles per minute **d** 0.75 miles per minute **e** 0.42 miles per minute **f** 0.7 miles per minute

8 **a** 36 mph **b** 72 mph **c** 27 mph **d** 360 mph

9 3120 km

10 **a** 76.5 miles **b** 38.25 miles **c** 12.75 miles **d** 68 miles **e** 136 miles **f** 191.25 miles

11 **a** 30 minutes **b** 2 hours **c** 48 minutes **d** 6 hours 40 minutes **e** 1 hour 5 minutes **f** 1 hour 13 minutes **g** 1 hour 12 minutes **h** 24 minutes

12 The less steep line as the truck would travel slower than the motorcycle.

13 **a** 45 km/h **b** 38.33 km/h **c** 48 mph **d** 85.71 km/h **e** 52.5 mph **f** 83.33 km/h **g** 53.33 mph **h** 30 km/h

14 **b** 12 noon

15 **b** 6:15 pm **c** 66.7 mph **d** 60 mph

IN FOCUS 14

1 **a** 56° **b** 68° **c** 61° **d** 47° **e** 13° **f** 51°

2 **a** 4.7 cm **b** 10.5 cm **c** 5.1 cm **d** 16.0 cm **e** 1.6 cm **f** 7.4 cm **g** 1.6 cm

3 **a** 61° **b** 37° **c** 68° **d** 13° **e** 45° **f** 34°

4 **a i** 64.3° **ii** 61.2° **iii** 32.1° **b i** 38.6° **ii** 35.0° **c i** 2.93 cm **ii** 145 cm **iii** 2.49 cm **d i** 21.0 cm **ii** 4.95 cm

5 A 27.30° B 30.47° C 25.62°

6 **a** 189 mm **b** 433 mm

IN FOCUS 15

1 **a** 6 **b** $p = t - 1$ **c** 2

2 **a** $x = \frac{P}{4}$ **b** 10.25 cm

3 **a** 270 mins **b** $W = \frac{(T - 20)}{20}$ **c** 5 pounds

4 **a** 8.75 pints **b** $l = \frac{4}{7p}$ **c** 4.86 litres

5 **a** $w = \frac{4}{l}$ **b** 6.2 cm

6 **a** $w = \frac{(p - 2l)}{2}$ **b** 4.2 cm

7 **a** $k = \frac{(V + 25)}{4}$ **b** $k = \frac{y}{3} + 1$ **c** $k = 7j$ **d** $k = \frac{4}{3}w$ **e** $k = 5 - 2v$
  f $k = b + 7$ **g** $k = \frac{5}{2}(d - 1)$ **h** $k = 8h - 3$ **i** $k = p - w$ **j** $k = \frac{(20 - q)}{5}$
  k $k = 3(13 - v)$ **l** $k = \frac{(v - s)}{t}$ **m** $k = bh - a$ **n** $k = \frac{h}{(2g)}$ **o** $k = \frac{(3 - 2y)}{5}$
  p $k = \frac{(9 + 2h)}{3}$

8 **a** $p = \sqrt{A}$ **b** $p = \sqrt{\left(\frac{v}{5}\right)}$ **c** $p = \sqrt{(g - 1)}$ **d** $p = \sqrt{\left(\frac{(q + 5)}{3}\right)}$ **e** $p = \sqrt{(4j)}$
  f $p = \sqrt{\left(\frac{3F}{2}\right)}$ **g** $p = \sqrt{(2 - z)}$ **h** $p = \sqrt{\left(\frac{K}{2\pi}\right)}$ **i** $p = \sqrt{\left(\frac{(A - 1)}{2}\right)}$
  j $p = \sqrt{\left(\frac{(H + 8)}{3}\right)}$

9 **a** 80, 45, 85, 28 **b** 35, 12, 22, 4 **c** 19.25, 2.25, 4, ⁻1.25

10 **a** 35 **b** 16 **c** ±7 **d** 38 **e** 225 **f** 11 **g** 1.1 **h** $\frac{7}{10}$

11 **a** $n^2 + 7n + 10$ **b** $f^2 + 11f + 10$ **c** $w^2 + 8w + 12$ **d** $2n^2 + 7n + 3$
  e $4d^2 + 19d + 21$ **f** $4y^2 + 28y + 45$ **g** $x^2 + 3x - 18$ **h** $v^2 + v - 6$
  i $b^2 - 2b - 15$ **j** $m^2 - 8m + 7$ **k** $h^2 - 6h + 5$ **l** $t^2 - 10t + 21$
  m $6p^2 + p - 1$ **n** $15x^2 + 8x - 12$ **o** $12y^2 - 24y - 15$
  p $12g^2 - 43g + 35$

12 **a** $(x + 3)(x + 4)$ **b** $(x + 2)(x + 6)$ **c** $(x + 1)(x + 6)$ **d** $(x + 2)(x + 3)$
  e $(x + 1)(x + 12)$ **f** $(x + 2)(x + 4)$

13 **a** $(x + 1)(x + 11)$ **b** $(x + 3)(x + 5)$ **c** $(x + 2)(x + 2)$ **d** $(x + 4)(x + 5)$
  e $(x + 4)(x - 1)$ **f** $(x + 5)(x - 3)$ **g** $(x + 6)(x - 2)$ **h** $(x + 2)(x - 3)$
  i $(x - 1)(x - 10)$ **j** $(x + 2)(x - 4)$ **k** $(x - 1)(x - 2)$ **l** $(x - 2)(x - 2)$

IN FOCUS 16

1 **a** bars with heights 1, 4, 1, 4, 5, 4, 2, 3
  b bars with heights 1, 2, 2, 2, 5, 7, 5

2 **a** 0.160– **b** 0.170–

4 **a** 0.07 seconds **b** 0.06 seconds

5 **a** 0.163 **b** 0.165

6 There are the same amount of men as women.

8 **a** 3 **b** 2

9 **a** 7 **b** 8

10 **a** 0.165 **b** 0.17

11 **a** 0.027 **b** 0.023

12 The men are on average faster, but the women were closer together

13 **a** bars with heights 7, 27, 20, 3, 2
  b bars with heights 2, 10, 7, 2, 2

14 Different totals of athletes

15 **a** 27.1 years **b** 28.3 years

16 **b** 20 minutes (C), 40 minutes (D), 7 minutes (E)
  c 86.2 minutes (C), 239 minutes (D), 45.1 minutes (E)
  e 85.5 minutes (C), 238 minutes (D), 45.4 minutes (E)
  f 5 minutes (C), 15 minutes (D), 2.6 minutes (E)

17 **a** 34, 33, 29, 28, 25, 22, 24, 27, 26

IN FOCUS 17

1 **a** 63.33% **b** 25.45% **c** 283.33% **d** 62% **e** 4.67% **f** 31.25%
  g 83.67% **h** 15.21% **i** 16.62% **j** 1.04%

2 **a** 29.5 kg **b** 2310 km **c** 3702.4 miles **d** 1822.5 yards **e** £27 183.70
  f 145 062.5 tonnes **g** 1.05 cm **h** 47.25 mm **i** £38.51 **j** 419.75 ml

3 **a** 426.8 ml **b** £36.6 **c** 22.75 mm **d** 0.74 cm **e** 3968.75 tonnes
  f £10 523.04 **g** 1267.2 yards **h** 2500.08 miles **i** 2322 km
  j 24.96 kg

4 **a** £218.37 **b** £13.74 **c** £52.71 **d** £25.56 **e** £315.55 **f** £187.99
  g £159.07 **h** £10.86 **i** £58.74 **j** £1761.33

5 £155.10

6 **a** Price busters **b** They are cheaper by 25p

7 **a** £405 **b** £4216 **c** £20.40 **d** £23.40 **e** £810 **f** £2300 **g** £3.75
  h £40.50 **i** £91.60 **j** £94.50

8 **a** £95.51 **b** £582.54 **c** £38.91 **d** £13.18 **e** £7011.2 **f** £3.96
  g £578.67 **h** £1355.71 **i** £32.45 **j** £70.58

9 **a** £255.31 **b** £11.06 **c** £38.29 **d** £8.50 **e** £63.83 **f** £76.55
  g £22.97 **h** £110.63 **i** £16.16 **j** £119.14

10 **a** £100.53 **b** £15.08

11 £3 157 895

IN FOCUS 18

2 **a** shape with vertices at (⁻2, 3), (2, 1), (2, ⁻1), (⁻4, ⁻1), (⁻4, 1)
  b shape with vertices at $(4\frac{1}{2}, 0), (5\frac{1}{2}, -\frac{1}{2}), (5\frac{1}{2}, -1), (4, -1), (4, -\frac{1}{2})$.

3 **b** **i** Rotation 90° anticlockwise about (1, 1)
  ii Reflection in $x = 5$
  c **i** C **ii** F **iii** E
  d Rotation 180° about (4, ⁻1)

4 **c** **i** Rotation 180° about (4, 0)
  ii Rotation 90° clockwise about (3, 0)
  iii Translation $\begin{pmatrix} 2 \\ 2 \end{pmatrix}$
  d Rotation 180° about (1, 0)

5 **b** **i** Rotation 180° about (⁻0.5, 1.5)
  ii Translation $\begin{pmatrix} {}^{-}14 \\ 0 \end{pmatrix}$

IN FOCUS 19

1 **a** 0.08 **b** 0.78 **c** 0.03 **d** 0.11

2 They represent all the outcomes.

3 **a** 0.92 **b** 0.18 **c** 0.11

4 No, peg down isn't as likely as peg up.

5 **a** 12 **b** 117 **c** 5

7 **b** 0.08 **c** 0.48 **d** 0.52

8 **a** Relative
  b Relative
  c Theoretical
  d Relative
  e Relative
  f Theoretical
  g Theoretical
  h Relative

9 **a** 0.19
  b No two outcomes can happen at the same time

10 **b** 15 and 30 are both multiples of 3 and 5

11 **a** 0.28 **b** 0.82 **c** 0.18 **d** 0.4

12 0.67

IN FOCUS 20

1 **a** $9 > {}^{-}4$ **b** ${}^{-}8 < {}^{-}3$ **c** $4 > {}^{-}2$ **d** ${}^{-}2 < 2$ **e** ${}^{-}5.6 < {}^{-}4.23$
  f ${}^{-}8 > {}^{-}11$

2 4, 71, ⁻3.2, ⁻0.33

3 **a** 4, 5, 6, 7, 8 **b** ⁻3, ⁻2, ⁻1 **c** ⁻5, ⁻4, ⁻3, ⁻2, ⁻1, 0 **d** ⁻6, ⁻5, ⁻4
  e none **f** 24 **g** ⁻14, ⁻15 **h** 1, 2, 3, 4, 5

5 The number of waiters has to be an integer, the weights do not

7 **a** $x \geqslant 9$ **b** $t < {}^{-}7$ **c** $p \leqslant 10.5$ **d** ${}^{-}11 \leqslant k \leqslant 11$ **e** $w < 2$ **f** $q > 9.3$
  g $t \geqslant 8$ **h** $h \leqslant 2$ **i** $f < 3$ **j** ${}^{-}9 < c < 9$ **k** $g \geqslant 4.5$ **l** $x > {}^{-}3$
  m $j < 19$ **n** $d \geqslant {}^{-}2$ **o** $v < 4\frac{1}{2}$ **p** $x \leqslant {}^{-}13$ **q** $s < 0$ **r** $u \leqslant {}^{-}3$

8 **a** 0 **b** 16

11 $x + y \leqslant 7, x < 2$

IN FOCUS 21

1 A: 2, B: 5

3 **a** 3 **b** 9

4 **b** 48 cm³ **c** 88 cm²

5 X: 78.6 cm³, Y: 18.4 cm³

6 *a* = 19.8 cm, *b* = 59.0 cm

7 **a** R: 3927.8 ml, S: 1247.7 ml, T: 74.8 ml, U: 1 231 504.3 ml
b R: 3.9 litres, S: 1.2 litres, T: 0.7 litres, U: 1231.5 litres

8 **a** length **b** area **c** area **d** volume **e** length **f** area
g none **h** volume **i** length **j** area **k** none **l** volume
m area **n** volume **o** area **p** length

IN FOCUS 22

1 **a** $^-3, ^-2$ **b** $^-5, 4$ **c** $^-2, 8$ **d** $^-9, 6$ **e** $^-5, 9$ **f** $^-8, 5$ **g** 2, 5
h $^-12, 3$ **i** $^-5, ^-3$ **j** 3, 6 **k** $^-12, 6$ **l** $^-3, 10$ **m** $^-9, 4$ **n** $^-12, 4$
o 3, 9 **p** $^-11, ^-7$ **q** $^-8, 9$ **r** $^-7, 6$ **s** $^-8, ^-7$ **t** $^-5, 12$

2 **a** 3.4 or –2.4 **b** 1.5 **c** 3.2 or –1.2 **d** 1.8 **e** 1.9 or –1.4 **f** 1.6

3 **a**

x	$^-2$	$^-1$	0	1	2	3	4	5
y	12	5	0	$^-3$	$^-4$	$^-3$	0	5

5 **a**

x	0	1	2	3	4	5
y	4	0	$^-2$	$^-2$	0	4

6 **a**

x	$^-5$	$^-4$	$^-3$	$^-2$	$^-1$	0	1	2	3
y	7	0	$^-5$	$^-8$	$^-9$	$^-8$	$^-5$	0	7

c $^-4, 2$

7 **b** $^-6, 3$

8 **b** 1, 5 **d** 0.6, 5.4

9 **b** $^-7, 3$ **d** $^-6.5, 2.5$ **e** $^-7.4, 3.4$

14 **c & d** $^-0.4, 2.4$

17 1.82

IN FOCUS 23

1 Not clear, leading, ambiguous.

7 **d** Moderate positive correlation

8 **a** £7300 **b** £4700

9 **a** 350 cc **b** 680 cc

10 **a** £8900 **b** 1160 cc

11 The vertical scale doesn't start at zero

12 The area of the large video box is about twice as large as the small one

13 **a** The number of injuries reduced steadily each year
d There were 17 or 18

IN FOCUS 24

1 **a** 3.04 cm **b** 30.1° **c** 5.73 cm **d** 6.07 cm

2 A: 6.51 cm, 2.04 cm², B: 14.52 cm, 12.3 cm²

3 **a i** 4.20 m **ii** 6.02 m **b** 19°

4 **a i** 5 km **ii** 2 km **b i** 135° **ii** 068° **iii** 277°
c i 5.39 km **ii** 22° **d** 4.24 km **e** 8.06 km

5 **a** 1.1 m
b 27.9°

IN FOCUS 25

1 **b** CD is a diameter

2 **b** 68°

3 22°

4 **a** 109° **b** 33°

5 **b** 96°

6 **b** 110°

7 **b** 75°

EXAM QUESTIONS

N1 **a** 64 p **b** £34.27 **c** 35p **d** 42p per kg

N2 **a** £61.80 **b** $(1.07)^3$ 1.225

N3 **a** 345 g, 225 g, 150 g, 3 eggs **b** 177°C

N4 **a** 8.19×10^{13} **b** 2.7×10^8

N5 **a** £4.80 **b i** 0.3 **ii** £3.60 **c** 7% **d** $\frac{1}{50}$

N6 **a** The answer should be smaller than 17.8
b 40, 60, 300

N7 **a i** £360 **ii** £230.40 **b** £375

N8 **a** £34.37 **b** £558.40

N9 **b**

Pattern	1	2	3	4	5	6	7
Number of dots	5	8	11	14	17	20	23

c 12 times 3, then add 2

N10 **a** £9.40 **b** 75 **c** 37.5%

N11 **a** £20 904 **b** 70, 300

N12 **a i** 0.125 **ii** 12.5% **b** £2815.75

N13 **a** 34 590 pesetas **b** £2.39

N14 **a i** 50, 30 **ii** 75 **iii** 79.36875 **b** 4.9×10^{11}

N15 **a** Ruth £100, Ben £80 **b** 60%

N16 **a** 10°C **b** −3°C

N17 **a** 163 **b** £3 200 000

N18 £48.50

N19 **a** £75 **b** £4.90 **c** Judy £10.50, Anna £14

N20 **a** £468 **b** 40%

N21 $\frac{1}{5}$

N22 **a** 36 **b** £564

N23 **a** 24 litres **b** 18p

N24 Tracy £4000, Wayne £3200

N25 32p

N26 **a** $\frac{7}{8}$, 0.8, $\frac{3}{4}$, 70%

N27 **a** 15, 9 **b** $4n - 3$

N28 **a** 3.629, 3.649, 3.662, 3.738, 3.747, 3.751 **b** 3.63

N29 £2720

N30 **i** $\frac{7}{10}$ **ii** $\frac{7}{20}$ **iii** $\frac{9}{20}$

N31 8 pounds

N32 York, Leeds, Newcastle, Scarborough, Hull, York or opposite
direction = 288 miles

N33 **a** ±5 **b** 64

N34 150 g, 120 g, 3 eggs, 135 g, 45 ml

N35 **a** 8 **b** 300 cm² **c** 240

N36 **a** 45% **b** 60

N37 **a** £735 **b** £196

N38 $\frac{3}{8}$, $\frac{5}{12}$ or $\frac{1}{3}$ for example

N39 **a** 1920 French francs **b** £2.50

N40 **a** 9472 **b** 25.8

N41 **a** £4800 **b** £5232 **c** £900

N42 0.45

N43 **a** 252 **b** £7840

N44 **b**

Numbr of pens	2	4	6	8	10	12	14	16
Number of fences	7	12	17	22	27	32	37	42

c 77

N45 2.5125×10^{-21} g

N46 7000 kg

N47 $n + 3$

N48 **a** 4200 **b i** 98.5 cm **ii** 97.5

N49 7.49

N50 **a** 5 : 2 **b** 40 years old **c** 14 years old

A1 **a** $80x$ **b** $80r + 60t$ **c** $80g + 120 = 1080$ **d** 12

A2 **b i** ⁻4.25 **ii** ⁻1.6, 2.6

A3 **b** 1.5, 1 **c** $3y = ⁻2x + 7$

A4 **a** ⁻2.7 **b i** $(x + 6)(x − 2)$ **ii** ⁻6, 2

A5 **a i** 2, ⁻1 **ii** subtract 3 each time **b** $20 − 3n$ **c** ⁻130

A6 **a** ⁻1, 0, 1, 2 **b i** $n > ⁻2.7$ **iii** −2

A7 **a** ⁻5, ⁻3, ⁻1, 1, 3, 5 **c i** ⁻3.8 **iii** 2.4

A8 3.53

A9 **a** 2900 litres **b** 8 rooms **c** 6 rooms **d** $R = \frac{(C - 2000)}{180}$

A10 **a** ⁻3, ⁻1, 1, 3, 5, 7 **c i** 6 **ii** ⁻1.75

A11 **a** 5 **b** 4 **c** 3.2 **d** −4.5 **e** $x \leqslant 2.5$

A12 **a** $2x^2 + x − 15$ **b i** $(x + 7)(x − 1)$ **ii** ⁻7, 1

A13 **a** £19 **b** 12 **c** $h = \frac{(C - 7)}{3}$

A14 **a** $P = bh + c$

A15 $8x + 14$ **c** 22.56 m²

A16 **a** 5 **b** ⁻0.5 **c** $\frac{1}{4}$

A17 **b** 35, 48 **c** $n^2 + 2n$

A18 **a** 7.6×10^3 **b** $M = \frac{(v^2R)}{G}$

A19 **a** $4x + 3$ **b** 6.5 **b** 7

A20 **a** $10x + 3$ **b** $6x^2 + 5xy − 4y^2$

A21 0.75, ⁻0.5

A22 **a** x^8 **b** y^4 **c** $4w^2$

A23 **a** $x^2 + 2x − 15$ **b** $3a(2a − 3b)$

A24 3.9

A25 **a** $t^5 − t^6$ **b** 30 **c** $3a$

A26 **a** $12xy$ **b** $15h + 10$

A27 **a** 9 **b** 8.5 **c** ⁻7

A28 3.66

A29 **a** 2 **b** 4.5

A30 **b** $v = \frac{(Pt)}{m}$

A31 **a** $x^2 + 2x − 24$ **b** 2, 3

A32 **a** 15.6°C **b** $F = \frac{9}{5}C + 32$

A33 2, ⁻1

A34 **a** 6.75 **b** 9

A35 **a** 6.63 **b** $u = \sqrt{(v^2 − 2as)}$

A36 $(x + 4)(x − 3) = 78$ **b ii** ⁻10 cm, 9 cm **iii** 13 cm, 6 cm

A37 **a** $C = 15n + 4$ **b** $n = \frac{(C - 4)}{15}$

A38 **a** $3x + 2y = 26, 4x + y = 28$ **b** £6

A39 **a** $x + y$ **b** $5(x + y)$

A40 **a** 17 **b** $d = 3n + 2$

A41 **a**

x	⁻3	⁻2	⁻1	0	1	2	3
y	4	⁻1	⁻4	⁻5	⁻4	⁻1	4

A42 **a**

x	⁻2	⁻1	0	1	2	3
y	⁻7	⁻4	⁻1	2	5	8

c i 1.5 **ii** ⁻5.5

S1 **a** 11 cm² **b i** 25.7 cm **ii** 39.3 cm²

S2 **a** 1.25 m **b** 40 cm **c** 140° **d** 118.5 cm

S3 **a** 20 cm² **b** 12 cm by 15 cm **c** 9

S4 **a** 18 000 cm³ **b** 8%

S5 **a** 2.6 m² **b** 3.9 m³ **c** 17420 kg

S6 40 000 cm³

S7 **a i** 5 km **ii** 11.20 am **iii** 10 minutes **c** 12 km/h

S8 **c** reflection in x-axis

S10 **a** 59° **b** 121°

S11 31.5 miles per gallon

S12 **a** 4.55×10^{-3} m³

S13 **a i** 145° **b** 83° **c** 35°

S14 13 cm²

S15 37 m²

S16 218 cm²

S17 1500 cm²

S18 17.7 cm

S19 **a** 63° **b i** 54° **ii** 63° **c** 54°

S20 **a** 2.5 m **b** 20 cm

S21 **a** 50 mph **b** 55 mph

S22 2.58 m²

S23 201 cm

S24 **a** 18 cm **b** 12 cm²

S25 **a** 86.64 cm² **b** 45.6 cm

S26 **a** 327 cm **b** 8495 cm²

S27 **a** 60° **b i** 9 cm **ii** 7.2 cm

S28 120 cm³

S29 **a** 13.5 cm **b** 4 cm **c** 60°

S30 **a** 42° **b** 69°

S31 **a** 4 cm **b** 12 cm

S32 **a** 13 cm

S33 30°

S34 **a** 5.29 cm **b** 41.4° **c** 12.4 cm

S35 37.9°

S36 116 cm

S37 **a** 109° **b** 24°

S38 13.4 m

S39 **a** 622 km **b** 600 km

S40 **a** corresponding **b** 80° **c i** 80° **ii** alternate

S41 295°

S42 24 m

S43 **a** rotation 180° about the origin or enlargement with scale factor ⁻1 and centre the origin.
  b (1, 2), (⁻2, 2), (1, 4)

S44 **a** vertices at (⁻6, 7), (⁻5, 7), (⁻5, 5)
  b reflection in the line $y = x$

S45 **a i** 1300 **ii** 20 km/h

S46 **a i** 11.00 am **ii** 30 minutes **b** 100 miles **c** 67 mph

S47 22 Deutschmarks

S48 **a** 109 yards **b** 9.1 metres

S49 **a** 11.4 m/s

S50 $3rl$: area, $\frac{2(r + l)^2}{h}$ length, $\frac{4\pi r^4}{3l}$: volume

D1 **a** 0.51 **b** 0.34

D2 **a** red **b i** 0.1 **ii** 0

D3 **a** 58 boxes **b** 0.078

D4 **a** 0.06 **b** 0.38

D5 **a** $\frac{1}{16}$ **b** $\frac{1}{6}$

D6   **a** **i** 0.14   **ii** 0   **ii** 1   **iv** 0.25   **c** 0.125

D7   **b** **i** $\frac{25}{144}$ **ii** $\frac{35}{72}$

D8   **i** 0.32   **ii** 0.76

D9   **b** $\frac{1}{6}$ **c** 10 times

D10   **a** 16 to 20   **b** 18

D11   **a** 16.2   **b** 20

D12   **a** 7   **b** 6
    c The boys did better as they had a higher median and a smaller range.

D13   **a** 29   **b** 9   **c** 6   **d** 8   **e** 6

D14   **a** Frequencies 6, 10, 7, 9, 8   **c** 16   **d** $10 \leqslant h < 15$

D15   **a** Frequencies 3, 7, 9, 4, 2   **b** $10 \leqslant t < 15$

D16   **a** £1060   **b** £736.95

D17   **i** Nippy sports: C, Tuff hatchback: A, Ace supermini: B, Mega estate: D   **ii** Nippy sports

D18   **a** 128°, 40°, 72°, 32°, 88°

D19   **a** 70   **b** 18 seconds   **b** 21%

D20   **b** **i** 157 cm   **ii** 33

D22   **b** 0.76   **c** 0.23

D23   304

D24   **a** **i** 0.2   **ii** 0   **b** 40

D25   **a** 0.91   **b** 3   **c** **i** 0.0081   **ii** 0.1638

D26   **i** $\frac{1}{6}$ **ii** $\frac{1}{4}$

D27   **i** $\frac{1}{11}$ **ii** $\frac{2}{11}$ **iii** $\frac{4}{11}$

D28   **a** 0.46   **b** 420

D29   **a** **i** 0.43   **ii** 0.74   **b** 13

D30   **i** $\frac{8}{13}$ **ii** $\frac{5}{13}$

D31   0.45

D32   **i** $\frac{6}{11}$ **ii** $\frac{5}{11}$

D33   £380

D34   38

D35   **a** Angles are 72°, 144°, 36°, 108°   **c** $\frac{2}{5}$

D36   **a** Angles are 140°, 16°, 56°, 94°, 54°
    b Frequencies 18, 36, 9, 27   **c** $\frac{7}{45}$

D37   **a** 90   **b** 40°   **c** 0.22   **d** $\frac{3}{25}$

D38   Angles are 30°, 90°, 72°, 168°

D39   73

D40   98.8 g

D41   **a** $0 \leqslant t < 10$

D42   1.2125 hours

D43   **a** 16.5 mm   **b** $\frac{5}{26}$ **c** 13 mm

PRACTICE NON-CALCULATOR PAPER

1   **a** £3.25   **b** $1.5 + 0.25x$

2   **a** 23   **b** $4x - 3$

4   **a** angles 75°, 129°, 108°, 48°   **b** 10%

5   **a** 55°   **b** 125°

6   **a** 13   **b** 6   **c** 6.5

7   **a** 33.3 mph   **b** 5h

8   **b** 21

9   2550 g

10   50

11   **a** $2 \times 2 \times 2 \times 3 \times 3$ **b** 8   **c** 24

13   **a** x^5 **b** x^2 **c** x^3

14   5

15   Savings Bond better

16 $10\pi + 44$

17   140

18 $x = 3, y = 1$

19 $y = \frac{1}{2}x + 2$

20   **a** 6×10^8 **b** 3×10^9

21   English results are less spreadout. Averages are the same.

22   **a** 65   **b** 115

PRACTICE CALCULATOR PAPER

1   £2.76

2   **a** $a + 6b$ **b** 7   **c** 4.8

3   **a** 444   **b** £139.86

4   **a** $4x$ **b** $16x$ **c** 9

6   Yes, 63 is bigger than 60

7   **b** 13.3 km/h

8   **b** positive correlation

9   130 cm^2

10   **a** $5 \times 6 = 6^2 - 6$ **b** $n \times (n + 1) = (n + 1)^2 - (n + 1)$

11   **a** 1.33   **b** 4.472

12   **a** $5 < T \leqslant 10$ **b** 11.9 minutes

13   6000

14   9

15   3.3

16   3400 m^2

17   **a** 5   **b** 1   **c** 3   **d** $\frac{1}{8}$ or 0.125

18   110°

19   23.3°

20   **a** $x(x + 4) = 45$ **b** $x^2 + 4x - 45 = 0$
    c $x = 5$, perimeter = 28   **d** 89.25

21   **a** 0.6   **b** **i** 0.6, 0.3, 0.7, 0.3, 0.7   **ii** 0.12

22   **a** 6   **b** 6

23 $x = \frac{c}{a - b}$